Lecture Notes in Computer Science 9787

Commenced Publication in 1973
Founding and Former Series Editors:
Gerhard Goos, Juris Hartmanis, and Jan van Leeuwen

More information about this series at http://www.springer.com/series/7407

Osvaldo Gervasi · Beniamino Murgante
Sanjay Misra · Ana Maria A.C. Rocha
Carmelo M. Torre · David Taniar
Bernady O. Apduhan · Elena Stankova
Shangguang Wang (Eds.)

Computational Science and Its Applications – ICCSA 2016

16th International Conference
Beijing, China, July 4–7, 2016
Proceedings, Part II

 Springer

Editors

Osvaldo Gervasi
University of Perugia
Perugia
Italy

Beniamino Murgante
University of Basilicata
Potenza
Italy

Sanjay Misra
Covenant University
Ota
Nigeria

Ana Maria A.C. Rocha
University of Minho
Braga
Portugal

Carmelo M. Torre
Polytechnic University
Bari
Italy

David Taniar
Monash University
Clayton, VIC
Australia

Bernady O. Apduhan
Kyushu Sangyo University
Fukuoka
Japan

Elena Stankova
Saint Petersburg State University
Saint Petersburg
Russia

Shangguang Wang
Beijing University of Posts
  and Telecommunications
Beijing
China

ISSN 0302-9743   ISSN 1611-3349   (electronic)
Lecture Notes in Computer Science
ISBN 978-3-319-42107-0   ISBN 978-3-319-42108-7   (eBook)
DOI 10.1007/978-3-319-42108-7

Library of Congress Control Number: 2016944355

LNCS Sublibrary: SL1 – Theoretical Computer Science and General Issues

Printed on acid-free paper

This Springer imprint is published by Springer Nature
The registered company is Springer International Publishing AG Switzerland

Preface

These multi-volume proceedings (LNCS volumes 9786, 9787, 9788, 9789, and 9790) consist of the peer-reviewed papers from the 2016 International Conference on Computational Science and Its Applications (ICCSA 2016) held in Beijing, China, during July 4–7, 2016.

ICCSA 2016 was a successful event in the series of conferences, previously held in Banff, Canada (2015), Guimares, Portugal (2014), Ho Chi Minh City, Vietnam (2013), Salvador, Brazil (2012), Santander, Spain (2011), Fukuoka, Japan (2010), Suwon, South Korea (2009), Perugia, Italy (2008), Kuala Lumpur, Malaysia (2007), Glasgow, UK (2006), Singapore (2005), Assisi, Italy (2004), Montreal, Canada (2003), (as ICCS) Amsterdam, The Netherlands (2002), and San Francisco, USA (2001).

Computational science is a main pillar of most present research as well as industrial and commercial activities and it plays a unique role in exploiting ICT innovative technologies. The ICCSA conference series has been providing a venue to researchers and industry practitioners to discuss new ideas, to share complex problems and their solutions, and to shape new trends in computational science.

Apart from the general tracks, ICCSA 2016 also included 33 international workshops, in various areas of computational sciences, ranging from computational science technologies to specific areas of computational sciences, such as computer graphics and virtual reality. The program also featured three keynote speeches and two tutorials.

The success of the ICCSA conference series, in general, and ICCSA 2016, in particular, is due to the support of many people: authors, presenters, participants, keynote speakers, session chairs, Organizing Committee members, student volunteers, Program Committee members, Steering Committee members, and many people in other various roles. We would like to thank them all.

We would also like to thank our sponsors, in particular NVidia and Springer for their very important support and for making the Best Paper Award ceremony so impressive.

We would also like to thank Springer for their continuous support in publishing the ICCSA conference proceedings.

July 2016

Shangguang Wang
Osvaldo Gervasi
Bernady O. Apduhan

Organization

ICCSA 2016 was organized by Beijing University of Post and Telecommunication (China), University of Perugia (Italy), Monash University (Australia), Kyushu Sangyo University (Japan), University of Basilicata (Italy), University of Minho, (Portugal), and the State Key Laboratory of Networking and Switching Technology (China).

Honorary General Chairs

Junliang Chen	Beijing University of Posts and Telecommunications, China
Antonio Laganà	University of Perugia, Italy
Norio Shiratori	Tohoku University, Japan
Kenneth C.J. Tan	Sardina Systems, Estonia

General Chairs

Shangguang Wang	Beijing University of Posts and Telecommunications, China
Osvaldo Gervasi	University of Perugia, Italy
Bernady O. Apduhan	Kyushu Sangyo University, Japan

Program Committee Chairs

Sen Su	Beijing University of Posts and Telecommunications, China
Beniamino Murgante	University of Basilicata, Italy
Ana Maria A.C. Rocha	University of Minho, Portugal
David Taniar	Monash University, Australia

International Advisory Committee

Jemal Abawajy	Deakin University, Australia
Dharma P. Agarwal	University of Cincinnati, USA
Marina L. Gavrilova	University of Calgary, Canada
Claudia Bauzer Medeiros	University of Campinas, Brazil
Manfred M. Fisher	Vienna University of Economics and Business, Austria
Yee Leung	Chinese University of Hong Kong, SAR China

International Liaison Chairs

Ana Carla P. Bitencourt	Universidade Federal do Reconcavo da Bahia, Brazil
Alfredo Cuzzocrea	ICAR-CNR and University of Calabria, Italy
Maria Irene Falcão	University of Minho, Portugal

Robert C.H. Hsu	Chung Hua University, Taiwan
Tai-Hoon Kim	Hannam University, Korea
Sanjay Misra	University of Minna, Nigeria
Takashi Naka	Kyushu Sangyo University, Japan
Rafael D.C. Santos	National Institute for Space Research, Brazil
Maribel Yasmina Santos	University of Minho, Portugal

Workshop and Session Organizing Chairs

Beniamino Murgante	University of Basilicata, Italy
Sanjay Misra	Covenant University, Nigeria
Jorge Gustavo Rocha	University of Minho, Portugal

Award Chair

Wenny Rahayu	La Trobe University, Australia

Publicity Committee Chair

Zibing Zheng	Sun Yat-Sen University, China
Mingdong Tang	Hunan University of Science and Technology, China
Yutao Ma	Wuhan University, China
Ao Zhou	Beijing University of Posts and Telecommunications, China
Ruisheng Shi	Beijing University of Posts and Telecommunications, China

Workshop Organizers

Agricultural and Environment Information and Decision Support Systems (AEIDSS 2016)

Sandro Bimonte	IRSTEA, France
André Miralles	IRSTEA, France
Thérèse Libourel	LIRMM, France
François Pinet	IRSTEA, France

Advances in Information Systems and Technologies for Emergency Preparedness and Risk Assessment (ASTER 2016)

Maurizio Pollino	ENEA, Italy
Marco Vona	University of Basilicata, Italy
Beniamino Murgante	University of Basilicata, Italy

Advances in Web-Based Learning (AWBL 2016)

Mustafa Murat Inceoglu	Ege University, Turkey

Bio- and Neuro-Inspired Computing and Applications (BIOCA 2016)

Nadia Nedjah   State University of Rio de Janeiro, Brazil
Luiza de Macedo Mourell   State University of Rio de Janeiro, Brazil

Computer-Aided Modeling, Simulation, and Analysis (CAMSA 2016)

Jie Shen   University of Michigan, USA and Jilin University,
                           China
Hao Chenina   Shanghai University of Engineering Science, China
Xiaoqiang Liun   Donghua University, China
Weichun Shi   Shanghai Maritime University, China
Yujie Liu   Southeast Jiaotong University, China

Computational and Applied Statistics (CAS 2016)

Ana Cristina Braga   University of Minho, Portugal
Ana Paula Costa Conceicao University of Minho, Portugal
  Amorim

Computational Geometry and Security Applications (CGSA 2016)

Marina L. Gavrilova   University of Calgary, Canada

Computational Algorithms and Sustainable Assessment (CLASS 2016)

Antonino Marvuglia   Public Research Centre Henri Tudor, Luxembourg
Mikhail Kanevski   Université de Lausanne, Switzerland
Beniamino Murgante   University of Basilicata, Italy

Chemistry and Materials Sciences and Technologies (CMST 2016)

Antonio Laganà   University of Perugia, Italy
Noelia Faginas Lago   University of Perugia, Italy
Leonardo Pacifici   University of Perugia, Italy

Computational Optimization and Applications (COA 2016)

Ana Maria Rocha   University of Minho, Portugal
Humberto Rocha   University of Coimbra, Portugal

Cities, Technologies, and Planning (CTP 2016)

Giuseppe Borruso   University of Trieste, Italy
Beniamino Murgante   University of Basilicata, Italy

Databases and Computerized Information Retrieval Systems (DCIRS 2016)

Sultan Alamri   College of Computing and Informatics, SEU,
                           Saudi Arabia
Adil Fahad   Albaha University, Saudi Arabia
Abdullah Alamri   Jeddah University, Saudi Arabia

Data Science for Intelligent Decision Support (DS4IDS 2016)

Filipe Portela   University of Minho, Portugal
Manuel Filipe Santos   University of Minho, Portugal

Econometrics and Multidimensional Evaluation in the Urban Environment (EMEUE 2016)

Carmelo M. Torre   Polytechnic of Bari, Italy
Maria Cerreta   University of Naples Federico II, Italy
Paola Perchinunno   University of Bari, Italy
Simona Panaro   University of Naples Federico II, Italy
Raffaele Attardi   University of Naples Federico II, Italy

Future Computing Systems, Technologies, and Applications (FISTA 2016)

Bernady O. Apduhan   Kyushu Sangyo University, Japan
Rafael Santos   National Institute for Space Research, Brazil
Jianhua Ma   Hosei University, Japan
Qun Jin   Waseda University, Japan

Geographical Analysis, Urban Modeling, Spatial Statistics (GEO-AND-MOD 2016)

Giuseppe Borruso   University of Trieste, Italy
Beniamino Murgante   University of Basilicata, Italy
Hartmut Asche   University of Potsdam, Germany

GPU Technologies (GPUTech 2016)

Gervasi Osvaldo   University of Perugia, Italy
Sergio Tasso   University of Perugia, Italy
Flavio Vella   University of Rome La Sapienza, Italy

ICT and Remote Sensing for Environmental and Risk Monitoring (RS-Env 2016)

Rosa Lasaponara   Institute of Methodologies for Environmental Analysis,
                            National Research Council, Italy
Weigu Song   University of Science and Technology of China, China
Eufemia Tarantino   Polytechnic of Bari, Italy
Bernd Fichtelmann   DLR, Germany

7th International Symposium on Software Quality (ISSQ 2016)

Sanjay Misra   Covenant University, Nigeria

International Workshop on Biomathematics, Bioinformatics, and Biostatisticss (IBBB 2016)

Unal Ufuktepe   American University of the Middle East, Kuwait

Land Use Monitoring for Soil Consumption Reduction (LUMS 2016)

Carmelo M. Torre	Polytechnic of Bari, Italy
Alessandro Bonifazi	Polytechnic of Bari, Italy
Valentina Sannicandro	University of Naples Federico II, Italy
Massimiliano Bencardino	University of Salerno, Italy
Gianluca di Cugno	Polytechnic of Bari, Italy
Beniamino Murgante	University of Basilicata, Italy

Mobile Communications (MC 2016)

Hyunseung Choo	Sungkyunkwan University, Korea

Mobile Computing, Sensing, and Actuation for Cyber Physical Systems (MSA4IoT 2016)

Saad Qaisar	NUST School of Electrical Engineering and Computer Science, Pakistan
Moonseong Kim	Korean Intellectual Property Office, Korea

Quantum Mechanics: Computational Strategies and Applications (QM-CSA 2016)

Mirco Ragni	Universidad Federal de Bahia, Brazil
Ana Carla Peixoto Bitencourt	Universidade Estadual de Feira de Santana, Brazil
Vincenzo Aquilanti	University of Perugia, Italy
Andrea Lombardi	University of Perugia, Italy
Federico Palazzetti	University of Perugia, Italy

Remote Sensing for Cultural Heritage: Documentation, Management, and Monitoring (RSCH 2016)

Rosa Lasaponara	IRMMA, CNR, Italy
Nicola Masini	IBAM, CNR, Italy Zhengzhou Base, International Center on Space Technologies for Natural and Cultural Heritage, China
Chen Fulong	Institute of Remote Sensing and Digital Earth, Chinese Academy of Sciences, China

Scientific Computing Infrastructure (SCI 2016)

Elena Stankova	Saint Petersburg State University, Russia
Vladimir Korkhov	Saint Petersburg State University, Russia
Alexander Bogdanov	Saint Petersburg State University, Russia

Software Engineering Processes and Applications (SEPA 2016)

Sanjay Misra	Covenant University, Nigeria

Social Networks Research and Applications (SNRA 2016)

Eric Pardede	La Trobe University, Australia
Wenny Rahayu	La Trobe University, Australia
David Taniar	Monash University, Australia

Sustainability Performance Assessment: Models, Approaches, and Applications Toward Interdisciplinarity and Integrated Solutions (SPA 2016)

Francesco Scorza	University of Basilicata, Italy
Valentin Grecu	Lucia Blaga University on Sibiu, Romania

Tools and Techniques in Software Development Processes (TTSDP 2016)

Sanjay Misra	Covenant University, Nigeria

Volunteered Geographic Information: From Open Street Map to Participation (VGI 2016)

Claudia Ceppi	University of Basilicata, Italy
Beniamino Murgante	University of Basilicata, Italy
Francesco Mancini	University of Modena and Reggio Emilia, Italy
Giuseppe Borruso	University of Trieste, Italy

Virtual Reality and Its Applications (VRA 2016)

Osvaldo Gervasi	University of Perugia, Italy
Lucio Depaolis	University of Salento, Italy

Web-Based Collective Evolutionary Systems: Models, Measures, Applications (WCES 2016)

Alfredo Milani	University of Perugia, Italy
Valentina Franzoni	University of Rome La Sapienza, Italy
Yuanxi Li	Hong Kong Baptist University, Hong Kong, SAR China
Clement Leung	United International College, Zhuhai, China
Rajdeep Niyogi	Indian Institute of Technology, Roorkee, India

Program Committee

Jemal Abawajy	Deakin University, Australia
Kenny Adamson	University of Ulster, UK
Hartmut Asche	University of Potsdam, Germany
Michela Bertolotto	University College Dublin, Ireland
Sandro Bimonte	CEMAGREF, TSCF, France
Rod Blais	University of Calgary, Canada
Ivan Blečić	University of Sassari, Italy
Giuseppe Borruso	University of Trieste, Italy
Yves Caniou	Lyon University, France

José A. Cardoso e Cunha	Universidade Nova de Lisboa, Portugal
Carlo Cattani	University of Salerno, Italy
Mete Celik	Erciyes University, Turkey
Alexander Chemeris	National Technical University of Ukraine KPI, Ukraine
Min Young Chung	Sungkyunkwan University, Korea
Elisete Correia	University of Trás os Montes e Alto Douro, Portugal
Gilberto Corso Pereira	Federal University of Bahia, Brazil
M. Fernanda Costa	University of Minho, Portugal
Alfredo Cuzzocrea	ICAR-CNR and University of Calabria, Italy
Florbela Maria da Cruz Domingues Correia	Intituto Politécnico de Viana do Castelo, Portugal
Vanda Marisa da Rosa Milheiro Lourenço	FCT from University Nova de Lisboa, Portugal
Carla Dal Sasso Freitas	Universidade Federal do Rio Grande do Sul, Brazil
Pradesh Debba	The Council for Scientific and Industrial Research (CSIR), South Africa
Hendrik Decker	Instituto Tecnológico de Informática, Spain
Adelaide de Fátima Baptista Valente Freitas	University of Aveiro, Portugal
Carina Soares da Silva Fortes	Escola Superior de Tecnologias da Saúde de Lisboa, Portugal
Frank Devai	London South Bank University, UK
Rodolphe Devillers	Memorial University of Newfoundland, Canada
Joana Dias	University of Coimbra, Portugal
Prabu Dorairaj	NetApp, India/USA
M. Irene Falcao	University of Minho, Portugal
Cherry Liu Fang	U.S. DOE Ames Laboratory, USA
Florbela Fernandes	Polytechnic Institute of Bragança, Portugal
Jose-Jesús Fernandez	National Centre for Biotechnology, CSIS, Spain
Mara Celia Furtado Rocha	PRODEB-Pós Cultura/UFBA, Brazil
Akemi Galvez	University of Cantabria, Spain
Paulino Jose Garcia Nieto	University of Oviedo, Spain
Marina Gavrilova	University of Calgary, Canada
Jerome Gensel	LSR-IMAG, France
Mara Giaoutzi	National Technical University, Athens, Greece
Andrzej M. Goscinski	Deakin University, Australia
Alex Hagen-Zanker	University of Cambridge, UK
Malgorzata Hanzl	Technical University of Lodz, Poland
Shanmugasundaram Hariharan	B.S. Abdur Rahman University, India
Tutut Herawan	Universitas Teknologi Yogyakarta, Indonesia
Hisamoto Hiyoshi	Gunma University, Japan
Fermin Huarte	University of Barcelona, Spain
Andrés Iglesias	University of Cantabria, Spain
Mustafa Inceoglu	Ege University, Turkey
Peter Jimack	University of Leeds, UK

Maurizio Pollino	Italian National Agency for New Technologies, Energy and Sustainable Economic Development, Italy
Alenka Poplin	University of Hamburg, Germany
Vidyasagar Potdar	Curtin University of Technology, Australia
David C. Prosperi	Florida Atlantic University, USA
Maria Emilia F. Queiroz Athayde	University of Minho, Portugal
Wenny Rahayu	La Trobe University, Australia
Jerzy Respondek	Silesian University of Technology, Poland
Ana Maria A.C. Rocha	University of Minho, Portugal
Maria Clara Rocha	ESTES Coimbra, Portugal
Humberto Rocha	INESC-Coimbra, Portugal
Alexey Rodionov	Institute of Computational Mathematics and Mathematical Geophysics, Russia
Jon Rokne	University of Calgary, Canada
Octavio Roncero	CSIC, Spain
Maytham Safar	Kuwait University, Kuwait
Chiara Saracino	A.O. Ospedale Niguarda Ca' Granda - Milano, Italy
Haiduke Sarafian	The Pennsylvania State University, USA
Jie Shen	University of Michigan, USA
Qi Shi	Liverpool John Moores University, UK
Dale Shires	U.S. Army Research Laboratory, USA
Takuo Suganuma	Tohoku University, Japan
Sergio Tasso	University of Perugia, Italy
Parimala Thulasiraman	University of Manitoba, Canada
Carmelo M. Torre	Polytechnic of Bari, Italy
Giuseppe A. Trunfio	University of Sassari, Italy
Unal Ufuktepe	American University of the Middle East, Kuwait
Toshihiro Uchibayashi	Kyushu Sangyo University, Japan
Mario Valle	Swiss National Supercomputing Centre, Switzerland
Pablo Vanegas	University of Cuenca, Equador
Piero Giorgio Verdini	INFN Pisa and CERN, Italy
Marco Vizzari	University of Perugia, Italy
Koichi Wada	University of Tsukuba, Japan
Krzysztof Walkowiak	Wroclaw University of Technology, Poland
Robert Weibel	University of Zurich, Switzerland
Roland Wismüller	Universität Siegen, Germany
Mudasser Wyne	SOET National University, USA
Chung-Huang Yang	National Kaohsiung Normal University, Taiwan
Xin-She Yang	National Physical Laboratory, UK
Salim Zabir	France Telecom Japan Co., Japan
Haifeng Zhao	University of California, Davis, USA
Kewen Zhao	University of Qiongzhou, China
Albert Y. Zomaya	University of Sydney, Australia

Reviewers

Abawajy, Jemal	Deakin University, Australia
Abuhelaleh, Mohammed	Univeristy of Bridgeport, USA
Acharjee, Shukla	Dibrugarh University, India
Andrianov, Sergei Nikolaevich	Universitetskii prospekt, Russia
Aguilar, José Alfonso	Universidad Autónoma de Sinaloa, Mexico
Ahmed, Faisal	University of Calgary, Canada
Alberti, Margarita	University of Barcelona, Spain
Amato, Alba	Seconda Universit degli Studi di Napoli, Italy
Amorim, Ana Paula	University of Minho, Portugal
Apduhan, Bernady	Kyushu Sangyo University, Japan
Aquilanti, Vincenzo	University of Perugia, Italy
Asche, Hartmut	Posdam University, Germany
Athayde Maria, Emlia Feijão Queiroz	University of Minho, Portugal
Attardi, Raffaele	University of Napoli Federico II, Italy
Azam, Samiul	United International University, Bangladesh
Azevedo, Ana	Athabasca University, USA
Badard, Thierry	Laval University, Canada
Baioletti, Marco	University of Perugia, Italy
Bartoli, Daniele	University of Perugia, Italy
Bentayeb, Fadila	Université Lyon, France
Bilan, Zhu	Tokyo University of Agriculture and Technology, Japan
Bimonte, Sandro	IRSTEA, France
Blecic, Ivan	Università di Cagliari, Italy
Bogdanov, Alexander	Saint Petersburg State University, Russia
Borruso, Giuseppe	University of Trieste, Italy
Bostenaru, Maria	"Ion Mincu" University of Architecture and Urbanism, Romania
Braga Ana, Cristina	University of Minho, Portugal
Canora, Filomena	University of Basilicata, Italy
Cardoso, Rui	Institute of Telecommunications, Portugal
Ceppi, Claudia	Polytechnic of Bari, Italy
Cerreta, Maria	University Federico II of Naples, Italy
Choo, Hyunseung	Sungkyunkwan University, South Korea
Coletti, Cecilia	University of Chieti, Italy
Correia, Elisete	University of Trás-Os-Montes e Alto Douro, Portugal
Correia Florbela Maria, da Cruz Domingues	Instituto Politécnico de Viana do Castelo, Portugal
Costa, Fernanda	University of Minho, Portugal
Crasso, Marco	National Scientific and Technical Research Council, Argentina
Crawford, Broderick	Universidad Catolica de Valparaiso, Chile

Cuzzocrea, Alfredo	University of Trieste, Italy
Cutini, Valerio	University of Pisa, Italy
Danese, Maria	IBAM, CNR, Italy
Decker, Hendrik	Instituto Tecnológico de Informática, Spain
Degtyarev, Alexander	Saint Petersburg State University, Russia
Demartini, Gianluca	University of Sheffield, UK
Di Leo, Margherita	JRC, European Commission, Belgium
Dias, Joana	University of Coimbra, Portugal
Dilo, Arta	University of Twente, The Netherlands
Dorazio, Laurent	ISIMA, France
Duarte, Júlio	University of Minho, Portugal
El-Zawawy, Mohamed A.	Cairo University, Egypt
Escalona, Maria-Jose	University of Seville, Spain
Falcinelli, Stefano	University of Perugia, Italy
Fernandes, Florbela	Escola Superior de Tecnologia e Gest ão de Bragança, Portugal
Florence, Le Ber	ENGEES, France
Freitas Adelaide, de Fátima Baptista Valente	University of Aveiro, Portugal
Frunzete, Madalin	Polytechnic University of Bucharest, Romania
Gankevich, Ivan	Saint Petersburg State University, Russia
Garau, Chiara	University of Cagliari, Italy
Garcia, Ernesto	University of the Basque Country, Spain
Gavrilova, Marina	University of Calgary, Canada
Gensel, Jerome	IMAG, France
Gervasi, Osvaldo	University of Perugia, Italy
Gizzi, Fabrizio	National Research Council, Italy
Gorbachev, Yuriy	Geolink Technologies, Russia
Grilli, Luca	University of Perugia, Italy
Guerra, Eduardo	National Institute for Space Research, Brazil
Hanzl, Malgorzata	University of Lodz, Poland
Hegedus, Peter	University of Szeged, Hungary
Herawan, Tutut	University of Malaya, Malaysia
Hu, Ya-Han	National Chung Cheng University, Taiwan
Ibrahim, Michael	Cairo University, Egipt
Ifrim, Georgiana	Insight, Ireland
Irrazábal, Emanuel	Universidad Nacional del Nordeste, Argentina
Janana, Loureio	University of Mato Grosso do Sul, Brazil
Jaiswal, Shruti	Delhi Technological University, India
Johnson, Franklin	Universidad de Playa Ancha, Chile
Karimipour, Farid	Vienna University of Technology, Austria
Kapcak, Sinan	American University of the Middle East in Kuwait, Kuwait
Kiki Maulana, Adhinugraha	Telkom University, Indonesia
Kim, Moonseong	KIPO, South Korea
Kobusińska, Anna	Poznan University of Technology, Poland

Korkhov, Vladimir	Saint Petersburg State University, Russia
Koutsomitropoulos, Dimitrios A.	University of Patras, Greece
Krishna Kumar, Chaturvedi	Indian Agricultural Statistics Research Institute (IASRI), India
Kulabukhova, Nataliia	Saint Petersburg State University, Russia
Kumar, Dileep	SR Engineering College, India
Laganà, Antonio	University of Perugia, Italy
Lai, Sen-Tarng	Shih Chien University, Taiwan
Lanza, Viviana	Lombardy Regional Institute for Research, Italy
Lasaponara, Rosa	National Research Council, Italy
Lazzari, Maurizio	National Research Council, Italy
Le Duc, Tai	Sungkyunkwan University, South Korea
Le Duc, Thang	Sungkyunkwan University, South Korea
Lee, KangWoo	Sungkyunkwan University, South Korea
Leung, Clement	United International College, Zhuhai, China
Libourel, Thérèse	LIRMM, France
Lourenço, Vanda Marisa	University Nova de Lisboa, Portugal
Machado, Jose	University of Minho, Portugal
Magni, Riccardo	Pragma Engineering srl, Italy
Mancini Francesco	University of Modena and Reggio Emilia, Italy
Manfreda, Salvatore	University of Basilicata, Italy
Manganelli, Benedetto	Università degli studi della Basilicata, Italy
Marghany, Maged	Universiti Teknologi Malaysia, Malaysia
Marinho, Euler	Federal University of Minas Gerais, Brazil
Martellozzo, Federico	University of Rome "La Sapienza", Italy
Marvuglia, Antonino	Public Research Centre Henri Tudor, Luxembourg
Mateos, Cristian	Universidad Nacional del Centro, Argentina
Matsatsinis, Nikolaos	Technical University of Crete, Greece
Messina, Fabrizio	University of Catania, Italy
Millham, Richard	Durban University of Technoloy, South Africa
Milani, Alfredo	University of Perugia, Italy
Misra, Sanjay	Covenant University, Nigeria
Modica, Giuseppe	Università Mediterranea di Reggio Calabria, Italy
Mohd Helmy, Abd Wahab	Universiti Tun Hussein Onn Malaysia, Malaysia
Murgante, Beniamino	University of Basilicata, Italy
Nagy, Csaba	University of Szeged, Hungary
Napolitano, Maurizio	Center for Information and Communication Technology, Italy
Natário, Isabel Cristina Maciel	University Nova de Lisboa, Portugal
Navarrete Gutierrez, Tomas	Luxembourg Institute of Science and Technology, Luxembourg
Nedjah, Nadia	State University of Rio de Janeiro, Brazil
Nguyen, Tien Dzung	Sungkyunkwan University, South Korea
Niyogi, Rajdeep	Indian Institute of Technology Roorkee, India

Oliveira, Irene	University of Trás-Os-Montes e Alto Douro, Portugal
Panetta, J.B.	Tecnologia Geofísica Petróleo Brasileiro SA, PETROBRAS, Brazil
Papa, Enrica	University of Amsterdam, The Netherlands
Papathanasiou, Jason	University of Macedonia, Greece
Pardede, Eric	La Trobe University, Australia
Pascale, Stefania	University of Basilicata, Italy
Paul, Padma Polash	University of Calgary, Canada
Perchinunno, Paola	University of Bari, Italy
Pereira, Oscar	Universidade de Aveiro, Portugal
Pham, Quoc Trung	HCMC University of Technology, Vietnam
Pinet, Francois	IRSTEA, France
Pirani, Fernando	University of Perugia, Italy
Pollino, Maurizio	ENEA, Italy
Pusatli, Tolga	Cankaya University, Turkey
Qaisar, Saad	NURST, Pakistan
Qian, Junyan	Guilin University of Electronic Technology, China
Raffaeta, Alessandra	University of Venice, Italy
Ragni, Mirco	Universidade Estadual de Feira de Santana, Brazil
Rahman, Wasiur	Technical University Darmstadt, Germany
Rampino, Sergio	Scuola Normale di Pisa, Italy
Rahayu, Wenny	La Trobe University, Australia
Ravat, Franck	IRIT, France
Raza, Syed Muhammad	Sungkyunkwan University, South Korea
Roccatello, Eduard	3DGIS, Italy
Rocha, Ana Maria	University of Minho, Portugal
Rocha, Humberto	University of Coimbra, Portugal
Rocha, Jorge	University of Minho, Portugal
Rocha, Maria Clara	ESTES Coimbra, Portugal
Romano, Bernardino	University of l'Aquila, Italy
Sannicandro, Valentina	Polytechnic of Bari, Italy
Santiago Júnior, Valdivino	Instituto Nacional de Pesquisas Espaciais, Brazil
Sarafian, Haiduke	Pennsylvania State University, USA
Schneider, Michel	ISIMA, France
Selmaoui, Nazha	University of New Caledonia, New Caledonia
Scerri, Simon	University of Bonn, Germany
Shakhov, Vladimir	Institute of Computational Mathematics and Mathematical Geophysics, Russia
Shen, Jie	University of Michigan, USA
Silva-Fortes, Carina	ESTeSL-IPL, Portugal
Singh, Upasana	University of Kwa Zulu-Natal, South Africa
Skarga-Bandurova, Inna	Technological Institute of East Ukrainian National University, Ukraine
Soares, Michel	Federal University of Sergipe, Brazil
Souza, Eric	Universidade Nova de Lisboa, Portugal
Stankova, Elena	Saint Petersburg State University, Russia

Sponsoring Organizations

ICCSA 2016 would not have been possible without the tremendous support of many organizations and institutions, for which all organizers and participants of ICCSA 2016 express their sincere gratitude:

Springer International Publishing AG, Switzerland
(http://www.springer.com)

NVidia Co., USA
(http://www.nvidia.com)

Beijing University of Post and Telecommunication, China
(http://english.bupt.edu.cn/)

State Key Laboratory of Networking and Switching Technology, China

University of Perugia, Italy
(http://www.unipg.it)

University of Basilicata, Italy
(http://www.unibas.it)

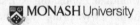

Monash University, Australia
(http://monash.edu)

Kyushu Sangyo University, Japan
(www.kyusan-u.ac.jp)

Universidade do Minho, Portugal
(http://www.uminho.pt)

Contents – Part II

Geometric Modeling, Graphics and Visualization

High Performance Computing and Networks

Parallel Sparse Matrix-Vector Multiplication Using Accelerators

Hiroshi Maeda[1] and Daisuke Takahashi[2(✉)]

[1] Graduate School of Systems and Information Engineering, University of Tsukuba,
1-1-1 Tennodai, Tsukuba, Ibaraki 305-8573, Japan
maeda@hpcs.cs.tsukuba.ac.jp
[2] Center for Computational Sciences, University of Tsukuba, 1-1-1 Tennodai,
Tsukuba, Ibaraki 305-8577, Japan
daisuke@cs.tsukuba.ac.jp

Abstract. Sparse matrix-vector multiplication (SpMV) is an essential computational kernel for many applications such as scientific computing. Recently, the number of computing systems equipped with NVIDIA's GPU and Intel's Xeon Phi coprocessor based on the MIC architecture has been increasing. Therefore, the importance of effective algorithms for SpMV in these systems is increasing. To the best of our knowledge, while previous studies have reported CPU and GPU implementations of SpMV for a cluster and MIC implementations for a single node, implementations of SpMV for the MIC cluster have not yet been reported. In this paper, we implemented and evaluated parallel SpMV on a GPU cluster and a MIC cluster. As shown by the results, the implementation for MIC achieved relatively high performance in some matrices with a single process, but it could not achieve higher performance than other implementations with 64 MPI processes. Therefore, we implemented and evaluated the single SpMV kernel to improve the performance of parallel SpMV.

Keywords: SpMV · Accelerator · GPU · MIC · Cluster

1 Introduction

Sparse matrix-vector multiplication (SpMV) is an important computational kernel for many applications such as scientific computing. For example, in physics and engineering simulations, accelerating SpMV is expected to reduce the execution time of existing applications. Algorithms for SpMV [3,6,7] have been proposed in several studies. However, their effects are largely dependent on the architecture of the computer and the nonzero structure of the matrix. Hence, the most appropriate algorithm depends on the matrix and the architecture of the computer. Recently, an increasing number of computing systems have been equipped with NVIDIA's GPU and Intel's Xeon Phi coprocessor based on the Many Integrated Core (MIC) architecture proposed by Intel in 2010. A system equipped with the Xeon Phi coprocessor was ranked first and a system equipped with Tesla K20X, which is NVIDIA's GPU, was ranked second in the TOP500

© Springer International Publishing Switzerland 2016
O. Gervasi et al. (Eds.): ICCSA 2016, Part II, LNCS 9787, pp. 3–18, 2016.
DOI: 10.1007/978-3-319-42108-7_1

list in November 2015. Many studies have applied a GPU to SpMV, such as GPU aware algorithms and sparse matrix formats [12,16], and several studies have also applied MIC to SpMV [10,15].

From these points of view, algorithms for SpMV using an accelerator effectively on clusters are required. To the best of our knowledge, while previous studies have reported CPU and GPU implementations of SpMV for a cluster [5,7] and MIC implementations with MPI/OpenMP [17], implementations of SpMV for the MIC cluster have not yet been reported. However, the MIC cluster is expected to achieve higher performance than CPU and GPU clusters. In this paper, we implement and evaluate algorithms for SpMV on a GPU cluster and a MIC cluster.

We previously published an evaluation of SpMV on GPU and MIC clusters [11], and we extend our evaluation in this paper. Mainly, we add the investigation of the single kernel.

The remainder of this paper is organized as follows. Section 2 describes SpMV. Section 3 explains the accelerator. Section 4 proposes an implementation of SpMV, and Sect. 5 gives the experimental results. Section 6 improves the single SpMV kernel for parallel SpMV. Finally, in Sect. 7, we provide concluding remarks.

2   Sparse Matrix-Vector Multiplication

A sparse matrix is a matrix populated primarily with zeros. SpMV is shown in Eq. (1) for an $M \times N$ sparse matrix A, vector x of length M, and vector y of length N.

$$y = Ax \tag{1}$$

This calculation is an essential operation in the CG method for solving linear equations, because SpMV consumes a large proportion of the execution time. Algorithms such as the CG method consist of iterative calculations, and the same matrix is repeatedly used. Therefore, a distribution method and preconditioning of the sparse matrix have been proposed under the condition that the matrix does not change across iterations.

Also, several formats of the sparse matrix have been proposed. For example, Compressed Row Storage (CRS) [3,14] is a standard sparse matrix format. It is specified by three arrays: val, which is the values of nonzero elements in the matrix; idx, which is a column index of the nonzero elements in matrix A; ptr, which is the row indices corresponding to val and idx. While the CRS format has advantages such as small memory space, load imbalance is a problem when the load per message-passing interface (MPI) process is decided by the number of rows.

2.1   Parallel SpMV

To parallelize SpMV, the distribution method and sparse matrix format must be decided. Several distribution methods that reduce the communication amount

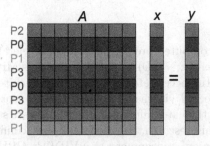

Fig. 1. Distribution of A, x, and y

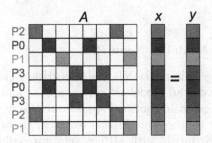

Fig. 2. Computation of each MPI process in parallel SpMV

and balances the load between MPI processes have been proposed. We adopted a hypergraph partitioning based one-dimensional distribution method [4] to distribute the sparse matrix and vectors.

Figure 1 is an example of the one-dimensional distribution of sparse matrix A and vectors x, y. P0–P3 mean the ranks of 4 MPI processes. In this distribution method, each MPI process holds a sparse matrix partitioned by the rows and vector elements that need to be multiplied by the diagonal elements of the sparse matrix. Each MPI process has the responsibility of calculating the elements in the rows. Figure 2 is an example of MPI processes that need to communicate with other MPI processes. Although the colored areas can be calculated in each MPI process locally without communication, no colored area causes communication with other MPI processes for vector x before the calculation to leave matrix A is distributed. This method minimizes the communication amount by communicating only the elements of vector x needed by each MPI process. In addition, this method can also overlap the communication and the calculation because local elements can be calculated without communication.

To decide the assignment between MPI processes and the rows of the sparse matrix, we used hypergraph partitioning based decomposition. In hypergraph partitioning based decomposition, we assume each row to be a vertex and each column to be a net. Hypergraph partitioning based decomposition can minimize the actual communication volume while maintaining the load balance.

3 Accelerator

The accelerator is a computer processor used to accelerate part of the functions of the CPU. Recently, NVIDIA's GPU and Intel's MIC architectures have become popular, and the accelerators for these two architectures are the same in that they are connected via the Peripheral Component Interconnect (PCI) Express bus. However, they are different in their hardware structure and method of programming. In this research, we used Compute Unified Device Architecture (CUDA), which is an environment of NVIDIA's GPU, and the Xeon Phi coprocessor, which is based on the MIC architecture. In this section, we explain only the programming model of Xeon Phi and omit the description of the GPU architecture and CUDA.

3.1 Programming Model of Xeon Phi

Xeon Phi uses two programming models: the offload model and the native model. Applications using the offload model are run on a CPU and the selected work is offloaded to the coprocessor. In contrast, applications using the native model are run exclusively on the coprocessor. In this research, we targeted a multi-node environment. On the Xeon Phi with MPI, several programming models can be used. In the host-only model, all MPI processes are run on the host CPUs. This model is a conventional model for CPU clusters. In the offload model, MPI processes are run on the host CPUs, while the offload region is executed on the Xeon Phi. In the symmetric model, MPI processes are run on the host CPUs and Xeon Phi. This model can exploit all resources. However, it is difficult to keep the load balance between the host CPU and Xeon Phi. In the MIC-only model, MPI processes are run on the Xeon Phi.

4 Implementation of SpMV

In this research, we implemented parallel SpMV that distributes the sparse matrix and vectors by using hypergraph partitioning based decomposition and performs computations by using the library for each CPU, GPU, and MIC cluster. Figure 3 gives the pseudocode of parallel SpMV with MPI.

To parallelize SpMV, we used the method explained in Sect. 2. We assigned a row in advance to each MPI process based on the hypergraph partitioning. First, each of the MPI processes begins asynchronous communication of an element of vector x needed by the MPI process. Next, the part of the SpMV that does not need communication, that is, the local elements of A, are computed with overlapping communication. Then, the MPI processes wait for completion of the communication and compute the rest of the elements of A. In the implementations for the GPU, the elements of vector x are transferred to the accelerator before computing SpMV in lines 26 and 30 of Fig. 3. After computation of SpMV in line 30 of Fig. 3, the elements of vector y are transferred to the host.

```
 1: inputs
 2:     Sparse matrix: A,
 3:     Vector: x,
 4:     number of the elements calculated locally of A: nLocal,
 5:     indices of elements of the vector x to be sent: elementsToSend,
 6:     total number of elements of the vector x to be sent: totalToBeSent,
 7:     numbers of elements to be sent in vector x: sendLength,
 8:     numbers of elements to receive in vector x: recvLength,
 9:     number of MPI processes: nProc,
10: outputs
11:     Vector: y
12:
13: for i = 0 to totalToBeSent − 1 do
14:     packed[i] ← x[elementsToSend[i]] {Packing (thread parallel)}
15: end for
16: nSend ← 0
17: for i = 0 to nProc − 1 do
18:     AsynchronousSend(packed[nSend],i) {Asynchronous Send to MPI pro-
        cess i}
19:     nSend ← nSend + sendLength[i]
20: end for
21: nRecv ← nLocal
22: for i = 0 to nProc − 1 do
23:     AsynchronousReceive(x[nRecv],i) {Asynchronous receive from MPI pro-
        cess i}
24:     nRecv ← nRecv + recvLength[i]
25: end for
26: y ← Ax {Computation of SpMV of local elements}
27: for i = 0 to nProc − 1 do
28:     WaitCommunication(i) {Wait until completion of communication with
        MPI process i}
29: end for
30: y ← y + Ax {Computation of SpMV of non-local elements}
31: return y
```

Fig. 3. Pseudocode of parallel SpMV with MPI

For the computational kernel, we used the CRS format for the sparse matrix and numerical libraries. To evaluate the implementations fairly, we used the available libraries provided by Intel Corporation and NVIDIA Corporation because they are considered to be efficiently optimized for each respective architecture. In the implementations for CPU and MIC, we used the Intel Math Kernel Library (MKL), which is the numerical library developed by the Intel Corporation. In the implementation for GPUs, we used the CUDA Sparse Matrix library (cuS-PARSE) developed by the NVIDIA Corporation.

5   Evaluation of Parallel SpMV

We evaluated the performance of the implementations on each cluster. We used the Highly Accelerated Parallel Advanced System for Computation Sciences/Tightly Coupled Accelerators (HA-PACS/TCA) as the GPU cluster and Cluster of Many-core Architecture processor (COMA) as the MIC cluster located at the Center for Computational Sciences, University of Tsukuba. The specifications of HA-PACS/TCA and COMA are described in Tables 1 and 2. We used square and unsymmetric matrices from The University of Florida Sparse Matrix Collection [2] as input matrices. We evaluated the implementations for the MIC-only model on the MIC cluster. Each process of the implementations consisted of the native model. We also evaluated the offload model. However, we exclude it in this paper because the performance was poorer than the others and our main purpose is to evaluate architectures. Also, we omitted the symmetric model because it is difficult to keep the load balance, such as the number of processes and threads between Xeon Phi and the host CPU.

In this evaluation, we measured the time needed to execute the code corresponding to Fig. 3. In the implementations for the GPU, we allocated memory for the matrix and transferred it to the GPU in advance. The measurement time comprises the time of transferring vectors x and y between the host CPU and the accelerator.

We evaluated the implementation for the GPU cluster on HA-PACS/TCA and that for the CPU and the MIC on COMA. The implementations are the

Table 1. Specifications of HA-PACS/TCA

CPU	Intel Xeon E5-2680v2 10 cores×2
Main memory	DDR3 1600 MHz 128 GB
GPU	NVIDIA Tesla K20X×4
Interconnection	Infiniband QDR×2
Compiler	Intel C++ Compiler 15.0.0
MPI	Intel MPI 5.0
CUDA Toolkit	6.5.14

Table 2. Specifications of COMA

CPU	Intel Xeon E5-2670v2 10 cores×2
Main memory	DDR3 1866 MHz 64 GB
MIC	Intel Xeon Phi 7110P 61 cores×2
Interconnection	Infiniband FDR×1
Compiler	Intel C++ Compiler 15.0.3
MPI	Intel MPI 5.1.1
Numerical library	Intel MKL 11.2.3

same as in our previous paper [11], however we changed the environments here. Each MPI process uses 5 CPU cores and 1 GPU, and each computation node is assigned 4 MPI processes in the implementation for the GPU cluster. Each MPI process is assigned 10 CPU cores, and each computation node is assigned 2 MPI processes in the implementation for the CPU cluster. Each MPI process is assigned 1 MIC, and each computation node is assigned 2 MPI processes, and we set the affinity of threads to compact and the number of threads to 240 in the implementation for the MIC cluster.

5.1 Experimental Results

The performance of each implementation is shown in Figs. 4 and 5. Figure 4 represents the single process and Fig. 5 represents the 64 MPI processes. The structure of the matrices affects the performance in the implementation of SpMV for the CPU cluster. However, the number of nonzero elements also affects the performances in the implementations of SpMV for the GPU and the MIC cluster.

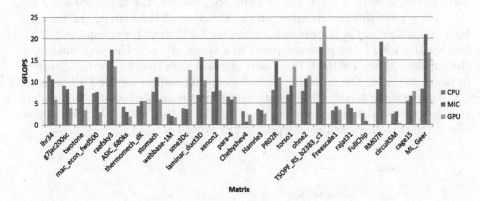

Fig. 4. Performance of each implementation (single process) (Color figure online)

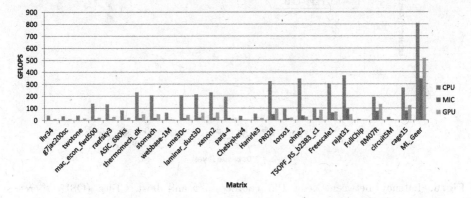

Fig. 5. Performance of each implementation (64 MPI processes) (Color figure online)

5.2 Performance of Each SpMV Implementation (single Process)

Figure 4 shows that the performance of the implementation for the CPU cluster is higher than others in matrices that have a small number of nonzero elements. However, the performance of the implementation for the MIC cluster becomes better than others in matrices that have a large number of nonzero elements. This is because communication is not needed with the single process and we could sufficiently exploit the parallelism of MIC. Also, the performance of the implementation for the GPU cluster is worse than that for the CPU cluster in matrices that have a small number of nonzero elements. However, the performance for the GPU cluster becomes better than that for the CPU in matrices that have a large number of nonzero elements.

5.3 Performance of Each SpMV Implementation (64 MPI Processes)

The result with the 64 MPI processes is significantly different from the result with the single process. First, the performance growth of the implementation for the CPU cluster differs depending on the structure of the matrix. On average, the performance of the implementation for the CPU cluster was improved 42.57 times with 64 MPI processes compared to a single process. The performance of the implementation for the GPU cluster is worse than that for the CPU cluster in many matrices. The reason is that the implementation for the GPU cluster transfers vector x to the GPU before executing the computation kernel and it transfers vector y, which is the result of SpMV, to the host CPU. This action affects the overall computation with 64 MPI processes.

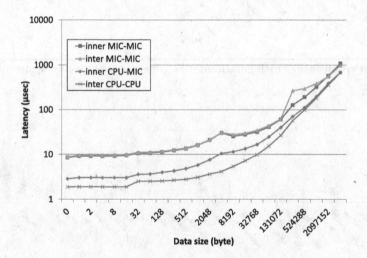

Fig. 6. Latency between Xeon Phi coprocessors and host CPUs (OSU Micro-Benchmark 4.3)

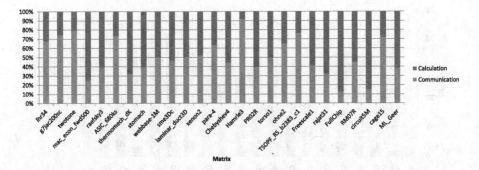

Fig. 7. Rate of calculation and communication time of the implementation for the MIC cluster with 64 MPI processes (Color figure online)

The performance of the implementation for the MIC cluster is worse than that for the CPU cluster in all of the matrices that we evaluated. Figure 6 compares the latency evaluated for osu_latency in the OSU Micro-Benchmark 4.3 [1] between the MICs and between the host CPUs on COMA. The word "inner" means the latency between MICs or between host CPUs in the same node. The latency of MPI communication between Xeon Phi is relatively large because it uses the PCI Express bus. Figure 7 shows the rate of calculation and the communication time of the implementation for the MIC cluster with 64 MPI processes. This figure indicates that not only communication but also calculation time is relatively large.

6   Investigation of SpMV Kernel for MIC

6.1   Existing Kernels and Their Problems

As shown in the previous section, the rate of calculation of the implementation for the MIC cluster was relatively high in some matrices (Fig. 7). The high rate is caused by two factors.

One factor is the decrease in problem size per MPI process due to the strong scaling. The effect of the decrease in problem size in the MIC is larger than that in the host CPU because a small problem cannot exploit MIC's parallelism sufficiently. Figure 8 shows the comparison of the performance of the CPU and the MIC when we changed the problem size. We used MKL with the CRS format. The horizontal axis is the matrix and the vertical axis is the performance in GFLOPS. We used dense matrices (e.g., dense-1000 means a 1000×1000 dense matrix). This figure shows the performance of the MIC is less than that of the CPU under dense-1500 because of the insufficiency of the problem size compared to the parallelism of the MIC.

The number of nonzeros in dense-1500 is 2.25×10^6. Hence, if the implementation for the MIC achieves higher performance than for the CPU, the matrix is supposed to need more than about $1.44 \times 10^8 (= 2.25 \times 10^6 \times 64)$ nonzero

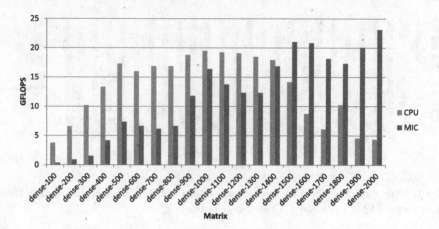

Fig. 8. Effect of the number of nonzero elements for CPU and MIC (Color figure online)

elements. However, ML_Geer, which has the most nonzero elements among the matrices that we measured has about 1.10×10^8 nonzero elements, which is fewer than 1.44×10^8.

Thus, the effect of the decrease in problem size is large when parallelizing SpMV with strong scaling. The main reason is that the MIC architecture needs much parallelism to exploit it sufficiently. However, it may be improved by changing the SpMV kernel.

Another problem is the unbalance of nonzero elements. Even though hypergraph partitioning takes account of the load balance between processes, the load balance between rows is ignored in each process. However, the CRS format, which is used in this paper, is normally parallelized by the rows. So, one thread in one process may cause the overall performance to be extremely low in matrices that have many nonzeros in a specific row.

Intel MKL offers some sparse matrix formats to calculate SpMV. However, the Coordinate (COO) and the Diagonal (DIA) formats cannot deal with an unbalance matrix without difficulty, and the Block Compressed Sparse Row (BSR) format can deal with the unbalance matrix. However the performance of the BSR format may extremely decline because it stores useless zero elements. This problem is expected to be improved by using the two-dimensional distribution method or changing the SpMV kernel.

In this section, we explain the approach by changing the SpMV kernel for these problems.

So far, ELLPACK Sparse Block (ESB) [10] and Compressed Sparse Row5 (CSR5) [9] have been proposed as new formats to improve SpMV on the MIC. CSR5 takes the load balance into account, whereas ESB does not. Therefore, we only evaluate CSR5 because the SpMV kernel needs to deal with the load balance to improve parallel SpMV.

$$A = \begin{Bmatrix} 0 & 1 & 0 & 0 & 2 & 3 \\ 4 & 5 & 6 & 7 & 8 & 9 \\ 10 & 0 & 0 & 0 & 0 & 0 \\ 0 & 11 & 0 & 12 & 0 & 0 \\ 0 & 0 & 0 & 0 & 0 & 13 \\ 0 & 14 & 0 & 0 & 15 & 0 \end{Bmatrix} \tag{2}$$

$$val = \begin{Bmatrix} 1 & 2 & 3 & 4 \\ 5 & 6 & 7 & 8 \\ 9 & 10 & 11 & 12 \\ 13 & 14 & 15 \end{Bmatrix} \tag{3}$$

$$idx = \begin{Bmatrix} 1 & 4 & 5 & 0 \\ 1 & 2 & 3 & 4 \\ 5 & 0 & 1 & 3 \\ 5 & 1 & 4 \end{Bmatrix} \tag{4}$$

$$index = \begin{Bmatrix} 0 & 1 & 2 & 0 \\ 1 & 2 & 3 & 4 \\ 5 & 0 & 0 & 1 \\ 0 & 0 & 1 \end{Bmatrix} \tag{5}$$

$$ptr = \begin{bmatrix} 0 & 3 & 9 & 10 & 12 & 13 & 15 \end{bmatrix} \tag{6}$$

6.2   Applying Segmented Scan to SpMV

The Segmented Scan method (SS) [13] divides the matrix into a fixed size vector, which is called the segment vector. SS performs load balancing well. This method stores Eq. (2) as Eqs. (3), (4), (5) and (6). In this example, the width of the segment vector is set to 4 elements. SpMV is performed as follows.

1. Multiply A and x by using Eqs. (3) and (4) (write the product to another space because the sparse matrix is repeatedly used)
2. Reduce to the head element of each row by using Eq. (5)
3. Write y by using Eq. (6)

Equation (5) is an index used in the reduction step, and it represents the number of nonzero elements from the head of the row. The index reduces the load unbalance by reduction, as shown in Fig. 9, because it can use multiple threads to process one row.

To the best of our knowledge, no previous work has applied the SS method to the MIC system. However, SS is expected to improve parallel SpMV because it takes into account the load balance and is simpler than CSR5. Therefore, we implemented SS for the MIC and evaluated it in the single process and in the 64 MPI processes.

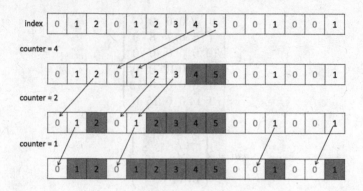

Fig. 9. Example of reduction using *index*

Implementation of SS. Figure 10 shows the simplified pseudocode of SS. In previous work [13], the reduction was done element by element, but the reduction of our implementation was done segment by segment to utilize single-instruction, multiple-data (SIMD) instructions. Also, we applied thread parallelism by using OpenMP in lines 4 and 13. Finally, we calculated the area that must be added before and we removed the branch in line 14.

```
 1: inputs
 2:      Matrix: A,
 3:      Vector: x,
 4: for i = 0 to H − 1 do
 5:     for j = 0 to W − 1 do
 6:         A.val_tmp[i][j] = A.val[i][j] × x[col_idx[i][j]]
 7:     end for
 8: end for
 9: nStep =int(ceil(log2(A.nMax + 1))) {A.nMax represents at most how many
    segments we need to sum up}
10: counter = 1 << nStep
11: for s = 0 to nStep − 1 do
12:     counter >>= 1
13:     for i = 0 to H − 1 do
14:         if counter ≤ A.segment_index[i] ∧ A.segment_index[i] ≤ counter × 2
            then
15:             for j = 0 to W − 1 do
16:                 A.val_tmp[i − counter][j]+ = A.val_tmp[i][j]
17:             end for
18:         end if
19:     end for
20: end for
21: ...
22: # We omit calculation of fractions and storing to vector y.
23: ...
```

Fig. 10. Pseudocode of SS

Performance Evaluation. We used dense, random, band, and unbalance matrices that we generated. In Figs. 11 and 12, the suffix of the matrices represents the number of rows and columns. We fixed the number of nonzero elements to $N \times \sqrt{N}$ except for dense matrices. To evaluate the performance of the MIC and the CPU, we used COMA (Table 2). First, we compared MKL (CRS format), CSR5 that deals with load balance, and SS in a single process. Even though ESB was proposed for the MIC, we did not include it because load balance is not a concern for ESB. For CSR5, we used published implementation [8]. Figures 11 and 12 show the comparisons of MKL, CSR5, and SS in a single process. Figure 11 shows the results of matrices that have fewer nonzero elements than band-8192, and Fig. 12 shows the results of matrices that have either more nonzero elements than band-8192 or nonzero elements equal to band-8192.

Reducing the width of the segment vector (W) of SS can easily keep the load balance. However, this procedure leads to performance degradation because of the increased overhead of assigning threads due to the increased number of steps in the reduction, and because of the increased cache misses due to assigning different threads to the same area.

Therefore, the optimal W depends on the matrices. However, we used $W = 1024$, which achieved relatively high performance in the matrices. The published implementation of CSR5 [8] failed in some matrices from band-1024 random-4096 and we could not evaluate it.

According to Fig. 11, the performance of SS is lower than that of MKL in most matrices. This is because SS calculates the multiplication and reduction parts separately, whereas MKL simply calculates each row.

In this evaluation, we fixed $W = 1024$. Hence, the advantage of SS is removed because reduction becomes a simple summation in matrices that have a small number of nonzero elements.

However, the performance of SS is greater than that of CSR5. This is because CSR5 uses the prefix sum scan. Though it keeps load balancing, it calculates

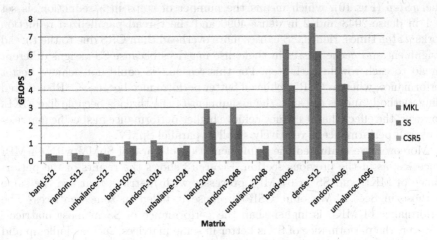

Fig. 11. Comparison of SS and MKL with a single process (for matrices that have a small number of nonzero elements) (Color figure online)

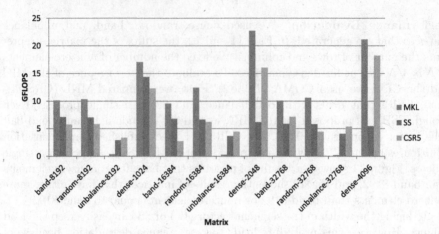

Fig. 12. Comparison of SS and MKL with a single process (for matrices that have a large number of nonzero elements) (Color figure online)

unused elements. When the number of nonzero elements is small, the reduction that is really needed is relatively small. Therefore, the performance of SS was higher than that of CSR5 because the effect of the unused reduction becomes large in CSR5.

The overall performance of MKL is high. However, the performance is extremely small in the unbalance matrices in Fig. 12. This is because MKL could not keep the load balance due to parallelizing in the unit of rows. The performances of SS and CSR5 are stable in these matrices. The results show that SS and CSR5 could show constant performances if the nonzero elements are extremely unbalanced.

Whereas the performances of MKL and CSR5 achieved higher performance in dense-2048 and dense-4096, the performance of SS in these matrices was low. The $nStep$ (Fig. 10), which means the number of steps in a reduction, is set to 1 in dense-2048 and 2 in dense-4096 and the thread parallelized reduction works $nStep$ times. Hence, SS causes more overhead than CRS due to the thread assignment and cache misses in the dense matrices because SS assigns different thread to each row in each step. For this reason, SS could not achieve better performance, whereas CSR5 achieved better performance because CSR5 adopted tiling method, and as a result, the assignment of the threads became like MKL. However, the effect that it cannot achieve higher performance in specific matrices to overall performance is relatively small in parallel SpMV.

Moreover, we evaluated the implementation of parallel SpMV with 64 MPI processes, as in the previous section. Figure 13 shows the results of the performance of MKL and SS in 64 MPI processes with matrices that are the same as those in Sect. 5. We omit CSR5 because it failed in some matrices. The performance of MKL is higher than the performance of SS in most matrices. However, the performance of SS is better in some matrices, such as Fullchip and

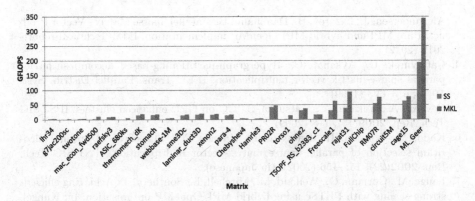

Fig. 13. Comparison of SS and MKL with 64 MPI processes (Color figure online)

circuit5M, because the nonzero elements are unbalanced in Fullchip and more than 8 % nonzero elements are in a single row.

7   Conclusion

In this paper, we presented a performance evaluation of parallel SpMV on CPU, GPU, and MIC clusters. On average, the performance of the implementation for the CPU cluster was improved 42.57 times with 64 MPI processes compared to a single process. However, the performance decreased in some matrices by parallelization because parallelization causes load unbalance and an increase of communication. Using an accelerator achieved higher performance than the implementations of parallel SpMV for the CPU cluster in matrices that have a large number of nonzero elements or a small number of MPI processes. However, the performance of the implementation of parallel SpMV for MIC cluster performance became lower as the number of MPI processes increased because of the increase of communication time and the decreased efficiency of the SpMV kernel.

Therefore, we proposed to apply SS to the implementation for the MIC cluster. As a result, although SS could not achieve higher performance in most matrices, it did achieve higher performance in the unbalanced matrices with 64 MPI processes.

Acknowledgments. This research was supported by Core Research for Evolutional Science and Technology (CREST) of Japan Science and Technology Agency (JST).

References

1. MVAPICH Benchmarks. http://mvapich.cse.ohio-state.edu/benchmarks/
2. Davis, T.: University of Florida Sparse Matrix Collection: sparse matrices from a wide range of applications. http://www.cise.ufl.edu/research/sparse/matrices/

3. Alexandersen, J., Lazarov, B., Dammann, B.: Parallel Sparse Matrix - Vector Product: Pure MPI and hybrid MPI-OpenMP implementation. IMM-Technical report-2012 (2012)
4. Catalyurek, U., Aykanat, C.: Hypergraph-partitioning-based decomposition for parallel sparse-matrix vector multiplication. IEEE Trans. Parallel Distrib. Syst. **10**(7), 673–693 (1999)
5. Cevahir, A., Nukada, A., Matsuoka, S.: CG on GPU-enhanced clusters. IPSJ SIG Tech. Rep. **2009**(15), 1–8 (2009)
6. Kudo, M., Kuroda, H., Katagiri, T., Kanada, Y.: The effect of optimal algorithm selection of parallel sparse matrix-vector multiplication. IPSJ SIG Tech. Rep. **2002**(22), 151–156 (2002). (in Japanese)
7. Lange, M., Gorman, G., Weiland, M., Mitchell, L., Southern, J.: Achieving efficient strong scaling with PETSc using hybrid MPI/OpenMP optimisation. In: Kunkel, J.M., Ludwig, T., Meuer, H.W. (eds.) ISC 2013. LNCS, vol. 7905, pp. 97–108. Springer, Heidelberg (2013)
8. Liu, W., Vinter, B.: bhSPARSEBenchmark SpMV using CSR5. https://github.com/bhSPARSE/Benchmark_SpMV_using_CSR5
9. Liu, W., Vinter, B.: CSR5: An Efficient Storage Format for Cross-Platform Sparse Matrix-Vector Multiplication. CoRR abs/1503.05032 (2015)
10. Liu, X., Smelyanskiy, M., Chow, E., Dubey, P.: Efficient sparse matrix-vector multiplication on x86-based many-core processors. In: Proceedings of the 27th International ACM Conference on International Conference on Supercomputing. ICS 2013, pp. 273–282. ACM (2013)
11. Maeda, H., Takahashi, D.: Performance evaluation of sparse matrix-vector multiplication using GPU/MIC cluster. In: 2015 Third International Symposium on Computing and Networking (CANDAR 2015). 3rd International Workshop on Computer Systems and Architectures (CSA 2015), pp. 396–399 (2015)
12. Monakov, A., Lokhmotov, A., Avetisyan, A.: Automatically tuning sparse matrix-vector multiplication for GPU architectures. In: Patt, Y.N., Foglia, P., Duesterwald, E., Faraboschi, P., Martorell, X. (eds.) HiPEAC 2010. LNCS, vol. 5952, pp. 111–125. Springer, Heidelberg (2010)
13. Ohshima, S., Sakurai, T., Katagiri, T., Nakajima, K., Kuroda, H., Naono, K., Igai, M., Itoh, S.: Optimized implementation of segmented scan method for CUDA. IPSJ Tech. Rep. 2010-HPC-**126**(1), 1–7 (2010). (in Japanese)
14. Pinar, A., Heath, M.T.: Improving performance of sparse matrix-vector multiplication. In: Proceedings of the 1999 ACM/IEEE Conference on Supercomputing. SC 1999. ACM (1999)
15. Saule, E., Kaya, K.: Performance evaluation of sparse matrix multiplication kernels on intel Xeon Phi. In: Wyrzykowski, R., Dongarra, J., Karczewski, K., Waśniewski, J. (eds.) Parallel Processing and Applied Mathematics. LNCS, vol. 8384, pp. 559–570. Springer, Heidelberg (2014)
16. Tang, W.T., Tan, W.J., Ray, R., Wong, Y.W., Chen, W., Kuo, S., Goh, R.S.M., Turner, S.J., Wong, W.: Accelerating sparse matrix-vector multiplication on GPUs using bit-representation-optimized schemes. In: Proceedings of the International Conference on High Performance Computing, Networking, Storage and Analysis. SC 2013, pp. 26:1–26:12 (2013)
17. Ye, F., Calvin, C., Petiton, S.G.: A study of SpMV implementation using MPI and OpenMP on intel many-core architecture. In: Daydé, M., Marques, O., Nakajima, K. (eds.) VECPAR 2014. LNCS, vol. 8969, pp. 43–56. Springer, Heidelberg (2015)

On the Cluster-Connectivity of Wireless Sensor Networks

H.K. Dai[1]([✉]) and H.C. Su[2]

[1] Computer Science Department, Oklahoma State University,
Stillwater, OK 74078, USA
dai@cs.okstate.edu
[2] Department of Computer Science, Arkansas State University,
Jonesboro, AR 72401, USA
suh@astate.edu

Abstract. Wireless sensor networks consist of sensor devices with limited energy resources and computational capabilities, hence network optimization and algorithmic development in minimizing the total energy or power to maintain the connectivity of the underlying network are crucial for their design and maintenance. We consider a generalized system model of wireless sensor networks whose node set is decomposed into multiple clusters, and show that the decision and the induced minimization problems of the cluster-connectivity of wireless sensor networks appear to be computationally intractable – completeness and hardness, respectively, for the nondeterministic polynomial-time complexity class. An approximation algorithm is devised to minimize the number of endnodes of inter-cluster edges within a factor of 2 of the optimum for the cluster-connectivity.

Keywords: Wireless sensor network · Connectivity · Nondeterministic polynomial-time complexity class · Approximation algorithm

1 Wireless Sensor Networks

Wireless sensor networks have wide range of applications involving monitoring, tracking, and controlling. A wireless sensor network consists of (numerous) sensor devices/nodes with limited energy resources, computational capabilities, and memory, in which broadcasting is an important communication mechanism. Consequently, a crucial design consideration for wireless sensor networks is to maximize their system energy efficiency [13,16].

1.1 System Model

Many studies consider a system model [15] of wireless sensor networks of stationary nodes with:

1. The availability of wide spectrum of bandwidth resources to alleviate contention for communication channels,

O. Gervasi et al. (Eds.): ICCSA 2016, Part II, LNCS 9787, pp. 19–33, 2016.
DOI: 10.1007/978-3-319-42108-7_2

2. The broadcast communication via omnidirectional antennas equipped for individual nodes, and

3. A finite set of node-power levels that provides, in practice, a discretized spectrum of energy-consumption requirements for communicating pairs of nodes to sustain their transmission links – as evidenced by the discrete node-power levels of commercially available sensors.

1.2   Graph-Theoretic Model

Most graph-theoretic definitions used in this paper are given in [1]. We will abbreviate "directed graph" to "digraph". For an undirected graph or digraph G, denote by $V(G)$ and $E(G)$ the node set and edge set of G, respectively. For a subset V' of nodes in G, the subgraph of G induced by V' is denoted by $G[V']$.

Denote by $\mathbb{R}_+$ the set of all positive reals. A wireless sensor network with a finite set P ($\subseteq \mathbb{R}_+$) of node-power levels can be modeled (as in [15]) by a digraph G in which $V(G)$ represents the set of sensor nodes and $E(G)$ represents the set of all possible unidirectional communication links between adjacent nodes.

The model is equipped with two functions:

1. A link-cost function $c : E(G) \rightarrow \mathbb{R}_+$ such that, for each directed arc/link $(u, v) \in E(G)$, its link-cost $c((u, v))$ yields the minimum power requirement for establishing/maintaining the link (u, v) – which depends on physical factors such as the node-pair distance and orientation of sender node, and environmental factors such as interference and noise, and

2. A node-power assignment function $p : V(G) \rightarrow P$ such that, for each sensor node $u \in V(G)$, $p(u)$ measures the transmission-power level assigned to u.

Hence, a transmission from a sender node u to a receiver node v of $V(G)$ via the link (u, v) entails that $p(u) \geq c((u, v))$.

Analogously, we can model a wireless sensor network based on an undirected graph [10], whose details are omitted.

1.3   Minimum Total Energy Topology Problems: Nondeterministic Polynomial-Time Completeness

Two core problems can be identified in designing and maintaining wireless sensor networks: network optimization and algorithmic development in minimizing the total energy/power to maintain the connectivity of the underlying network/graph.

Based on the link directionality (unidirectional/bidirectional), graph connectivity (strongly digraph/weakly graph/higher-order node- or edge-connectivity) and the topological constraints on connected node/cluster-spanning subgraphs, variations of these optimization and algorithmic problems have been formulated and studied in the literature [2–4, 10, 11, 15].

Generally, the source of complexity is in two forms: the number of possible topologically-constrained connected node/cluster-spanning subgraphs, subject to prescribed allowable discrete node-power assignments. The complexity of most minimum total energy/power topology problems appear to be computationally intractable – completeness or hardness for the nondeterministic polynomial-time ($\mathcal{NP}$) complexity class (see studies in [3,12,15] and references therein).

2   Clustered Wireless Sensor Networks

We consider a generalized system model of wireless sensor networks whose node set is decomposed into multiple clusters [14]. Wireless multi-radio sensor networks [5,6,8,9,12] are realization of such model.

2.1   System Model

The sensors of a clustered wireless sensor network are (often hierarchically) organized into clusters. The aggregation of sensor nodes may be based on their intra-cluster versus inter-cluster spatial proximity, service attributes, etc.

The communication protocol usually works in two phases:

1. Intra-cluster broadcasting: a source/sender node broadcasts with usually low/moderate node-power consumption to all other nodes of the same cluster, and
2. Inter-cluster cooperative transmission: all receiver nodes collaborate the transmission to the destination node with the same node-power level.

2.2   Graph-Theoretic Model

For our complexity study, it suffices to consider a dual node-power levels: $P = \{p_{\text{low}}, p_{\text{high}}\}$ with $p_{\text{low}} < p_{\text{high}}$.

We model the studied clustered wireless sensor networks with:

1. A connected undirected graph G in which $V(G)$ represents the set of sensor nodes and $E(G)$ represents the sets of all possible bidirectional links between adjacent nodes, and
2. A non-empty finite set $\mathcal{C}$ of non-empty connected clusters that partitions $V(G)$: (1) every cluster in $\mathcal{C}$ spans a connected subgraph of G, (2) for all $C, C' \in \mathcal{C}$, either $C \cap C' = \emptyset$ or $C = C'$ (but not both), and (3) $\cup_{C \in \mathcal{C}} C = V(G)$.

The edge set $E(G)$ is partitioned into two categories: inter-cluster edges versus intra-cluster ones. A boundary node of G is an endnode of an inter-cluster edge of G.

In the context of cluster-connectivity, two distinct clusters A and B are: (1) cluster-adjacent if there exist boundary nodes $a \in V(A)$ and $b \in V(B)$ such that $\{a, b\}$ is an inter-cluster edge, and (2) cluster-connected if either A and B are cluster-adjacent or $A, C_1, C_2, \ldots, C_n, B$ is a sequence of all-distinct successively cluster-adjacent clusters for some clusters $C_1, C_2, \ldots, C_n$ of $\mathcal{C}$ where $n \geq 1$ (note that any two successive clusters may not have a common boundary noes in the "intermediate" cluster).

The link-cost function $c : E(G) \to \mathbb{R}_+$ associates: for each intra-cluster edge/link $\{u, v\}$ (within a connected cluster), $c(\{u, v\}) = p_{\text{low}}$, and for each inter-cluster edge/link $\{u, v\}$, $c(\{u, v\}) = p_{\text{high}}$.

2.3   Connectivity Problems of Clustered Wireless Sensor Networks

The simplicity/duality of the node-power level set P of G translate the optimization of energy topology to the connectivity of the underlying clusters of G as follows. We consider two connected node-spanning topologies: spanning tree and Hamiltonian path. The complexities of the Spanning Tree and Hamiltonian Path problems for general graphs are deterministic polynomial-time solvable and nondeterministic polynomial-time complete, respectively.

Note that in the context of each of the following cluster-connectivity problems, we may disregard the intra-connectedness of each cluster and assume that each cluster is an empty subgraph (with void edge set).

1. Clustered-Graph Spanning Tree
   - Problem Instance: An undirected graph G with a non-empty finite set $\mathcal{C}$ of non-empty clusters of G that partitions $V(G)$.
   - Decision: Is there a set $R \subseteq V(G)$ of cluster-representatives (that is, for each $C \in \mathcal{C}$, $|V(C) \cap R| = 1$) such that the subgraph of G induced by R, $G[R]$, is connected (or equivalently, $G[R]$ embeds a node-spanning tree)?
2. Clustered-Graph Hamiltonian Path
   - Problem Instance: An undirected graph G with a non-empty finite set $\mathcal{C}$ of non-empty clusters of G that partitions $V(G)$.
   - Decision: Is there a set $R \subseteq V(G)$ of cluster-representatives (that is, for each $C \in \mathcal{C}$, $|V(C) \cap R| = 1$) such that the subgraph of G induced by R, $G[R]$, is a Hamiltonian path?

A related topology-control problem of hierarchically structured wireless networks is studied in [12], in which the source of complexity lies in finding a node-spanning tree with constrained number of non-pendant/non-leaf nodes (cluster-heads) with the minimum total node-power over a prescribed set of discrete node-power levels:

k-Distinct-r Strong Minimum Energy Hierarchical Topology

- Problem Instance: An n-node complete undirected graph K_n, an edge/link-cost function c with a discrete cardinality-k set P of node-power levels (that is, $c : E(K_n) \to P$), and two threshold constants: a positive integer r for the number of non-pendant nodes, and $M \in \mathbb{R}_+$ for the total node-power.
- Decision: Is there a node-spanning tree T of K_n such that T has at most r non-pendant nodes and $\sum_{u \in V(T)} \max_{\{u, v\} \in E(T)} c(\{u, v\})$ (total node-power of T) $\leq M$?

The above topology-control k-Distinct-r Strong Minimum Energy Hierarchical Topology problem is proven to be $\mathcal{NP}$-complete in [12], accompanied with, when the threshold constant r is fixed, an $\frac{r+1}{2}$-approximation algorithm for the problem with arbitrary k.

3   Clustered-Graph Connectivity Problems: Nondeterministic Polynomial-Time Completeness

We prove the $\mathcal{NP}$-completeness of the Clustered-Graph Spanning Tree problem via a deterministic polynomial-time reduction from the 3-Dimensional Matching problem, which is known to be $\mathcal{NP}$-complete [7]:

3-Dimensional Matching

- Problem Instance: Three mutually-disjoint sets X, Y, and Z, all with the same cardinality n, and a set $T \subseteq X \times Y \times Z$ of ordered triples.
- Decision: Is there a cardinality-n subset $M \subseteq T$ of ordered triples such that M is a 3-dimensional matching saturating X, Y, and Z: for all ordered triples $(x_1, y_1, z_1), (x_2, y_2, z_2) \in M$, if $(x_1, y_1, z_1) \neq (x_2, y_2, z_2)$ then $x_1 \neq x_2$ and $y_1 \neq y_2$ and $z_1 \neq z_2$?

Theorem 1. *The Clustered-Graph Spanning Tree problem is $\mathcal{NP}$-complete.*

Proof. We first show that the Clustered-Graph Spanning Tree problem is in $\mathcal{NP}$ with a deterministic polynomial-time verifier for the problem. Given an instance of the problem: an undirected graph G with a non-empty finite set $\mathcal{C}$ of N non-empty clusters $C_1, C_2, \ldots, C_N$ of G partitioning $V(G)$, a certificate corresponds to a cardinality-N subset $R \subseteq V(G)$, and the verification algorithm checks that R is a set of cluster-representatives (that is, for each $i = 1, 2, \ldots, N$, $|V(C_i) \cap R| = 1$) and the subgraph $G[R]$ of G induced by R is connected – both of which can be decided in polynomial time.

To prove that the Clustered-Graph Spanning Tree problem is $\mathcal{NP}$-hard, we construct below a desired deterministic polynomial-time reduction that transforms a problem instance of the 3-Dimensional Matching problem to one of the Clustered-Graph Spanning Tree problem.

Consider a problem instance $(X, Y, Z; T)$ of the 3-Dimensional Matching problem: three mutually-disjoint sets X, Y, and Z, all with the same cardinality n, and a set $T \subseteq X \times Y \times Z$ of ordered triples. Enumerate the sets X, Y, and Z as $(x_1, x_2, \ldots, x_n)$, $(y_1, y_2, \ldots, y_n)$, and $(z_1, z_2, \ldots, z_n)$, respectively. Construct the corresponding undirected graph G with its vertex set $V(G)$ partitioned into the family $\mathcal{C}$ of clusters in four ranks and edge set $E(G)$ consisting of inter-cluster edges between successive ranks as follows.

1. Partition of $V(G)$ into $\mathcal{C} = \mathcal{X} \cup \mathcal{Y} \cup \mathcal{Z} \cup \mathcal{W}$:
   (a) Rank $\mathcal{X}$ of n singleton clusters: for $\eta = 1, 2, \ldots, n$,

   $$\mathcal{X}_\eta = \{x_\eta\}.$$

   (b) Rank $\mathcal{Y}$ of n order-n clusters: for $\eta = 1, 2, \ldots, n$,

   $$\mathcal{Y}_\eta = \{y_{\eta,\kappa} \mid \kappa = 1, 2, \ldots, n\}.$$

   (c) Rank $\mathcal{Z}$ of n order-n clusters: for $\eta = 1, 2, \ldots, n$,

   $$\mathcal{Z}_\eta = \{z_{\eta,\kappa} \mid \kappa = 1, 2, \ldots, n\}.$$

(d) Rank $\mathcal{W}$ of a singleton cluster:

$$\mathcal{W} = \{w\}.$$

2. Structure of $E(G)$:
   (a) For every ordered triple $(x_i, y_j, z_k) \in T$ where $i, j, k \in \{1, 2, \ldots, n\}$, include the two inter-cluster edges $\{x_i, y_{j,i}\}$ and $\{y_{j,i}, z_{k,i}\}$ in $E(G)$.
   (b) For all $\eta \in \{1, 2, \ldots, n\}$ and $\kappa \in \{1, 2, \ldots, n-1\}$, include the inter-cluster edges $\{z_{\eta,\kappa}, x_{\kappa+1}\}$ in $E(G)$.
   (c) For all $\eta \in \{1, 2, \ldots, n\}$, include the inter-cluster edges $\{z_{\eta,n}, w\}$ in $E(G)$.

Figure 1 illustrates an example problem instance of the 3-Dimensional Matching and its corresponding problem instance of the Clustered-Graph Spanning Tree.

Note that:

1. The node set $V(G)$ is of cardinality $2n^2 + n + 1$ and is partitioned into $3n + 1$ clusters in the four ranks, and the edge set $E(G)$ is of cardinality at most $2|T| + (n-1)n + n \, (= 2|T| + n^2)$, and
2. The above correspondence, denoted by f, from a problem instance $(X, Y, Z; T)$ of the 3-Dimensional Matching to the problem instance $f((X, Y, Z; T)) = (G, \mathcal{C})$ of the Clustered-Graph Spanning Tree is a deterministic polynomial-time transformation.

We show that the deterministic polynomial-time transformation is indeed a reduction from the 3-Dimensional Matching to the Clustered-Graph Spanning Tree below.

Lemma 1. *The deterministic polynomial-time transformation f is a reduction from the 3-Dimensional Matching problem to the Clustered-Graph Spanning Tree problem.*

Proof. Firstly, assume that a problem instance $(X, Y, Z; T)$ of the 3-Dimensional Matching embeds a 3-dimensional matching in T, that is, there exists a sequence of n ordered triples in T, which is enumerated as:

$$(x_1, y_{j_1}, z_{k_1}), (x_2, y_{j_2}, z_{k_2}), \ldots, (x_n, y_{j_n}, z_{k_n})$$

such that the two index sets $\{j_1, j_2, \ldots, j_n\}$ and $\{k_1, k_2, \ldots, k_n\}$ are permutations of $\{1, 2, \ldots, n\}$.

Consider the corresponding problem instance under the above transformation f: $f((X, Y, Z; T)) = (G, \mathcal{C})$. The transformation f yields that, for each $i = \{1, 2, \ldots, n\}$, the ordered triple (x_i, y_{j_i}, z_{k_i}) gives rise to two inter-cluster edges in G: $\{x_i, y_{j_i,i}\}$ (joining the clusters $\mathcal{X}_i$ and $\mathcal{Y}_{j_i}$) and $\{y_{j_i,i}, z_{k_i,i}\}$ (joining the clusters $\mathcal{Y}_{j_i}$ and $\mathcal{Z}_{k_i}$).

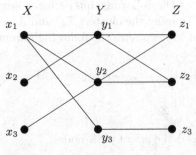

$$X = \{x_1, x_2, x_3\}, Y = \{y_1, y_2, y_3\}, Z = \{z_1, z_2, z_3\}$$
$$T = \{\, (x_1, y_1, z_1), (x_1, y_2, z_1), (x_1, y_2, z_2), (x_1, y_3, z_3),$$
$$(x_2, y_1, z_1), (x_2, y_1, z_2),$$
$$(x_3, y_2, z_1), (x_3, y_2, z_2)\}$$

(a) problem instance $(X, Y, Z; T)$ of the 3-Dimensional Matching

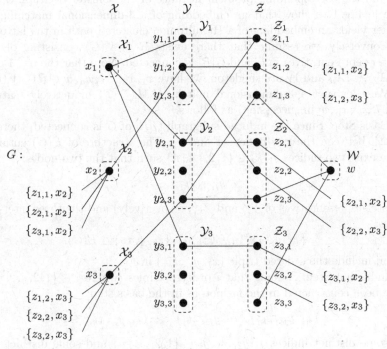

(b) corresponding problem instance $(G, \mathcal{C})$ of
the Clustered-Graph Spanning Tree

Fig. 1. A deterministic polynomial-time reduction from the 3-Dimensional Matching to the Clustered-Graph Spanning Tree: (a) example problem instance of the 3-Dimensional Matching; (b) the corresponding problem instance of the Clustered-Graph Spanning Tree.

With the inclusion of the following inter-cluster edges in G: for each $i = 1, 2, \ldots, n$, $\{z_{k_i,i}, x_{i+1}\}$ (joining the clusters $\mathcal{Z}_{k_i}$ and $\mathcal{X}_{i+1}$) and $\{z_{k_n,n}, w\}$ (joining the clusters $\mathcal{Z}_{k_n}$ and $\mathcal{W}$), we have the following (Hamiltonian) path saturating all the clusters of $\mathcal{C}$:

$$\mathcal{X}_1 \xrightarrow{\{x_1,y_{j_1,1}\}} \mathcal{Y}_{j_1} \xrightarrow{\{y_{j_1,1},z_{k_1,1}\}} \mathcal{Z}_{k_1} \xrightarrow{\{z_{k_1,1},x_2\}} \mathcal{X}_2 \xrightarrow{\{x_2,y_{j_2,2}\}} \mathcal{Y}_{j_2} \xrightarrow{\{y_{j_2,2},z_{k_2,2}\}} \mathcal{Z}_{k_2} \xrightarrow{\{z_{k_2,2},x_3\}}$$

$$\mathcal{X}_3 \xrightarrow{\{x_3,y_{j_3,3}\}} \cdots \xrightarrow{\{y_{j_n,n},z_{k_n,n}\}} \mathcal{Z}_{k_n} \xrightarrow{\{z_{k_n,n},w\}} \mathcal{W}$$

Denote by $R\,(\subseteq V(G))$ the set of path-nodes:

$$\{x_1, y_{j_1,1}, z_{k_1,1}, x_2, y_{j_2,2}, z_{k_2,2}, \ldots, x_n, y_{j_n,n}, z_{k_n,n}, w\}.$$

Since the index sets $\{j_1, j_2, \ldots, j_n\}$ and $\{k_1, k_2, \ldots, k_n\}$ are permutations of $\{1, 2, \ldots, n\}$, R is a set of $3n + 1$ cluster-representatives in G, that is, for every cluster C of G, $|V(C) \cap R| = 1$. Clearly, the subgraph (path) induced by R, $G[R]$, is connected, as desired.

Figure 2 employs the example problem instance of the 3-Dimensional Matching and the corresponding problem instance of the Clustered-Graph Spanning Tree in Fig. 1 to show that an embedding of a 3-dimensional matching in the former yields an embedding of a Hamiltonian clustered-path in the latter.

Conversely, we assume that there exists $R \subseteq V(G)$ consisting of $3n + 1$ cluster-representatives in G and $G[R]$ is connected. Note that the $n + 1$ clusters $\mathcal{X}_1, \mathcal{X}_2, \ldots, \mathcal{X}_n$ and $\mathcal{W}$ are singleton, we have $x_1, x_2, \ldots, x_n, w \in R\,(\subseteq V(G[R]))$.

We construct a 3-dimensional matching $M\,(\subseteq T)$ inductively (saturating $x_1, x_2, \ldots, x_n, w$ in succession) as follows.

Basis Step: Since the induced subgraph $G[R]$ of G is connected, there exists a path between the two nodes x_1 and x_2. The structure of $E(G)$ entails that there exist two indices $j_1, k_1 \in \{1, 2, \ldots, n\}$ such that the two nodes:

$$y_{j_1,1}, z_{k_1,1} \in R$$

(cluster-representatives of $\mathcal{Y}_{j_1}$ and $\mathcal{Z}_{k_1}$, respectively) and the three inter-cluster edges:

$$\{x_1, y_{j_1,1}\}, \{y_{j_1,1}, z_{k_1,1}\}, \{z_{k_1,1}, x_2\} \in E(G).$$

Then, include the ordered triple (x_1, y_{j_1}, z_{k_1}) in M.

Induction Step: Assume that i ordered triples (where $i \in \{1, 2, \ldots, n - 1\}$) have been constructed in the fashion as in the basis step:

$$(x_1, y_{j_1}, z_{k_1}), (x_2, y_{j_2}, z_{k_2}), \ldots, (x_i, y_{j_i}, z_{k_i}) \in M$$

for some distinct indices $j_1, j_2, \ldots, j_i \in \{1, 2, \ldots, n\}$ and some distinct indices $k_1, k_2, \ldots, k_i \in \{1, 2, \ldots, n\}$ such that the $2i$ nodes:

$$y_{j_1,1}, z_{k_1,1}, y_{j_2,2}, z_{k_2,2}, \ldots, y_{j_i,i}, z_{k_i,i} \in R$$

(cluster-representatives of $\mathcal{Y}_{j_1}, \mathcal{Z}_{k_1}, \mathcal{Y}_{j_2}, \mathcal{Z}_{k_2}, \ldots, \mathcal{Y}_{j_i}, \mathcal{Z}_{k_i}$, respectively) and the $3i$ inter-cluster edges:

$$\{x_1, y_{j_1,1}\}, \{y_{j_1,1}, z_{k_1,1}\}, \{z_{k_1,1}, x_2\}, \ldots, \{z_{k_i,i}, x_{i+1}\} \in E(G[R])\,(= E(G)).$$

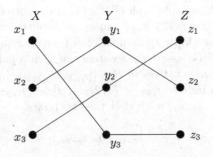

$$X = \{x_1, x_2, x_3\}, Y = \{y_1, y_2, y_3\}, Z = \{z_1, z_2, z_3\}$$
$$T = \{(x_1, y_1, z_1), (x_1, y_2, z_1), (x_1, y_2, z_2), (x_1, y_3, z_3),$$
$$(x_2, y_1, z_1), (x_2, y_1, z_2),$$
$$(x_3, y_2, z_1), (x_3, y_2, z_2)\}$$

(a) problem instance $(X, Y, Z; T)$ of the 3-Dimensional Matching embeds
   a 3-dimensional matching in T: $\{(x_1, y_3, z_3), (x_2, y_1, z_2), (x_3, y_2, z_1)\}$

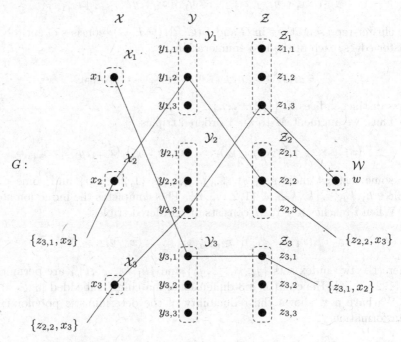

(b) corresponding problem instance $(G, \mathcal{C})$ of
   the Clustered-Graph Spanning Tree embeds a Hamiltonian clustered-path:
   $x_1, y_{3,1}, z_{3,1}, x_2, y_{1,2}, z_{2,2}, x_3, y_{2,3}, z_{1,3}, w$

Fig. 2. With the example problem instance of the 3-Dimensional Matching and the corresponding problem instance of the Clustered-Graph Spanning Tree in Fig. 1, an embedding of a 3-dimensional matching in the problem instance gives rise to a Hamiltonian clustered-path in the corresponding problem instance.

Now, since the induced subgraph $G[R]$ of G is connected, there exists a path in $G[R])$ between the two nodes x_{i+1} and x_{i+2} (note that x_{i+2} denotes w when $i = n - 1$). The membership requirement of R (cluster-representativeness in G) and the structure of $E(G)$ assert the existence of such a path in $G[R]$ from x_{i+1} to x_{i+2} via cluster-representatives in $R - \{y_{j_1,1}, z_{k_1,1}, y_{j_2,2}, z_{k_2,2}, \ldots, y_{j_i,i}, z_{k_i,i}\}$; hence there exist two indices $j_{i+1} \in \{1, 2, \ldots, n\} - \{j_1, j_2, \ldots, j_i\}$ and $k_{i+1} \in \{1, 2, \ldots, n\} - \{k_1, k_2, \ldots, k_i\}$ such that the two nodes:

$$y_{j_{i+1},i+1}, z_{k_{i+1},i+1} \in R - \{y_{j_1,1}, z_{k_1,1}, y_{j_2,2}, z_{k_2,2}, \ldots, y_{j_i,i}, z_{k_i,i}\}$$

(cluster-representative of $\mathcal{Y}_{j_{i+1}}$, $\mathcal{Z}_{k_{i+1}}$, respectively) and the three inter-cluster edges:

$$\{x_{i+1}, y_{j_{i+1},i+1}\}, \{y_{j_{i+1},i+1}, z_{k_{i+1},i+1}\}, \{z_{k_{i+1},i+1}, x_{i+2}\} \in E(G[R]) (= E(G)).$$

Since the $2i + 2$ nodes:

$$y_{j_1,1}, z_{k_1,1}, y_{j_2,2}, z_{k_2,2}, \ldots, y_{j_i,i}, z_{k_i,i}, y_{j_{i+1},i+1}, z_{k_{i+1},i+1}$$

are cluster-representatives in G and $E(G[R]) (= E(G))$ consists of entirely inter-cluster edges, each of the two enumerations:

$$j_1, j_2, \ldots, j_i, j_{i+1} \text{ and } k_1, k_2, \ldots, k_i, k_{i+1}$$

gives distinct indices in $\{1, 2, \ldots, n\}$.

Thus, we augment M to $i + 1$ ordered triples:

$$(x_1, y_{j_1}, z_{k_1}), (x_2, y_{j_2}, z_{k_2}), \ldots, (x_i, y_{j_i}, z_{k_i}), (x_{i+1}, y_{j_{i+1}}, z_{k_{i+1}})$$

for some distinct indices $j_1, j_2, \ldots, j_i, j_{i+1} \in \{1, 2, \ldots, n\}$ and some distinct indices $k_1, k_2, \ldots, k_i, k_{i+1} \in \{1, 2, \ldots, n\}$. This completes the induction step.

When i reaches $n - 1$, M consists of n ordered triples:

$$(x_1, y_{j_1}, z_{k_1}), (x_2, y_{j_2}, z_{k_2}), \ldots, (x_n, y_{j_n}, z_{k_n})$$

where the two index sets $\{j_1, j_2, \ldots, j_n\}$ and $\{k_1, k_2, \ldots, k_n\}$ are permutations of $\{1, 2, \ldots, n\}$. Hence M is a 3-dimensional matching embedded in T.

We have now shown the reducibility of the deterministic polynomial-time transformation f.   □

This completes the proof of the $\mathcal{NP}$-completeness of the Clustered-Graph Spanning Tree problem.   □

The complexity of the Clustered-Graph Hamiltonian Path problem is $\mathcal{NP}$-completeness – when considering its general-graph version. In fact, the deterministic polynomial-time reduction in proving Theorem 1 shows the $\mathcal{NP}$-completeness of the Clustered-Graph Hamiltonian Path problem in Theorem 2.

Theorem 2. *The Clustered-Graph Hamiltonian Path problem is $\mathcal{NP}$-complete.*

4   Minimization of Clustered-Graph Connectivity: Nondeterministic Polynomial-Time Hardness, and Approximation Algorithm

The Clustered-Graph Connectivity decision problem induces an optimization problem: Minimum-Boundary Clustered-Graph Spanning Tree as follow. Analogous to the context of the Clustered-Graph Spanning Tree and Hamiltonian Path problems, we may assume that each cluster of an underlying graph is an empty subgraph (with void edge set).

Minimum-Boundary Clustered-Graph Spanning Tree

- Problem Instance: An undirected graph G with a non-empty finite set $\mathcal{C}$ of non-empty clusters of G that partitions $V(G)$.
- Computation: Determine a minimum-cardinality boundary set $R \subseteq V(G)$ of boundary nodes that can support the cluster-connectivity (connecting every pair of distinct clusters) of G.

The Minimum-Boundary Clustered-Graph Spanning Tree abstracts the networks of dual-band radio devices with a low-power/short-range and a high-power/long-range radios – commercially available off-the-shelf. The optimization problem captures a practical setting of such networks in which: (1) all the low-power/short-range radios are always active by default (represented by the connected clusters of the underlying graph), and (2) the number of high-power/long-range radios is minimized to maintain the global connectivity (represented by a minimum-cardinality boundary set in the underlying graph).

We cast the Minimum-Boundary Clustered-Graph Spanning Tree problem as a related decision problem by imposing an upper bound on the cardinality of a boundary set supporting the cluster-connectivity in the underlying graph. In our context, the decision problem related to the Minimum-Boundary Clustered-Graph Spanning Tree problem is:

Boundary Clustered-Graph Spanning Tree

- Problem Instance: An undirected graph G with a non-empty finite set $\mathcal{C}$ of non-empty clusters of G that partitions $V(G)$, and an integer b.
- Decision: Is there a boundary set $R \subseteq V(G)$ of boundary nodes that can support the cluster-connectivity (connecting every pair of distinct clusters) of G with $|R| \leq b$?

Theorem 3. *The Boundary Clustered-Graph Hamiltonian Path problem is $\mathcal{NP}$-complete.*

Proof. To see that the Boundary Clustered-Graph Spanning Tree problem is in $\mathcal{NP}$, we describe a deterministic polynomial-time verifier for the problem. Given an instance of the problem: an undirected graph G with a non-empty finite set $\mathcal{C}$ of non-empty clusters of G that partitions $V(G)$, and an integer b, we show that a certificate corresponding to a subset $R \subseteq V(G)$ consisting of boundary nodes that can support the cluster-connectivity of G with $|R| \leq b$

can be verified in deterministic polynomial time. The verification algorithm can achieve the following tasks in deterministic polynomial time:

1. Deriving a clustered-graph G' corresponding to G such that: (1) $V(G')$ is the set of $\mathcal{C}$ of all non-empty clusters of G, and (2) $E(G')$ represents the cluster-adjacency of G in which $\{C_1, C_2\} \in E(G')$ if and only if C_1 and C_2 are cluster-adjacent (that is, there exist boundary nodes $u_1 \in V(C_1)$ and $u_2 \in V(C_2)$ such that $\{u_1, u_2\}$ is an inter-cluster edge of G), and
2. Checking that: (1) R is a set of boundary nodes supporting the cluster-connectivity of G – via the connectivity of G', and (2) $|R| \le b$.

We prove the $\mathcal{NP}$-hardness of the problem with a deterministic polynomial-time reduction that transforms a problem instance of the Clustered-Graph Spanning Tree problem ($\mathcal{NP}$-complete according to Theorem 1) to a problem instance of the Boundary Clustered-Graph Spanning Tree problem. The reduction transforms a problem instance of the Clustered-Graph Spanning Tree problem: an undirected graph G with a non-empty finite set $\mathcal{C}$ of non-empty clusters of G that partitions $V(G)$ to the problem instance of the Boundary Clustered-Graph Spanning Tree problem: the undirected graph G with the non-empty finite set $\mathcal{C}$ of non-empty clusters of G that partitions $V(G)$, and the integer $b = |\mathcal{C}|$. We note that the following statements are equivalent:

1. There exists a set $R \subseteq V(G)$ of cluster-representatives (that is, for every $C \in \mathcal{C}$, $|V(C) \cap R| = 1$) such that the subgraph of G induced by R, $G[R]$, is connected.
2. There exists a set $R \subseteq V(G)$ such that for every $C \in \mathcal{C}$, $|V(C) \cap R| = 1$, and every pair of distinct clusters in G are cluster-connected.
3. There exists a set $R \subseteq V(G)$ of boundary nodes that can support the cluster-connectivity of G with $|R| \le |\mathcal{C}|$ $(= b)$.

The transformation and the verification via the equivalence can be accomplished in deterministic polynomial time, which give the $\mathcal{NP}$-hardness of the Boundary Clustered-Graph Spanning Tree problem.   □

To cope with the $\mathcal{NP}$-hardness of the Minimum-Boundary Clustered-Graph Spanning Tree problem, we present below a polynomial-time approximation algorithm Approx-Boundary that computes near-optimal solutions – a boundary set whose cardinality is guaranteed to be less than twice the cardinality of an optimal boundary set.

The input to the approximation algorithm Approx-Boundary is an undirected graph G with a non-empty finite set $\mathcal{C}$ of non-empty clusters of G that partitions $V(G)$ (as noted above, we assume that each cluster is an empty subgraph with void edge set – disregarding any intra-cluster edges).

The algorithm incrementally maintains: (1) a set R of selected boundary nodes, (2) the family $\mathcal{C}(R)$ of hosting clusters of the boundary nodes of R, and (3) the set $e(R)$ of selected inter-cluster edges incident with the boundary nodes of R, such that $C(R)$ always form a single tree of cluster-connected clusters via the inter-cluster edges of $e(R)$.

All the inter-cluster edges between (distinct) clusters of $C(R)$ are considered as inactive. The algorithm initializes the statistics of the cluster-neighborhood and cluster-degree of a node u, denoted by $\Gamma(u)$ and $d(u)$ respectively, and incrementally updates the statistics with inactive inter-cluster edges: (1) for all inter-cluster edges between u and endnodes of a common cluster, exactly one endnode is arbitrarily included in $\Gamma(u)$, and (2) $d(u) = |\Gamma(u)|$.

Algorithm. Approx-Boundary (G, C)

Require: An undirected graph G with a non-empty finite set C of non-empty clusters of G that partitions $V(G)$.

Ensure: A boundary set $R \subseteq V(G)$ of boundary nodes that supports the cluster-connectivity of G via a tree of cluster-connected clusters formed by the inter-cluster edges in $e(R)$.

1: $R := \emptyset$; $C(R) := \emptyset$; $e(R) := \emptyset$;
2: For every node $u \in V(G)$, compute $\Gamma(u)$ and $d(u)$;

3: Let $u_{\max}$ be a node of $V(G)$ with the maximum cluster-degree:
   $u_{\max} := \arg\max\{d(v) \mid v \in V(G)\}$ (breaking ties arbitrarily);
4: $R := \{u_{\max}\} \cup \Gamma(u_{\max})$;
   $C(R) := \{C \in C \mid V(C) \cap R \neq \emptyset\}$;
   $e(R) := \{\{u_{\max}, v\} \mid v \in \Gamma(u_{\max})\}$;

5: Inactivate all the inter-cluster edges between (distinct) clusters in $C(R)$;
   For every $u \in \cup_{C \in C} V(C)$, update $\Gamma(u)$ and $d(u)$:
   $\Gamma(u) := \Gamma(u) - \{v \in \Gamma(u) \mid v \in \cup_{C \in C(R)} V(C)\}$;
   $d(u) := |\Gamma(u)|$;

6: while $C(R) \neq C$ loop
7:    Let $u_{\max}$ be a node of $\cup_{C \in C(R)} V(C)$ with the maximum cluster-degree
      via active inter-cluster edges:
      $u_{\max} := \arg\max\{d(v) \mid v \in \cup_{C \in C(R)} V(C)\}$ (breaking ties arbitrarily);
8:    $R := R \cup (\{u_{\max}\} \cup \Gamma(u_{\max}))$;
      $C(R) := C(R) \cup \{C \in C \mid V(C) \cap \Gamma(u_{\max}) \neq \emptyset\}$;
      $e(R) := e(R) \cup \{\{u_{\max}, v\} \mid v \in \Gamma(u_{\max})\}$;

9:    Inactivate all the inter-cluster edges between (distinct) clusters in $C(R)$;
      For every $u \in \cup_{C \in C(R)} V(C)$, update $\Gamma(u)$ and $d(u)$:
      $\Gamma(u) := \Gamma(u) - \{v \in \Gamma(u) \mid v \in \cup_{C \in C(R)} V(C)\}$;
      $d(u) := |\Gamma(u)|$;

10: return R and $e(R)$;

The algorithm adopts a greedy strategy and operates as follows. The selection of boundary nodes of R starts with a node with the maximum cluster-degree (breaking ties arbitrarily) and the inclusion of its cluster-neighborhood; the sets $C(R)$ and

$e(R)$ are initialized accordingly. The augmentations of R, $\mathcal{C}(R)$, and $e(R)$ inactivate all the inter-cluster edges between (distinct) clusters in $\mathcal{C}(R)$, which entail the updates of the statistics of $\Gamma(u)$ and $d(u)$ for all the nodes u possibly affected by the inactivation, that is, $u \in \cup_{C \in \mathcal{C}(R)} V(C)$. Then, the algorithm incrementally grows the boundary set R and the tree of cluster-connected clusters in $\mathcal{C}(R)$ by: (1) adding to R a boundary node $u_{\max} \in \cup_{C \in \mathcal{C}(R)} V(C)$ that contributes the maximum cluster-adjacency to $\mathcal{C}(R)$ via active inter-cluster edges, that is,

$$u_{\max} = \arg \max\{d(v) \mid v \in \cup_{C \in \mathcal{C}(R)} V(C)\} \text{ (breaking ties arbitrarily)},$$

together with its cluster-neighborhood $\Gamma(u_{\max})$ and augmentations of $\mathcal{C}(R)$ and $e(R)$ accordingly, and (2) performing the necessary inactivation of the inter-cluster edges and updates of Γ- and d-statistics as described above for the initialization. The algorithm terminates when $\mathcal{C}(R) = \mathcal{C}$.

Theorem 4. *Approx-Boundary is a polynomial-time approximation algorithm with performance ratio less than 2 for the Minimum-Boundary Clustered-Graph Spanning Tree problem.*

Proof. Omitted due to space constraint. □

5 Concluding Remarks

Similar to most minimum total energy/power topology problems, the cluster-connectivity of wireless sensor networks, modeled as Clustered-Graph Spanning Tree and Hamiltonian Path problems, appear to be computationally intractable – completeness for the nondeterministic polynomial-time complexity class. Its induced decision and minimization problems, modeled as Boundary Clustered-Graph Spanning Tree and Minimum-Boundary Clustered-Graph Spanning Tree problems, are also $\mathcal{NP}$-complete and -hard, respectively.

Some of the major methods to cope with the intractability include approximation and randomized algorithms, and heuristic techniques. We have devised a polynomial-time approximation algorithm with performance ratio less than 2 for the Minimum-Boundary Clustered-Graph Spanning Tree problem. Our work in progress includes a comparative empirical study based on the current implementation of the algorithm Approx-Boundary versus other greedy Kruskal- and Prim-like approximation algorithms.

References

1. Bondy, J.-A., Murty, U.S.R.: Graph Theory. Graduate Texts in Mathematics. Springer, New York, London (2007)
2. Chen, W.-T., Huang, N.-F.: The strongly connecting problem on multihop packet radio networks. IEEE Trans. Commun. **37**(3), 293–295 (1989)
3. Cheng, X., Narahari, B., Simha, R., Cheng, M.X., Liu, D.: Strong minimum energy topology in wireless sensor networks: NP-completeness and heuristics. IEEE Trans. Mob. Comput. **2**(3), 248–256 (2003)

4. Clementi, A.E.F., Penna, P., Silvestri, R.: Hardness results for the power range assignment problem in packet radio networks. In: Hochbaum, D.S., Jansen, K., Rolim, J.D.P., Sinclair, A. (eds.) RANDOM 1999 and APPROX 1999. LNCS, vol. 1671, pp. 197–208. Springer, Heidelberg (1999)
5. Cui, S., Goldsmith, A.J., Bahai, A.: Energy-efficiency of MIMO and cooperative MIMO techniques in sensor networks. IEEE J. Sel. Areas Commun. 22(6), 1089–1098 (2004)
6. Cui, S., Madan, R., Goldsmith, A.J., Lall, S.: Cross-layer energy and delay optimization in small-scale sensor networks. IEEE Trans. Wireless Commun. 6(10), 3688–3699 (2007)
7. Karp, R.M.: Reducibility among combinatorial problems. In: Miller, R.E., Thatcher, J.W. (eds.) Complexity of Computer Computations, pp. 85–103. Plenum Press, New York (1972)
8. Laneman, J.N., Wornell, G.W.: Distributed space-time-coded protocols for exploiting cooperative diversity in wireless networks. IEEE Trans. Inf. Theory 49(10), 2415–2425 (2003)
9. Li, X.: Space-time coded multi-transmission among distributed transmitters without perfect synchronization. IEEE Signal Process. Lett. 11(12), 948–952 (2004)
10. Lloyd, E.L., Liu, R., Marathe, M.V., Ramanathan, R., Ravi, S.S.: Algorithmic aspects of topology control problems for ad hoc networks. ACM Mobile Netw. Appl. 10(1–2), 19–34 (2005)
11. Panda, B.S., Shetty, D.P.: A local search based approximation algorithm for strong minimum energy topology problem in wireless sensor networks. In: Hota, C., Srimani, P.K. (eds.) ICDCIT 2013. LNCS, vol. 7753, pp. 398–409. Springer, Heidelberg (2013)
12. Panda, B.S., Shetty, D.P., Pandey, A.: k-distinct strong minimum energy topology problem in wireless sensor networks. In: Natarajan, R., Barua, G., Patra, M.R. (eds.) ICDCIT 2015. LNCS, vol. 8956, pp. 187–192. Springer, Heidelberg (2015)
13. Singh, S., Raghavendra, C.S., Stepanek, J.: Power-aware broadcasting in mobile ad hoc networks. In: Proceedings of IEEE International Symposium on Personal, Indoor and Mobile Radio Communications (PIMRC), 12–15 September 1999, Osaka, Japan, pp. 22–31 (1999)
14. Stathopoulos, T., Lukac, M., McIntire, D., Heidemann, J.S., Estrin, D., Kaiser, W.J.: End-to-end routing for dual-radio sensor networks. In: Proceedings of IEEE INFOCOM 2007: 26th IEEE International Conference on Computer Communications, Joint Conference of the IEEE Computer and Communications Societies, 6–12 May 2007, Anchorage, Alaska, USA, pp. 2252–2260 (2007)
15. Čagalj, M., Hubaux, J.-P., Enz, C.: Minimum-energy broadcast in all-wireless networks: NP-completeness and distribution issues. In: Proceedings of the Eighth Annual International Conference on Mobile Computing and Networking (MOBICOM), 23–28 September 2002, Atlanta, Georgia, USA, pp. 172–182 (2002)
16. Wieselthier, J.E., Nguyen, G.D., Ephremides, A.: Algorithms for energy-efficient multicasting in static ad hoc wireless networks. ACM Mobile Netw. Appl. 6(3), 251–263 (2001)

A PBIL for Load Balancing in Network Coding Based Multicasting

Huanlai Xing[1](✉), Ying Xu[2], Rong Qu[3], and Lexi Xu[4]

[1] School of Information Science and Technology,
Southwest Jiaotong University, Chengdu, People's Republic of China
hxx@home.swjtu.edu.cn
[2] College of Computer Science and Electronic Engineering,
Hunan University, Changsha, People's Republic of China
[3] School of Computer Science, University of Nottingham, Nottingham, UK
[4] Department of Network Optimisation and Management,
China Unicom Network Technology Research Institute,
Beijing, People's Republic of China

Abstract. One of the most important issues in multicast is how to achieve a balanced traffic load within a communications network. This paper formulates a load balancing optimization problem in the context of multicast with network coding and proposes a modified population based incremental learning (PBIL) algorithm for tackling it. A novel probability vector update scheme is developed to enhance the global exploration of the stochastic search by introducing extra flexibility when guiding the search towards promising areas in the search space. Experimental results demonstrate that the proposed PBIL outperforms a number of the state-of-the-art evolutionary algorithms in terms of the quality of the best solution obtained.

Keywords: Load balancing · Multicast · Network coding · Population based incremental learning

1 Introduction

With the popularity of the Internet, network traffic has been dominated by a dramatically increasing number of multimedia applications, e.g. online games, IPTV, VoD, remote education, and video conferencing. It has been reported that over 90 % of the Internet traffic comes from multimedia. Multicast is one of the efficient technologies developed for supporting one-to-many multimedia applications with stringent quality-of-service (QoS). Therefore, this technology has drawn significant amount of research attention from both academia and industry [1]. Unfortunately, multicast with store-and-forward strategy cannot guarantee that a theoretical maximal throughput is always obtained. Network coding allows intermediate nodes to perform mathematical operations to incoming information if necessary. When incorporated into multicast, it can always guarantee the theoretical maximal throughput and hence becomes an ideal technique for point-to-multipoint data transmission [2].

© Springer International Publishing Switzerland 2016
O. Gervasi et al. (Eds.): ICCSA 2016, Part II, LNCS 9787, pp. 34–44, 2016.
DOI: 10.1007/978-3-319-42108-7_3

Load balancing is an issue in network resource allocation. Network service providers are eager to make use of the network infrastructure as fully as possible so as to accommodate more users with different QoS guarantees. Undoubtedly, a more balanced traffic load helps to make use of the remaining network resources to function more efficiently. Hence, load balancing has been a hot spot in the field of communications for many years [3].

There are some research efforts dedicated to the load balancing in network coding based multicast (NCM). A NCM algorithm was presented and compared with two traditional multicast routing algorithms with respect to the achievable throughput and load balancing in [4]. A reliable data dissemination protocol, which adapts network coding for decreasing the broadcast traffic in code updates, was developed in [5]. In [6], the problem of exploiting the abilities of next generation terminals in satellite systems with network coding was investigated. In multi-hop wireless networks, a flexible energy-efficient multicast routing algorithm with network coding was put forward [7]. However, all work above assumes that coding recombination has to be executed at all coding-possible nodes within a network, which would consume serious network computational and buffering resources, as coding incurs expensive computational overhead [8–10]. Since coding might only be necessarily performed at a limited number of coding-possible nodes, it is of vital importance to consider the load balancing issue in NCM, where coding is to be performed when needed. However, this issue has received little attention.

Population based incremental learning (PBIL) is an estimation of distribution algorithm, incorporating competitive learning concept into evolutionary algorithm. PBIL builds a probability model and evolves it to lead the search towards promising areas in the search space. Due to the simplicity and efficiency, the algorithm has been applied to a wide range of optimization problems, including the stabilizer design problem in power system [11], the dynamic optimization problem [12], the robot soccer system optimization [13], the antenna design problem [14], and the network coding resource minimization problem [15, 16], etc.

In this paper, a load balancing optimization problem in the context of NCM is formulated, where coding is performed only when necessary. A modified PBIL with a new probability vector update scheme is adopted to optimize the problem. The new scheme maintains a set of best-so-far samples obtained during the evolution. At each generation, a proportion of samples in the set are randomly selected and used to update the probability vector, which to a certain extent helps improve the global exploration. Performance comparisons show that the proposed PBIL is superior to a number of state-of-the-art evolutionary algorithms (EAs) with respect to the solution quality.

2 Problem Formulation

We represent a communications network by a directed graph $G = (V, E)$, where V and E are sets of nodes and links, respectively. The number of links in E is denoted by $|E|$. Assume every link in G is numbered and let $e_i \in E$ denote link i, where e_i is associated with a maximum bandwidth B_i^{max} and a currently consumed bandwidth B_i^{csmd}, where

$B_i^{max} \geq B_i^{csmd}$. There is a source node $s \in V$, a set of receivers $T = \{t_1, \ldots, t_d\}$, $t_k \in V, k = 1, \ldots, d$, and an expected bandwidth $R_{s \rightarrow T}$ from s to each receiver in T [8, 9].

We call an intermediate node in G as a *merging* node if it is non-source, non-receiver and has multiple incoming links. Only merging nodes can perform coding operations. So, all coding-possible nodes in G are merging nodes. Given a NCM request, the task is to find a connected subgraph (i.e. sub-network) in G to deliver NCM data traffic [9]. We refer to such subgraph as NCM subgraph and denote it by $G_{s \rightarrow T}$. A NCM subgraph consists of multiple paths, where each path originates from source s and terminates at a receiver. We refer to *link-disjoint* paths as paths which do not have any common link. More details can be found in [9].

In this paper, identical bandwidth consumption, denoted by $B_{s \rightarrow T}$, incurs in each link occupied by the NCM subgraph. So, R link-disjoint paths to the same receiver will incur $R \cdot B_{s \rightarrow T}$ bandwidth consumption, where R is an integer. Let ω_i be the bandwidth utilization ratio in $e_i \in E$, Φ be the number of link-disjoint paths to each receiver in $G_{s \rightarrow T}$, and $\Omega_z(s \rightarrow t_k)$ be path z from s to t_k in $G_{s \rightarrow T}$, $z = 1, \ldots, \Phi$, respectively. Let $\rho_z(s \rightarrow t_k)$ and $r(s \rightarrow t_k)$ denote the link set of $\Omega_z(s \rightarrow t_k)$ and the achievable bandwidth from s to $t_k \in T$ in $G_{s \rightarrow T}$, respectively.

In a communications network, each link is associated with a bandwidth utilization ratio (BUR) which reflects the percentage of how much bandwidth has been used. The task of this work is to find a NCM subgraph in G, with the average BUR minimized and a number of constraints met. A smaller average BUR implies a more balanced traffic load in G.

Minimize:

$$\bar{\omega} \tag{1}$$

where,

$$\bar{\omega} = \left(\sum\nolimits_{i \in |E|} \omega_i \right) / |E| \tag{2}$$

$$\omega_i = \left(B_{s \rightarrow T} \cdot c_i + B_i^{csmd} \right) / B_i^{max} \tag{3}$$

$$c_i = \begin{cases} 1, & e_i \in G_{s \rightarrow T} \\ 0, & otherwise \end{cases} \tag{4}$$

Subject to:

$$B_{s \rightarrow T} + B_i^{csmd} \leq B_i^{max}, \quad \forall i = \{1, \ldots, |E|\} \tag{5}$$

$$r(s \rightarrow t_k) = \Phi \cdot B_{s \rightarrow T}, \quad \forall t_k \in T \tag{6}$$

$$\rho_m(s \rightarrow t_k) \cap \rho_n(s \rightarrow t_k) = \emptyset, \quad \forall m, n \in \{1, \ldots, \Phi\}, m \neq n \tag{7}$$

The objective of the problem concerned is to minimize the average BUR, which is shown in Eq. (1). Equation (2) defines the average BUR. The BUR value associated with each link is calculated based on Eq. (3). Equation (4) defines the coefficient c_i.

Constraint (5) explains the bandwidth constraint. The achievable bandwidth $r(s \rightarrow t_k)$ is Φ times larger than $B_{s \rightarrow T}$ because Φ link-disjoint paths from the source to each receiver are to be constructed, which is shown in Constraint (6). Constraint (7) reflects that any two paths in $G_{s \rightarrow T}$ cannot have common link, as long as they terminate at the same receiver.

3 The Proposed PBIL

This section first introduces the new probability vector (PV) update scheme and then describes the overall procedure of the proposed PBIL in detail.

3.1 New PV Update Scheme

In the original PBIL, best-so-far sample (i.e. solution/individual) is used to update PV. But this update scheme may cause serious prematurity very easily, which is because depending on a single sample might lead the search to local optima quickly. A number of improved PV update schemes are thus developed including the famous Hebbian-inspired rule [17]. This rule defines that PV should be updated by a set of samples obtained during the evolution rather than a single sample and to some extent helps maintain a relatively high level of diversity.

The new PV update scheme is based on the Hebbian-inspired rule. Different from other variants, the proposed scheme operates an external population (EP) which only records the best-so-far samples obtained during the evolution. At each generation, a subset of samples is randomly selected from EP and their statistical information is then extracted and used to update PV.

Let $P(t)$ and N denote the PV at generation t and the number of samples generated at each generation, respectively. Let $SS_{EP} = \{B_1, ..., B_M\}$ be the subset of samples randomly selected from EP, where M is smaller than N and B_i is the i-th selected sample, $i = 1, ..., M$. Denote the probability distribution of all samples in SS_{EP} and the learning rate by P_{SS} and α, respectively. The update of $P(t)$ is defined in Eqs. (8) and (9).

$$P(t) = (1.0 - \alpha) \cdot P(t-1) + \alpha \cdot P_{SS} \tag{8}$$

$$P_{SS} = \frac{1}{M} \sum_{k=1}^{M} B_k \tag{9}$$

By introducing extra uncertainty to the PV update process, the proposed scheme helps guide the search exploring unknown areas in the search space as much as possible where optimum may reside. With a certain level of diversity preserved, it helps to prevent the search getting stuck at local optima and hence improves global exploration.

3.2 The Overall Procedure of the Proposed PBIL

In the literature, the binary link state (BLS) individual representation has been widely adopted when tackling network coding related optimization problems [9, 10, 15, 16, 18, 19]. The proposed PBIL is also based on BLS encoding. Details can be found in [9, 10].

Assume there is an individual (sample) X. In terms of fitness evaluation, we first check the feasibility of X. We call X feasible if it results into a valid NCM subgraph $G_{s \to T}(X)$ and infeasible otherwise. If X is feasible, its fitness value is set to the average BUR in G; otherwise, its fitness value is set to a sufficiently large number (100 % in this paper since BUR is in the range [0, 100 %]). Note that a feasible individual requires more time than infeasible ones for evaluation since those feasible not only undergo feasibility checking but also NCM subgraph construction. And constructing a NCM subgraph consumes more time than feasibility checking.

```
Initialization
1.Set t = 0
2.Set P(t) = {0.7,...,0.7}
3.Generate N samples by sampling P(t) and put them in set H(t)
4.Put the N samples into EP
repeat
5.Set t = t + 1
6.Evaluate all samples in H(t-1)
7.Update EP by finding N best samples from H(t-1) and EP
8.Randomly select M samples from EP and put them into SS_EP
9.Update P(t) by using the proposed PV update scheme in Sub-
    section 3.1
10.Mutate P(t) by Eq.(10)
11.Generate N samples by P(t)
until termination condition is met
```

Fig. 1. Overall procedure of the proposed PBIL

The overall procedure of the proposed PBIL is shown in Fig. 1. PV $P(t)$ is initialized as $\{0.7,\ldots,0.7\}$, meaning the probability of generating '1' at each position in $P(t)$ is 0.7. This helps guide the search to explore promising areas in the search space. Set $H(t)$ is the sampling set generated by PV at each generation. Note that an all-one individual is inserted into $H(t)$ to guarantee that the search begins with a feasible sample [10]. The initial N samples are copied into **EP** to form the external population. In the main loop, each sample in $H(t-1)$ is evaluated and associated with a fitness value. Then, **EP** is updated. Later on, M random samples are picked up from **EP** and used to update $P(t)$ (M = $N/2$ in this paper).

Mutation operation is implemented to add probability disturbance to PV so that prematurity is alleviated. Let the mutation probability and the amount of mutation at each position in PV, P_k, denoted by p_m and σ, respectively. Let RND_k be a random number (either 0.1 or 1.0, both with a probability of 0.5). If $RND_k < p_m$, P_k is mutated by Eq. (10)

$$P_k = (1.0 - \sigma) \cdot P_k + RND_k \cdot \sigma \qquad (10)$$

The termination criterion is that the evolution reaches a predefined number of generations.

4 Performance Evaluation

This section evaluates the proposed PBIL by comparing it with a number of state-of-the-art EAs with respect to the solution quality.

4.1 Test Instances

First of all, test instances are introduced briefly. We consider 12 benchmark instances including 2 fixed and 10 randomly generated directed networks. These instances have been widely adopted for performance evaluation when handling with network coding related optimization problems [8–10, 15, 16, 18–20]. The two fixed networks are F1 (i.e. 7-copy) and F2 (i.e. 15-copy) networks, where each copy is a modified butterfly network and x-copy network is constructed by cascading a number of identical copies [18]. As for the random instances, the network scale is from 20 to 60 nodes.

Table 1 shows all instances for performance comparison. In all experiments, each link is with an identical maximum bandwidth, i.e. 100 Mbps. The consumed bandwidth of link $e_i \in E$ prior to NCM is randomly generated in the range [1, 50] Mbps with a uniform distribution. The bandwidth consumption of each link in NCM subgraph is set to 30 Mbps. As mentioned in Sect. 2, the single objective of the problem concerned in the paper is to minimize the average BUR. A smaller average BUR represents a better network load balancing performance. All experiments were run on a Windows XP computer with Intel(R) Core(TM) E8400 3.0 GHz, 2G RAM. For performance comparison, each algorithm is run 20 times on each instance.

4.2 Overall Performance Evaluation

In this subsection, performance comparison is carried out among a number of state-of-the-art EAs on 12 benchmark instances. The following lists all algorithms for comparison.

- **GA1**: GA with BLS representation [9].
- **GA2**: GA with block transmission state (BTS) representation [10].
- **UMDA**: univariate marginal distribution algorithm (UMDA) in [21]. Different from PBIL, UMDA utilizes statistics of the last generation to generate a new generation of samples.
- **QEA1**: Quantum-inspired evolutionary algorithm (QEA) [22]. Based on BLS representation, QEA1 adopts rotation angle step (RAS) and quantum mutation probability (QMP) to update individuals, where RAS is randomly generated and QMP is based on the current fitness value of the associated individual.

Table 1. Benchmark instances for performance comparison

Instances	Nodes	Links	Receivers	Rate
F1	57	84	8	2
F2	121	180	16	2
R1	20	37	5	3
R2	20	39	5	3
R3	30	60	6	3
R4	30	69	6	3
R5	40	78	9	3
R6	40	85	9	4
R7	50	101	8	3
R8	50	118	10	4
R9	60	150	11	5
R10	60	156	10	4

– **QEA2**: Another variant of QEA [23]. Different from QEA1, QEA2 modifies the values of RAS and QMP according to the current and previous fitness values of the associated individual.
– **PBIL1**: PBIL devised for the network coding resource minimization problem [16].
– **PBIL2**: The proposed PBIL with the novel PV update scheme in Sect. 3.

All EAs above uses BLS representation except GA2. The population size and the number of iterations are set to 20 and 200 for each algorithm, respectively. We adopt suggested parameter settings for GA1, GA2, UMDA, QEA1, QEA2, and PBIL1 [9, 10, 16, 21–23]. In PBIL2, we set the learning rate $\alpha = 0.1$, the mutation probability $p_m = 0.02$, and the probability variance at each position $\sigma = 0.05$, respectively. The number of random samples selected at each generation from EP is 10. All results are collected by running each algorithm 20 times.

Table 2 shows the results of mean value (%) and standard deviation (SD). It is clearly seen that the proposed algorithm, PBIL2, performs the best if considering all test instances. For each instance, PBIL2 obtains the smallest mean value and promising SD, indicating that PBIL2 has a stabilized and outstanding optimization performance in

Table 2. Results of mean value (%) and SD (Best results are in bold)

Instances	GA1	GA2	UMDA	QEA1	QEA2	PBIL1	PBIL2
F1	45.33 (0.34)	45.29 (0.32)	**45.11** (0.00)	45.17 (0.13)	45.35 (0.29)	45.31 (0.22)	**45.11** (0.00)
F2	46.20 (0.24)	46.41 (0.28)	46.37 (0.15)	46.57 (0.09)	46.74 (0.08)	46.74 (0.04)	**46.10** (0.06)
R1	44.25 (0.86)	44.49 (1.06)	44.54 (0.68)	43.78 (0.41)	43.64 (0.25)	43.69 (0.40)	**43.43** (0.00)
R2	41.73	42.24	41.58	41.36	41.19	40.87	**40.52**

(*Continued*)

Table 2. (*Continued*)

Instances	GA1	GA2	UMDA	QEA1	QEA2	PBIL1	PBIL2
	(1.06)	(1.25)	(0.52)	(0.37)	(0.33)	(0.43)	(0.00)
R3	39.95	40.32	39.73	39.33	39.25	39.26	**38.89**
	(1.22)	(1.48)	(0.46)	(0.47)	(0.42)	(0.41)	(0.26)
R4	34.88	34.56	34.82	34.45	34.54	34.46	**34.42**
	(0.62)	(0.40)	(0.37)	(0.23)	(0.25)	(0.25)	(0.19)
R5	43.12	43.38	43.03	43.09	42.80	43.09	**42.40**
	(1.70)	(2.16)	(0.38)	(0.53)	(0.43)	(0.42)	(0.22)
R6	42.69	42.83	42.97	42.62	42.66	42.47	**42.40**
	(0.49)	(0.63)	(0.41)	(0.27)	(0.31)	(0.15)	(0.05)
R7	38.73	38.43	38.09	38.14	38.18	38.26	**37.09**
	(1.72)	(1.79)	(0.52)	(0.62)	(0.45)	(0.39)	(0.41)
R8	40.89	41.59	40.76	41.56	41.51	41.39	**40.28**
	(0.93)	(1.64)	(0.44)	(0.41)	(0.47)	(0.56)	(0.24)
R9	39.38	39.62	39.48	39.79	39.85	39.92	**39.26**
	(0.17)	(0.36)	(0.22)	(0.20)	(0.20)	(0.19)	(0.09)
R10	36.32	36.45	36.20	36.50	36.36	36.66	**36.05**
	(0.89)	(0.95)	(0.29)	(0.31)	(0.37)	(0.30)	(0.14)

finding near-optimal solutions. By building an evolving probabilistic model, PBIL2 can generate promising samples from the PV at a relatively high probability, which to a certain extent guides the search towards promising areas in the search space. The novel PV update scheme helps PBIL2 gain better global exploration ability and avoid prematurity since the PV update is no longer dependent on a single sample but a random set of samples at each generation. With global exploration enhanced, PBIL2 has more opportunity to reach the global optima and thus obtains the best performance. Figure 2 shows the box plots of the seven algorithms in six selected instances including F2, R1, R3, R5, R7, and R9. One can find that PBIL2 always outperforms the rest of the EAs for comparison.

The average computational time (ACT) is another important performance indicator when evaluating EAs. Table 3 shows ACT values of all algorithms in all instances. On the one hand, it can be observed that PBIL2 obtains a relatively large ACT, which means it takes long time for navigating each evolutionary search. This is because PBIL2 does well in global exploration and feasible samples are generated at a high probability. As mentioned in Subsect. 3.2, feasible samples consume more time than infeasible ones. If more feasible samples are generated during the evolution, an algorithm definitely incurs more computational cost and hence more ACT. With the novel PV update scheme integrated, PBIL2 generates significant number of feasible samples and hence consumes relatively large amount of ACT. On the other hand, PBIL2 is not the one with the largest ACT either, compared with the others. One may dramatically reduce ACT of PBIL2 by using parallel computation techniques.

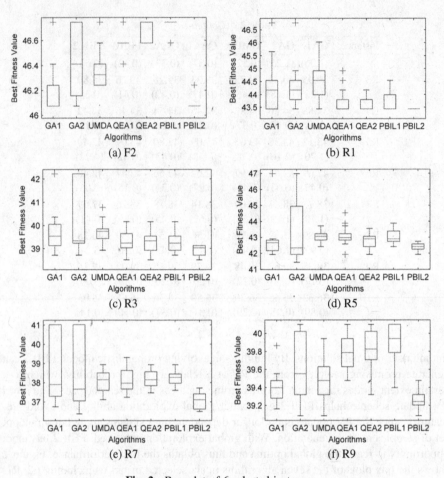

(a) F2

(b) R1

(c) R3

(d) R5

(e) R7

(f) R9

Fig. 2. Box plot of 6 selected instances

Table 3. Results of ACTs (sec.)

Instances	GA1	GA2	UMDA	QEA1	QEA2	PBIL1	PBIL2
F1	2.2	2.6	2.9	2.2	1.7	0.7	2.6
F2	15.4	12.3	18.1	6.1	4.4	3.1	15.2
R1	0.9	1.0	1.0	1.3	1.0	0.6	1.0
R2	0.9	0.8	1.1	1.3	1.0	0.5	1.1
R3	2.1	1.9	2.5	2.6	2.2	1.1	2.5
R4	2.5	2.8	2.7	3.4	2.6	1.5	2.7
R5	3.6	3.4	3.9	3.0	2.8	1.1	3.4
R6	2.9	2.5	2.9	2.1	1.5	1.2	2.6
R7	4.8	4.9	5.9	5.2	4.6	2.3	5.7
R8	8.8	5.8	8.7	5.3	4.7	2.2	7.2
R9	14.9	10.4	14.5	7.6	7.3	4.4	13.0
R10	13.4	12.6	15.4	14.3	12.8	7.5	15.1

5 Conclusions

This paper formulates a load balancing optimization problem in network coding based multicast (NCM), where the objective is to keep the network traffic load as balanced as possible when supporting data transmission of NCM. Average bandwidth utilization ratio is used to measure to what extent network traffic load is balanced. To handle with the problem above, we present a modified PBIL with a novel probability vector (PV) update scheme. This scheme introduces additional flexibility in guiding the search by updating the PV by a number of random best samples, which helps enhance the global exploration capability and prevent the search from local optima. Experimental results show that the proposed PBIL gains the best optimization performance compared with a number of state-of-the-art evolutionary algorithms regarding the quality of the solution obtained. Due to the advantages of the proposed algorithm, it can be applied to a number of optimization problems in wireless communications, e.g. traffic control and network configuration in Long Term Evolution (LTE) cellular systems [24, 25].

Acknowledgements. This research was supported in part by NSFC (No.61401374), the Fundamental Research Funds for the Central Universities (No. 2682014RC23), the Project-sponsored by SRF for ROCS, SEM, P. R. China and University of Nottingham, UK.

References

1. Benslimane, A.: Multimedia Multicast on the Internet. ISTE, Norwood (2007)
2. Li, S.Y.R., Yeung, R.W., Cai, N.: Linear network coding. IEEE Trans. Inform. Theory **49**(2), 371–381 (2003)
3. Wang, N., Pavlou, G.: Traffic engineered multicast content delivery without MPLS overlay. IEEE Trans. Multimedia **9**(3), 619–628 (2007)
4. Chi, K., Yang, C., Wang, X.: Performance of network coding based multicast. IEE Proc. Commun. **153**(3), 399–404 (2006)
5. Hou, I.H., Tsai, Y.E., Abdelzaher, T.F., Gupta, I.: AdapCode: adaptive network coding for code updates in wireless sensor networks. In: Proceedings of the INFOCOM (2008)
6. Vieira, F., Lucani, D.E., Alagha, N.: Codes and balances: multibeam satellite load balancing with coded packets. In: Proceedings of the ICC (2012)
7. Jiang, D., Xu, Z., Li, W., Chen, Z.: Network coding-based energy-efficient multicast routing algorithm for multi-hop wireless networks. J. Syst. Softw. **104**, 152–165 (2015)
8. Kim, M., Ahn, C.W., Médard, M., Effros, M.: On minimizing network coding resources: an evolutionary approach. In: Proceedings of the NetCod (2006)
9. Kim, M., Médard, M., Aggarwal, V., O'Reilly, V., Kim, W., Ahn, C.W., Effros, M.: Evolutionary approaches to minimizing network coding resources. In: Proceedings of the INFOCOM (2007)
10. Kim, M., Aggarwal, V., O'Reilly, V., Médard, M., Kim, W.: Genetic representations for evolutionary minimization of network coding resources. In: Proceedings of the EvoCOMNET (2007)
11. Folly, K.A.: Multimachine power system stabilizer design based on a simplified version of genetic algorithms combined with learning. In: Proceedings of the ISAP2005 (2005)

12. Yang, S., Yao, X.: Population-based incremental learning with associative memory for dynamic environments. IEEE Trans. Evolut. Comput. **12**(5), 542–561 (2008)

13. Kim, J.H., Kim, Y.H., Choi, S.H., Park, I.W.: Evolutionary multi-objective optimization in robot soccer system for education. IEEE Comput. Intell. Mag. **4**(1), 31–41 (2009)

14. Ho, S.L., Yang, S., Bai, Y., Huang, J.: A robust metaheuristic combing clonal colony optimization and population-based incremental learning methods. IEEE Trans. Magn. **50**(2) (2014). DOI:10.1109/TMAG.2013.2283886

15. Xing, H., Qu, R.: A population based incremental learning for network coding resources minimization. IEEE Commun. Lett. **15**(7), 698–700 (2011)

16. Xing, H., Qu, R.: A population based incremental learning for delay constrained network coding resource minimization. In: Di Chio, C., et al. (eds.) EvoApplications 2011, Part II. LNCS, vol. 6625, pp. 51–60. Springer, Heidelberg (2011)

17. Gonzalez, C., Lozano, J.A., Larranaga, P.: Analyzing the population based incremental learning algorithm by means of discrete dynamical systems. Complex Syst. **12**, 465–479 (2000)

18. Ahn, C.W., Yoo, J.C.: Multi-objective evolutionary approach to coding-link cost trade-offs in network coding. Electron. Lett. **48**(25), 1595–1596 (2012)

19. Xing, H., Qu, R.: A nondominated sorting genetic algorithm for bi-objective network coding based multicast routing problems. Inform. Sci. **233**, 36–53 (2013)

20. Xing, H., Qu, R., Bai, L., Ji, Y.: On minimizing coding operations in network coding based multicast: an evolutionary algorithm. Appl. Intell. **41**(3), 820–836 (2014)

21. Lozada-Chang, L.V., Santana, R.: Univariate marginal distribution algorithm dynamics for a class of parametric functions with unitation constraints. Inform. Sci. **181**(11), 2340–2355 (2011)

22. Xing, H., Ji, Y., Bai, L., Sun, Y.: An improved quantum-inspired evolutionary algorithm for coding resource optimization based network coding multicast scheme. AEUE **64**(12), 1105–1113 (2010)

23. Ji, Y., Xing, H.: A memory-storable quantum-inspired evolutionary algorithm for network coding resource minimization. In: Kita, E. (Ed.) Evolutionary Algorithm, InTech, pp. 363–380 (2011)

24. Xu, L., Chen, Y., Chai, K.K., Schormans, J., Cuthbert, L.: Self-organising cluster-based cooperative load balancing in OFDMA cellular networks. Wiley Wirel. Commun. Mobile Comput. **15**(7), 1171–1187 (2015)

25. Xu, L., Cheng, X., Chen, Y., Chao, K., Liu, D., Xing, H.: Self-optimised coordinated traffic shifting scheme for LTE cellular systems. In: Proceedings of the ICSON2015 (2015)

A Proposed Protocol for Periodic Monitoring of Cloud Storage Services Using Trust and Encryption

Alexandre Pinheiro[1(✉)], Edna Dias Canedo[2(✉)], Rafael Timóteo de Sousa Jr.[1(✉)], and Robson de Oliveira Albuquerque[1(✉)]

[1] Electrical Engineering Department, University of Brasília - UnB, Brasília, Brazil
tenpinheiro@dct.eb.mil.br, desousa@unb.br, robson@redes.unb.br
[2] Faculdade UnB Gama, University of Brasília – UnB, Brasília, Brazil
ednacanedo@unb.br

Abstract. The advantages of using cloud computing include scalability, availability and a virtually 'unlimited' storage capacity. However, it is challenging to build storage services that are at the same time safe from the customer point-of-view and that run in public cloud infrastructures managed by service providers that can not be fully considered trustworthy. Owners of large amounts of data have to keep their data in the cloud for a long period of time without the need to keep copies of the original data or to access it. In such cases, questions of Integrity, availability, privacy and trust are still challenges in the adoption of Cloud Storage Services to ensure security, especially when losing or leaking information can bring significant damage, be it legal or business-related. With such concerns in mind, this paper proposes a protocol to monitor the information stored in the cloud and the behaviour of contracted storage services periodically. The proposed protocol, which is based on trust and encryption, is validated by analysis and simulation that demonstrate its utilization of computing resources compared to its results regarding cloud storage protection that are achieved over time.

Keywords: Protocol · Trust · Cloud data storage · Integrity · Data monitoring

1 Introduction

Companies, institutions and government agencies generate large amounts of information in digital format, such as documents, projects, transactions records, etc., every day. For legal or business reasons, this information needs to remain stored for a long period of time and this has become an issue for IT managers.

The use of cloud services for storing sensitive information started to gain relevance, along with the popularization, cost reductions and an ever-growing supply of Cloud Storage Services. However, the question of ensuring integrity and confidentiality still has to be evaluated in such services in order to protect the stored information.

Cloud services for data storage are fast, cheap, and almost infinitely scalable. However, reliability can be an issue, as even the best services sometimes fail [1]. Additionally a considerable number of organizations consider security and privacy as obstacles to the acceptance of public cloud services [2].

© Springer International Publishing Switzerland 2016
O. Gervasi et al. (Eds.): ICCSA 2016, Part II, LNCS 9787, pp. 45–59, 2016.
DOI: 10.1007/978-3-319-42108-7_4

Data integrity is defined as the accuracy and consistency of stored data, i.e., proprieties indicating that the data has not changed and has not been broken [2]. Cloud services should provide mechanisms to confirm data integrity, while still ensuring user privacy.

Considering these requirements and constraints, this paper proposes a protocol to provide customers of Cloud Storage Services with the sustained assurance of the existence and integrity of their files stored in the cloud without the need to keep copies of the original files or to expose their contents. The proposed protocol was developed and tested within storage services from both private and public clouds.

This paper is structured as follows: Sect. 2 reviews of concepts and definitions on cloud computing, encryption, and trust. Section 3 presents works related to data integrity in the cloud. Section 4 comprises the proposed protocol, presenting its trust classification process for Cloud Storage Services, allowing the permanent monitoring of the stored data. A detailed analysis of the proposed protocol is given in Sect. 5 while Sect. 6 is devoted to simulations that characterize the expected behaviour for the trust level assigned to the Cloud Storage Service over time. Section 7 closes this paper with some conclusions and future work outlining.

2   Background

Cloud computing is being gradually adopted in different business scenarios, to obtain flexible and reliable computing environments, with several supporting solutions available in the market. Since it uses different technologies (e.g., virtualization, utility computing, grid computing, and service-oriented architecture) and constitutes a completely new computational paradigm, cloud computing requires high-level management activities which include: (a) service provider selection, (b) virtualization technology selection, (c) virtual resources allocation, and (d) monitoring and auditing to comply with Service Level Agreements (SLAs) [3].

A solution in cloud computing consists of several elements such as clients, data canters, and distributed servers. These elements form the three parts of a cloud solution [3, 4]. Each element has a purpose and a specific role in delivering working application based on the cloud.

The cloud computing architecture is structured in layers, each one dealing with a particular aspect for making application resources available. Basically, there are two main layers, namely, a lower and a higher resource layer. The lower layer comprises the physical infrastructure and is responsible for the virtualization of storage and computational resources. The higher layer provides specific services, such as Software as a Service (SaaS), Platform as a Service (PaaS), and Infrastructure as a Service (IaaS). These layers may have their own management and monitoring systems, independent from one another, thus improving flexibility, reuse and scalability [5, 6].

Considering to the intended access methods and availability, there are different models of deployment for cloud computing services that include private cloud, public cloud, community cloud, and hybrid cloud environments [7].

2.1 Encryption

Encryption is the process of converting a plaintext message into cipher-text that can be deciphered back to the original message. An encryption algorithm along with a key is used in the encryption and decryption of data. There are several data encryption methods that form the basis of network security, employing schemes with block or stream ciphers.

The type and length of the keys used depend on the encryption algorithm and the needed security level. In conventional symmetric encryption, a single key is used. With this key, the sender can encrypt a message and a recipient can decrypt the ciphered message but the security of the key becomes an issue since at least two copies of the key exist, one in the sender, and another in the recipient. In asymmetric encryption, the encryption key and the decryption key are correlated but different. One is a public key of the recipient that the sender can use to encrypt the message and the other is a recipient private key allowing the recipient to decrypt the message [8].

2.2 Trust

Trust is a common reasoning for humans to face the complexities of the world and to think sensibly about the possibilities of everyday life. Trust is strongly linked with expectancies in something, which implies a degree of uncertainty and optimism. It is the choice of putting something on the hands of others, considering the other's behaviour to determine how to act in a given situation [9].

Trust can be considered as particular level of subjective probability in which an agent believes that another agent will perform a certain action, which is subject to monitoring [10]. Also, trust can be represented as an opinion so that situations involving trust and trust relationships can be modelled. Thus, positive and negative feedback on a specific member can be accumulated and used to calculate its future behaviour [11].

A scale for trust or distrust was proposed in [9] with values that are defined in the interval $[-1, 1]$, where -1 represents complete distrust and 1 represents blind trust. Blind trust is a trust level indicating no need for service monitoring whereas complete distrust is a level indicating that the service will probably always fail.

2.3 Hash

Hash value or hash code or simply hash, is the result of applying a mathematical function that takes a string of any size as data source and returns a relatively small and fixed-length string. Any bit changed in the source string alters dramatically the resulting hash code [12]. These functions are designed to make it very difficult to deduce from a hash value the string which was used to calculate this hash. Also, it is required that it is extremely difficult to find two strings whose hash codes are the same, i.e., a hash collision.

3 Related Work

In order to try to guarantee the integrity of data stored in cloud services, many research works suggest solutions with both advantages and disadvantages regarding the domain analysed in this paper. A protocol is proposed in [13] to enable the Cloud Storage Services to prove that a file subjected to verification is not corrupted. To that end a formal and secure definition of proof of retrievability is presented and the paper introduces the use of sentinels which are special blocks, hidden in the original file prior to encryption and then used to challenge the cloud service. In [14], another scheme based on [13] is presented where one does not need to encrypt all the data, but only a few bits per data block.

George and Sabitha [15] propose a two parts solution to improve privacy and integrity. The first part, called 'anonymization' is used to identify fields in records that could identify their owners. Anonymization then uses techniques such as generalisation, suppression, obfuscation, and addition of anonymous records to enhance data privacy. The second part, called 'integrity checking', uses public and private key encryption techniques to generate a tag for each record on a table. Both parts are executed with the help of a trusted third party called 'enclave' that saves all the data generated that will be used for deanonymisation and integrity verification.

An encrypted integrity verification method is proposed by [16]. The proposed method uses a new hash algorithm, the Dynamic User Policy Based Hash Algorithm. Hashes on data are calculated for each authorised cloud user. For data encryption, an Improved Attribute-Based Encryption algorithm is used. Encrypted data and its hash value are saved separately in cloud storage. Data integrity can be verified only by an authorized user and it is necessary to retrieve all the encrypted data and its hash.

Another proposal to simultaneously achieve data integrity verification and privacy preserving is found in [17] which proposes the use of two encryption algorithms for every data upload or download transaction. The Advanced Encryption Standard (AES) algorithm is used to encrypt a client's data which will be saved in a Cloud Storage Service, and a RSA-based partial homomorphic encryption technique is used to encrypt AES encryption keys that will be saved in a third party together with a hash of the file. Data integrity is verified only when a client downloads one file.

In [18], a data integrity auditing protocol is proposed to allow the fast identification of corrupted data using homomorphic cipher-text verification and a recoverable coding methodology. Checking the integrity of outsourced data is done periodically by a trusted or untrusted entity. Due to the adopted methodology, both the total auditing time and the communication cost could be reduced.

In the work of [19], a security model is presented of public verification for storage correctness assurance that supports dynamic data operation. The model guarantees that no challenged file blocks should be retrieved by the verifier during the verification process and no state information should be stored at the verifier side between audits. A Merkle Hash Tree (MHT) is used to save the hashes of authentic data values and both the values and positions of data blocks are authenticated by the verifier.

In [20], an approach is presented that allows inserting large volumes of encrypted data in non-relational databases hosted in the cloud and afterwards performs queries on inserted data without the use of a decryption key. Although it is not the main focus of the work, this approach could be used to verify the integrity of content stored in the cloud through the evaluation of responses to queries with previously calculated results.

4   File Integrity Monitoring Protocol

This section presents our proposed protocol whose central objective is to enable the provision of an outsourced storage service that clients can continuously verify regarding the integrity of their stored files. By using the proposed protocol to interact with the storage services provider, the clients are not required to keeping copies of the original files, or revealing the content of these files.

4.1   Protocol Requirements

One of the main requirements of this protocol is to prevent the cloud provider from offering and charging a client for a safe storage service that in practice is not being provided. Also the protocol must operate with low bandwidth consumption, and provide rapid identification of a misbehaving service, strong defences against fraud, ensuring data confidentiality and also giving utmost predictability to the Integrity Check Service while avoiding the overloading of Cloud Storage Services.

4.2   Protocol Operating Principle

The protocol execution involves three interacting entities: the storage client, de storage provider and a third party that performs file integrity verification, as shown in Fig. 1. The basic operating procedure required by the protocol is the encryption of the original file, followed by dividing the encrypted file into 4096 small chunks, which in turn are randomly permuted and then grouped into data blocks, each one with distinct 16 file chunks. Hashes are generated from these data blocks. Each hash, together with the addresses of the chunks forming the corresponding data block, is sent to the Integrity Check Service.

The selection and distribution of chunks used to assemble the data blocks is done in cycles. The number of cycles will vary according to the file storage period. Each cycle generates 256 data blocks without repeating chunks. The data blocks generated in each cycle contain all of the chunks of the encrypted file ($256 * 16 = 4096$).

Each hash and its corresponding chunk addresses will be used only once by the Integrity Check Service to send an integrity verification challenge to the Cloud Storage Service provider. On receiving a challenge with the chunk addresses, the Cloud Storage Service reads the chunks from the stored file, assembles the data block, generates a hash from data block and sends this hash as answer to the Integrity Check Service.

To finalize, the answer hash and the original hash are compared by the Integrity Check Service. If the hashes are equal, it means that the content of chunks in the stored file is intact.

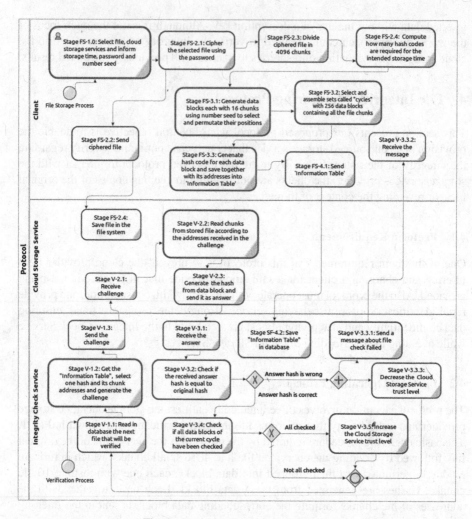

Fig. 1. Proposed Protocol Architecture

4.3 Protocol Architecture

Formally the protocol architecture is based in three components: (i) clients; (ii) storage services in the cloud; and (iii) an integrity check service. The interaction between the architectural members is carried out through an exchange of asynchronous messages.

The protocol comprises two distinct execution processes as shown in Fig. 1. The first one is called 'File Storage Process' and runs on demand having the client as its starting entity. The second is the 'Verification Process' which is instantiated by an Integrity Check Service and executed continuously to verify one Cloud Storage Service. An Integrity Check Service can simultaneously verify more than one Cloud Storage Services through parallel instances of the Verification Process.

4.3.1   File Storage Process

The File Storage Process is responsible for preparing the file to be sent to the Cloud Storage Service and for generating the information needed for its verification by the Integrity Check Service. In Fig. 1, each stage of 'File Storage Process' is identified with the prefix 'Stage FS-' followed by the stage number and, if necessary, by its sub-process number.

The process initiates with the user selecting a file, a password, a number seed, and the time in years for which she/he intends to maintain the file integrity monitoring. The password will serve to generate the secret key used to encrypt the chosen file while the number seed will add extra entropy to the process, since a random seed is used to warrant an unpredictable selection and distribution of the data that is source of the hash codes needed to check file integrity.

One or more storage cloud services should also be selected. Considering the need to ensure the recoverability of files, it is important to use more than one provider to provide redundancy, since customers are not required to keep copies of original files.

The second stage consists of several parallel file encryption tasks, while the encrypted files are sent to the chosen storage services. Also, in a parallel process, a logical division of an encrypted file into 4096 chunks is also run. When the file size is not a multiple of 4096 bytes, the padding to complete it is taken from the beginning of the file. The last step in this stage is to calculate the number of hash codes needed to perform the verification of the file, without repeating it throughout the requested time period. The result should always be a multiple of 256.

In the next stage, data blocks are generated from a random union of 16 chunks of the original file. For this, after the split, each file chunk receives an address code between 0 and 4095, represented in hexadecimal format (000 - FFF).

An algorithm developed for this purpose selects the chunks and performs permutation of their positions for each data block. The algorithm uses the seed provided by the client to create the random seed which will be used to add entropy to the pseudo-random sequence generator that chooses the sequences of address codes. The same algorithm is also responsible for grouping data blocks in sets called 'cycles'. Each cycle consists of 256 data blocks with all the 4096 chunks of the original file. This stage comprises sub-processes 'FS-3.1' to 'FS-3.3' in Fig. 1.

The last stage is devoted to building a data structure named 'information table' containing a record header made up by a file identifier, the chunk size in bytes, the amount of body records, the total number of cycles, and a checksum. Each record in the table represents a data block and contains fields with the address codes of the 16 file chunks, their cycle number and hash code. This table is sent to the Integrity Check Service while the corresponding file is sent to the Cloud Storage Service.

4.3.2   Verification Process

This process, whose stages are prefixed with 'Stage V-' in Fig. 1, is intended to periodically perform an integrity check of client files saved in Cloud Storage Services and to rate these services according to a reliability scale for the verified file integrity.

The first stage of the process verifies what next file should have its integrity checked in a given storage service, and then performs in parallel the file verification in each other registered storage service. After selecting the file, its information table is read and the number of the last checking cycle is retrieved. When the file is new or when a last cycle has already been completed, a new, not yet checked, cycle is randomly chosen.

Next, the process fetches from the information table the not-yet-verified next record, belonging to the selected cycle, and containing the hash code and address codes of the data block that will be verified. The challenge is then assembled with the address codes obtained from the selected record, the chunk length, and the file identifier retrieved from the information table header. An identifier is assigned to this challenge and the challenge package is sent to the service storage. Since there possibly are several parallel sent challenges the process registers their identifiers in a local challenge pool, sets a timer for the arrival of each response, and then waits for responses corresponding to each challenge.

In the second stage, the Cloud Storage Service receives the challenge and concurrently retrieves from the stored file all the chunks defined in the challenge, considering the length and each address code received. All retrieved chunks are grouped into a data block whose hash is produced. The resulting hash is packaged with the challenge identifier and sent in response to the Integrity Check Service.

Thus, in the last stage, the Integrity Check Service receives the response, finds the corresponding challenge in the pool, reads the original record from the information table and compares the received hash with the hash that gave rise to the challenge. If they do not match, a message is sent to the file owner (client) warning on the failed file verification. Whenever a file verification process fails, the trust level assigned to the failing storage service is immediately downgraded as shown in the sub-processes 'V-3.1'to 'V-3.6' in Fig. 1. If the Integrity Check Service does not receive a response from the Cloud Storage Service regarding a challenge, after the wait time expiration, the original challenge is re-sent and the wait time is squared. After the 10th unsuccessful attempt, the process considers that the challenge failed, consequently downgrading the trust level as well.

If the compared hash codes are equal, then a flag is saved in the information table record, indicating that the data block represented by that hash code was successfully verified. After that, this record will be checked as to whether it is the last in the current cycle, meaning that the whole data block from the original file has already been successfully verified. In this case, the Cloud Storage Service will receive a positive assessment. Thus, ending the stage, the trust level reclassification of the storage service is done. Upon completion of this stage, the process is re-executed from the first stage.

4.3.3   Trust Level Classification Process

As explained, whenever a verification process fails, the trust level assigned to the verified storage service is downgraded. When the current trust level is greater than zero, it is set to zero and, when the trust value is between 0 and −0.5, it is reduced by 15 %. Otherwise, a value of 2.5 % of the difference between the current trust level value and −1 is calculated and the result is subtracted from the trust level value.

Conversely, whenever a checking cycle is completed without failures (all the data blocks of a file are checked without errors), the Cloud Storage Service trust level is raised. If the current trust level value is less than 0.5, then the trust level value is raised by 2.5 %. Otherwise, a value of 0.5 % of the difference between 1 and the current trust level value is calculated and the result is added to the trust level value.

The trust level assigned to a cloud storage service will affect the verification periodicity of stored files and the time needed to get a complete file cycle. When the verification of all data blocks in a cycle is successfully concluded, this means that all chunks of the original file were tested.

Table 1 presents the minimum percentages of stored files that will be verified per day, as well as the minimum percentage of data blocks that will be checked by file and per day, according to the trust level assigned to a storage service. The values 1 and −1, respectively representing blind trust and complete distrust as described in Sect. 2.2, are incompatible with the object of classification and are not considered.

Table 1. Classification of the trust levels

Trust level	Value range	Files verified by day	% from file verified by day	Data Blocks verified by day
Very high trust	]0.9, 1[	15 %	~0.4 %	1
High trust	]0.75, 0.9]	16 %	~0.8 %	2
High medium trust	]0.5, 0.75]	17 %	~1.2 %	3
Low medium trust	]0.25, 0.5]	18 %	~1.6 %	4
Low trust	]0, 0.25]	19 %	~2.0 %	5
Low distrust	]−0.25, 0]	20 %	~2.4 %	6
Low medium distrust	]−0.5, −0.25]	25 %	~3.2 %	8
High medium distrust	]−.75, −0.5]	30 %	~ 4.0 %	10
High distrust	]−0.9, −0.75]	35 %	~ 4.8 %	12
Very high distrust	]−1, 0.9]	50 %	~5.6 %	14

Whenever the trust level value attains zero, a fact which commonly occurs when a service has not yet been rated, or when its trust level has been reduced due to faults being identified, a small fixed value is assigned to determine the initial trust. Thus, if the last check resulted in a 'positive assessment', a value of +0.1 is assigned for the trust level, otherwise the assigned value is −0.1.

5   Protocol Analysis

The analysis criteria are derived from the requirements for the proposed protocol, including: low consumption of network bandwidth, predictability and economy in consumption of the Integrity Check Service resources, fast identification of misbehaving services, privacy, resistance against fraud, and no overloading of the storage service.

The proposed logical division of the original file into 4096 chunks, grouped into blocks of 16 chunks, aims at minimizing the storage service overhead by reducing the amount of data to be read for each verification, also enabling the parallel execution of searching and recovering tasks for each data chunk.

The need for fast identification of misbehaving services also contributed to determining operational parameters. For instance, the protocol uses a random selection of 16 file chunks in the data block to allow checking the integrity of various parts of the file in a single verification step.

Privacy is attained with the use of 256-bit hash codes to represent each data block, regardless of their original size. Hash codes allow the Integrity Check Service to verify files hosted in storage services, without necessarily knowing the file contents.

Furthermore, the use of hash codes in combination with a fixed amount of data blocks provides predictability and low usage of the network bandwidth. Also, it is possible to pre-determine the computational cost required to verify the integrity of a file, the total time foreseen for its storage, regardless of its size. Figure 2 shows the daily consumption of network bandwidth (minimum and maximum) by the proposed protocol, for each file hosted in a Cloud Storage Service, considering the sending of challenges and reception of responses for each proposed trust level.

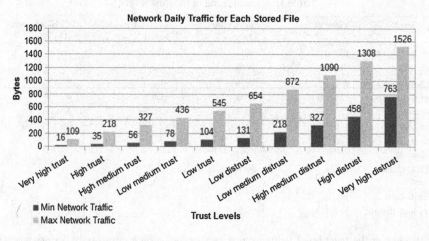

Fig. 2. Daily consumption of network bandwidth for each file hosted in cloud. (Color figure online)

There is also the possibility to predict the total number of data blocks, as it is a function of the time predicted for the file storage, since each hash code and the chunk addresses that form the data are used only once, uniquely and exclusively as a challenge to the Cloud Storage Service. Calculations below were made based on a worst-case scenario, i.e. the hypothetical situation where the Cloud Storage Service remains rated as 'Very High Distrust' throughout the whole file storage period.

According to Table 1, in a Cloud Storage Service rated as 'Very High Distrust', it is necessary to check at least 14 data blocks of each file a day. As the data blocks generation is performed in cycles with 256 blocks each, to determine the total number of data blocks to be generated (Eq. 2), it is necessary to first calculate the total number of cycles (Eq. 1).

$$Cycles = ROUND\left(\frac{(14 * 366) * Years}{256}, 0\right) \tag{1}$$

$$DataBlocks = Cycles * 256 \tag{2}$$

Table 2 shows the number of data blocks for each file, generated according to the needed storage time.

Table 2. Number of data blocks necessary according to the storage time.

Storage time (Years)	Foreseen file cycles	Data blocks per file
5	101	25856
10	203	51968
15	304	77824
20	406	103936
30	608	155648
50	1014	259584
100	2028	519168

From the definition of the number of blocks, it is possible to determine the size of the 'information table' and, therefore, the computing cost to transfer and store it in an Integrity Check Service. Figure 3 shows the size variation of the 'Information Table' according to the number of years planned for its storage.

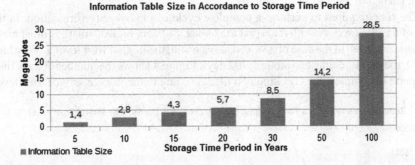

Fig. 3. The 'information table' size variation per number of years planned for its storage.

Finally, fraud resistance is obtained by means of a selection and swapping algorithm that assigns the entropy needed to render impracticable any attempt to predict which chunks are in each data block, as well as the order in which these chunks were joined. A brute force attack, generating hash codes for all possible combinations of data blocks, is not feasible as the number of possible combinations for the arrangement of 4096 file chunks in blocks with 16 chunks each is about 6.09×10^{57} blocks (Eq. 3). Consequently, to generate and store 256-bit hash codes for all possible combinations of data blocks would require about 1.77×10^{47} TB in disk space (Eq. 4).

$$A_{n,p} = \frac{n!}{(n-p)!} \rightarrow A_{4096,16} = \frac{4096!}{(4096-16)!} \cong 6.09 \times 10^{57} \tag{3}$$

$$Disk\ Space(TB) = \frac{(6.09 \text{x } 10^{57}) * 256}{8 * 1024^4} \cong 1.77 \times 10^{47} \qquad (4)$$

6 Simulations

In order to demonstrate when and how changes occur in the trust level assigned to the Cloud Storage Service, mathematical simulations were performed. The results refer to a scenario where one Integrity Check Service is being used by one client to checking files on one Cloud Storage Service. The first simulation was done to determine the maximum number of days required to complete a file check cycle (all its data blocks being checked at least once) on each trust level. To obtain the results we considered the percentages defined for each trust level on Table 1.

The time to complete the file check cycle can vary between a minimum and a maximum time according to the number of files being checked in the same Cloud Storage Service. However, file size should not influence the time spent for its assessment, considering that the number of data blocks checked daily represents a percentage of the size of each file verified. The maximum time set can only be expanded if either the Integrity Check Service or the Cloud Storage Service has no processing power or bandwidth needed to meet its daily requests.

The time required to perform a complete cycle in file integrity verification, in the case of a valued service with the worst trust level, of 'Very high distrust', is of, at most, 36 days, whereas, in the case of a valued service with the highest trust level, ('Very high trust'), the required time can attain 1,707 days. Figure 4 shows the number of days (min and max) to complete the checking of a file according to the storage service trust level.

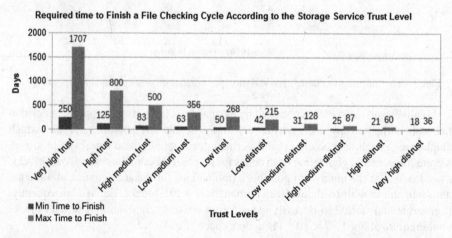

Fig. 4. Increase in the number of days to complete checking a file as per trust level. (Color figure online)

This time variation is intended to reward services with lessen failures, minimizing consumption of resources such as processing and bandwidth. Moreover, it allows prioritizing the protocol to checking files that are stored in services that have already failed, reducing the time required to determine if there are any corrupted or lost files.

The second simulation aims at determining the number of file check cycles completed successfully as needed, without any intermediate failure record, for a Cloud Storage Service to be rated with the highest trust levels. Thus, starting from the situation not evaluated, when the trust value is equal to zero, 67 cycles are needed to reach an intermediate classification that can already be considered quite reliable, the 'High medium trust'. To achieve the highest trust level, 'Very high trust', one needs 384 cycles. Figure 5 shows the maximum number of checking cycles without failure for a Cloud Storage Service to receive repeated raises until getting the maximum trust level.

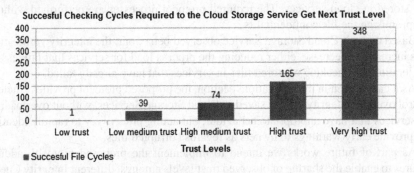

Fig. 5. The number of checking cycles without failures to raise a storage service trust level.

Finally, the third simulation evaluates the trust level decline of a Cloud Storage Service when integrity failures are identified in checking the data blocks of monitored files. The simulation starts with a fault identified in a storage service, so far classified at any level considered reliable. The trust levels for a reliable service are specified in Table 1 between 'Very high trust' and 'Low trust', i.e., a range where the trust level value is greater than zero.

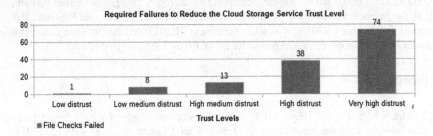

Fig. 6. Number of failures identified needed to downgrade the storage service trust level.

In this case, the trust level value for the service immediately receives value zero, being reclassified to a 'Low distrust' trust level that can also be regarded as the lowest distrust level. Having identified over 74 flaws in data blocks, without concluding at the same time a successful file checking cycle, the Cloud Storage Service is reclassified as 'Very high distrust', the highest distrust level. Figure 6 shows the number of failures identified that are needed to reduce the Cloud Storage Service trust level.

Hence, the proposed protocol pace to build trust in a service provider is slower than the pace to react to failures of this provider, consequently mistrusting its service.

7   Conclusions

In this paper a protocol is proposed to warranty the integrity of files saved in Cloud Storage Services. Through an Integrity Check Service hosted by a third party and by using trust levels, the protocol allows a qualified and continued integrity monitoring of files stored in cloud services. The required protocol processes are performed without compromising the file confidentiality.

Based on each Cloud Storage Service observed behaviour, the integrity verification rates increase or decrease, either reducing the process load for services that fail less or accelerating the checking of stored files in services that have already failed.

As it can be seen in the simulations section, the proposed protocol provides efficient control over the integrity of files saved in Cloud Storage Services, without overloading the services that have appropriate behaviour, but acting proactively to rapidly identify and provide early warnings to owners as regards corrupted files.

As part of future works we intend to implement the proposed protocol, adding features to enable the sharing of observed trust levels amongst different Integrity Check Services. Also, it is interesting to choose and test a hash algorithm that is safe and presents adequate performance for devices with low processing power such as tablets and smartphones. Furthermore, we want to add a mechanism to ensure that, only when authorized by a client, the Integrity Check Services will be authorized to send challenges to Cloud Storage Services.

Acknowledgments. The authors wish to thank the Brazilian research, development and innovation Agencies CAPES (Grant FORTE 23038.007604/2014-69), FINEP (Grant RENASIC/ PROTO 01.12.0555.00) and the Research Support Foundation of the Federal District FAPDF, as well as the Science and Technology Department of the Brazilian Army, the Brazilian Public Administration School ENAP (Cooperation Agreement 02/2013) and the Brazilian Union Public Defender DPGU (Cooperation Agreement 30/2014), for their support to this work.

References

1. Tandel, S.T., Shah, V.K., Hiranwal, S.: An implementation of effective XML based dynamic data integrity audit service in cloud. Int. J. Soc. Appl. Comput. Sci. **2**(8), 449–553 (2014)
2. Dabas, P., Wadhwa, D.: A recapitulation of data auditing approaches for cloud data. Int. J. Comput. Appl. Technol. Res. (IJCATR) **3**(6), 329–332 (2014)
3. Miller, M.: Cloud Computing Web-Based Applications that Change the Way You Work and Collaborate online. Que Publishing, Pearson Education, Canada (2008)

4. Velte, T., Velte, A., Elsenpeter, R.: Cloud Computing, A Practical Approach. McGraw-Hill, Inc., New York (2009)
5. Zhou, M., Zhang, R., Zeng, D., Qian, W.: Services in the cloud computing era: a survey. In: 4th International Universal Communication, Symposium (IUCS), pp. 40–46. IEEE Shanghai, China (2010)
6. Jing, X., Jian-jun, Z.: A brief survey on the security model of cloud computing. In: Ninth International Symposium on Distributed Computing and Applications to Business, Engineering and Science (DCABES), pp. 475–478. IEEE Hong Kong (2010)
7. Mell, P., Grance, T.: The NIST definition of cloud computing. Technical report, National Institute of Standards and Technology (2009)
8. Bellare, M.: Public-Key Encryption in a Multi-user Setting: Security Proofs and Improvements. Springer, Heidelberg (2000)
9. Marsh, S.P.: Formalizing trust as a computational concept. Department of Computing Science and Mathematics, University of Stirling, Doctorate Thesis (1994)
10. Gambetta, D.: Can we trust trust. In: Gambetta, D. (ed.) Trust: Making and Breaking Cooperative Relations, Chap. 13, Electronic edn., pp. 213–237. Department of Sociology, University of Oxford (2008)
11. Jøsang, A., Knapskog, S.J.: A metric for trusted systems. In: Global IT Security, pp. 541–549 (1998)
12. Bose, R.: Information Theory, Coding and Cryptography, 2nd Edn., pp. 297–298. Tata McGraw-Hill, New Delhi (2008)
13. Juels, A., Kaliski, B.S.: Pors: proofs of retrievability for large files. In: 14th ACM Conference on Computer and Communication Security (CCS), Alexandria, VA, pp. 584–59 (2007)
14. Kumar, R.S., Saxena, A.: Data integrity proofs in cloud storage. In: Third International Conference on Communication Systems and Networks (COMSNETS), Bangalore (2011)
15. George, R.S., Sabitha, S.: Data anonymization and integrity checking in cloud computing. In: Fourth International Conference on Computing (ICCCNT), Communications and Networking Technologies, Tiruchengode (2013)
16. Kavuri, S.K.S.V.A., Kancherla, G.R., Bobba, B.R.: Data authentication and integrity verification techniques for trusted/untrusted cloud servers. In: International Conference on Advances in Computing, Communications and Informatics (ICACCI), New Delhi, pp. 2590–2596 (2014)
17. Al-Jaberi, M.F., Zainal, A.: Data integrity and privacy model in cloud computing. In: International Symposium on Biometrics and Security Technologies (ISBAST), Kuala Lumpur, pp. 280–284 (2014)
18. Kay, H., et al.: An efficient public batch auditing protocol for data security in multi-cloud storage. In: 8th ChinaGrid Annual Conference (ChinaGrid), Changchun, pp. 51–56 (2013)
19. Wang, Q., Wang, C., Li, J., Ren, K., Lou, W.: enabling public verifiability and data dynamics for storage security in cloud computing. In: Backes, M., Ning, P. (eds.) ESORICS 2009. LNCS, vol. 5789, pp. 355–370. Springer, Heidelberg (2009)
20. Jordão, R., Martins, V.A., Buiati, F., Sousa Jr., R.T., Deus, F.E.: Secure data storage in distributed cloud environments. In: IEEE International Conference on Big Data (IEEE BigData), Washington DC, USA, pp. 6–12 (2014)

Implementation of Multiple-Precision Floating-Point Arithmetic on Intel Xeon Phi Coprocessors

Daisuke Takahashi[✉]

Center for Computational Sciences, University of Tsukuba,
1-1-1 Tennodai, Tsukuba, Ibaraki 305-8573, Japan
daisuke@cs.tsukuba.ac.jp

Abstract. In this paper, we propose an implementation of multiple-precision floating-point addition, subtraction, multiplication, division and square root on Intel Xeon Phi coprocessors. Using propagated carries in multiple-precision floating-point addition is a major obstacle to vectorization and parallelization. By using the carry skip method, the operation of performing propagated carries in the multiple-precision floating-point addition can be vectorized and parallelized. A parallel implementation of floating-point real FFT-based multiplication is presented, as multiplication is a fundamental operation in fast multiple-precision arithmetic. The experimental results of multiple-precision floating-point addition, multiplication, division and square root operations on an Intel Xeon Phi 5110P are then reported.

1 Introduction

Numerical computations on modern computers are generally performed with arithmetic operations of the elementary *float* and *double* datatypes. The precision of the 32-bit *float* datatype is approximately 6 decimal digits and the precision of the 64-bit *double* datatype is approximately 15 decimal digits in IEEE 754 representation. Although quad precision is available on computers which can support the necessary software, its precision is at most 33 decimal digits.

However in the fields of scientific and engineering computing, a higher accuracy is often required. Furthermore, floating-point operations also have round-off errors which tend to compound in large-scale computations to the point where these errors cannot be ignored. To reduce these rounding errors and increase accuracy, we can use multiple-precision arithmetic instead.

Several software packages are available for multiple-precision floating-point arithmetic on CPUs [2,3,7,8,18]. GMP [7] is probably the most widely used of these packages due to its greater functionality and efficiency. Although GMP supports single instruction multiple data (SIMD) instructions, there is generally not much support for propagating this sort of carry [7].

Also, there are several implementations of multiple-precision floating-point arithmetic on graphics processing units (GPUs) [15,16]. For example, the CUMP

© Springer International Publishing Switzerland 2016
O. Gervasi et al. (Eds.): ICCSA 2016, Part II, LNCS 9787, pp. 60–70, 2016.
DOI: 10.1007/978-3-319-42108-7_5

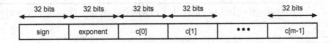

Fig. 1. Multiple-precision floating-point number representation

library [16] contains functions which allow operations involving large objects such as vectors or matrices.

The Intel Many Integrated Core Architecture (Intel MIC Architecture) has emerged as an important computational accelerator in high-performance computing systems. An implementation of multiple-precision integer arithmetic on the Intel Xeon Phi coprocessors was presented in [12]. However, to the best of our knowledge, an implementation of multiple-precision floating-point arithmetic on Intel Xeon Phi coprocessors has not yet been reported. The Intel Xeon Phi 5110P has 60 cores and 512-bit SIMD instructions, making both vectorization and parallelization particularly important to fully utilize the capabilities of the Intel Xeon Phi coprocessors.

In this paper, we propose an implementation of multiple-precision floating-point addition, subtraction, multiplication, division and square root operations on the Intel Xeon Phi coprocessors.

The remainder of this paper is organized as follows. Section 2 describes a multiple-precision floating-point number representation. Section 3 describes a vectorization and a parallelization of multiple-precision floating-point addition, subtraction and multiplication. Section 4 describes a vectorization and a parallelization of multiple-precision floating-point division and square root. Section 5 presents results comparing the performance of our implementation with the GMP package. In Sect. 6, we provide some concluding remarks.

2 Multiple-Precision Floating-Point Number Representation

Intel Initial Many Core Instructions (Intel IMCI) supports vpaddd instruction for 32-bit integer addition, and vpmulhud and vpmulld instructions for 32-bit × 32-bit → 32-bit integer multiplication [9].

However, the Intel IMCI does not support the Intel Advanced Vector Extensions 512 (Intel AVX-512) vpaddq instruction for 64-bit integer addition, or the Intel AVX-512 vpmuludq instruction for 32-bit × 32-bit → 64-bit integer multiplication [10]. Thus, we use a 32-bit integer array to represent multiple-precision floating-point numbers.

A multiple-precision floating-point number is represented with a sign and a signed exponent and 32 bits of mantissa with a 32-bit integer array as shown in Fig. 1. Sign s (1 or -1) is stored in the first element of the array, and exponent q $(-2^{31} \leq q \leq 2^{31} - 1)$ is stored in the second element. Mantissa $c = \sum_{i=0}^{m-1} c_i \times 10^{-8(i+1)}$ is stored from the third to the $m+2$-th element of the array using the big endian format (radix $= 10^8$).

Algorithm 1. Multiple-precision parallel addition with the carry skip method [20]

Input: $X = \sum_{i=0}^{n-1} x_i B^i$, $Y = \sum_{i=0}^{n-1} y_i B^i$
Output: $Z = X + Y := \sum_{i=0}^{n-1} z_i B^i$
1: **for** $i = 0$ to $n - 1$ **do in parallel**
2: $\quad z_i \leftarrow x_i + y_i$
3: **while** $\max\limits_{0 \le i \le n-2} (z_i) > B$ **do**
4: $\quad c_0 \leftarrow 0$
5: $\quad$ **for** $i = 0$ to $n - 2$ **do in parallel**
6: $\quad\quad c_{i+1} \leftarrow z_i$ div B
7: $\quad$ **for** $i = 0$ to $n - 2$ **do in parallel**
8: $\quad\quad z_i \leftarrow (z_i \bmod B) + c_i$
9: $\quad z_{n-1} \leftarrow z_{n-1} + c_{n-1}$
10: **while** $\max\limits_{0 \le i \le n-2} (z_i) = B$ **do**
11: $\quad$ **for** $i = 0$ to $n - 1$ **do in parallel**
12: $\quad\quad$ **if** $(z_i = B)$ **then**
13: $\quad\quad\quad p_i \leftarrow i$
14: $\quad\quad$ **else**
15: $\quad\quad\quad p_i \leftarrow n - 1$
16: $\quad l \leftarrow \min\limits_{0 \le i \le n-1} (p_i)$
17: $\quad$ **for** $i = l + 1$ to $n - 1$ **do in parallel**
18: $\quad\quad$ **if** $(z_i < B - 1)$ **then**
19: $\quad\quad\quad p_i \leftarrow i$
20: $\quad\quad$ **else**
21: $\quad\quad\quad p_i \leftarrow n - 1$
22: $\quad m \leftarrow \min\limits_{l+1 \le i \le n-1} (p_i)$
23: $\quad z_l \leftarrow z_l - B$
24: $\quad$ **for** $i = l + 1$ to $m - 1$ **do in parallel**
25: $\quad\quad z_i \leftarrow z_i - (B - 1)$
26: $\quad z_m \leftarrow z_m + 1$
27: **return** Z.

With this representation multiple-precision floating-point numbers can be represented as $s \times c \times 10^q$. For example, the value of π is stored in a 32-bit integer array x as x[0] = 1, x[1] = 1, x[2] = 31415926, x[3] = 53589793, and so on.

3   Vectorization and Parallelization of Multiple-Precision Floating-Point Addition, Subtraction and Multiplication

The operation counts for n-digit multiple-precision floating-point sequential addition, subtraction and multiplication by a single-precision integer are clearly $O(n)$. However, parallelization of these operations is difficult when propagated carries and borrows are used.

One obstacle to parallelization is the repeating of the propagation of carries, as in the case of $0.99999999\cdots 9 + 0.00000000\cdots 1$, which can be handled with the carry skip method [14].

Algorithm 1 is for multiple-precision parallel addition with the carry skip method [20]. We assume that the input data has been normalized from 0 to $B-1$ and is stored in arrays X and Y. Here, c_i $(0 \le i \le n-1)$ and p_i $(0 \le i \le n-1)$ are working arrays to store carries and indices, respectively. In Algorithm 1, multiple-precision addition without the propagation of carries is performed at lines 1 and 2. Then, incomplete normalization is performed from 0 to B in lines 3 to 9. Note that the while loop in lines 3 to 9 is repeated at most twice. The range for carry skipping is determined in lines 11 to 22. Finally, carry skipping is performed in lines 23 to 26.

Because carry skips occur intermittently in the case of $0.998998998\cdots +$ $0.0100100100\cdots$, this is one of the most complicated examples for use with the carry skip method [20]. For this "worst case", the carry look-ahead method is effective. However, the carry look-ahead method requires $O(\log P)$ steps on parallel computers that have P processors. Since we assumed that such a worst case rarely occurs, the less computationally expensive carry skip method was considered here to be sufficient for conducting multiple-precision parallel addition. The same methods can be applied to multiple-precision parallel subtraction and multiplication with a single-precision integer.

In Algorithm 1, each **for** loop can be vectorized and parallelized. Also, the computation of the maximum values in lines 3 and 10 are parallelized by the OpenMP reduction clause. In lines 6 and 8, integer division and modulo operations are needed. When the radix is a power of two, the integer division and modulo operations can be performed by right shift and AND operations, respectively. However, in this case, a binary-to-decimal radix conversion is needed. For ease of implementation, we use a radix $= 10^8$ in the implementation of the multiple-precision floating-point arithmetic.

Multiple-precision floating-point multiplication of n-digit numbers can be performed in $O(n \log n \log \log n)$ operations with the Schönhage-Strassen algorithm [17], which uses the fast Fourier transform (FFT) [5]. However, the Schönhage-Strassen algorithm may not be able to take full advantage of computers with fast floating-point hardware. Therefore, in this paper we use the multiple-precision multiplication algorithm which uses floating-point real FFT for fast multiplication that was presented in [3].

For the floating-point real FFT-based multiplication, we can use the "balanced representation" [6], which tends to yield reduced errors for the convolutions we wish to calculate. In this technique, an n-digit multiplicand $X = \sum_{i=0}^{n-1} x_i B^i$ with radix-B is represented as follows:

$$X = x_0' + x_1'B + x_2'B^2 + \cdots + x_{n-1}'B^{n-1}, \ |x_i'| \le \left\lfloor \frac{B}{2} \right\rfloor. \tag{1}$$

In the floating-point real FFT-based multiplication, the radix $B = 10^4$ is used. Each input data word (32-bit) is split into two words upon entry to the FFT-based multiplication.

The key operations in FFT-based multiple-precision floating-point multiplication are the FFT and normalization operations, on which a significant proportion of the total computation time is spent. Therefore, we can reduce computing time with an efficient parallel FFT routine in the FFT-based multiple-precision parallel multiplication.

The normalization of results in FFT-based multiple-precision parallel multiplication is essentially the same as the incomplete normalization in Algorithm 1. Thus, the incomplete normalization can also be vectorized and parallelized. In the multiple-precision parallel multiplication, the while loop in lines 3 to 9 in Algorithm 1 may be repeated more than twice due to the larger values of z_i in line 2.

The Intel C/C++ and Fortran compilers also support automatic vectorization of floating-point loops using Intel IMCI. In this paper, this automatic vectorization was used. For the Intel MIC Architecture, memory movement is optimal when the data starting address lies on 64-byte boundaries. Thus, we specified a directive "!dir$ attributed align : 64" for arrays.

4   Vectorization and Parallelization of Multiple-Precision Floating-Point Division and Square Root

The computation time of multiple-precision division and square root operations is considerably longer than that of addition, subtraction, or multiplication. Several methods have been proposed to perform division and square root operations [4,11,13]. It is well known that multiple-precision division and square root operations can be reduced to multiple-precision addition, subtraction and multiplication by Newton iteration [13].

4.1   Newton Iteration

We denote the number of operations used to multiply two n-digit numbers as $M(n)$. Newton iteration requires $O(M(n))$ operations to perform n-digit division and square root operations.

In division, the quotient of a/b is computed as follows. First, the following Newton iteration is used that converges to $1/b$:

$$x_{k+1} = x_k + x_k(1 - bx_k), \tag{2}$$

where the multiplication of x_k and $(1 - bx_k)$ can be performed with only half of the precision of x_{k+1} [3].

The final iteration is performed as follows [11]:

$$a/b \approx (ax_k) + x_k \left\{ a - b(ax_k) \right\}, \tag{3}$$

Table 1. Specification of the machine

Platform	Intel Xeon Phi coprocessor
Number of CPUs	1
Number of cores	60
CPU Type	Intel Xeon Phi 5110P 1.053 GHz
L1 Cache (per core)	I-Cache: 32 KB
	D-Cache: 32 KB
L2 Cache (per core)	512 KB
Main Memory	GDDR5 8 GB
OS	Linux 2.6.38.8 + mpss3.6

where the multiplication of a and x_k, and the multiplication of x_k and $\{a - b(ax_k)\}$ can be performed with only half of the final level of precision.

Square roots are computed by the following Newton iteration that converges to $1/\sqrt{a}$:

$$x_{k+1} = x_k + \frac{x_k}{2}(1 - ax_k^2), \tag{4}$$

where the multiplication of x_k and $(1 - ax_k^2)/2$ can also be performed with only half of the precision of x_{k+1}.

The final iteration is performed as follows [11]:

$$\sqrt{a} \approx (ax_k) + \frac{x_k}{2}\left\{a - (ax_k)^2\right\}, \tag{5}$$

where the multiplication of a and x_k, and the multiplication of x_k and $\left\{a - (ax_k)^2\right\}/2$ can be performed with only half of the final level of precision.

These iterations are performed by doubling the precision for each iteration.

4.2 Vectorization and Parallelization

To perform the Newton iteration in parallel, it is necessary to parallelize the multiple-precision parallel addition, subtraction and multiplication. For these operations, the parallel algorithms given in Sect. 3 can be applied. These operations can also be vectorized.

The half operation in Eq. (5) is a multiple-precision division by a single-precision integer. The multiple-precision division can be parallelized [19].

5   Performance Results

To evaluate the performance of the multiple-precision floating-point arithmetic, we compared it against the GMP 6.1.0 [7].

We averaged the elapsed times obtained from 10 executions of the multiple-precision floating-point addition ($\pi + \sqrt{2}$), multiplication ($\pi \times \sqrt{2}$), division

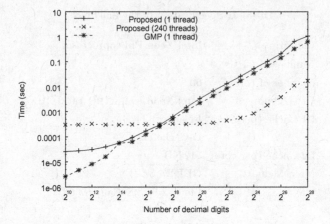

Fig. 2. Performance of the multiple-precision floating-point addition for the calculation of $\pi + \sqrt{2}$ (Intel Xeon Phi 5110P, 240 threads)

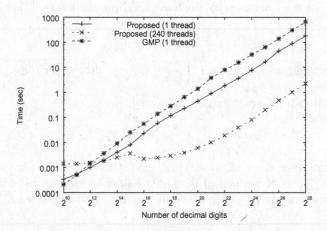

Fig. 3. Performance of the multiple-precision floating-point multiplication for the calculation of $\pi \times \sqrt{2}$ (Intel Xeon Phi 5110P, 240 threads)

($\sqrt{2}/\pi$) and square root ($\sqrt{\pi}$). We note that the respective values of n-digit π and $\sqrt{2}$ were prepared in advance. The selection of these operations has no particular significance here, but was convenient to establish definite test cases, the results of which were used as randomized test data.

The specification for the platform used is shown in Table 1. The compiler used was the Intel Fortran Compiler (`ifort`, version 16.0.0.109) for the implemented multiple-precision floating-point arithmetic. The compiler options used were specified as "`ifort -O3 -mmic -no-prec-div -openmp`". The compiler used was the Intel C Compiler (`icc`, version 16.0.0.109) for the GMP, with the compiler options "`icc -O3 -mmic`".

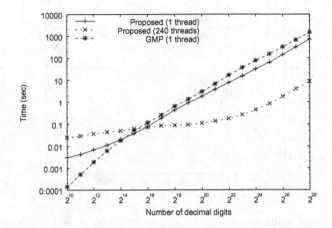

Fig. 4. Performance of the multiple-precision floating-point division for the calculation of $\sqrt{2}/\pi$ (Intel Xeon Phi 5110P, 240 threads)

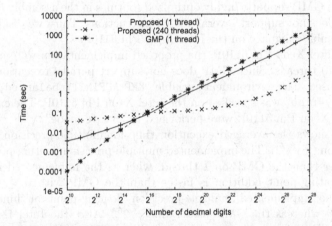

Fig. 5. Performance of the multiple-precision floating-point square root for the calculation of $\sqrt{\pi}$ (Intel Xeon Phi 5110P, 240 threads)

The compiler option "-O3" is for maximizing the speed and enabling more aggressive optimizations. The compiler option "-mmic" is for generating instructions for the Intel MIC Architecture. The compiler option "-no-prec-div" is for enabling the division-to-multiplication by reciprocal optimization. The compiler option "-openmp" is for enabling the compiler to generate multi-threaded code based on the OpenMP directives. The executions on the Intel Xeon Phi 5110P were performed in "native mode". The FFT routine from the FFTE library (version 6.0) [1] with optimization for the Intel Xeon Phi was used for the implemented multiple-precision floating-point arithmetic.

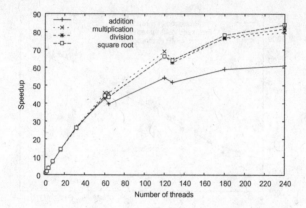

Fig. 6. Speedup for multiple-precision floating-point addition, multiplication, division and square root, $n = 2^{28}$ decimal digits (Intel Xeon Phi 5110P, 240 threads)

Although GMP supports highly optimized routines in the assembly language, Intel IMCI does not support several x86_64 instructions. Thus, we disabled routines in assembly language on the Intel Xeon Phi 5110P.

On the Intel Xeon Phi 5110P, the proposed implementation were used with from 1 to 240 threads. Since GMP does not support parallel execution, only 1 thread was used. The environment variable "KMP_AFFINITY=balanced, granularity=fine" was specified on the Intel Xeon Phi 5110P. The executions on the Intel Xeon Phi 5110P were performed in "native mode".

Figure 2 shows the averaged execution time for multiple-precision floating-point addition ($\pi + \sqrt{2}$). The implemented multiple-precision floating-point addition is slower than the GMP on 1 thread, whereas the implemented multiple-precision floating-point addition is faster than the GMP with $n \geq 2^{18}$ (240 threads). The implemented multiple-precision floating-point arithmetic uses radix $= 10^8$, whereas the GMP uses radix $= 2^{64}$. Also, the Intel IMCI does not support SIMD instruction for integer division. This is the reason the GMP is faster than the implemented multiple-precision floating-point addition on 1 thread.

Figure 3 shows the averaged execution time for multiple-precision floating-point multiplication ($\pi \times \sqrt{2}$). The implemented multiple-precision floating-point multiplication is faster than the GMP for the cases of $n \geq 2^{12}$ (1 thread) and $n \geq 2^{13}$ (240 threads). A possible reason for this is the performance of the floating-point real FFT is faster than the GMP's FFT, and the normalization operation is vectorized.

Figure 4 shows the averaged execution time for multiple-precision floating-point division ($\sqrt{2}/\pi$). The implemented multiple-precision floating-point division is faster than the GMP for $n \geq 2^{15}$ (1 thread) and $n \geq 2^{16}$ (240 threads).

Figure 5 shows the averaged execution time for multiple-precision floating-point square root ($\sqrt{\pi}$). The implemented multiple-precision floating-point square root is faster than the GMP for $n \geq 2^{16}$ (1 thread and 240 threads).

For $n \leq 2^{15}$ (240 threads), we can clearly see that the parallelization over-head dominates the execution time in Figs. 4 and 5. This is because the Newton iteration of division and square root operations performs small digit addition, subtraction and multiplication.

Figure 6 shows the speedup for multiple-precision floating-point addition $(\pi + \sqrt{2})$, multiplication $(\pi \times \sqrt{2})$, division $(\sqrt{2}/\pi)$ and square root operations $(\sqrt{\pi})$ of $n = 2^{28}$ decimal digits. The speedup of multiple-precision floating-point addition is less than that of multiple-precision floating-point multiplication, division and square root operations. This is because the multiple-precision floating-point addition is memory-intensive operation. For the multiple-precision floating-point multiplication, division and square root operations, the speedup is more than 70 on 240 threads. These results indicate the implemented multiple-precision floating-point arithmetic utilizes the hyper-threading on the Intel Xeon Phi 5110P.

6   Conclusion

We presented an implementation of multiple-precision floating-point addition, subtraction, multiplication, division and square root operations on Intel Xeon Phi coprocessors.

In FFT-based multiple-precision floating-point multiplication, FFTs and normalization are the elementary operations, on which a significant proportion of the total computation time is spent.

The operations for performing propagated carries and borrows in multiple-precision floating-point addition, subtraction and multiplication were parallelized using the carry skip method. Similar to multiple-precision floating-point addition and subtraction, some parts of the normalization of results in the multiple-precision multiplication can be vectorized and parallelized.

We have succeeded in vectorizing and parallelizing the basic routines for addition, subtraction, multiplication, division and square root operations on the Intel Xeon Phi coprocessors. The results of our performance evaluation demonstrate that the implemented multiple-precision floating-point arithmetic can better exploit the capabilities of the Intel Xeon Phi coprocessors.

Acknowledgments. This research was partially supported by Core Research for Evolutional Science and Technology (CREST), Japan Science and Technology Agency (JST).

References

1. FFTE: A Fast Fourier Transform Package. http://www.ffte.jp/
2. Bailey, D.H.: High-Precision Software Directory. http://www.davidhbailey.com/dhbsoftware/
3. Bailey, D.H.: Algorithm 719: multiprecision translation and execution of FORTRAN programs. ACM Trans. Math. Softw. **19**, 288–319 (1993)

4. Brent, R., Zimmermann, P.: Modern Computer Arithmetic. Cambridge University Press, Cambridge (2010)
5. Cooley, J.W., Tukey, J.W.: An algorithm for the machine calculation of complex Fourier series. Math. Comput. **19**, 297–301 (1965)
6. Crandall, R., Fagin, B.: Discrete weighted transforms and large-integer arithmetic. Math. Comput. **62**, 305–324 (1994)
7. Free Software Foundation Inc.: The GNU MP Bignum Library. https://gmplib.org/
8. Free Software Foundation Inc.: The GNU MPFR Library. http://www.mpfr.org/
9. Intel Corporation: Intel Xeon Phi coprocessor instruction set architecture reference manual (2012). https://software.intel.com/sites/default/files/forum/278102/327364001en.pdf
10. Intel Corporation: Intel architecture instruction set extensions programming reference (2015). https://software.intel.com/sites/default/files/managed/07/b7/319433-023.pdf
11. Karp, A.H., Markstein, P.: High-precision division and square root. ACM Trans. Math. Softw. **23**, 561–589 (1997)
12. Keliris, A., Maniatakos, M.: Investigating large integer arithmetic on Intel Xeon Phi SIMD extensions. In: Proceedings of 9th International Conference on Design & Technology of Integrated Systems in Nanoscale Era (DTIS 2014), pp. 1–6 (2014)
13. Knuth, D.E.: The Art of Computer Programming. Seminumerical Algorithms, vol. 2, 3rd edn. Addison-Wesley, Reading (1997)
14. Lehman, M., Burla, N.: Skip techniques for high-speed carry propagation in binary arithmetic units. IRE Trans. Electron. Comput. **EC–10**, 691–698 (1961)
15. Lu, M., He, B., Luo, Q.: Supporting extended precision on graphics processors. In: Proceedings of Sixth International Workshop on Data Management on New Hardware (DaMoN 2010), pp. 19–26 (2010)
16. Nakayama, T., Takahashi, D.: Implementation of multiple-precision floating-point arithmetic library for GPU computing. In: Proceedings of 23rd IASTED International Conference on Parallel and Distributed Computing and Systems (PDCS 2011), pp. 343–349 (2011)
17. Schönhage, A., Strassen, V.: Schnelle Multiplikation großer Zahlen. Computing **7**, 281–292 (1971)
18. Smith, D.M.: Algorithm 693: a FORTRAN package for floating-point multiple-precision arithmetic. ACM Trans. Math. Softw. **17**, 273–283 (1991)
19. Takahashi, D.: A parallel algorithm for multiple-precision division by a single-precision integer. In: Lirkov, I., Margenov, S., Waśniewski, J. (eds.) LSSC 2007. LNCS, vol. 4818, pp. 729–736. Springer, Heidelberg (2008)
20. Takahashi, D.: Parallel implementation of multiple-precision arithmetic and 2,576,980,370,000 decimal digits of π calculation. Parallel Comput. **36**, 439–448 (2010)

Towards a Sustainable Architectural Design by an Adaptation of the Architectural Driven Design Method

Luis Villa[1], Ivan Cabezas[2(✉)], Maria Lopez[2], and Oscar Casas[2]

[1] R&D Department, SIO, Cali, Colombia
lvilla@sio.com.co
[2] Facultad de Ingeniería, Universidad de San Buenaventura Cali, Cali, Colombia
{imcabezas,mtlopez,ocgarcia}@usbcali.edu.co

Abstract. Sustainability is a global concern. It must be addressed by different sectors of society, even by the information technology sector. Moreover, such effort should not be only focused on direct impacts of technology over the environment, but also on the software engineering discipline by itself, which is facing now other dimensions of sustainability. In this paper, an adaptation of the Attribute-Driven Design method including sustainability as a driver is introduced. The proposal is based on sustainability guidelines established by the Karlskrona Manifesto. It involves a multidimensional sustainability analysis considering three levels of impacts and opportunities for each one of the architectural components. The adaptation is motivated by the design of a sustainable architecture of a cloud-based personal health record. The designed architecture is termed Health Catalogue Repository. It offers cloud services, allowing interoperability and timely access to patients' clinical information. Energy consumption and resources optimization are contemplated like fundamental sustainability characteristics in the architectural design. The achieved design offers a better trade-off among quality attributes and sustainability constraints. The achieved design by using the proposal reduces the long-term impacts whilst increases the sustainability opportunities in architectural components.

Keywords: Sustainability · Architecture software design · PHR · e-Health · Cloud computing

1 Introduction

Global population as well as economic systems inherent to society are growing at a fast pace. It has increased the consumption of resources, in an indiscriminate manner. Moreover, most of economic systems are focused on achieving short-term goals, ignoring long-term effects, which may be irreversible in many cases [28]. For instance, technological advances help indeed to improve quality of life, but at the same time, they have an impact over the environment [3]. These impacts are related to social, economic and environmental changes in the well-being and health of people, and may be reflected by

O. Gervasi et al. (Eds.): ICCSA 2016, Part II, LNCS 9787, pp. 71–86, 2016.
DOI: 10.1007/978-3-319-42108-7_6

phenomena such as displacement of communities, poverty, malnutrition, climate disorders, greenhouse effect or even extreme weather, among others [25].

Sustainability issues such as carbon emissions, global warming, energy, water and natural resources consumption should be a priority for organizations and governmental institutions [16]. In practice, these issues and their relevance may be understood by the Software Engineering (SE) community. However they may be not properly discussed, since it requires considering multiple elements and dimensions. Moreover, a definition of sustainability can be adjusted from the perspective from which it is analyzed. Among the different sectors and initiatives on which sustainability must be a top priority concern, public health and e-health are fundamental ones.

Public health programs should be continuously addressed by government institutions and international organizations in order to enhance its coverture as well as improving quality of provided services. It has a direct impact on the people's well-being [31]. In practice, a collaborative environment among health institutions is required for providing an adequate service. Such collaboration, as well as an optimization of resources can be made possible by e-health initiatives.

To support the creation of a collaborative environment of the health system, a technological platform allowing information interoperability and integration between electronic medical records and health systems is required. The Electronic Health Record (EHR) is the digital information related to the management and delivery of health services. EHR management involves a very large volume of information generated in different formats such as text files, structured and semi-structured documents, images, sound or video, which must be stored and preserved according to conditions and times established by each country's law [9, 18].

From author's viewpoint, properly tackling sustainability issues is not only a must but also a challenge for SE researchers and practitioners. It requires a participation of the whole SE community in order to propose new models of production, consumption and life [8]. The task of designing sustainable architectural systems is the focus of authors in this paper. Sustainability should be treated as a driver for designing architectural systems involving multiple stakeholders and impacting on other system requirements. In this regard, many architectural designs are guided by ISO 25010 quality attributes [17]. However, it does not include sustainability as a quality attribute of software. Thus if in practice, sustainability is sometimes not considered during a design process it cannot be associated exclusively or mainly to a lack of consciousness, but to a lack of the proper tools or guidelines for accomplish such endeavor.

In this paper, an adaptation of the Attribute-Driven Design method (ADD) is presented. It allows achieving a sustainable architectural design. The proposed method, termed as Sustainable Attribute-Driven Design method (SADD) includes sustainability drivers as one of the inputs with higher priority. In the architectural decisions is important to consider fundamental sustainability characteristics as highest-priority architectural drivers in order to design a sustainable system. Moreover, an analysis of sustainability is document based on a multidimensional impact matrix of the architectural components considered during the design phase. The architectural sustainability analysis of the Catalogue Health Repository (HCR) architecture was conducted based on the proposed method.

The paper is structured as follows. The problem statement and solution approach are summarized in Sect. 2. The background and related works are described in Sect. 3. The SADD method is proposed in Sect. 4. The architectural sustainability analysis of the HCR is discussed in Sect. 5 by the proposed method. Finally, conclusions are stated in Sect. 6.

2 Problem Statement and Solution Approach

Many architectural designs rely on standards such as the ISO 25010, in order to define quality attributes [17]. Nevertheless, the ISO 25010 standard does not include sustainability as a software quality attribute. Moreover, sustainability is not a trivial concept, requiring a multidimensional and early analysis. In practice, sustainability has to be considered during the architectural design process, in order to obtain a sustainable system. However, software architects may lack of proper tools and guidelines to achieve such endeavor. The solution approach presented in this paper consists in a sustainable architectural design method – SADD –, by an adaptation to the ADD method proposed by the SEI. It considers sustainability like a restriction in the design inputs. Thus, sustainability as a design driver may imply impacts or opportunities over other system requirements and multiple stakeholders. In this way, the SADD method aims to be a tool for analyzing sustainability impacts and opportunities of an architecture, during the design process.

3 Background

As we already pointed out, sustainability can be defined according to the perspective from which it is analyzed. From a social perspective, sustainability can be defined as the policies and institutions impacting on the integration of the various groups and cultural practices in a fair and equitable manner [24]. From a development perspective, a sustainable development is such one satisfying present needs, without compromising the ability of future generations to satisfy their own needs [27]. Regarding to projects, sustainability is the capacity to continuously provide generated benefits during an extended period. From a social perspective, the sustainability of a system is intended to ensure equity in the distribution and opportunity on delivery of social services such as health and education [13]. This section is focused on sustainability from a technological perspective. In this regard, technological sustainability initiatives are motivated by the introduction of environmental policies with a global reach in order to protect natural resources. For example, initiatives of big technology companies to optimize resource consumption, to generate healthy environments and to improve the delivery of services for increasing the quality of life for people. These initiatives are related to strategies of sustainability [12]. Each country's policies have a direct impact on citizens' quality of life as improve conditions and life expectancy should be a priority for governments [15]. Sustainability can be analyzed as an architectural quality attribute. It requires an analysis

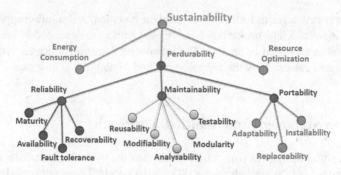

Fig. 1. Sub-characteristics for sustainability

from different dimensions (i.e. environmental, economic, social, technical and individual) in order to establish the impacts of the system against other established requirements.

3.1 Sustainability and the Karlskrona Manifesto

Two types of sustainability can be discussed in software engineering: relative and absolute. Relative sustainability implies that a function will be preserved for a specific time. Absolute sustainability implies that the software or service will help to preserve the environment and human well-being [21]. According to the guidelines of the Karlskrona Manifesto [2] the level of compliance of sustainability should not be focused exclusively on architecture direct impacts. Thus, it is appropriate to analyze the different levels of impact: direct (primary), indirect (secondary) and long term (tertiary) [14], which are described in Table 1.

Table 1. Sustainability impacts levels

Level	Description
First Level: Primary Effects	Effects of the physical existence of ICT: environmental impacts of the production, use, recycling and disposal of hardware
Second Level: Secondary Effects	Indirect environmental effects of ICT due to its power to change processes in a modification of their environmental impacts
Third Level: Tertiary Effects	Environmental effects of the medium or long-term adaptation of behavior or economic structures due to the stable availability of ICT and the services it provides

In [6] it is established that an analysis of sustainability as a quality attribute from a perspective of software architecture, requires an establishment of the sub-characteristics related to the requirements that the architecture should satisfy. The architecture should ensure that the software product is energy-efficient when it works, uses resources appropriately, and that it endures over time. Sustainability sub-characteristics and their relation to quality attributes are depicted in Fig. 1.

To analyze the perdurability as a characteristic of sustainable software, it can be related to software quality attributes, as defined in ISO 25010 [17]. To analyze sub-characteristics such as energy consumption and resource optimization, acceptance criteria for sustainability are required. Definitions of the sustainability sub-characteristics are described below. Energy consumption refers to the amount of necessary energy for a system component to perform its functions and satisfy its requirements. Resource optimization is the level of compliance with sustainability for the component during duties execution in terms of quantities and types of used resources, such as other software, hardware, and materials. Perdurability ensures that the software product will be retained and will last over time. It allows modifiability and adaptation to new circumstances without losing their functionality nor other quality features. In [20] the impacts of software development on the environment and the relation among sustainability characteristics and software quality attributes in order to develop environmentally sustainable software products are analyzed. If sustainability is considered as a quality attribute, it will be competing with other quality attributes during decision making about design and prioritization of drivers. Sustainability is a multidimensional concept that aims at sustainable development. Five sustainability dimensions are considered: Environmental, social, economic, technical and individual [11, 22]. Each dimension addresses a specific topic that can be affected by system implementation. Details of each dimension are presented in Table 2.

Table 2. Sustainability dimensions

Dimension	Description	Topics
Environmental	Long-term effects of human activities on natural systems	Ecosystems, Raw resources, Climate change, Food production, Water, Pollution and Waste
Social	Social communities and the factors that erode trust in society	Social equity, Justice, Employment, Democracy
Economic	Assets, capital and value added	Wealth creation, Prosperity, Profitability, Income
Technical	Longevity of information, systems and infrastructure	Maintenance, Innovation, Obsolescence, Data integrity
Individual	Well-being of humans as individuals	Mental/physical well-being, Education, Self-respect, Skills, Mobility

3.2 Architecture Driven Design - ADD

The Architecture Driven Design (ADD) is a method for defining a software architecture proposed by the Software Engineering Institute (SEI) [29]. In architectural design methods such as the ADD, requirements and quality attributes are prioritized by stakeholders. ADD follows a iterative process decomposing a system or system element by applying architectural tactics and patterns that satisfy its driving quality attribute requirements. The ADD method is illustrated in Fig. 2.

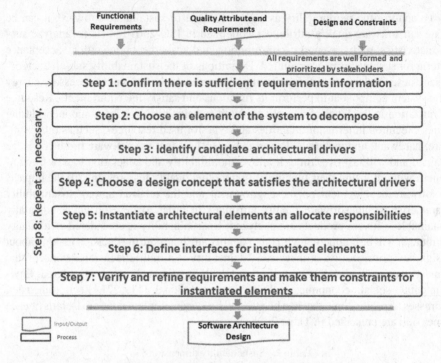

Fig. 2. Architecture driven design model

SUSTAINABILITY IMPACT MATRIX OF <<COMPONENT>>						
<<COMPONENT>>	First Order (Design and		Second Order (Application and Use)		Third Order (Over Time Use)	
Dimension	Impacts	Opportunities	Impacts	Opportunities	Impacts	Opportunities
Environmental	<<Impact>>	<<Opportunity>>	<<Impact>>	<<Opportunity>>	<<Impact>>	<<Opportunity>>
Social	<<Impact>>	<<Opportunity>>	<<Impact>>	<<Opportunity>>	<<Impact>>	<<Opportunity>>
Technical	<<Impact>>	<<Opportunity>>	<<Impact>>	<<Opportunity>>	<<Impact>>	<<Opportunity>>
Economic	<<Impact>>	<<Opportunity>>	<<Impact>>	<<Opportunity>>	<<Impact>>	<<Opportunity>>
Individual	<<Impact>>	<<Opportunity>>	<<Impact>>	<<Opportunity>>	<<Impact>>	<<Opportunity>>

Fig. 3. Sustainability impacts and opportunities matrices templates

3.3 Sustainable Health Services on Cloud

It is not common to find consolidated medical records. Usually, the EHR is fragmented into different systems of Electronic Medical Records (EMR) [1], belonging to different institutions providing health services. Such fragmentation hampers both timely accesses to EHR, and evidence-based decision making, which are fundamental to provide adequate medical services. Currently, an ubiquitous access to medical records in a timely manner from anywhere and at any time [19] is a need for government institutions. Several countries have implemented projects for promoting the development of sustainable technology platforms, and improved health services delivery, based on health records bank models and e-health cloud services [30]. Providing EHR information on cloud enables integration, consolidation, availability and timely access to patients' medical records. In addition, cloud computing provides flexibility and scalability, which is favorable for system sustainability. Information processing on cloud has also to be taken under consideration due to the sensitivity of EHR data, as well as the multiple actors around it: medical and nursing staff, care institutions and regulatory institutions. Granted EHR access should be regulated for ensuring data reliability and integrity. It is also important to consider associated risks of implementing a cloud based system. National and international institutions have issued guidelines regarding the treatment of health information of individuals, which those responsible for the EHR shall meet to protect information confidentiality and privacy. Interoperability mechanisms are required for information sharing among health actors. Sharing EHRs supports the sustainability of the health system from different dimensions: environmental, social, economic, technical and individual.

The system architecture shall ensure the functioning of the EHR as a lasting over time solution, in order to optimize the use of resources, comply the guidelines established by health services regulatory agencies, and satisfy users' care expectations. Currently, environmental impacts of technology are a global concern. SE should support this purpose by ensuring the construction of sustainable system designs.

It should be highlighted that a timely access to EHR is required to support decision-making based on evidence by e-health. It requires of interoperability and integration mechanisms among different actors in the health ecosystem in order to consolidate clinical patient information, under a sustainable model. In this regard, the HCR is a repository for storage, management and custody of the EHR based on a regulated model of Personal Health Record (PHR) as a cloud service. The HCR is related to community cloud, Personal Health Record (PHR), Health Record Bank (HRB) [10], Health as a Service (eHaaS) [4] and SOA concepts [7, 26]. A regulated PHR consolidates information from EMR of health institutions into a single PHR system managed by the user aiming to achieve a unique and complete EHR. An HCR enables the consolidation of a baseline for performed medical care consultations. The HCR distributes the costs of storage, infrastructure, security and maintenance among various health agencies that store information in the HRB. The HCR exposes healthcare functionalities as a service and is based on SOA reference architecture [23].

4   Proposal Solution

System quality attributes should be defined and refined during the architectural design process. Requirements engineering is the key to identify sustainability issues. It translates the needs, expectations and concerns from users within a specific context into technical requirements that can be implemented into a software system. Nevertheless, identifying requirements does not guarantee a per se sustainability of a system.

From authors' point of view, sustainability of a software architecture shall be addressed at the very beginning of its design process. It can be tackled by a multidimensional analysis on the technical characteristics and attributes of the system. Such analysis will arise only if the architectural design is guided by sustainability drivers. However, a software architect may lack of the proper tools in order to analyze sustainability issues. Taking this into account, the proposal SADD includes a modification of step 4: an activity to be performed for verifying and refining requirements. It consists in analyzing and documenting sustainability impacts and opportunities of each architectural component regarding to the five dimensions of sustainability – environmental, economic, social, technical, and individual – against the three levels considered by the Karlskrona Manifesto – direct, indirect and long term.

Such analysis produces a set of sustainability impacts and opportunities matrices as deliverable. Such set of matrices will be the input for another deliverable termed Architectural Sustainability Analysis. This document is, in essence, a multidimensional holistic analysis of the architectural design, regarding to sustainability. It contains the justifications, as well as the inherent tradeoff, in order to declare an architectural design as sustainable. Moreover, an evaluation of the sustainability matrix is considered. The goal of such evaluation is to identify whether a component is aligned with the architecture sustainability. In this regard, a quantitative evaluation assigning numerical values (i.e. weights) to each dimension and level of sustainability implies great effort and requires a deeper multidimensional analysis [5]. Thus, a qualitative assessment is considered by the proposal in order to determine whether or not an architectural component is sustainable. The component is sustainable if the opportunities are most relevant for the impact sustainability. In summary, the SADD method considers the below mentioned adaptations, which are illustrated in Fig. 4.

- Inputs: Design and Constrains: Sustainability shall be considered as the architectural constraint with the highest priority of drivers.
- Process:
  - Step 4: The original step "Choose a design concept that satisfies the architectural drivers" is divided now in the four following steps.

    Step 4.1. Choose a design concept.

    Step 4.2. Build the sustainability impacts and opportunities matrix: Fill the impact matrix of the design concept for each component. It involves identifying the impacts and opportunities of the design concept concerning sustainability, taking into account to the five dimensions of sustainability, and the three levels proposed on the Karlskrona Manifesto. For each dimension and level, identify sustainability impacts and opportunities provided by the component. The sustainability impacts are negative effects generated by the adoption of the selected component. The sustainability opportunities are positive effects of the component included in the architecture. The impacts and opportunities are defined in terms of physical, human and environmental resources, technical specifications and quality attributes required by the component and these are related with the sustainability of the architecture in different dimensions. The template for sustainability impact matrix is illustrated in Fig. 3.

    Step 4.3. Evaluate sustainability: Evaluate compliance of sustainability constraints and adequacy of chosen design concepts. The previously documented sustainability impacts and opportunities matrix, should be analyzed as a tool for validating the chosen design concept. The selection of a component that is aligned with the sustainability of architecture is the result of a trade-off among impacts and opportunities for each dimension/level. If the matrix component opportunities (positive aspects) on each sustainability dimension and each level are more and more relevant than generated impacts (negative aspects) by the component, it meets with the sustainability driver. During the analysis, are impacts and opportunities of the third level (long-term), are considered as more relevant than impacts of the second and the first levels. Thus, if multiple components are being evaluated, the selected component should offers more opportunities and less generated impacts.

Step 4.4. Repeat as necessary: If in the sustainability validation, the trade-off among impacts versus opportunities is favoring impacts, then go back to the step 4.1. Otherwise, continue to step 5.

- Outputs: An architectural Sustainability Analysis will be documented after finishing the design process, explaining why, the architectural design can be, in a holistic way, defined as sustainable, and considering a set of sustainability impacts and opportunities matrices during the choose design concept phase for each one of the architectural elements, and for each sustainability dimension.

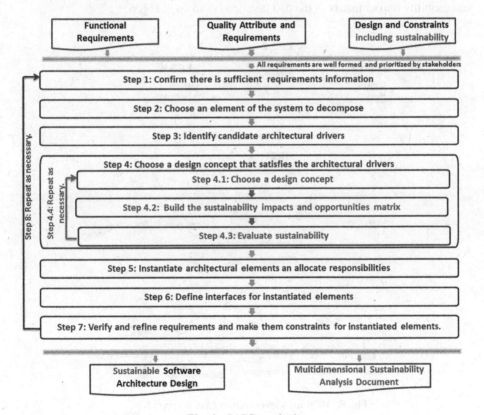

Fig. 4. SADD method

5   Architectural Sustainability Analysis of the HCR

The HCR model proposed in [26] is taken here as case study. It was designed using the conventional ADD method as a distributed architecture – without considering sustainability issues. Its high level architecture is presented in Fig. 5. In that architecture the HRB is distributed among several nodes of health promoter entities (EPS - *Entidades Promotoras de Salud*). EPS are responsible for registering people in the health system and ensuring health care services to its members. Thus, a reengineering process of the HCR architecture was performed for incorporating and analyzing sustainability as follows. Steps 1 to 7 were run iteratively. In step 1, drivers are reviewed including sustainability as the constraint with highest priority. In Step 2 the repository component of HCR is selected. In step 3 the sustainability, performance, security and Availability are identified as priority drivers. This repository could increase the computation efficiency, reduce computational overhead, as well as enhance fault recovery and information privacy. In step 4, concepts for repository are analyzed. In Step 4.1 HRB distributed concept is selected. In Step 4.2 Sustainability impact matrix of repository component of the HCR distributed architecture is developed. Impacts and opportunities for each impact level on the five dimensions of sustainability are identified and evaluated. The obtained sustainability impact matrix in the first iteration is shown in Fig. 6.

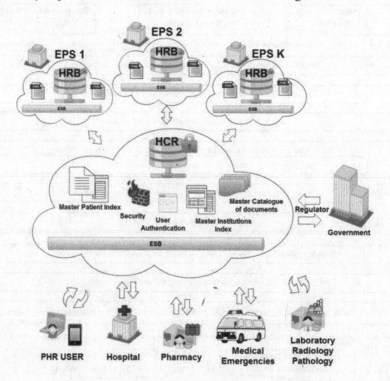

Fig. 5. HCR high level architectural diagram [26]

DISTRIBUTED HEALTH RECORD BANK Dimension	First Order (Design and Production)		Second Order (Application and Use)		Third Order (Over Time Use)	
	Impacts	Opportunities	Impacts	Opportunities	Impacts	Opportunities
Environmental	Increased waste generation		Increased energy consumption for each EPS node		Increased carbon footprint	
Social	Requires a high skills to development team		Requires Regulatory laws for HRB use / Requires SLA for information sustainability	Allows to accessible Health Information		Offers a better health attention
Technical	Involves complex designs, difficult administration, management and maintenance	Allows a distributed storage	Implies duplicated health information		Implies on site platform obsolescence	
	Limited capacity of HBR in each EPS node		Implies distributed personal health information		Perdurability of health documents depends of EPS Nodes	
Economic	Implies an expensive design process / Implies a high investment on infrastructure for each EPS node		Requires the investment on licenses for storage for eache EPS node	Allows the generation of new incomes by incorporated services	Requires of a sustained financial support	Allows new business models
Individual	Requires a user centered design	Allows privacy, confidentiality and protection of health information	Requires of technological skills	Offers on-line accessibility to Health information	Requires of responsibility on information security	

Fig. 6. Sustainability impact and opportunities matrix for distributed HRB component

In the step 4.3, the HRB component sustainability is evaluated. The HRB is distributed among the EPS' nodes. The HRB sustainability trade-off analysis for each dimension resulted in an array of impacts and opportunities, where the impacts on sustainability are more and more relevant than the opportunities and benefits it generates. In conclusion, a distributed HRB is not sustainable since it implies a lot of energy consumption for each HRB EPS node. In addition each EPS would incur in significant costs of investment in infrastructure in order to support the level of transactional requests made by the central HCR. Moreover, a distributed HRB does not achieve an optimized use of physical resources. Thus, a return to the step 4.1 is necessary.

In the second iteration of step 4.1, a centralized repository concept was selected. In Step 4.2 sustainability impact matrix of HRB centralized architecture is developed. The obtained sustainability impact matrix is shown in Fig. 7. In Step 4.3, centralized HRB sustainability is evaluated. It is observed that an HRB in a centralized architecture provides greater opportunities for the sustainability of HCR. In fact, a centralized HRB optimizes resource usage, reduces energy consumption and waste reduction. As the trade-off shows a component sustainability, the design process continues with the remainder steps.

SUSTAINABILITY IMPACT MATRIX OF HEALTH RECORD BANK IN HCR CENTRALIZED						
HEALTH RECORD BANK	**First Order** (Design and Production)		**Second Order** (Application and Use)		**Third Order** (Over Time Use)	
Dimension	Impacts	Opportunities	Impacts	Opportunities	Impacts	Opportunities
Environmental		Avoids waste generation		Allows a cost-effective use of natural resources		Allows a low carbon footprint
Social		Requires basic skills to development team	Requires Regulatory laws for HRB use / Requires SLA for information sustainability	Allows to accessible Health Information		Offers a better health attention
Technical	Implies complex designs	Allows a distributed storage, with reusability of components And flexible services by demand	Requires secure storage mechanisms	Allows a timely access to consolidated personal health information		Avoids on site platform obsolescence
	Requires cloud development capabilities	Offers an high capacity and scalabilty		Avoids duplicated health information		Allows perdurable storage of health documents
Economic	Implies an expensive design process	Allows a low investment on infrastructure		Allows the generation of new incomes by incorporated services / Avoids the investment on licenses for storage	Requires of a sustained financial support	Allows new business models
Individual	Requires a user centered design	Allows privacy, confidentiality and protection of health information	Requires of technological skills	Offers on-line accessibility to Health information	Requires of responsibility on information security	Allows ubiquity information

Fig. 7. Sustainability impact and opportunities matrix for centralized HRB component

The architectural design of the HCR was finished by using the presented proposal for each component. Once applied the SADD method, the obtained HCR is a centralized architecture. Centralized components of HCR optimize resources usage and allow energy saving, which are fundamental to system sustainability. In Fig. 8 the sustainable high level architecture of HCR is presented. Then sustainability analysis for each dimension according to the results of the sustainability impacts and opportunities matrices is performed.

5.1 Environmental Sustainability

As a strategy for environmental sustainability, leading technology providers have proposed the use of cloud computing. The use of such technology seeks to have the lowest possible impact on the environment. With the implementation of architectural systems based on cloud services computational storage, consumption of power, cooling and ventilation systems are optimized in order to minimize its environmental impact. Therefore the HCR can be declared as environmentally sustainable since it is based on cloud architecture. Moreover, the cloud-service-providers selection must favor those

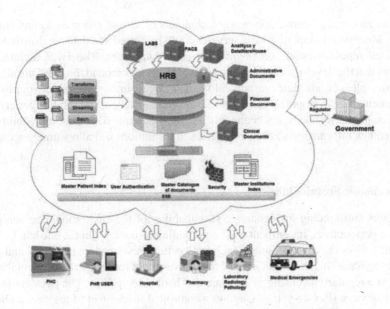

Fig. 8. High level architecture diagram of sustainable HCR

who are committed to protect the environment. Suppliers should be selected considering criteria such as adopted measures to reduce the impact on the carbon footprint, the percentage of renewable energy generators used, and the level of optimization based routing. The HCR implementation encourages the use of documents in electronic format as a mechanism for information sharing. Thereby, it reduces paper consumption and use of printing materials avoiding deforestation and environmental pollution rates. Since the HCR provides remote health services to patients, it prevents unnecessary journeys. Avoiding unnecessary travel has a positive environment impact, since it reduces the rate of generation of carbon dioxide, the carbon footprint and greenhouse effect. Thus, fuel consumption and transportation costs are reduced.

5.2 Social Sustainability

The architecture of the HCR system seeks to improve quality of health services. It promotes a collaborative environment among institutions and different actors in the health system, enabling an optimized use of the resources, as well as a timely and safety access to patients' clinical information, according to government laws for treatment of patients' health information. Moreover, the HCR allows users to perform management on their own PHRs.

5.3 Technical Sustainability

The architecture of HCR on cloud provides access to advanced technology. It allows a flexible and scalable growth of HCR system. Cloud computing increases the capacity of the physical infrastructure available for processes execution, optimizing physical resources and energy consumption, by techniques of virtualization and consolidated storage. Moreover, components and services could be reused and data are distributed in specialized repositories in agreement with used data types. The HCR architecture considers that the development of applications and services around the PHR, are aligned with principles of absolute sustainability. The HCR implementation requires the involvement of a development team with advanced knowledge, skills and experience of cloud computing technologies. Standards adoption is required for interoperability, in order to allow information sharing among health institutions and allow timely access to PHR.

5.4   Economic Sustainability

The factors surrounding the economic sustainability of the HCR should be analyzed from two perspectives, investment costs and funding resources. On the one hand, cloud computing allows to have technological infrastructure, development platforms and software as services. This avoids an initial capital investment for infrastructure acquisition, as well as a regularly licensing of software and hardware updates. The payment model for used services (Pay-as-you-go), allows a controlled investment of resources. On the other hand, cloud services allow focusing on the business value in order to generate a greater profitability by offering new services. Additional incomes may be obtained by offering HCR services such as information query for scientific and academic research purposes, merchandising of apps for interacting with services and processes arising from medical care such as authorizations, billing, payment or delivery of medicaments.

5.5   Individual Sustainability

The HCR architecture aims to improve the quality of experience perceived by each user. It allows multiple e-Health services such as telemedicine, virtual assistance and electronic prescribing, enabling and facilitating and ubiquitous patients care. Moreover, the HCR allows users to perform management on their own medical information by PHRs.

6   Conclusions

In this paper, the problem of how to design a sustainable architecture is addressed and discussed. The main contribution consists in an adaptation of the ADD method, in order to achieve a sustainable architectural design. Such adaptation is termed as the SADD. It modifies mainly the required inputs as well as the step 4 of the conventional method. Moreover it brings a clear guideline to software architects. The guideline can be stated in a very concise way: sustainability must be the highest priority constraint. Such guideline can be fulfilled by two activities. The first activity consists in analyzing each architectural component according to the five sustainability dimensions, regarding the three impact levels considered in the Karlskrona Manifesto. This activity has a deliverable termed as a sustainability impacts and opportunities matrix. The matrix allows to a software architect determine if he/she should repeat the fourth step, due to sustainability unfulfilled issues, or if he/she could moves forward with the design. The second activity is performed when an architect considers he/she has accomplished a sustainable design. It consists in an Architectural Sustainability Analysis document. Such document allows a discussion as well as a comprehension by the stakeholders on the multidimensional sustainability of an architectural design. The proposal was applied and exemplified in the context of the HCR. The value of the proposal relies on providing to the software architects concrete tools to conduct a very complex, challenging and abstract task, within a context which historically has only considered the direct impacts of an architectural design.

References

1. Alliance, T.N., et al.: Defining Key Health Information Technology Terms (2008)
2. Becker, C., et al.: Sustainability design and software: the karlskrona manifesto. In: ICSE 2015: 37th International Conference on Software Engineering (2015)
3. Benjumea Llorente, F.: Caso de éxito en empresa: la innovación como modelo de negocio. In: Cajamar, F. (ed.) Innovación y desarrollo económico, pp. 183–200 (2010)
4. Black, A.S., Sahama, T.: eHealth-as-a-Service (eHaaS): the industrialisation of health informatics, a practical approach. In: 2014 IEEE 16th International Conference on e-Health Networking, Applications and Services (Healthcom), pp. 555–559 (2014)
5. Cabezas, I., Trujillo, M.: A method for reducing the cardinality of the pareto front. In: Alvarez, L., Mejail, M., Gomez, L., Jacobo, J. (eds.) CIARP 2012. LNCS, vol. 7441, pp. 829–836. Springer, Heidelberg (2012)
6. Calero, C., et al.: Towards a software product sustainability model. J. Sustain. 25010, 4 (2013)
7. Castrillón, H.Y., et al.: Modelo Arquitectónico para Interoperabilidad entre Instituciones Prestadoras de Salud en Colombia. Univ. del cauca, programa Maest. en Comput (2012)
8. Castro, E.V.: Tecnologías de información que contribuyen con las prácticas de Green IT. Ingenium 8(19), 11–26 (2014)
9. CMS: Health insurance reform: security standards. Fed. Regist. 68(34), 8334 (2003)
10. Gold, J.D., Ball, M.J.: The Health Record Banking imperative: A conceptual model (2007)
11. Goodland, R., Bank, W.: Sustainability: human, social economic and environmental. Soc. Sci. 6(11), 220–225 (2002)
12. Greenpeace International: Make It Green - Cloud Computing and its Contribution to Climate Change. Forbes. 12 (2010)
13. Harris, J.: A Survey of Sustainable Development: Social and Economic Dimensions. Island Press, Washington, D.C (2001)
14. Hilty, L.M., et al.: The relevance of information and communication technologies for environmental sustainability - a prospective simulation study. Environ. Model Softw. 21(11), 1618–1629 (2006)
15. Indarte, S., Pazos Gutiérrez, P.: Estándares e interoperabilidad en salud electrónica: requisitos para una gestión sanitaria efectiva y eficiente. In: CEPAL (2012)
16. Info-Tech Research Group: Green IT: ¿Por qué las medianas empresas están invirtiendo ahora? (2009)
17. International Organization For Standarization ISO: ISO/IEC 25010:2011 (2011). http://www.iso.org
18. Ireland, D.P.C. of EU Directive 95-46-EC - Chapter 1 - Data Protection Commissioner – Ireland. https://www.dataprotection.ie/docs/EU-Directive-95-46-EC-Chapter-1/92.htm

19. Kim, T.W., Kim, H.C.: A healthcare system as a service in the context of vital signs: proposing a framework for realizing a model. In: Computers and Mathematics with Applications, pp. 1324–1332 (2012)
20. Koçak, S.A., et al.: Integrating Environmental Sustainability in Software Product Quality (2015)
21. Lago, P., et al.: Exploring initial challenges for green software engineering. ACM SIGSOFT Softw. Eng. Notes. **38**(1), 31 (2013)
22. Penzenstadler, B., Femmer, H.: A generic model for sustainability with process- and product-specific instances. In: GIBSE 2013 – Proceedings of the 2013 Workshop on Green In Software Engineering Green by Software Engineering, June 2015, pp. 3–7 (2013)
23. Philip, B., et al.: Evaluating a Service-Oriented Architecture. http://resources.sei.cmu.edu/library/asset-view.cfm?assetid=8443
24. Polèse, M., Stren, R.E.: The Social Sustainability of Cities: Diversity and the Management of Change (2000)
25. St Louis, M.E., Hess, J.J.: Climate change: impacts on and implications for global health. Am. J. Prev. Med. **35**(5), 527–538 (2008)
26. Villa, L., et al.: Electronic health record as an eHaaS. In: 2015 10th Computing Colombian Conference (10CCC) (2015)
27. World Commission on Environment and Development (WCED): Our Common Future. Oxford University Press, Oxford (1987)
28. WEO: World Energy Outlook. Int. Energy Agency, 690 pp. (2014)
29. Wojcik, R.: Attribute-Driven Design Method – ADD. http://www.sei.cmu.edu/architecture/tools/define/add.cfm
30. World Health Organization: WHO | WHO health systems strategy (2007). http://www.who.int/healthsystems/strategy/en/
31. Yang, A., et al.: "Sustainability" in global health. Glob. Public Health. **5**(2), 129–135 (2010)

The Design and Implementation on the Android Application Protection System

Cui Haoliang[1], Huang Ruqiang[2(✉)], Shi Chengjie[1],
and Niu Shaozhang[1]

[1] Beijing Key Lab of Intelligent Telecommunication Software and Multimedia,
Beijing University of Posts and Telecommunications, Beijing 100876, China
{cui.haoliang,happy_chengjie}@163.com,
szniu@bupt.edu.cn
[2] International School, Beijing University of Posts and Telecommunications,
Beijing 102209, China
hrqhrq2012@163.com

Abstract. As an open-source mobile platform, Android is facing with the severe problems of security and then the applications that running on this platform also confront with the same threats. This paper concludes the secure problems with which android applications are facing and gives a research on the current defense solutions. A security reinforcement system based on the Dex protection is proposed in order to defense the dynamic monitoring and modification. This system combines the static defense solution and dynamic defense solution, implements the purpose to tamper-proofing, anti-debugging for Android applications and improves the reliability and security of the software.

Keywords: Security threats · Secondary-loading · Tamper-proofing · Anti-debugging

1 Introduction

Recently, open-ended Android platform, complicated application sources and extensive software give rise to the higher frequency of the security vulnerabilities, which makes users become more and more worried about the security of mobile phone and personal data.

Traditional protections of software mainly focus on improving the capability of defensing hostile attack specifically which includes that the software not being maliciously modified, core data and algorithm not being inversed and software not being used unauthorized. Besides these basic secure requirements, software protection must defense dynamic debugging and process injection.

Software reinforcement technology of Android application is too hard to implement due to the characteristic of Android system. There are many security methods that only aim at a specific attack point rather than generate a comprehensive and effective solution. One of the typical Android security system areas is to provide a transparent, compatible and effective security reinforcement system.

© Springer International Publishing Switzerland 2016
O. Gervasi et al. (Eds.): ICCSA 2016, Part II, LNCS 9787, pp. 87–105, 2016.
DOI: 10.1007/978-3-319-42108-7_7

The research result of this paper plays the important theory and practical value to improve the security reinforcement working during the process of Android software developments. The result helps developers to eliminate worrying about lacking of security knowledge and provide security premise of developing the software.

This paper continues as follows. Section 2 introduces the background on this topic, such as two kinds of security threats and some current security reinforcement methods. Section 3 makes an analysis about static reinforcement solution and explains the mechanisms of these measures. Section 4 demonstrates the design of the dynamic defense solution and explains its working theory. Section 5 explains the whole architecture of reinforcement system and its specific mechanism. Section 6 makes the security analysis about the reinforcement system. Section 7 gives a conclusion about this paper.

2 Related Background

2.1 Security Threats

According to different analysis method, the security threats with which Android software is facing can be separate into static attack and dynamic attack.

Static attack. Static attack is to decompile the APK package of software and then tamper the program code maliciously or obtain the core code of the program [1]. Furthermore, static analysis is one of common presentation of static attack. In particular, static analysis is a technology that making use of some technical method such as Lexical analysis, syntax analysis, control flow and data flow to scan the program file and then generate disassembly code without running the code. Attackers will get knowledge about the program by reading the disassembly code, which makes the attack becomes easily [2]. Based on this method's principle, some tools also help to analyze the dependent relationship between different packages, obtain the invoking process of classes in order to make in-depth analysis. Attackers should locate critical code among a large amount of code by some special method, such as information feedback method, Eigen function method and sequential view method.

Dynamic attack. The mechanism of dynamic attack is to debug the program while the software is running or inject into the core function to generate the running log and then analyze the running logic of the program [3]. The main steps of locating critical code in the process of dynamic debug is to run the program, observe the output of program to determine the key point of program. There are some main methods. They include jog output method, stack trace method and analytic method.

2.2 Security Reinforcement Measure

At present, the defense technologies of static attack includes code obfuscation technology, software encryption technology and software tamper-proofing technology.

(1) Code obfuscation technology is to reorganize and process the program on the premise that not change function of code and running logic in order to protect the code. Christian Collerg separated the code obfuscation into four types: data obfuscation, layout obfuscation, control obfuscation and preventive obfuscation [4]. Obfuscating the code in different layers increases the difficulty of decompiling largely. Although attackers are able to decompile the code, this technology increases the difficulty of code analysis.

(2) Software encryption technology is a technical method that encrypt the binary code of software, in order to prevent attackers from obtaining binary plaintext.

(3) The main purpose of software tamper-proofing technology is to protect the integrity of software, protect program from tampering, distributing and using unlawfully [5]. This technology consists of two forms. Software-based form mainly contains checksum, software aging, assertion checking and key protection technology [6]. Hardware-based form mainly coordinates with the hardware equipment to protect the integrity of software, such as the hardware-based digital rights management method.

3 Static Defense Solution

3.1 DEX Shell Mechanism

Android uses Dalvik virtual machine as the running environment of the software. This VM design the executable file named DEX (Dalvik VM executes) for Android platform exclusively. Attackers is able to use decompile tools to transfer the DEX file into smali or jar file which is used to analyze the code logic of software and taper with code. Thus, the major object of static protection to application software is DEX file. DEX shell technology includes several steps to protect the application with shell program. The first step is to encrypt the DEX file and embed the shelling program into the new DEX file. When the program is running, the shelling program decrypt the original DEX file, and finally execute the DEX file. Dynamic defense mainly focuses on protecting from debugging dynamically. The mechanism of dynamic defense should comply with the debug tools working theory, which helps to find the defense solution. JDWP protocol debug tools and Ptrace debug are the main methods of dynamic debugging to the Android application at present.

3.2 DEX Shell Technology

At present, software shell technology is one of the most popular software protection technology. Shell technology on the PC platform is relative mature, which appears many general shell tools. Because of the particularity of Android system, Android does not provide official shell method as IOS AND WinPhone system. However, there are lots of APK shell program in the application markets. In addition, those reinforcement products on Android platform are not as high generality as PC platform.

The core purpose of software shell technology is to help application to defense decompiling activities of attackers and avoid that the code has exposed in front of unauthorized person. The shell program always change the entrance of the program's execution. The shell program becomes the real entrance of program. The technology also encrypts the binary data of the original program. When the shell program execute, the code of shelling is running at first, and next, the decryption, decompression and dynamic loading the original program through this shell program. Besides these steps, shell program need to check the executing environment of the software. It also looks over whether there is any debug software tries to debug the application dynamically. If there is any, the program should be terminated for protect itself. The shell program prevents attackers analyzing the real executed code of the application through the binary file of software. The shell program increase the capability of decompiling and unauthorized modifying in a great extent. This shows the shell program mechanism of Android application in the Fig. 1 below.

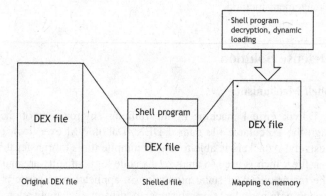

Fig. 1. Shell program mechanism

3.3   Class Loading Mechanism of Dalvik

The class loading mechanism of virtual machine is the whole process that loading the defined class data from class file into memory through the virtual machine. The life-cycle of the class in virtual machine contains seven stages: loading, verify, check, prepare, analyze, initialize, use and uninstall. In java code, the first step of running the program is to loading the class, java VM use the class loader to load the java class file into VM. This shows the method of using java class below: java project is import into java file suffixed with .java, compile the java file and get the class file suffixed with . class (Java byte file). Class loader is responsible for transferring the byte file into an instance of java.lang.Class. Each java class is represented in this initialization. Creating an object of this class can use newInstance() method to accomplish.

Dalvik VM is the same as java VM, which need to load the corresponding class into memory at the first step of running the program. There are different places between these two class loading mechanism. Developer is able to define its own class loader as the child class of ClassLoader, and then use defineClass method to load class from binary file. However, this method cannot apply to Android platform. Android mainly use DexClassLoader and PathClassLoader to achieve the loading of class, these two Loaders are both loading the class through DexFile.

Because DEX file is the executable file of Dalvik, Dalvik VM only recognizes DEX file and file suffixed with .apk or .jar, rather than Class file. Furthermore, .apk and .jar file need to contain DEX file. DexClassLoader obtains DEX objects through static method named loadDex(). PathClassLoader generates DexFile object by the function named DexFile(path). There is one major difference between two Loaders. When using DexClassLoader, it need to provide one optimized Directory to store the optimized DEX file. As for PathClasssLoader, DEX file cannot be decompressed from .apk and . jar package. It support operating the DEX format file directly.

The Classloader of Dalvik adopts the parents' entrusted mode, which means that parents' loader will start to load at the very first. Only parents' loader cannot finish the class loading, it will entrust original class loader to finish the work. If it has finished the loading, it will return success.

3.4 Design of Static Defense Solution

Secondary-loading technology means loading the DEX file dynamically when the software is executing. This solution is based on secondary-loading technology. Firstly, it encrypts the DEX file of the application, and then decrypt the file with native code at the entrance of the program. Finally, the file will be loaded dynamically with ClassLoader.

The static defense solution which is based on the secondary-loading technology acts as an important part of the whole security reinforcement system. The whole reinforce system is divided into two parts: PC side and Android side. The section of shelling is executed at the PC side. It mainly includes four steps: shell program, encryption, DEX integrity check and the APK signature. The structure of the static defense solution is shown in the Fig. 2.

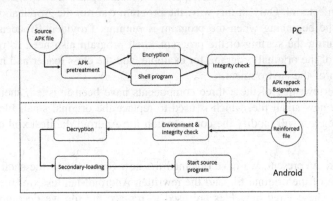

Fig. 2. Static defense mechanism

Encryption. Encryption means that the DEX file of the original program will be encrypted according to the encrypt algorithm, which generates the binary file. In the reinforcement system, the protection object is the DEX file of original program. Shelling process always encrypt the DEX file in order to avoid the DEX file and the plaintext of the byte file obtaining by the attackers. Encrypt algorithm always adopts the symmetric algorithm. However, in order to protect the details about encrypt algorithm from being analysis by attackers, the program should try to invoke the algorithm code in the open project, instead of the encrypt function that system provides. It shows the encryption process in the Fig. 3.

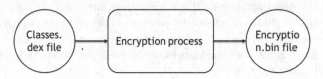

Fig. 3. Encryption process

Shell program. Shell program is the core part of the whole static defense solution. It is responsible for generating the DEX file of the shell program according to the component information of original program and replacing the smali file of original program with the smali file transferred from DEX file.

The first problem of running the shell program which has to be solved is how to load all component of original program. The shell program code will be executed at the very first. Before the DEX file of original program is loaded dynamically, all components are not able to be loaded, otherwise there will throw a loading exception. In the whole process of starting program, the initialization of components, such as Application, Provider, Broadcast, is determined by system. Application is the context of the whole program and system will start this application of the program at first. In addition, the developer always is used to making the initialized work in the Application. Thus, the shell program need to rewrite the Application, and replace the Application of the AndroidMainfest.xml. Only in this way, the program can run the Application of shell program at the beginning when the program is running. Provider, Broadcast will be initialized during the starting of the program, shell program also need to rewrite the components of the original program, for example, add the class loader and modify the configuration of the environment.

When the rewriting of these three components have been finished, shell program will generate new smali file, which is used to replace the original smali file. Furthermore, the program will modify the configuration file AndroidManifest.xml of original program.

APK repack. At present, we have obtained the new Smali file of the shell program, encrypted file of the original file and the rewritten AndroidManifest.xml file from the steps above. These three new files are used to replace the file we decompiled from original program, and then pack them together and sign them.

Decryption. Decryption means the bin file of original program should be decrypted in order to finish the loading work later. The code of decryption always uses java code to accomplish. In order to increase the security of code, we need to confound the java code in order to increase the difficulty of decompiling.

There is one another method that use the Native code to provide protection, which is more secure than the code confounding. Native code is too difficult to decompile for attackers, and it requires well knowledge about compilation to decompile the Native code.

Secondary-loading technology. The next step is to load the DEX file after we get the plaintext of DEX file. The general measure of shelling is to use DexClassLoader to accomplish the dynamic loading of apk file. DexClassLoader have to extend the loader class of shell program. The Classloader of Dalvik adopts the parents entrusted mode, that is, when original program start to load the class, parents classloader(class loader of shell program) will work at first. Its own ClassLoader starts to load when the shell program does not load successfully. When the shell program wants to invoke the method of the original program, it can use the method: Class.forName(String class-Name, boolean initializeBoolean, ClassLoader classLoader). In this method, Dex-ClassLoader is one parameter of the method. There will not exist a failure in the process of loading in this way.

Because of the initial work in the Application of the original program, shell program has rewritten the Application, even in the file of AndroidMainfest.xml, the initial Application is the Application of Shell program. In this situation, the system will execute the Application of shell program when the program starts to run. In order to ensure the logic of the original program, the Application in the file of AndroidMainfest.xml should be replaced with the original Application when the code of original program has been loaded successfully. The shell program should set the configuration of Application object when it has run over. The specific work is to configure the loading environment dynamically, modify the Application of original program by revoking and finally start the original program.

4 Dynamic Defense Solution

4.1 Dynamic Attack Mechanism

Dynamic defense mainly focuses on protecting from debugging dynamically. The mechanism of dynamic defense should comply with the debug tools working theory, which helps to find the defense solution. JDWP protocol debug tools and Ptrace debug are the main methods of dynamic debugging to the Android application at present.

(1) JDWP is the communicating protocol between the debug tools and java virtual machine. Many debug tools are implemented based on this protocol. Because Android java virtual machine provides the ports of debugging, debugger establishes the socket connection to debug program through these ports, daemon will poll the socket connections that the system have established and check that whether characteristics of connection accords with debug condition. If any debugger connection is discovered, the program will be terminated immediately.

(2) Ptrace is the parent process that Linux system provide can monitor other process. Making use of Ptrace technology will debug the Android software and inject code [7, 8]. The mechanism of defense solution to the Ptrace attack judges whether the process is in the traced mode by detecting the status of process.

4.2 Dynamic Defense Mechanism

JDWP debug defense. According to the last part, dynamic debug use the debugger based on JDWP to hook the software and obtain the data of running the program. One of the effective method to defense dynamic attack is to add code of checking debugger into the source code. Setting the value of android: debuggable to be "false", in order to avoid the debugging to the program. During the whole process of running program, it is important to check that whether the value has been changed. The common defense solution is to aborting the program when the value has been changed.

The defense solution that this project proposal is embarked from the debug mechanism of JDWP. Android virtual machine provides the ports to be debugged and debugger establish the socket connection with these ports to debug the program. Thus, the demon program is responsible for checking whether the attribute value accord with the debug feature by monitoring the socket connection that system establishes. If there is any connection from debugger, the program should be stopped or disconnect.

The interface monitor module will monitor the socket port dynamically, if a connection exists, the module sends the monitor information to the anti-debugging module when some program is planning to debug the program through the JDWP debug interface. The anti-debugging module mainly applies to terminate the JDWP debug process, when this module receives the message from the monitor module, it will end the thread to prevent the program from being debugged.

Ptrace debug defense. The defense mechanism of Ptrace attack is to check the state of process. The solution will terminate the process when the process is determined as "traced" state. according to this principle, this section introduces a defense solution which used the process-ring technology to prevent the application from being debugged dynamically.

The processes-ring program use three processes to defense the dynamic debugging. Each process will monitor the other two processes that if there is any process is being traced, the process will be terminated. The purpose of using three processes is thinking of the performance balance point. If only two processes monitor each other, attackers also take two processes to monitor them, the demon process is possible to be killed and the program can be suspended to be debugged. In other word, the more processes monitor each other, the more difficult attackers make the Ptrace debug. However, many processes will run out of performance and produce error during the program executes. Thus, the form of three processes is the best choice to monitor each other. On the one hand, it defenses the dynamic debug thoroughly, on the other hand, it will not get an impact on the execution of the program.

5 System Design and Implementation

According to the Android security threat and current defense method that previous content mentioned, this paper proposes a secure and comprehensive reinforcement system, this system combines the static defense solution and dynamic defense solution, which makes the file that reinforced by this system defense some static and dynamic attacks.

Current security reinforcement system just uses multiple different defense solutions that focus on different attack patterns [9]. This mechanism would make the whole reinforcement process become more complex. When different solutions become incompatible with each other, some troubles will appear in the process. The system of this paper will avoid the situation that file would be reinforced many times.

5.1 Design

In order to solve the static and dynamic attack that Android applications are facing with, we adopt the method that combines the static defense and dynamic defense together and embeds the security defense solution into the lifetime of generation and publish. The security reinforcement program contains two parts: PC and Android. PC side is responsible for encrypting and shelled-protecting the application, embedding the safe boot code into the application to make sure that the application is protected by the shell during running time. Android is working during the initial stage of booting, it will start the shell program, monitor the running environment, check the integrity, decrypt the code file and start the daemon in order to ensure the security of the boot and running environment and prevent program from being sniffed and debugged dynamically.

The security reinforcement system consists of two parts: PC side and Android side. PC side is responsible for using script language to generate Java file of shell program, which includes encryption, shell program, signature and repack. Android side is responsible for checking the running environment, Integrity check and decryption. In order to protect the code, Android make use of Native Code to implement.

PC side and Android side both adopt MVC module, MVC module is the abbreviation of model-view-controller, which is a classic structure module of software. MVC module divides the application into three different roles. Each module has its own work and assignment. Model: manage the logic of program, data structure and state. View: present the statement and data through user interface. Control: situate between Model and View, Control module send the data from View module to Model module. This framework make the system can be reused and flexible.

5.2 Architecture of System

This paper implements a security reinforcement system based on the reinforcement program we proposed. The system is composed of PC side and Android side. PC side is responsible for shelling program with automated scripts and Android side aims at shell processing and dynamic defense function.

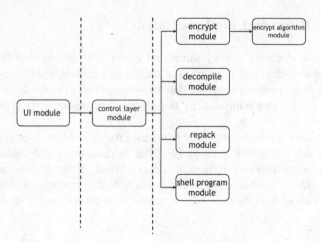

Fig. 4. Static defense mechanism

The static defense mechanism based on the secondary loading technology is an important part of the security reinforcement system. Figure 4 shows the PC side structure of the static defense mechanism.

PC side. PC side includes UI module, control layer module, decompile module, repack module, shell program module, encrypt module, encrypt module and algorithm module. UI module is the whole interface of the reinforcement system, UI module receive the APK program which need to be reinforced, control layer will send the file to APK processing module.

(1) UI module presents the function of the whole reinforcement system: obtain one APK file that needs to be reinforced and presents reinforced APK file.

(2) Control module is a kind of isolation layer, which separate the UI layer apart from processing module. When Control module receive the request from UI module, it sends the APK file to processing module to finish the reinforcement work. The underlying module sends the result to control module, and finally presents the result in UI module.

(3) Decompiling module mainly uses script language to encapsulate the decompiling function of ApkTool to provide the invocation to other module. ApkTool is a common decompile tool that transfers APK file into smali file and decrypts AndroidManifest.xml.

(4) The task of repack module is to pack the new generated DEX file, and produce the new reinforced program. This module repack the new smali file which produced by shell program with ApkTool.

(5) Shell program is the core module of the whole reinforcement system. Shell program generates the reinforced DEX file. It analyzes AndroidManifest.xml file of original program, obtain Application, Content Provider, Broadcast of original program and state specific processing mode of each file.

(6) Encryption module invokes the encrypt algorithm module and encrypt the new generated DEX file. The core part of the shell program is to encrypt the protection object, in order to avoid exposing the plaintext of the original program.

(7) Encryption algorithm module contains all the basic encrypt algorithm involved in the reinforcement system, such as AES, 3DES. Other module can invoke these algorithms by using the interface of this module. AES is the abbreviation of Advanced Encryption Standard, which is one of the most common Symmetric key algorithms. Triple DES means triple data encryption algorithm keyword. This algorithm encrypts the database in three times, which increases the difficulty of cracking by using longer key length.

Android side. In the Android side, there are unpacking program and dynamic defense solution. Android side includes JNI interface module, control module, DEX processing module, file operation module, encryption algorithm module, JDWP module and Ptrace module. JNI module uses java code to invoke the code of local interface; control module is responsible for controlling the invocation of all interface in Android side; encryption module achieves some kinds of encrypt algorithm, such as: AES, 3DES. The Android side structure of the static defense mechanism is shown in Fig. 5.

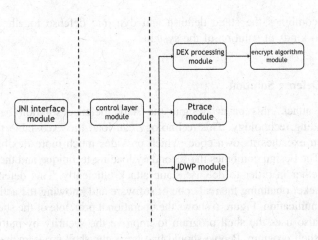

Fig. 5. Static defense mechanism

(1) JNI module is similar with JNI Onload function in the .so library file. In the Native Code programming, the Dalvik execute system.loadLibrary in the C component, the Onload() method will execute at the first place. This module becomes the entrance of invoking local code and make some initialization work.

(2) The control module separates the JNI module and other function module. Android side mainly focus on Native Code, and control layer provides all invoking interface of other module APIs. Java code layer can connect these APIs to accomplish specific purpose.

(3) The major task of DEX processing module is to decrypt DEX file and load the plaintext of DEX.

(4) Encryption algorithm module contains all the basic encrypt algorithm involved in the reinforcement system. Other module can invoke this module to implement specific function. Hackers always analyze the encrypt algorithm of software by searching the specific encryption method. The code obfuscation increases the difficulty to analyze the encrypt method.

Android side also has the JDWP and Ptrace module to implements the dynamic defense.

(1) This module is used to prevent software from being debugged by Ptrace function. The processing-ring program monitors whether the software is traced. It interrupts the software if the program finds out that the software is being debugged, which prevents the Ptrace debugging.
(2) JDWP module can monitor the debug port dynamically to judge any debugging request from client. The connection will close up when the request appears, which avoids debugging in the java layer happened during the whole lifecycle.

6   Implement of Defense Solution

This system combines the static defense and dynamic defense together. This part introduces two kinds of solution of the system.

6.1   Static Defense Solution

As for static attack, this paper designs the static defense solution based on the secondary-loading technology. This technology can load the executable file dynamically, and then executes its own code, which provides much more flexibility for the applications. The design combines the secondary loading technique and the encryption technique together in order to defense static attack efficiently. This defense solution avoids the attacker obtaining the real code of software and knowing the activity pattern and data communication. Figure 6 shows the operational principle of the shell program. The solution also uses the shell program to improve the security by putting the real code into the shell program. People should first break the shell program and get access to the key in order to obtain the running code.

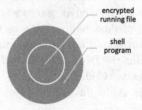

Fig. 6. Shell program

6.2   Dynamic Defensing Solution

As for dynamic attack, this paper designs port monitoring defense mechanism and process-ring defense mechanism and increases the difficulty of dynamic debugging to, the program under the white-box environment. Dalvik virtual machine accomplishs Java Debug Wire Protocol, thus, Dalvik vm supports lots of source code debugging in the different developing environment and is able to use any debugger based on JDWP. The defense technology we proposed is derived from JDWP debug principle. Daemon can poll the socket that system established and monitor the attribute of the socket. If there is any connection from debugger, the ending program will start to run or cut the connection off.

We design the process-ring technology to defense dynamic attack based on Ptrace. This method uses the monitoring processes to monitor each other, that is, if any process is traced, this process will be terminated. In the design, three processes monitor each other, which keep a capability balance among three processes. The more processes get involved, the stronger to defense the debug attack. However, more processes will consume the capability of system to a large extend. Figure 7 shows the mechanism of the processing-ring program.

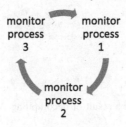

Fig. 7. Processing-ring

7   Security Analysis

In order to prove the security of the reinforcement system, this section adopts the contrast experiment to verify it, that is, comparing the general APK file with the APK file reinforced by the system to analyze the security of the system. The content of the analysis is divided into static security analysis and dynamic security analysis.

7.1   Static Security Analysis

In this experiment, *test.apk* is the file without reinforcement and *testsecurity.apk* is the reinforced file. This experiment make use of Apktool to decompile these two APK files and analyze the result of decompiling (Fig. 8).

The steps are as follows:

(1) Use **apktool d test.apk** command to decompile

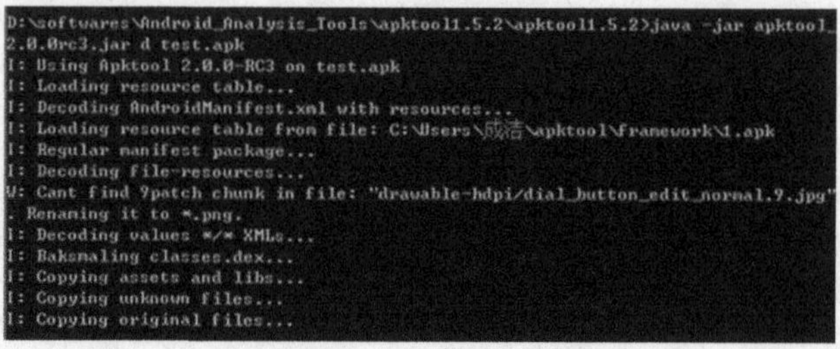

Fig. 8. Using Apktool to decompile the APK file

(2) Check whether the decompiled folder contains bytecode file of the APK file (Fig. 9)

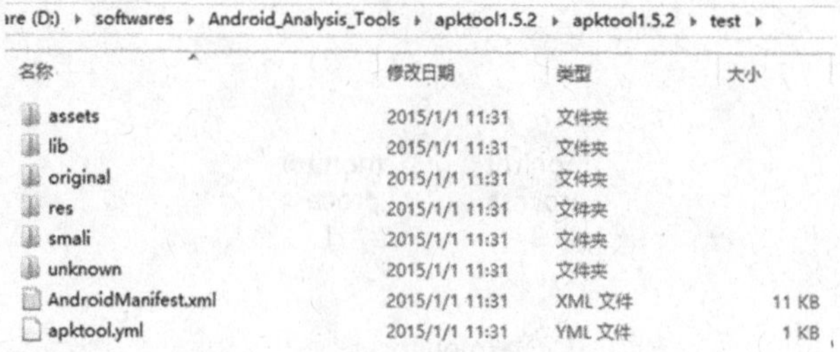

Fig. 9. The result of decompiling APK file

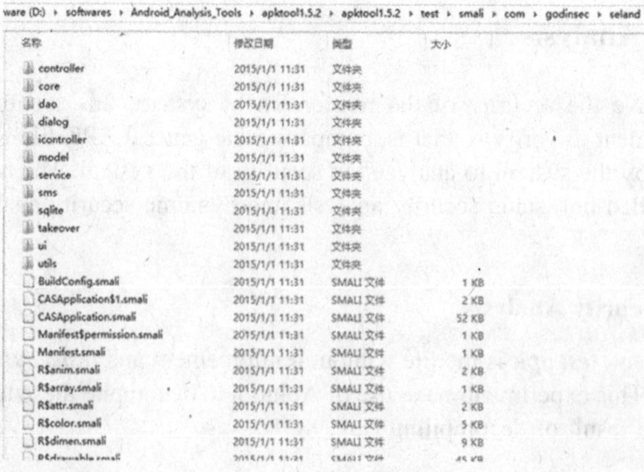

Fig. 10. Smali file of APK file without reinforcement

(3) Use ApkTool decompile testsecurity.apk file (Fig. 10)

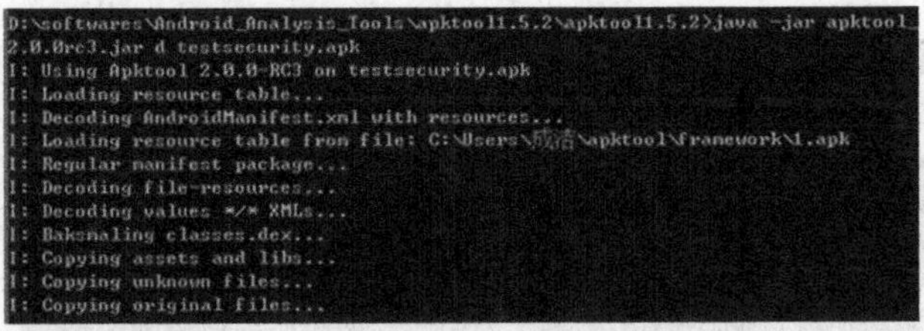

Fig. 11. Decompile the reinforced APK file

(4) Check whether the decompiled folder contains bytecode file of the APK file
(Fig. 11)

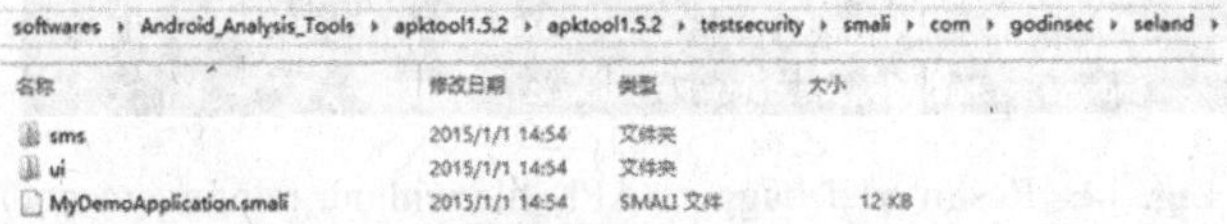

Fig. 12. Result of decompiling reinforced file

According to the experiment, we can conclude that after the reinforcement, although using ApkTool to decompile, it cannot obtain the bytecode file of the APK file and all the bytecode files are covered. It verifies that this reinforcement system improves the capability of defensing static attack to a large extend (Fig. 12).

7.2 Dynamic Security Analysis

Attackers always use Andbug, Ida Pro to debug the Android software dynamically, we will analyze the dynamic security of the program that had been reinforced through these two tools (Fig. 13).

Andbug debug analysis steps:

(1) Running the APK file without reinforcement, use *adb shell ps* to check the process *pid* of APK file without reinforcement

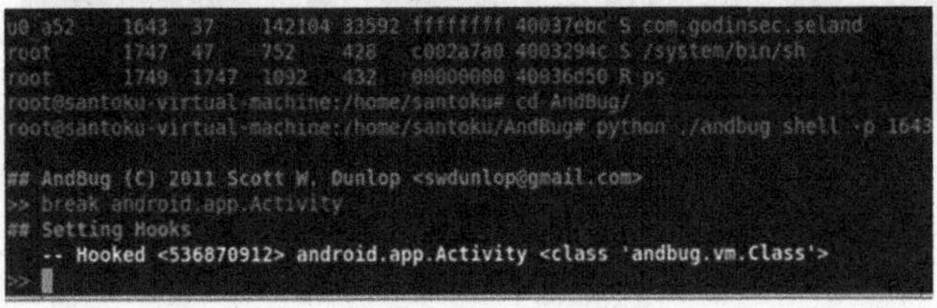

Fig. 13. Running process diagram of APK file without reinforcement

(2) Use *./andbug shell –p pid* to enter dynamic debug environment, use "*break android.app.Activity*" command to invoke Activity class break point.

When the program is entered into break point, the stack calling information is shown in the Fig. 14.

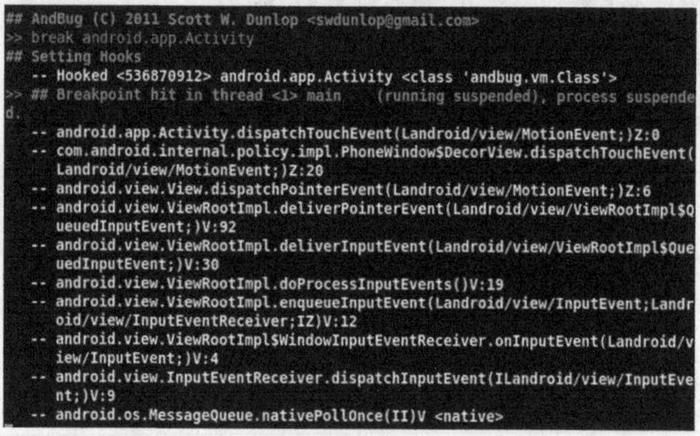

Fig. 14. Result of debugging APK file without reinforcement

(3) Run the APK reinforced file, use *adb shell ps* to check the process *pid* of APK reinforced file.
(4) Use *./andbug shell –p pid* to enter dynamic debug environment, use "*break android.app.Activity*" command to invoke Activity class break point

When we tried to make the break points to the APK reinforced file, there happened a debug error, which is shown in Fig. 15.

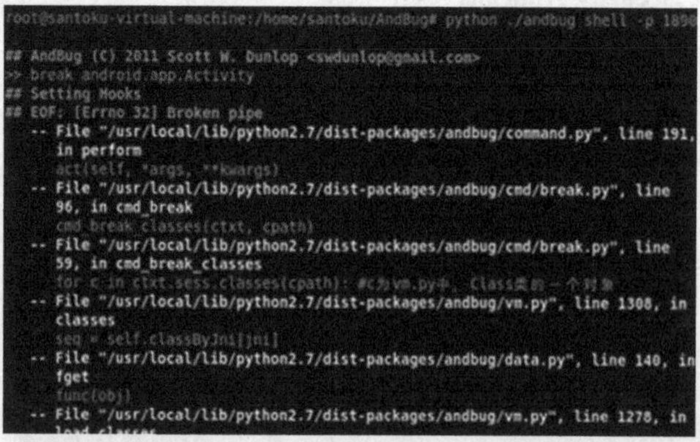

Fig. 15. Result of debugging APK reinforced file

Ida Pro debug analysis steps:

(1) Run the Ida Pro debugger, use **Debugger- > attach** to start debugging to the program without reinforcement.

Fig. 16. IDA PRO chooses the process to debug

Choose the process of APK, debug it, step into break point

The APK file enter into the break point and go on the debugging, which is shown in the Fig. 16. The IDA is running regularly (Figs. 17, 18, 19).

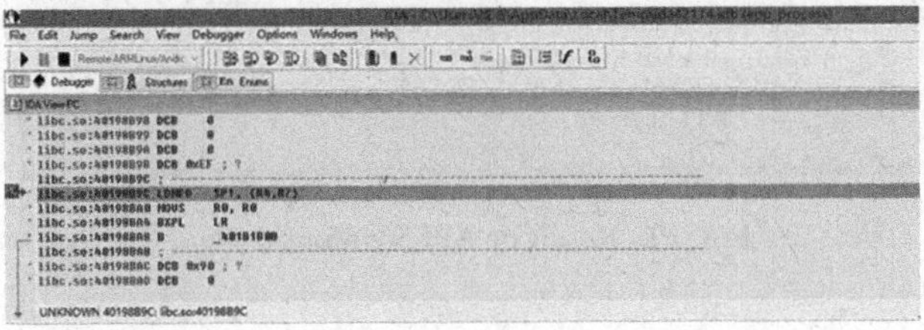

Fig. 17. Result of APK file without reinforcement

(2) Running the APK reinforced file, use ***Debugger- > attach*** to debug, there come three processes out after running.

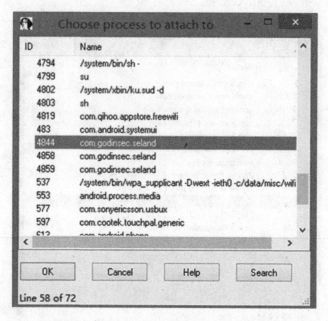

Fig. 18. Choose one of the three process to debug

(3) Choose one of them to debug it.

It is not able to debug the APK reinforced file and an error occurs when the process is debugged. The result is shown in Fig. 19. IDA has occurred an error during the whole process of debugging.

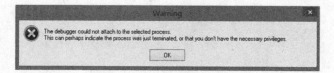

Fig. 19. Result of APK reinforced file

8 Conclusions

This paper mainly studies the research and implementation of the Android application software protection. Started from basic knowledge about Android, this paper analyzes current static attack and defense technology, dynamic attack and its defense technology. This paper proposes a universal security reinforcement scheme in order to provide

Android software protection. The security reinforcement system only applies to the application program running on the Dalvik virtual machine. The next major work is to study the static and dynamic attack of Android operation system that is running on the ART virtual machine and its defense mechanism.

Acknowledgment. This work was supported by National Natural Science Foundation of China (No. 61370195, U1536121, 61070207), Beijing Natural Science Foundation (No. 4132060) and National Cryptography Development Foundation of China (MMJJ201201002).

References

1. 2015 IEEE/ACM 37th IEEE International Conference on Software Engineering (ICSE), pp. 725–728. IEEE (2015)
2. Ji H.: The Research and Implement on Operating System of Mobile Phone on Android Platform, Southeast University (2010)
3. Qiu L, Zhang Z, Shen Z, et al.: AppTrace: Dynamic trace on Android devices. In: 2015 IEEE International Conference on Communications (ICC). IEEE (2015)
4. Collerg, C., Thomborson, C., Low, D., et al.: A taxonomy of obfuscating transformations. Department of Computer Science, The University of Auckland, New Zealand, July 1997
5. Yanfei, L., Sha, F., et al.: Survey of software tamper proofing technique. Modern Telecommun. Tech. **45**(1), 66–70 (2015). doi:10.3969/j.issn.1002-5316.2015.01.013. 74
6. Haibo, S., Yuda, S., et al.: Tamper-proofing technology of computer software. Modern Comput. Prof. **2**, 45–48 (2005)
7. Bin, L.: The research and implement on software activity analysis system based on Android sandbox, Beijing University of Posts and Telecommunications (2012)
8. Zheng, M., Sun, M., Lui, J.C.S.: DroidTrace: a ptrace based Android dynamic analysis system with forward execution capability. In: 2014 International Wireless Communications and Mobile Computing Conference (IWCMC), pp. 128–133 (2014)
9. Feng, X., Lin, J., Jia, S.: The research on security reinforcement of Android applications. In: 2015 4th International Conference on Mechatronics, Materials, Chemistry and Computer Engineering. Atlantis Press (2015)

Extending the ITU-T G.1070 Opinion Model to Support Current Generation H.265/HEVC Video Codec

Debajyoti Pal[(✉)], Tuul Triyason, and Vajirasak Vanijja

School of Information Technology,
King Mongkut's University of Technology Thonburi,
Toongkru, Bangkok, Thailand
debajyoti.pal@mail.kmutt.ac.th,
{tuul.tri,vachee}@sit.kmutt.ac.th

Abstract. Online video streaming is one of the most promising applications that is being widely used today. Such streaming videos at high definition (HD) resolution or up consume a large network bandwidth. Current generation video codecs like H.265/High Efficiency Video Coding (HEVC) and VP9 are expected to reduce this bandwidth requirement while providing an excellent viewing quality. ITU-T has developed a standardized parametric opinion model called G.1070 that tries to assess the Quality of Experience (QoE) of any multimedia content. The model outputs an overall multimedia quality M_q which is a combination of the video quality V_q and speech quality S_q. The function V_q has to be validated for different video codecs and formats by carrying out subjective experiments. In this paper we propose for the first time a set of coefficients that enables us to extend the G.1070 opinion model to support the H.265 video codec at full-HD resolution.

Keywords: QoE · H.265 · MOS · Video Quality Assessment

1 Introduction

As per a recent report published in [1] the demand for online video streaming services has increased to a great extent in the past few years. Majority of the video contents are being watched on small form factor screens ranging between 5 in.- 10 in. typically found in mobile and tablet devices either over mobile networks or Wi-Fi [2]. This has resulted in a greater demand for network bandwidth which has further led to the emergence of newer codecs like VP9 from Google and H.265/HEVC from Joint Collaborative Team on Video Coding (JCT-VC) [3]. A correct estimation of the QoE for these newer codecs is required to assess their benefits and advantages over older generation of codes like H.264/AVC. ITU-T Recommendation G.1070 is a standardized parametric planning model that is useful as a QoE/QoS planning tool for assessing the combined effects of variations in several video and speech parameters that affect the QoE [4]. The model is actually a combination of three parts. The first part provides us with a speech quality estimation function (Sq) which is a simplified version of the E-model [5]. The second part provides us with a video quality estimation function

© Springer International Publishing Switzerland 2016
O. Gervasi et al. (Eds.): ICCSA 2016, Part II, LNCS 9787, pp. 106–116, 2016.
DOI: 10.1007/978-3-319-42108-7_8

(Vq) that depends upon application and network level parameters like bit rate (BR, frame rate (FR) and packet loss rate (PLR) of the encoded video. Both of these functions have been designed carefully from various subjective experiments [6–8]. The third part which is known as the multimedia quality estimation function (Mq) takes into account the Mean Opinion Scores (MOS) of both Sq and Vq and also accounts for any introduced delay (absolute audiovisual delay or audiovisual media asynchrony) to estimate the overall video quality.

Vq is estimated from a combination of factors that comes from the coding impairment and the packet loss impairment of the encoded video. It is based on the main assumption that for a particular experimental setup (BR, FR, video codec type, video format and video display size) the obtained video quality follows a Gaussian distribution. This assumption is for the coding impairment only and does not take into account the effect of the network. The effect that the network has during online video streaming on the final output is taken care of by a packet loss degradation factor that follows a decaying exponential pattern. The overall video quality is thus a combination of both and depends on a set of coefficients (v_1 to v_{12}) that needs to be trained for a particular experimental setup. In particular the coefficients depend upon the codec type, video format and the video display size. The experiments that have been carried out so far allows us to validate the G.1070 model for MPEG4-Part2, H.264 and MPEG2 video codec ranging from CIF, QCIF, QVGA, VGA to HD display resolutions [6–9]. In this paper we have carried out a subjective video quality experiment for the H.265 video codec at full HD resolution (following the ITU-T Recommendation P.910) and extract the optimized set of coefficients for this particular configuration for the G.1070 model. We further analyze the validity and robustness of the model from the data gathered.

2 Experimental Setup

We have conducted a subjective video quality experiment on 24 subjects. The subjects were aged between 18–30 and balanced in gender. They were non-experts and were not concerned with multimedia quality as a part of their work. ITU-T Recommendation P.910 has been used as the reference while conducting this experiment [10]. Four reference videos have been carefully selected from the publicly available SVT High Definition Multi Format Test Set maintained by the Video Quality Experts Group (VQEG) [11]. All the reference videos have different type of video contents each having different levels of spatial information and temporal information. The authors in paper [12] have shown that the perceived video quality depends on the type of video content under certain conditions. Thus, we estimated that it would be wise to include a wide variety of video content to capture all possible variations. The details of the video sequences have been given in Table 1. Each of the sequences are of length 10 s and contain 500 frames each. We have used 5 different bit rates and frame rates for each sequence under the condition of no packet loss. Thus, we get a total of 100 test conditions per user for the first case. In the case of packet loss we selected 5 specific combinations of bit rate and frame rate (explained in later section) combined with 4 different values of packet loss to obtain 80 test conditions per user for the second case. The experiment details have been shown in Table 2.

Table 1. Details of selected video sequences

Sequence no	Name	Description	Content feature
1	CrowdRun	Start of a crowd running	Multiple fast moving objects with high spatial detail
2	DucksTakeOff	Ducks are on water and take off	Fast random motion and camera panning
3	OldTown	Sky view of Stockholm	Slow smooth motion and camera panning and zooming
4	ParkJoy	People are running in front of trees during a left to right camera movement	Random moving objects with camera zoon in and high contrast

Table 2. Experiment details

Video Codec	H.265/HEVC
Encoder Version	HM14
Video Format	Full HD (1080p)
Video Display Size (inch)	5.7
Video Bit Rate (kbps)	512, 1000, 2000, 4000, 8000
Video Frame Rate (fps)	8, 15, 24, 30, 60
Packet Loss Rate (%)	0.5, 1, 5, 10
Packet Loss Pattern	Random
Video Sequences	CrowdRun, DucksTakeOff, OldTown, ParkJoy

The video sequences were presented on a Samsung Galaxy Note 5 having 64 GB of internal storage, 4 GB ram, running the latest version of Android (Marshmallow 6.0.1) and capable of displaying up to 2 K resolution. The videos were presented to the users using a third party application (MX Player free version). We have selected this device because not only it supports resolutions of up to 2 K, but it can also decode H.265 sequences smoothly without dropping any frames or visual latency. As all the videos were preloaded into the mobile, hence we turned on the flight mode while carrying out the experiment.

The subjective test was carried out in a controlled laboratory environment as recommended by ITU-T P.910. The viewing distance was about 80 cm and the viewing angle about 30°. The participants were left alone while completing their tasks to minimize the unwanted effects of being supervised [13]. Before starting the actual experiment, the subjects were put through a training phase during which they saw and evaluated a set of 20 stimuli that was representative of the entire set of video sequences that they would face during the test.

We used a 9 point Absolute Category Rating (ACR) scale as prescribed in Recommendation P.910 to gather the subjective scores instead of a 5 point scale since it is considered to be more accurate. The results were then linearly mapped to the 5 point ACR scale.

3 Result Analysis

3.1 Video Quality Affected by Coding

Our aim is to use the subjective data that we gathered from the experiment to extract the coefficients v_1 to v_{12} as described by the G.1070 model. The video quality estimation function V_q as proposed by the G.1070 model has been described by Eqs. (1) to (7).

$$V_q = 1 + I_{Coding} \exp\left(-\frac{P_{pl}}{D_{P_{pl}}}\right) \tag{1}$$

Where, P_{pl} represents the packet loss %, $D_{P_{pl}}$ represents the degree of video quality robustness due to packet loss and I_{coding} represents the basic video quality affected by the coding distortion. In the event of no packet loss Eq. (1) reduces to:

$$V_q = 1 + I_{Coding} \tag{2}$$

Where, I_{coding}-is expressed as:

$$I_{Coding} = I_{O_{fr}} \exp\left\{-\frac{\left(\ln(Fr) - \ln\left(o_{fr}\right)\right)^2}{2D_{fr^2}}\right\} \tag{3}$$

Where, I_{ofr} represents the maximum video quality at each video bit rate B_r, F_r represents the frame rate, O_{fr} represents the optimal frame rate which maximizes the video quality for a particular bit rate B_r and D_{fr} is the degree of video quality robustness due to the frame rate F_r. O_{fr} is expressed as:

$$O_{fr} = v_1 + v_2 \times B_r, \quad 1 \le O_{fr} \le 30 \tag{4}$$

I_{ofr} is expressed as:

$$I_{O_{fr}} = v_3 - \frac{v_3}{1 + \left(\frac{B_r}{v_4}\right)^{v_5}}, \quad 0 \le I_{O_{fr}} \le 4 \tag{5}$$

And D_{fr} is expressed as:

$$D_{fr} = v_6 + v_7 \times B_r \tag{6}$$

And $D_{P_{pl}}$ is expressed as:

$$D_{P_{pl}} = v_{10} + v_{11} \exp\left(-\frac{F_r}{v_g}\right) + v_{12} \exp\left(-\frac{B_r}{v_g}\right), \quad 0 < D_{P_{pl}} \tag{7}$$

We start the process by finding out v_1 to v_7 first as described by Eqs. (4) to (6) for the case of no packet loss rate. By performing the first curve fitting to the subjective

Table 3. I_{Ofr}, O_{fr} and D_{fr} values for different bit rates

Variables	512 kbps	1000 kbps	2000 kbps	4000 kbps	8000 kbps
I_{Ofr}	2.00	2.68	2.96	3.38	3.87
O_{fr}	24.02	24.19	23.78	22.83	20.91
D_{fr}	1.87	1.80	1.86	1.91	1.99

data that we have gathered from the experiment, the values of O_{fr}, I_{Ofr} and D_{fr} for each bit rate are calculated and shown in Table 3. Applying these values to Eqs. (4) to (6) we can extract the coefficients v_1 to v_7.

Figure 1 shows the plot between O_{fr} and the bit rate B_r. There is no cut-off limit of the optimal frame rate beyond which it saturates. Instead, it decreases in a linear fashion towards higher bit rates. The model shows this trend, except for a fluctuation at a bit rate of 512 kbps where it underestimates the MOS. Also, the condition laid down by G.1070 model about the O_{fr} range to lie between 0 to 30 is proven to be valid for our experimental setup.

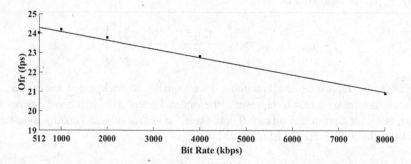

Fig. 1. Optimal frame rate (O_{fr}) vs. bit rate (B_r)

Figure 2 shows the graph of I_{ofr} vs. the bit rate B_r. There is a perfect fit for I_{ofr} as described by Eq. (5) except for the lower bit range region (512 kbps and 1000 kbps). Also, we find that $0 \leq I_{ofr} \leq 4$.

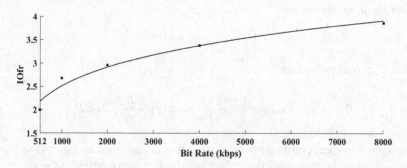

Fig. 2. I_{ofr} affected by the Bit rate (B_r)

Figure 3 shows the relationship between the degree of video quality robustness (D_{fr}) and the bit rate (B_r). We observe that D_{fr} increases linearly with the bit rate as predicted by Eq. (6) except at 512 kbps and 1000 kbps. Following this method we are able to extract the value of the coefficients v_1 to v_7 for our particular experimental setup. The coefficients have been listed in Table 5.

Figure 4 shows the variation of MOS with frame rate. From Eqs. (2) and (3) it is clear that the video quality which is affected by the coding distortion only is subjected to a Gaussian distribution. This fact is confirmed by Fig. 4. For every bit rate there is a threshold frame rate where we get the maximum MOS, following which the MOS decreases.

We can conclude from this part that for no packet loss, the video quality estimation function V_q predicted by the G.1070 model fits quiet well to our set of data except for a lower bit rate range of 512 kbps–1000 kbps.

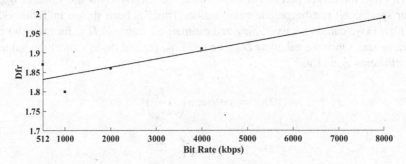

Fig. 3. Degree of video quality robustness (D_{fr}) vs. Bit rate (B_r)

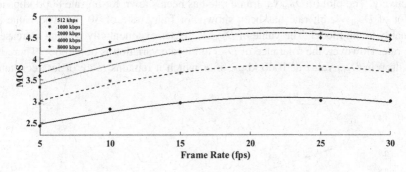

Fig. 4. Perceptual video quality for different bit rates

3.2 Video Quality Affected by Packet Loss

In order to limit the number of testing conditions of the subjective test, we chose 4 values of packet loss combined with 5 bit rate and frame rate combinations because this is sufficient to carry out the data fitting with regards to the number of unknown parameters. The test conditions have been obtained by combining the highest bit rate

Table 4. Bit rate (B_r)/Frame rate (F_r) and Packet loss rate (P_{lr}) combination

Bit rate/Frame rate combination	Packet loss rate (%)	I_{Coding}
1000 kbps, 30 fps	0.5, 1, 5, 10	2.72
2000 kbps, 30 fps	0.5, 1, 5, 10	3.20
8000 kbps, 15 fps	0.5, 1, 5, 10	3.57
8000 kbps, 25 fps	0.5, 1, 5, 10	3.59
8000 kbps, 30 fps	0.5, 1, 5, 10	3.54

i.e. 8000 kbps with three values of frame rate (15, 25 and 30) and also by combining the frame rate of 30 fps with three values of bit rate (1000, 2000, 8000). This is illustrated in Table 4.

The Video Quality estimation function of G.1070 in case of packet loss is given by Eq. (1). From the earlier part of our experiment, we already know the value of I_{Coding} for our specific bit rate/frame rate combination. This has been shown in Table 4. So, from Eq. (1) we can do a curve fitting and estimate the value of D_{Ppl} for every bit rate and frame rate. Once we calculate D_{Ppl} we use Eqs. (8) and (9) to extract the value of the coefficients v_8 and v_9.

$$D_{P_{pl}} = a + b\exp\left(-\frac{F_r}{v_g}\right) \qquad (8)$$

$$D_{P_{pl}} = c + d\exp\left(-\frac{B_r}{v_g}\right) \qquad (9)$$

The plot of D_{Ppl} vs. frame rate and bit rate has been shown in Figs. 5 and 6 respectively. The plot of D_{Ppl} vs. frame rate has been shown for bit rate 8000 kbps and the plot of D_{Ppl} vs. bit rate has been shown for frame rate of 30 fps. The values fit reasonably to the model. The value of D_{Ppl} decreases exponentially with an increase in frame rate. However, the tolerance of D_{Ppl} to bit rate variation is quiet poor. There is a sharp drop in D_{Ppl} value at 1000 kbps, after which it remains more or less constant.

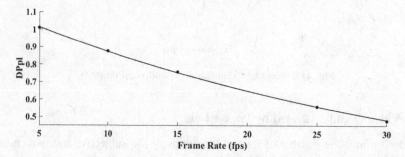

Fig. 5. Variation of D_{Ppl} with Frame rate

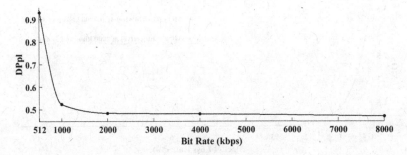

Fig. 6. Variation of D_{Ppl} with Bit rate

Table 5. List of our coefficients for H.265/HEVC codec at full HD resolution

Coefficients	Value
v_1	24.52
v_2	-4.44×10^{-4}
v_3	0.4456
v_4	13.29
v_5	0.2133
v_6	1.821
v_7	2.105×10^{-5}
v_8	41.26
v_9	0.42
v_{10}	-0.16
v_{11}	1.32
v_{12}	49.39

After finding out the values of v_8 and v_9 we again do a curve fitting for finding out the coefficients v_{10} to v_{12} as per Eq. (7). List of all the coefficients from v_1 to v_{12} has been shown in Table 5.

Figures 7, 8, 9, 10 and 11 show the video quality variation with packet loss rate for the bit rate/frame rate combinations that we have selected.

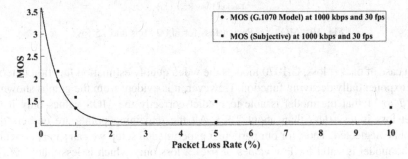

Fig. 7. MOS vs. Packet loss for 1000 kbps and 30 fps

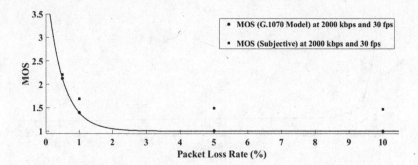

Fig. 8. MOS vs. Packet loss for 2000 kbps and 30 fps

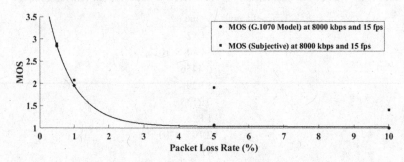

Fig. 9. MOS vs. Packet loss for 8000 kbps and 15 fps

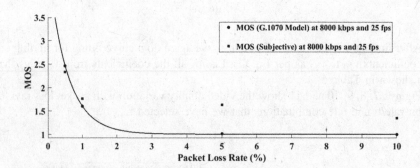

Fig. 10. MOS vs. Packet loss for 8000 kbps and 25 fps

In case of packet loss, G.1070 models the video quality estimation function in terms of an exponentially decaying function. However, it is evident from the graphs shown in Figs. 7 to 11 that the model is able to predict correctly the MOS values only if the packet loss is less (less than about 2 %). All the remaining points are outliers that cannot be neglected. Thus, for our particular experimental setup we can safely conclude that the model is valid for low values of packet loss only which is less than 2 %.

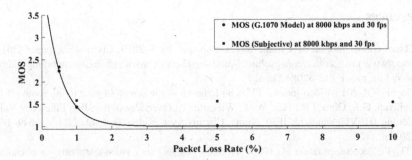

Fig. 11. MOS vs. Packet loss for 8000 kbps and 60 fps

3.3 Accuracy of the Model

Figure 12 shows the prediction accuracy of the G.1070 model. We calculate the model accuracy separately under the conditions of without packet loss and with packet loss. The R^2 value is 0.792 for the condition of no packet loss, 0.151 for the condition of packet loss and 0.784 taking into account both the conditions. Similarly the Pearson correlation coefficients for the condition of no packet loss, packet loss and overall are 0.89, 0.389 and 0.885 respectively. We have calculated the correlation using the same data that we used for finding out the coefficients. So, we need to verify the accuracy of the model further by testing it with some other set of data.

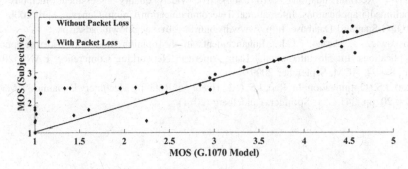

Fig. 12. Prediction accuracy of the G.1070 model for H.265 codec

4 Conclusion

We have carried out this experiment to extract a new set of coefficients that can extend the ITU-T G.1070 model for supporting current generation codec like H.265/HEVC at full HD resolution. The model gives a fairly good overall accuracy rate for our set of extracted parameters. However, in the case of packet loss the G.1070 model could only estimate within a range of 0 % to 2 %. Hence, our extracted coefficients are valid only within this range.

References

1. Cisco. Global mobile data traffic forecast update, 2014–2019. Cisco white paper (2015). http://www.cisco.com/c/en/us/solutions/collateral/serviceprovider/visual-networking-index-vni/white_paper_c11-520862.html
2. Ooyala. Q1 2013 Video Index - TV is no longer a single screen in your living room (2013)
3. Sullivan, G.J., Ohm, J.R., Han, W.-J., Wiegand, T.: Overview of the High Efficiency Video Coding (HEVC) standard. IEEE Trans. Circuits Syst. Video Technol. **22**(12), 1649–1668 (2012)
4. ITU-T Recommendation G.1070. Opinion model for video-telephony applications. International Telecommunication Union, Geneva, July 2012
5. ITU-T Recommendation G.107. The E-model, a computational model for use in transmission planning. International Telecommunication Union, Geneva, March 2005
6. ITU-T Contribution COM 12-C34-E. Verification of video quality estimation model for videophone services. Source: NTT, Japan, ITU-T Study Group 12 Meeting, Geneva, January 2007
7. Yamagishi, K., Hayashi, T.: Verification of video quality opinion model for videophone services. In: 2nd ISCA Tutorial and Research Workshop on Perceptual Quality of Systems, pp. 143–148, September 2006
8. Yamagishi, K., Hayashi, T.: QRP08-1: opinion model for estimating video quality of videophone services. In: Global Telecommunications Conference, GLOBECOM 2006. IEEE, San Francisco, CA, pp. 1–5 (2006)
9. Belmudez, B., Möller, S.: Extension of the G.1070 video quality function for the MPEG2 video codec. In: Second International Workshop on Quality of Multimedia Experience (QoMEX), Trondheim, pp. 7–10 (2010)
10. ITU-T Recommendation P.910. Subjective video quality assessment methods for multimedia applications. International Telecommunication Union, Geneva, April 2008
11. VQEG Standard Database. http://www.its.bldrdoc.gov/vqeg/downloads.aspx
12. Joskowicz, J., Ardao, J.C.L.: Enhancements to the opinion model for video-telephony applications. In: 5th International Latin American Networking Conference, LANC 2009, pp. 87–94. ACM, September 2009
13. Rao, J.S.: Optimization. In: Rao, J.S. (ed.) History of Rotating Machinery Dynamics. HMMS, vol. 20, pp. 341–351. Springer, Heidelberg (2011)

New Advantages of Using Chains in Computing Multiple $s - t$ Probabilistic Connectivity

Alexey S. Rodionov[✉] and Denis A. Migov

Institute of Computational Mathematics and Mathematical Geophysics SB RAS,
Prospect Akademika Lavrentjeva 6, 630090 Novosibirsk, Russia
alrod@sscc.ru, mdinka@rav.sscc.ru
http://www.sscc.ru/

Abstract. We consider the problem of a network reliability calculation
for a network with unreliable communication links and perfectly reliable nodes. For such networks, we study two different reliability indices:
network average pairwise connectivity and average size of a connected
subgraph that contains some special vertex. The problem of precise calculation of both these characteristics is known to be NP-hard. Both indices
may be calculated or estimated through complete or partial enumeration
of pairs of vertexes and calculation of their pairwise reliability. Methods
for speeding up this process in the case when there are chains in a graph
structure are presented in the paper.

Keywords: Network reliability · Pairwise connectivity · Random
graph · Algorithm

1 Introduction

Random graphs are widely used for analysing different indices of a network
reliability (see [4,7,10] and many others). As a rule, network reliability is defined
as some connectivity measure and the associated problems are NP-hard [12].

In most of researches it is assumed that vertexes are reliable and edges fail
statistically independently. This model corresponds to real networks where the
reliability of nodes is much higher than reliability of communication links. Transport and wireless networks are good examples.

Depending on goals, various reliability measures are used in practice. In this
paper we consider two of them: (1) average pairwise connectivity (APC) and (2)
average size of a connected subgraph that contains some special node (central
node, CN) (ASCS). APC is a subject of a number of recent researches (see [8,9],
for example, and ours [3]). APC characterizes a network's reliability from point
of view of possibility to establish an arbitrary pairwise connection that is of
most importance in P2P networks, for example. ASCS is almost unexplored in

D.A. Migov—This research was supported by Russian Foundation for Basic Research
under grant 16-37-00345 and by grant of the Program of basic researches of the
Presidium of Russian Academy of Science.

© Springer International Publishing Switzerland 2016
O. Gervasi et al. (Eds.): ICCSA 2016, Part II, LNCS 9787, pp. 117–128, 2016.
DOI: 10.1007/978-3-319-42108-7_9

general case, some estimations of a size of largest connected component can be found concerning Erdös-Rènyi model [1], but not in connection with CN. ASCS characterizes monitored part of a network, which is of great importance in censor networks, for example.

We research algorithms for obtaining exact values and bounds for both of these indices.

In this paper we denote nodes by their numbers, that is s means s-th node of a graph. Edges are denoted as e_{ij} (edge connected nodes i and j) or e_i (i-th edge in a chain), depending on context. Reliability of e_{ij} we designate as p_{ij}, $q_{ij} = 1 - p_{ij}$.

While the task of finding $s - t$ probabilistic connectivity (pairwise reliability) is well explored and effective methods with use of sequential reduction of chains, when all chains that do not contain nodes s or t as inner ones are substituted by edges with equivalent reliability, had been proposed long ago [2,5], the case when one or both of these nodes lay inside a chain is unexplored. Such case is treated in primitive way: a chain is divided by a pivot node and is considered as pair of chains each of which can be substituted by an edge with equivalent reliability.

In the current paper we show how to calculate $s - t$ pairwise reliability R_{st} in this case more efficiently and how to organize multiple calculation of $s - t_i$ pairwise reliabilities between node s and nodes t_i of some chain. Such multiple calculation are needed for calculation of APC ($\bar{R}(G)$) and ASCS ($C(G)$):

$$\bar{R}(G) = \sum_{i=1}^{n-1} \sum_{j=i+1}^{n} R_{ij}/C_n^2; \tag{1}$$

$$C(G) = \sum_{i=1}^{n} R_{1i}w_i. \tag{2}$$

When calculating ASCS we assume that a c-node has number one. A weight w_i corresponds to expected number of nodes merged into the node i after applying different reducing techniques [3].

This equations are obvious and seem to be non-effective as they require complete enumeration of all pairs of nodes in the first case and pairs containing a c-node in the second, but they are effective when parallelize calculation of R_{ij}.

Moreover, knowing part of pairwise reliabilities we can obtain upper and lower bounds of $\bar{R}(G)$ and $C(G)$ as it will be shown later.

Hereafter we show how to calculate sum of several R_{ij} in one step if there is at least one chain in a graph's structure, that is very common and almost inevitable case in real networks.

2 Factoring Method for Network Reliability Calculation

Factoring method is the most widely used method for calculating different network reliability measures [6]. This technique partitions the probability space into two sets, based on the success or failure of one network particular element (node or link). The chosen element is called factored element. So we obtain two graphs, in one of which the factored element is absolutely reliable and in the second one factored element is absolutely unreliable that is, is removed. The probability of the first event is equal to the reliability of factored element; the probability of the second event is equal to the failure probability of factored element. Thereafter obtained graphs are subjected to the same procedure. The law of total probability gives expression for the network reliability. For $s - t$ probabilistic connectivity of graph with unreliable edges the corresponding formula takes the following form:

$$R_{st}(G) = p_e R_{st}(G/e) + (1 - p_e)R_{st}(G\backslash e), \qquad (3)$$

where p_e — reliability of edge e, G/e — is a graph obtained by contracting edge e from G, $G\backslash e$ — is a graph obtained by deleting e from G. Recursions continue until either disconnected graph is obtained, or until a graph for which the probabilistic connectivity can be calculated directly is obtained — it can be a graph of a special type or small dimension graph.

Note, that if a pivot node initially has a degree not equal to 2, then it may obtain such degree during factoring process and possible graph's reductions.

3   One of Pivot Nodes is Inside a Chain

Let i be an inner node of some chain Ch with terminal nodes x and y. Without loss of generality we can assume that nodes of a chain have numbers from x to y in increasing order and $x < i < y$. Let edges of a Ch be $e_x = (x, x+1), \dots, e_{x+k-1} = (x + k - 1, y)$. Let A be an event that some node s ($s \notin [x, \dots, y]$) and i are connected, B and E — that pairs of nodes $s - x$, and $s - y$ are connected in $G\backslash Ch$. We can obtain a probability of A as:

$$P(A) = P(BE)P(A|BE) + P(B\bar{E})P(A|B\bar{E}) + P(\bar{B}E)P(A|\bar{B}E). \qquad (4)$$

It is easy to see that $P(B \cup E) = R_{s,\langle x,y \rangle}(G/Ch)$, where $\langle x, y \rangle$ is a number of a node that is obtained by contracting nodes x and y. From this and using meaning of events B and E, we obtain that

$$P(B\bar{E}) = R_{s,\langle x,y \rangle}(G/Ch) - R_{sx}(G\backslash Ch);$$
$$P(\bar{B}E) = R_{s,\langle x,y \rangle}(G/Ch) - R_{sy}(G\backslash Ch);$$
$$P(BE) = R_{sy}(G\backslash Ch) + R_{sx}(G\backslash Ch) - R_{1,\langle x,y \rangle}(G/Ch).$$

Thus we have that

$$R_{si} = \left[R_{1y}(G\backslash Ch) + R_{sx}(G\backslash Ch) - R_{s,\langle x,y \rangle}(G/Ch) \right] \left(\prod_{j=x}^{i-1} p_j + \prod_{j=i}^{y-1} p_j - \prod_{j=x}^{t-1} p_j \right) +$$

$$\left[R_{s,\langle x,y\rangle}(G/Ch) - R_{sy}(G\backslash Ch)\right] \prod_{j=i}^{y-1} p_j +$$

$$\left[R_{s,\langle x,y\rangle}(G/Ch) - R_{sx}(G\backslash Ch)\right] \prod_{j=x}^{i-1} p_j$$

$$= R_{sx}(G\backslash Ch) \prod_{j=i}^{y-1} p_j + R_{sy}(G\backslash Ch) \prod_{j=x}^{i-1} p_j -$$

$$\left[R_{sy}(G\backslash Ch) + R_{sx}(G\backslash Ch) - R_{s,\langle x,y\rangle}(G/Ch)\right] \prod_{j=x}^{t-1} p_j. \tag{5}$$

Thus, for obtaining total income into $C(G,p)$ of all pairs $1-i$ where i is a node of a chain we need to obtain $R_{1x}(G\backslash Ch, p)$, $R_{1y}(G\backslash Ch, p)$, and $R_{1,\langle x,y\rangle}(G/Ch, p)$ - only:

$$S(G, Ch) = R_{1y}(G\backslash Ch) \sum_{i=x}^{t} w_i \left[\prod_{j=x}^{i-1} p_j - \prod_{j=x}^{y-1} p_j\right] +$$

$$R_{1x}(G\backslash Ch) \sum_{i=x}^{y-1} w_i \left[\prod_{j=i}^{y-1} p_j - \prod_{j=x}^{y-1} p_j\right] + R_{1,\langle x,y\rangle}(G/Ch) \prod_{j=x}^{y-1} p_j \sum_{i=x}^{y} w_i. \tag{6}$$

When calculating income of this pairs into $\bar{R}(G)$ we use the same equation with all w_i equal to 1.

Now let us consider the case when a c-node is an inner node of some chain. Let nodes of this chain be numbered from 1 to k and corresponding edges from 1 to $k-1$ and let s, $1 < s < k$ be a c-node.

For obtaining reliability of connection between c-node and some other node of a chain (terminal nodes included) we consider pseudo-cycle that consists of a chain Ch and pseudo-edge e_{1k} whose reliability is $R_{1k}(G\backslash Ch)$. Thus total income of all pairs $v_s - v_i$, $i \in \overline{1,k}\backslash s$ into $C(G,p)$ is

$$Y(G, Ch) = \sum_{\substack{i=1 \\ i \neq s}}^{k} w_i \left\{\prod_{j=i}^{s-1} p_j + R_{1k}(G\backslash Ch) \left[\prod_{j=1}^{i-1} p_j \prod_{j=s}^{k-1} p_j - \prod_{j=1}^{k-1} p_j\right]\right\}. \tag{7}$$

Using the same pseudo-cycle we calculate an income of all pairs of nodes in the chain into APC:

$$\Delta = \sum_{i=1}^{k-1} \sum_{j=i+1}^{k} \left\{\prod_{r=i}^{j-1} p_r + R_{1k}(G\backslash Ch) \left[\prod_{r=1}^{i-1} p_j \prod_{r=j}^{k-1} p_j - \prod_{j=1}^{k-1} p_j\right]\right\}. \tag{8}$$

4 Both Pivot Nodes are Inside Different Chains

In this section we assume that graph G under consideration consists of subgraph G_0 and two adjoining to G_0 chains Ch_1 and Ch_2 (see Fig. 1). Notations for

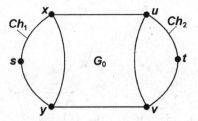

Fig. 1. Graph with 2 chains

end vertices of the first chain are x and y. For the second chain notations of end vertices are u and v. By p_{sx} we denote the product of reliabilities of edges between s and x in Ch_1. Values of p_{sy}, p_{ut}, p_{vt} are defined similarly.

We denote by G_0' graph G_0 with merged vertices x, y; by G_0'' – graph G_0 with merged vertices u, v. In the first case new vertex is denoted by $\langle x, y \rangle$, in the second case — by $\langle u, v \rangle$. Graph G_0 with merged vertices x, y and merged vertices u, v is denoted by G_0'''. These graphs are presented in the Fig. 2. We use similar notations for any other graph which completely includes G_0. We designate reliable edge e_{ij} as $\mathbf{e}_{ij}$.

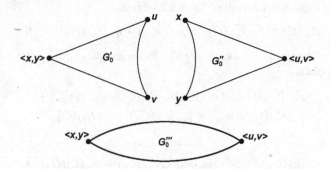

Fig. 2. Graphs derived from G_0 by merging some nodes

Theorem 1. *The following equation holds*

$$R_{st}(G) = p_{sx}p_{sy}(p_{ut}p_{tv}R_{\langle u,v\rangle,\langle x,y\rangle}(G_0''') + p_{ut}q_{tv}R_{u,\langle x,y\rangle}(G_0') + q_{ut}p_{tv}R_{v,\langle x,y\rangle}(G_0')) +$$
$$p_{sx}q_{sy}(p_{ut}p_{tv}R_{\langle u,v\rangle,x}(G_0'') + p_{ut}q_{tv}R_{ux}(G_0) + q_{ut}p_{tv}R_{vx}(G_0)) +$$
$$q_{sx}p_{sy}(p_{ut}p_{tv}R_{\langle u,v\rangle,y}(G_0'') + p_{ut}q_{tv}R_{uy}(G_0) + q_{ut}p_{tv}R_{vy}(G_0)). \quad (9)$$

Proof. In the preliminary step we perform parallel-series transformation to Ch_1 and Ch_2. As a result, chain Ch_1 is reduced in edges $e_{x,s}$ and $e_{y,s}$ with reliabilities p_{sx} and p_{sy} respectively. Chain Ch_2 is reduced in edges e_{ut} and e_{vt} with reliabilities p_{ut} and p_{vt} respectively. Let us denote graph $G_0 \cup e_{ut} \cup e_{vt}$ by G_1. Now we start factoring procedure as is shown in the Fig. 3. For better understanding of

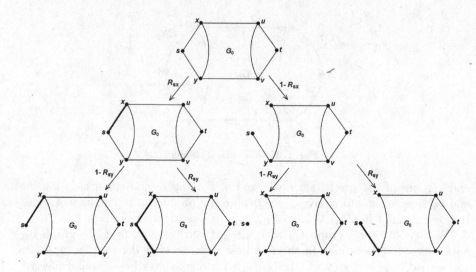

Fig. 3. Factoring procedure applied to the reduced graph G.

the procedure we draw reliable edges as thick ones, not merging their ends. First we apply the factoring procedure (3) with edge e_{sx}:

$$R_{st}(G) = p_{sx}R_{st}(G_1 \cup e_{sy} \cup \mathbf{e}_{sx}) + q_{sx}R_{st}(G_1 \cup e_{sy}).   \qquad (10)$$

For these two terms we again use the factoring procedure with edge e_{sy} as pivot. So we obtain:

$$R_{st}(G_1 \cup e_{sy} \cup \mathbf{e}_{sx}) = p_{sy}R_{st}(G_1 \cup \mathbf{e}_{sx} \cup \mathbf{e}_{sy}) +$$
$$q_{sy}R_{st}(G_1 \cup \mathbf{e}_{sx}) = p_{sy}R_{zt}(G_1') + q_{sy}R_{xt}(G_1).   \qquad (11)$$

$$R_{st}(G_1 \cup e_{sy}) = p_{sy}R_{st}(G_1 \cup \mathbf{e}_{sy}) + q_{sy}R_{st}(G_1)$$
$$= p_{sy}R_{yt}(G_1) + 0.   \qquad (12)$$

Thus Eq. (10) becomes the following:

$$R_{st}(G) = p_{sx}p_{sy}R_{\langle x,y \rangle,t}(G_1') + p_{sx}q_{sy}R_{xt}(G_1) + q_{sx}p_{sy}R_{yt}(G_1).   \qquad (13)$$

Expressions for each term of the provided equation can be found in the same way. For this purpose we consider another two-node cut $\{u, v\}$ instead $\{x, y\}$ and graph G_0 instead G_1. In other worlds we substitute in (13) s by t, x by u, y by v. t should be substituted by x, y, z for obtaining $R_{xt}(G_1)$, $R_{yt}(G_1)$, $R_{zt}(G_1)$ respectively. So we have the following:

$$R_{xt}(G_1) = p_{ut}p_{tv}R_{\langle u,v \rangle,x}(G_0'') + p_{ut}q_{tv}R_{ux}(G_0) + q_{ut}p_{tv}R_{vx}(G_0),   \qquad (14)$$
$$R_{yt}(G_1) = p_{ut}p_{tv}R_{\langle u,v \rangle,y}(G_0'') + p_{ut}q_{tv}R_{uy}(G_0) + q_{ut}p_{tv}R_{vy}(G_0),$$
$$R_{zt}(G_1') = p_{ut}p_{tv}R_{\langle u,v \rangle,\langle x,y \rangle}(G_0''') + p_{ut}q_{tv}R_{u,\langle x,y \rangle}(G_0') + q_{ut}p_{tv}R_{v,\langle x,y \rangle}(G_0').$$

Substitution of terms in (13) by expressions from (14) directly leads to Eq. (9) □.

Now we transfer to computing multiple pairwise reliabilities in the case under consideration (note that our reasoning is easily expanded on the case of more than 2 chains in a graph, one can simply consider pairs of chains in order). First let us assume that nodes in the Ch_1 are numbered from 1 to k and those in the Ch_2 – from $n - d + 1$ to n.

From (7) and (9) we obtain that total income of pairs of nodes from these chains into numerator of (1) is:

$$
\begin{aligned}
D = & \sum_{s=1}^{k-1} \sum_{i=s+1}^{k} \left[\prod_{r=s}^{i-1} p_r + R_{1k}(G \backslash Ch) \left(\prod_{j=1}^{s-1} p_j \prod_{j=i}^{k-1} p_j - \prod_{j=1}^{k-1} p_j \right) \right] + \\
& \sum_{s=n-d+1}^{n} \sum_{\substack{i=n-d+1 \\ i \neq s}}^{n} \left[\prod_{r=s}^{i-1} p_r + R_{n-d+1,k}(G \backslash Ch) \left(\prod_{j=n-d+1}^{s-1} p_j \prod_{r=i}^{n-1} p_j - \right. \right. \\
& \left. \left. \prod_{j=n-d+1}^{n-1} p_j \right) \right] + \sum_{s=1}^{k} \sum_{t=n-d+1}^{n} \left[R_{\langle u,v \rangle, \langle x,y \rangle}(G_0''') \prod_{i=1}^{k-1} p_i \prod_{i=n-d+1}^{n-1} p_i + \right. \\
& R_{u, \langle x,y \rangle}(G_0') \prod_{i=1}^{k-1} p_i \prod_{j=n-d+1}^{t-1} p_i \left(1 - \prod_{j=t}^{n-1} p_i \right) + \\
& R_{v, \langle x,y \rangle}(G_0') \prod_{i=1}^{k-1} p_i \left(1 - \prod_{i=n-d+1}^{t-1} p_i \right) \prod_{i=t}^{n-1} p_i + \\
& R_{\langle u,v \rangle, x}(G_0'') \prod_{i=1}^{s-1} p_i \left(1 - \prod_{i=s}^{k-1} p_i \right) \prod_{i=n-d+1}^{n} p_i + \\
& R_{ux}(G_0) \prod_{i=1}^{t-1} p_i \left(1 - \prod_{i=t}^{n-1} p_i \right) \prod_{i=n-d+1}^{t-1} p_i \left(1 - \prod_{i=t}^{n-1} p_i \right) + \\
& R_{vx}(G_0) \prod_{i=1}^{t-1} p_i \left(1 - \prod_{i=t}^{n-1} p_i \right) \left(1 - \prod_{i=n-d+1}^{t-1} p_i \right) \prod_{i=t}^{n-1} p_i + \\
& R_{\langle u,v \rangle, y}(G_0'') \left(1 - \prod_{i=1}^{s-1} p_i \right) \prod_{i=s}^{k-1} p_i \prod_{i=n-d+1}^{n-1} p_i + \\
& R_{uy}(G_0) \left(1 - \prod_{i=1}^{s-1} p_i \right) \prod_{i=s}^{k-1} p_i \prod_{i=n-d+1}^{t-1} p_i \left(1 - \prod_{i=t}^{n-1} p_i \right) + \\
& R_{vy}(G_0) \left(1 - \prod_{i=1}^{t-1} p_i \right) \prod_{i=t}^{n-1} p_i \left(1 - \prod_{i=n-d+1}^{t-1} p_i \right) \prod_{i=t}^{n-1} p_i \right].
\end{aligned}
\tag{15}
$$

5 Obtaining Bounds for $\bar{R}(G)$ and $C(G)$ Through Known Part of R_{ij}

We use cumulative updating of bounds of reliability indices (LB – lower bound, and UB – upper one) similar to [11]. One of significant particularities of such bounds is that at the end they reach an exact value. In the Fig. 4 we present behavior of APC bounds on example of lattice 4×4 with equally reliable edges ($p = 0.5$).

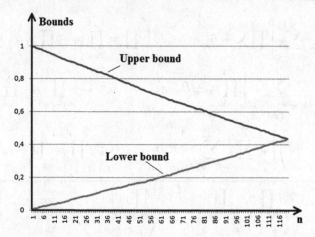

Fig. 4. Cumulative updating of bounds of APC on example of lattice 4×4, all $p_{ij} = 0.5$

Assume that we know some pairwise reliabilities in a graph with n nodes, say k of them. For short we denote them as R_i, $i = 1, \ldots, k$. In this case, assuming that in the best case all rest pairs are connected, and in the worst – disconnected, we can obtain LB and UB of APC as

$$LB_{APC} = \frac{\sum\limits_{i=1}^{k} R_i}{C_n^2}, \tag{16}$$

$$UB_{APC} = 1 - \frac{\sum\limits_{i=1}^{k} R_i}{C_n^2}. \tag{17}$$

After obtaining any new R_{ij} bounds are updated. Using (15) allows updating by larger steps thus leading to faster decision making about a graph's reliability.

Bounds for ASCS can be obtained in a similar way: we assume that some k of R_{1i} are known already. Using the same idea that in the best case all rest pairs are connected, and in the worst – disconnected, we have:

$$LB_{ASCS} = 1 + \sum_{i=1}^{k} R_{1i} w_i, \tag{18}$$

$$UB_{ASCS} = n - \sum_{i=1}^{k} R_{1i} w_i. \tag{19}$$

Using (6) and, if a c-node is inside some chain, (7) again allows larger steps of updating bounds.

6 Example

Let us consider random graph presented in the Fig. 5. We use common approach to calculation R_{ij} using reduction of chains thus obtaining for each pair of nodes new graph with different edges' reliabilities and not more than 4 nodes thus allowing simple, but tiresome if handmade, calculation of pairwise reliabilities that are presented in the Table 1.

From this we simply obtain that $\bar{R}(G) = 0.979618723$ and $C(G) = 7.805386414$.

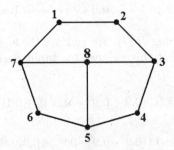

Fig. 5. Random graph with 8 nodes and 9 edges. All edges are equally reliable with $p = 0.9$

Table 1. Pairwise reliabilities for the graph in the Fig. 5

(i,j)	R_{ij}	(i,j)	R_{ij}
(1,2)	0.9777268548	(3,5)	0.9907266168
(1,3)	0.9740900268	(3,6)	0.9801808137
(1,4)	0.9635859306	(3,7)	0.9890491068
(1,5)	0.9721756647	(3,8)	0.9932902758
(1,6)	0.9653880996	(4,5)	0.9856073448
(1,7)	0.9777268548	(4,6)	0.9732547386
(1,8)	0.9746929827	(4,7)	0.9801808137
(2,3)	0.9777268548	(4,8)	0.9842371317
(2,4)	0.9653880996	(5,6)	0.9856073448
(2,5)	0.9721756647	(5,7)	0.9907266168
(2,6)	0.9635859306	(5,8)	0.9946753758
(2,7)	0.9740900268	(6,7)	0.9856073448
(2,8)	0.9746929827	(6,8)	0.9842371317
(3,4)	0.9856073448	(7,8)	0.9932902758

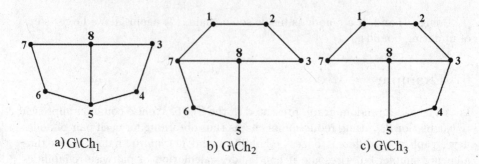

a) $G\backslash Ch_1$ b) $G\backslash Ch_2$ c) $G\backslash Ch_3$

Fig. 6. Graph G after deletion of chains

Now let us use results of the paper. We have three chains in the graph's structure: Ch_1=7-1-2-3, Ch_2=3-4-5, and Ch_3=5-6-7. Each node but 8th belongs to one or more chains.

First we discuss obtaining $\bar{R}(G)$. We can use (8) for pairs inside chains and (15) for the rest of pairs. For this we need to obtain $R_{37}(G\backslash Ch_1)$, $R_{35}(G\backslash Ch_2)$, and $R_{57}(G\backslash Ch_3)$ (see Fig. 6):

$$R^{(1)} = R_{37}(G\backslash Ch_1) = 0.9(0.9 + 0.9^2 - 0.9^3)^2 + 0.1(0.9^2 + 0.9^4 - 0.9^6)$$
$$= 0.9595908;$$
$$R^{(2)} = R_{35}(G\backslash Ch_2) = 0.1(0.9^5 + 0.9^2 - 0.9^7) + 0.9(0.9^3 + 0.9 - 0.9^4) \times$$
$$(0.9^2 + 0.9 - 0.9^3) = 0.0.95119272;$$
$$R^{(3)} = R_{57}(G\backslash Ch_3) = 0.1(0.9^5 + 0.9^2 - 0.9^7) + 0.9(0.9^3 + 0.9 - 0.9^4) \times$$
$$(0.9^2 + 0.9 - 0.9^3) = 0.0.95119272.$$

So total income of all pairs of nodes inside chains (there are 12 such pairs) into the numerator of (1) is:

$$S_1 = R_{(1)} \cdot (3 \cdot 0.9^2 + 2 \cdot 0.9 + 1 - 6 \cdot 0.9^3) + 3 \cdot 0.9 + 2 \cdot 0.9^2 + 0.9^3 +$$
$$R^{(2)}(1 + 2 \cdot 0.9 - 3 \cdot 0.9^2) + 0.9^2 + 2 \cdot 0.9 +$$
$$R^{(3)}(1 + 2 \cdot 0.9 - 3 \cdot 0.9^2) + 0.9^2 + 2 \cdot 0.9 = 11.7942923376.$$

This income immediately gives us the following bounds for APC: LB=0,4212247 and UB=0,9926533.

For the rest of pairs we consider pairs of nodes between chains using the Theorem 1. First we consider chains Ch_1 and Ch_2. Note that they have one common node – 3. Thus we find the total income of pairs 1-4,1-5,2-4, and 2-5 into the numerator. For this we must obtain graphs G_0, G_0', G_0'', and G_0''', and find all pairwise reliabilities for these graphs that are mentioned in (9). Graphs are presented in the Fig. 7. In the terms of the Theorem $x = 7$, $y = v = 3$ and $u = 5$.

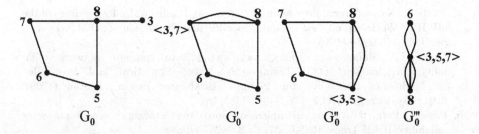

Fig. 7. Graph, derived from G according the Theorem 1.

Needed pairwise reliabilities can be obtained trivially and they are:

$$R_{\langle u,v\rangle,\langle x,y\rangle}(G_0''') = R_{v,\langle x,y\rangle}(G_0') = R_{\langle u,v\rangle,y}(G_0'') = R_{vy}(G_0) = 1;$$
$$R_{u,\langle x,y\rangle}(G_0') = R_{\langle u,v\rangle,x}(G_0'') = 0.97929;$$
$$R_{ux}(G_0) = 0.9639; \quad R_{uy}(G_0) = R_{vx}(G_0) = 0.87561.$$

Now we can easily obtain total income of all mentioned pairs into our numerator. After considering all chains obtain as all pairwise reliabilities, as their total income into reliability indices under consideration.

7 Conclusion

In this paper we have introduced new results about $s - t$ pairwise reliability that allows efficient calculation of reliability of pairs in which at least one node belongs to a chain. Most effective our equations are in the case when one need calculate pairwise connectivity between all pairs of nodes or between one special node and rest of them. Note that these equations allow natural parallelization. Results are fresh and we have executed few experiments till now, but we can say that calculation of pairwise reliability between opposite corners of a lattice 5×5 was executed more than 3 time faster with our method than by common factoring method with series-parallel reduction. It is obvious also that efficiency of our approach increases with increasing of average chains' length.

References

1. Bollobas, B.: Random Graphs, 2nd edn. Cambridge University Press, Cambridge (2001). http://dx.doi.org/10.1017/CBO9780511814068
2. Colbourn, C.J.: The Combinatorics of Network Reliability. Oxford University Press Inc., New York (1987)
3. Gadyatskaya, O., Rodionov, A.S., Rodionova, O.K.: Using EDP-polynomials for network structure optimization. In: Gervasi, O., Murgante, B., Laganà, A., Taniar, D., Mun, Y., Gavrilova, M.L. (eds.) ICCSA 2008, Part II. LNCS, vol. 5073, pp. 1061–1076. Springer, Heidelberg (2008)

4. Jereb, L.: Network reliability: models, measure and analysis. In: Proceedings of the 6th IFIP Workshop on Performance Modeling and Evaluation of ATM Networks, pp. T02/1–T02/10 (1998)
5. Lucet, C., Manouvrier, J.F.: Exact methods to compute network reliability. In: Ionescu, D.C., Limnios, N. (eds.) Statistical and Probabilistic Models in Reliability, pp. 279–294. Birkhäuser Boston, Boston (1999). http://dx.doi.org/10.1007/978-1-4612-1782-4_20
6. Page, L., Perry, J.: A practical implementation of the factoring theorem for network reliability. IEEE Trans. Reliab. **37**(3), 259–267 (1988)
7. Petingi, L.A.: Reliability study of mesh networks modeled as random graphs. In: Proceedings of the 2010 International Conference on Mathematical Models for Engineering Science, MMES 2010, Stevens Point, Wisconsin, USA, pp. 85–93. World Scientific and Engineering Academy and Society (WSEAS) (2010). http://dl.acm.org/citation.cfm?id=1965658.1965688
8. Potapov, A., Goemann, B., Wingender, E.: The pairwise disconnectivity index as a new metric for the topological analysis of regulatory networks. BMC Bioinform. **9**, 227 (2008). http://dx.doi.org/10.1186/1471-2105-9-227
9. Sun, F., Shayman, M.: On the average pairwise connectivity of wireless multihop networks. In: 2005 Global Telecommunications Conference, GLOBECOM 2005, vol. 3, p. 5. IEEE, November 2005
10. Tauro, S.L., Palmer, C.R., Siganos, G., Faloutsos, M.: A simple conceptual model for the internet topology. In: 2001 Proceedings of the Global Telecommunications Conference, GLOBECOM 2001, 25–29 November 2001, San Antonio, TX, USA, pp. 1667–1671 (2001). http://dx.doi.org/10.1109/GLOCOM.2001.965863
11. Won, J.M., Karray, F.: Cumulative update of all-terminal reliability for faster feasibility decision. IEEE Trans. Reliab. **59**(3), 551–562 (2010)
12. Xie, M., Dai, Y., Poh, K.: Computing Systems Reliability: Models and Analysis. Kluwer, Dordrecht (2004)

A Delay-Driven Switching-Based Broadcasting Scheme in Low-Duty-Cycled Wireless Sensor Networks

Dung T. Nguyen[1], Vyacheslav V. Zalyubovskiy[2], Thang Le-Duc[1],
Duc-Tai Le[1], and Hyunseung Choo[1(✉)]

[1] College of Information and Communication Engineering, SKKU, Suwon, Korea
{ntdung,ldthang,ldtai,choo}@skku.edu
[2] Sobolev Institute of Mathematics, Novosibirsk, Russia
slava@math.nsc.ru

Abstract. Wireless Sensor Networks (WSNs) are proven to be an important part of the everyday modern life. In that, broadcasting essentially delivers network-wide configurations, code updates, or route finding requests to the sensor nodes. Addressing the delay performance of tree-based broadcasting in low-duty-cycled WSNs, this paper proposes a novel switching-based scheme to enhance the overall broadcasting delay given one sink node sending out the packet. Simulation results show that the proposed algorithm significantly improves delay compared to the well-known schemes, while maintaining a comparable number of transmissions.

1 Introduction

Wireless Sensor Networks (WSNs) nowadays appear in various applications in real life, including surveillance, environmental and industrial monitoring, and healthcare. With very limited power and computation resources, communications in WSNs face many challenges: reliability, complexity, synchronization, energy consumption and so on [1]. One of the major research categories on WSNs is broadcasting in which a broadcasting packet is sent out from a source node, and it is expected to reach all the sensor nodes in an efficient way [2]. Being a primitive common tasks in WSNs, broadcasting is generally evaluated by delay and energy cost [3].

WSNs exist in the form of an ad-hoc network, where nodes wirelessly communicate with each other [4]. Since the communication range is limited, a node may have to be a forwarder in a multi-hop connection. As the common nature of wireless communications, communicating links between the nodes are naturally unreliable [5], i.e. several retransmissions may be performed to deliver one packet. Unreliability, hence, poses serious degradation of the delay and energy consumption performance of WSNs.

Duty cycling has been introduced to tackle the energy consumption problem in WSNs, where events occur unpredictably, and hence idle listening will rapidly drain the nodes' batteries [6]. In the duty-cycled mode, sensor nodes only wake up for short periods of time to send or receive packets, and then switch to the dormant state. Such a mechanism can prolong the network lifetime, but in exchange, it becomes a problem for delay-sensitive tasks/applications.

© Springer International Publishing Switzerland 2016
O. Gervasi et al. (Eds.): ICCSA 2016, Part II, LNCS 9787, pp. 129–139, 2016.
DOI: 10.1007/978-3-319-42108-7_10

A recent approach to improve the energy efficiency of broadcasting is to build a schedule-based broadcasting tree, in which all the children of a non-leaf node wake up at the same time to receive the packet simultaneously [7]. This approach allows a parent to wake up one time per working period instead of several times (depending on the number of child nodes). However, delay efficiency is still an open problem.

To overcome the challenges, in this work, we develop a broadcasting scheme running on a given schedule-based tree, which improves the overall network's broadcasting delay performance. The algorithm allows nodes, which have failed to receive a packet from the parent, to receive it from their neighbor if the switches are beneficial. To the best of our knowledge, the contribution of this paper is of three-fold:

- A switching decision is made focusing on the delay efficiency,
- A local sequencing of broadcasting timeslots among child nodes of a parent is made based on the global knowledge, i.e. nodes with large delay-ahead to broadcast to their descendants are given more chances to switch and get an earlier packet delivery,
- A collision-free broadcasting timeslots assignment, in which nodes take its broadcasting timeslots based on a sequence driven by their subtree's delay-ahead.

The organization of the paper is as follows. Section 2 reviews the related works. In Sect. 3, we represent the system model, assumptions and formulate the problem. Sections 4 and 5 respectively describe the proposed scheme, and the performance compared with the well-known algorithms. In Sect. 6, we conclude our work, and present future direction.

2   Related Work

Broadcasting in low-duty-cycled WSNs has been well-studied. A number of efficient tree structures were proposed that considered unreliable nature of wireless communications and duty-cycling. Guo et al. [8] introduced an Energy Optimal Tree which minimizes the total number of transmissions per broadcasting packet, Niu et al. [9] aimed to provide a minimum delay broadcasting tree. Multi-pipeline scheduling [10] and Correlated Flooding [7] schedule the child nodes of a common parent to wake up and receive a packet simultaneously, so that the parent does not wake up multiple times in a working period. Such the schemes are referred to as *schedule-based duty cycle*. DSRF [11] describes a switching-based flooding algorithm for scheduled-based WSNs, to let nodes, which have failed to receive a packet from parents, utilize transmissions from the neighboring nodes. The algorithm addresses both the energy consumption and delay performance for multiple broadcasting packets. In [12], Im et al. proposed a criticality-awareness algorithm to consider the latency-ahead of a node as a primary factor. The goal is to preferentially take care of the transmissions to the high latency-ahead nodes, even if it causes collisions elsewhere, to achieve a smaller broadcasting delay. The authors in [13] proposed an energy-efficient flooding by flexibly adjusting the tree structure, while guaranteeing the overall expected delay constraint. To ensure the reliability of a transmission, collecting ACKs from the receivers is a must as presented in [14].

3 Preliminaries

3.1 System Model and Assumptions

A WSN contains a number of sensor nodes deployed arbitrarily on a field, in which links between them are unreliable. Basically, link quality is an average number of transmissions to get one success [8]. Link quality between any pair of nodes can be measured beforehand, for example by a wireless link estimation described in [9]. Each node maintains a table of neighboring nodes, as well as link qualities to the neighbors. We assume that the tree is given, and all the children of a parent will be scheduled to periodically wake up at the broadcasting timeslot of the parent to receive the broadcasting packet. Those children of a common parent are called *siblings*. Note that two siblings might not be neighbors.

Time is entirely divided into working periods of equal length. Each working period, in turn, consists of a number of timeslots. A timeslot is sufficient for a round trip of DATA-ACK exchanges. A completion of sending through a link is confirmed by returning an ACK. Local clocks are assumed to be synchronized, so that given the schedule, a node knows when to wake up synchronously with others. Such a synchronization can be achieved by various protocols [15, 16].

3.2 Timeslot Design

As illustrated in Fig. 1, a timeslot's duration is sufficient for transmitting the data packet plus receiving up to M acknowledgements (M subsequent sub-slots). The total length of the sub-slots is relatively small compared to the data period. Considering one parent, each sub-slot is assigned to specific children following a predefined order. Therefore, any node can have at most M children in the broadcast tree. After a parent sends a data packet, any child node that receives successfully will reply with an ACK, in its assigned sub-slot. A node that has not yet gotten the packet, by overhearing the ACKs from the neighboring siblings, will get to know that it has failed to receive from the parent (called *FRP node*).

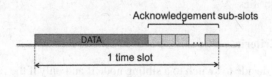

Fig. 1. A timeslot design

3.3 Problem Formulation

A source node has the broadcasting packet. Every node is scheduled to wake up at its parent's broadcasting timeslot to receive the broadcasting packet through unreliable links and confirm its reception status within the same timeslot by replying with an ACK.

Multiple children are given different ACK priorities (i.e. sending ACKs at different sub-slots), so that their replies to a common parent do not collide. The broadcasting delay is defined as the time when the last leaf node in the network receives the packet. Rather than focus on a broadcasting structure design, our goal is to design a dynamic algorithm to improve the broadcasting delay of a given tree with a collision-free duty-cycling schedule.

4 Proposed Scheme

4.1 Motivation

A major shortcoming of tree-based broadcasting is that a packet strictly follows the tree structure. In schedule-based duty cycle, because all child nodes wake up at the same time to listen to a broadcasting packet and ACK if successful, there are chances for the *FRP nodes* to overhear and perceive the reception results of other siblings. Hence, the *FRP node* may choose to switch to receive from one of the neighboring siblings rather than waiting for the parent in the next working periods. The existing switching algorithm, DSRF [11], aims to provide energy-efficient switches. In this work, an *FRP node* chooses a neighboring sibling that provides a maximal energy savings. Hence, even switching to a node via a weaker link can be considered, incurring a long delay to the *FRP node*. To boost up the broadcasting performance, the algorithm schedules the wake-up sequence of nodes in a sense that individually optimizes the reception probability for each child nodes set of a parent. Consequently, all child nodes are treated equally regardless of how big the subtree is, which causes an ineffective network-wide delay performance.

The objective of this paper, firstly, is to formulate a delay-driven criteria to perform a switch. Based on the scheduled-based broadcasting tree, for each packet, an *FRP node* will look for other neighboring siblings who received the packet already. Among them, it selects one node to switch to if the switch brings a gain in delay. Secondly, a sending scheduling priority is investigated to prioritize child nodes based on the expected delay to broadcast to its subtree. Last but not least, a collision-free schedule is built with respect to the obtained priorities.

4.2 Switching Criteria

An *FRP node* will decide to switch to a sibling node if and only if the following criteria hold:

- A sibling node already received the packet
- The broadcasting timeslot of the sibling node must be ahead of the *FRP node*'s one
- The sibling node must be one of the neighbors of the *FRP node*, and there must remain at least one ACK slot to allocate to the *FRP node*
- The switching must be beneficial in terms of delay

After a parent broadcasts a data packet, nodes which successfully received will return the ACKs within their pre-allocated acknowledgement sub-slots. An ACK message conveys: *(i)* the node's ID, *(ii)* the reception status (successful), *(iii)* the number of child nodes it is accommodating. Other *FRP nodes* can overhear the ACKs, and know the reception results of those siblings. It then looks for the neighboring sibling whose broadcasting timeslot is ahead of it, and can provide shorter delay than the parent can.

Figure 2 depicts an example of a switch. Node s is the source node that originates the broadcasting packet. Assume that in the first attempt, node s successfully delivers the packet to node a, but not node b. Node a's broadcasting timeslot is prior to node b, and node b decides to switch to node a, which means that it will wake up at timeslot t_a to receive the packet. Once decided to switch to receive from node a, node b will keep waiting for the packet from node a in the subsequent working periods until a success.

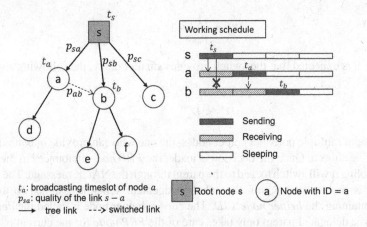

Fig. 2. An example of switching to a sibling

The advantages of switching are: *(i)* an *FRP node* gets one more transmission before its broadcasting timeslot, meaning a certain probability to be able to broadcast in its immediate timeslot; *(ii)* a non-leaf sibling will send to the sibling's children, regardless of the switch, which means that the *FRP node* gets some free transmissions from the sibling; and *(iii)* if all the child nodes of a parent receive successfully, or switch, then the parent does not have to broadcast in the next working period.

4.3 Switching Decision

Upon each failed reception, an *FRP node* will compare the expected delay if it stays receiving from the parent, or switches to a sibling node. Those sibling nodes to be considered are called *helper nodes*. In *schedule-based duty cycling*, if a node can receive the packet before its broadcasting timeslot of a current working period, it then can send to its child nodes within the current working period. Thus, from the *FRP node*'s perspective, receiving from any node, whose broadcasting timeslot is ahead of its, does not matter. If it stays with the parent, the first retransmission will be in the next working period. Therefore, the remaining expected number of working periods until it successfully receives the packet is simply inversely proportional to the link quality to the parent.

For example in Fig. 2, expected remaining working periods if node b stays with the parent s is:

$$D_b^{par} = \frac{1}{p_{sb}} \tag{1}$$

$\frac{1}{p_{sb}}$ is also called the expected transmissions count from node s to node b (noted as $EXT(s, b)$). In case node b switches to node a, because in the current working period, the helper node has not transmitted yet, the expected remaining working periods until one successful transmission is:

$$D_{ab}^{sw} = \frac{1}{p_{ab}} - 1 \tag{2}$$

Since it is expected that the switch provides smaller delay, the following should be satisfied:

$$D_{ab}^{sw} < D_b^{par} \tag{3}$$

In case of multiple potential helper nodes, the one who can provide maximum delay gain will be chosen. Once the decision is made, the *FRP node* announces to the neighboring sibling it will switch to, and to the parent through the NACK message. The NACK message is broadcasted within the acknowledgement sub-slot assigned to the *FRP node*, containing the *helper node's ID*. The chosen helper node is called the *delegated parent*. The delegated parent only takes care of the *FRP node* for the current packet.

If an *FRP node* does not know about a sibling's reception status, it will assume that the sibling switches to the same selected helper node. Then if the expected number of children after switching exceeds the limit M, the *FRP node* will cancel its decision, and look for another helper node. An effective way to alleviate the loss of ACKs and overcome the problem that the *FRP node* cannot overhear from the non-neighboring siblings is to use M tail bits of an ACK to represent the reception status of all siblings [11]. If a node overhears the ACK from other nodes, it will encode those results in the associated bits, and then do its acknowledgement with both its own result, and the overheard results.

4.4 Sending Schedule Priority

An *FRP node* should switch to a helper node which is ahead of it in the sending sequence to get a certain chance to start broadcasting in the current working period. With different sending schedule priority, the probability of switching for each node is different. For a non-leaf child node, i.e. there is a sub-tree rooted at the node, its reception failure causes all of its descendants to wait for some additional working periods. Apparently, among the child nodes, the one with a bigger sub-tree (in terms of delay to broadcast to all the descendants) should have more chances to get the packet early.

We introduce a term '*EXT_ahead*' for non-leaf nodes. Between a node m and one of its descendants n, let's say the tree path to go from m to n is $R(m, n)$. $R(m, n)$ consists of a set of tree links that the packet has to pass. Assuming the set of descendant nodes of node m is $Des(m)$, then the *EXT_ahead* of node m is defined as:

$$ETX\_ahead(m) = \max_{n \in Des(m)} ETX(m, n) \tag{4}$$

In which:

$$ETX(m, n) = \sum_{(u,v) \in (m,n)} ETX(u, v) \tag{5}$$

In order to give larger *EXT_ahead* node more chances to switch, the algorithm assigns it a latter position in a sending sequence. For example, if $EXT\_ahead(b) > EXT\_ahead(a)$, the broadcasting timeslot of node a should be ahead of that of node b. The sequence of the broadcasting timeslots is same as the sequence of ACKs.

4.5   Collision-Free Scheduling

The schedule should avoid both the primary and secondary collisions [17]. Let us assume the network has a part as in Fig. 3, and the algorithm is allocating a broadcasting timeslot for node n at hop-count $\mathcal{L}$. Obviously, all the nodes in the previous hops are scheduled. The neighboring nodes of n include $\{m, l, p, q, z, u, v, y\}$. On one hand, the broadcasting timeslot of n should be different from the receiving timeslots of the one-hop neighbors p, q and z, otherwise the transmissions to its children will cause collisions with the reception at p and q. On the other hand, the broadcasting timeslot of n should not be duplicated with broadcasting timeslots of the neighbors and siblings to avoid potential collisions (i.e. nodes p, r). Node n does not have to consider its children, because the children are only allocated broadcasting timeslots after their parent gets one.

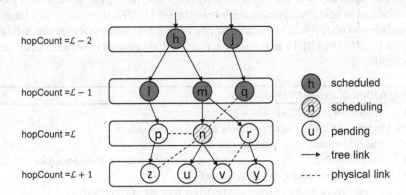

Fig. 3.   Scheduling broadcasting timeslot for node n

The scheduling starts from the source node, and is spread hop-by-hop as in Algorithm 1. In each hop, the algorithm examines each node one-by-one. A node's broadcasting timeslot should be scheduled after its parent, and after all of the siblings with a lower *EXT_ahead*.

Algorithm 1. TimeslotScheduling

1	**Procedure** *TimeslotScheduling*
2	**Input**: *all information of the network*
3	**Output:** *a collision-free schedule*
4	V: set of all nodes in the network, $Q \leftarrow \emptyset$
5	$bcTS(Source) \leftarrow 0$ //broadcasting timeslot of source node is initialized to be 0
6	Calculate the ETX-ahead of all nodes in V
7	$hc \leftarrow 0$
8	**while** $V \neq \emptyset$:
9	**for** $k \in V$ **&** $hopCount(k) == hc$:
10	**move** k **from** V **to** Q
11	**end**
12	**for** $m \in Q$ **&** $children(m) \neq \emptyset$:
13	sort m's non-leaf children in increasing order of ETX-ahead and stored in the
14	list $sortedChildren(m)$
15	**for** n $\in sortedChildren(m)$:
16	$assignTS(n)$
17	**end**
18	**end**
19	$hc \leftarrow hc + 1$
20	$Q \leftarrow \emptyset$
21	**end**

In Algorithm 1, for each node in the *sortChildren(m)*, the algorithm calls a *assignTS()* function (line 16) to find a suitable broadcasting timeslot for the child node. The procedure *assignTS()* is described as in Algorithm 2. Considering a child node n of node m, the algorithm will check all the neighbors and siblings of node n, and stores those allocated sending/receiving timeslots in the set *UsedTS* (line 5–7). The timeslot to be assigned to node n will be the closest one behind the parent m's timeslot, which is not in the *UsedTS* (line 8). Here we assume that T is long enough to achieve a collision-free schedule.

Algorithm 2. *assignTS (n)*

1	**Procedure** *assignTS(n)*
2	**Input:** *all information of neighbors and siblings of node n*
3	**Output:** *a timeslot assigned to node n*
4	$UsedTS \leftarrow \emptyset$ //the set of timeslots that n should not take
5	**for** $r \in neighbors(n)$ **or** $r \in siblings(n)$:
6	**add** broadcasting timeslot and receiving timeslot of r to $UsedTS$
7	**end**
8	$bcTS(n) \leftarrow$ the closest timeslot $\in [0, T-1]$ behind the $bcTS(m)$ and is not in $UsedTS$

The timeslot scheduling algorithm (Algorithm 1) performs $|V|$ iterations (line 12). For each node m, the sorting procedure (line 13 and 14) takes a constant time because each node has at most M children. Next, the function *assignTS(n)* inside each iteration (line 16) examines all the *neighbors* and *siblings* of node n (Algorithm 2). Although the number of neighbors may vary with each node, the total number of times the algorithm scans the neighbors throughout $|V|$ iterations is at most $O(|E|)$ (E: a set of all physical links in the network). Since each node has no more than $M - 1$ siblings, the total number of times scanning the siblings in $|V|$ iterations is $O(|V|)$. Therefore, after combining all the running times, the computational complexity of the scheduling algorithm is $O(|V| + |E|)$.

5   Performance Evaluation

We assume that in the initialization phase, all network nodes collects local information (e.g. neighbors, siblings, link qualities). Then the EXT_ahead value is estimated. After finishing the scheduling, the network turns into the duty-cycling mode, and starts broadcasting.

5.1   Simulation Environment

We generate 50 random topologies for each setting. To form the communication edges between nodes, only links with a quality higher than 0.3 are considered. For each topology, 20 broadcasting packets are scheduled individually. All nodes are identical and with a communication range of 40 m. The source node is placed at (0 m, 0 m). The number of allowed child nodes is $M = 7$.

In order to demonstrate the delay performance of the proposed idea, we use the broadcasting tree EOT [8] as the backbone structure. The proposed scheme is compared with the two following baselines:

- EOT-based broadcasting: The packet is broadcasted through the EOT, with no switching allowed.
- DSRF: the Dynamic Switching-based Reliable Flooding proposed by [11]. The scheme considers energy efficiency while making switching decisions.

All the schemes apply the collision-free scheduling presented in Sect. 4.5. We compare the average broadcasting delay, and the average number of transmissions to complete a broadcasting packet.

5.2   Simulation Results

Impact of Network Density. We keep the network side length of the area at 200×200 m, and vary the number of nodes from 100 to 300. A working period consists of 100 timeslots. The proposed scheme is 6–10 % better than DSRF in delay, and ~20 % better than EOT-based broadcasting. In terms of number of transmissions, the proposed scheme requires slightly less transmissions than the DSRF and EOT-based schemes. When the number of nodes increases, the network density increases, so that average number of child nodes increases. Consequently, there are more chances to switch and achieve better delay. In return, because there are more nodes, the algorithm has to accommodate with more transmissions for each broadcasting packet (Fig. 4).

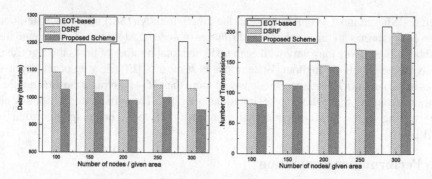

Fig. 4. Performance on different network densities.

Impact of Network Size. In this experiment, we vary the network's side length, while keeping the network density constant. The number of nodes in each case is changed accordingly. A working period has 100 timeslots. With a small size network, the proposed scheme and the DSRF provide similar performances. When the network size becomes larger, the performance gap increases up to 6 %. This is reasonable, since a small-sized network does not have as many switches as a large one. Across all the experiments, the proposed scheme is comparable to the DSRF, and is 7–9 % better than the EOT-based broadcasting in terms of the number of transmissions (Fig. 5).

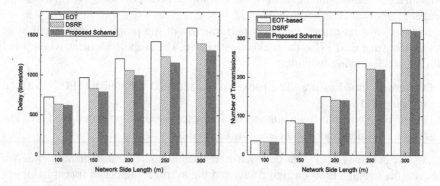

Fig. 5. Performance on different network sizes.

6   Conclusion

In this paper, we proposed a delay-driven broadcasting scheme which utilizes transmissions from neighboring sibling nodes for better delay. The nodes are scheduled based on the size of the subtree they have to accommodate. Through simulation results, the proposed scheme has up to 10 % shorter delay than the DSRF, and remarkably dominates the tree-based flooding without the help of a switching mechanism. In the future, we plan to study a tree structure that maximizes the performance of the proposed scheme, as well as work on a distributed approach.

Acknowledgement. This research was supported by the MSIP, Korea, under the G-ITRC support program (IITP-2015-R6812-15-0001) supervised by the IITP, Priority Research Centers Program through the National Research Foundation of Korea (NRF) funded by the Ministry of Education, Science and Technology (NRF-2010-0020210), and by Institute for Information & communications Technology Promotion (IITP) grant funded by the Korea government (MSIP) (No. 10041244, Smart TV 2.0 Software Platform).

References

1. Chen, Q., Cheng, S., Gao, H., Li, J., Cai, Z.: Energy-efficient algorithm for multicasting in duty-cycled sensor networks. J. Sens. **15**(12), 31224–31243 (2015)
2. Akyildiz, I.F., Su, W., Sankarasubramaniam, Y., Cayirci, E.: Wireless sensor networks: a survey. J. Comput. Netw. **38**(4), 393–422 (2002)
3. Xu, L., Zhu, X., Dai, H., Wu, X., Chen, G.: Towards energy-fairness for broadcast scheduling with minimum delay in low-duty-cycle sensor networks. J. Comput. Commun. **75**(1), 81–96 (2016)
4. Gorce, J.M., Zhang, R., Parvery, H.: Impact of radio link unreliability on the connectivity of wireless sensor networks. EURASIP J. Wirel. Commun. Netw. (2007)
5. Gu, Y., He, T.: Data forwarding in extremely low-duty-cycled sensor network with unreliable communication links. In: Proceedings of SenSys 2007, pp. 321–334 (2007)
6. Choe, J., Ha, N.P.K., Hong, J., Choo, H.: Fast and reliable data forwarding in low-duty-cycle wireless sensor networks. In: Proceedings of ICCSA 2012, pp. 324–338 (2012)
7. Guo, S., Kim, S.M., Zhu, T., Gu, Y., He, T.: Correlated flooding in low-duty-cycle wireless sensor networks. In: Proceedings of ICNP 2011, pp.383–392 (2011)
8. Guo, S., Gu, Y., Jiang, B., He, T.: Opportunistic flooding in low-duty-cycle wireless sensor networks with unreliable links. In: Proceedings of INFOCOM 2009, pp. 2787–2802 (2009)
9. Niu, J., Cheng, L., Gu, Y., Jun, J., Zhang, Q.: Minimum-delay and energy efficient flooding tree in asynchronous low-duty-cycle wireless sensor networks. In: Proceedings of WCNC 2013, pp.1261–1266 (2013)
10. Guo, S., He, T.: Robust Multi-pipeline scheduling in low-duty-cycle wireless sensor networks. In: Proceedings of INFOCOM 2012, pp.361–369 (2012)
11. Cheng, L., Gu, Y., He, T., Niu, J.: Dynamic switching-based reliable flooding in low-duty-cycle wireless sensor networks. In: Proceedings of INFOCOM 2013, pp. 1393–1401 (2013)
12. Im, G., Le, T., Choo, H., Kim, D.S.: Critical-path aware broadcast scheduling in duty-cycled wireless sensor networks. In: Proceedings of ICOIN 2015, pp. 410–411 (2015)
13. Wu, S., Niu, J., Cheng, L., Chou, W.: Energy efficient flooding under minimum delay constraint in synchronous low-duty-cycle wireless sensor networks. In: Proceedings of ComComAp 2014, pp.121–126 (2014)
14. Zhu, T., Zhong, Z., He, T., Zhang, Z.: Exploring link correlation for efficient flooding in wireless sensor networks. In: Proceedings of NSDI 2010, pp. 1–15 (2010)
15. Su, W., Akyildiz, I.F.: Time-diffusion synchronization protocol for wireless sensor networks. ACM Trans. Netw. **13**(2), 384–397 (2005)
16. Wu, Y.C., Chaudhari, Q., Serpedin, E.: Clock synchronization of wireless sensor networks. In: IEEE signal processing magazine 2011, pp. 124–138 (2011)
17. Ha, N.P.K., Zalyubovskiy, V., Choo, H.: Delay-efficient data aggregation scheduling in duty-cycled wireless sensor networks. In: Proceedings of RACS 2012, pp. 203–208 (2012)

Memory-Aware Scheduling
for Mixed-Criticality Systems

Zheng Li[1]([⊠]) and Li Wang[2]

[1] School of Computer Sciences, Western Illinois University, Macomb, IL, USA
z-li2@wiu.edu
[2] College of Engineering and Information Science, DeVry University,
Chicago, IL, USA

Abstract. In this paper, by taking both memory-access and computation time cost into consideration, a two-phase execution, i.e. memory-access phase first to fetch the instructions and required data, and then computation, is proposed to model mixed criticality tasks. Based on the proposed task model, a fixed-priority based scheduling algorithm is developed to schedule the mixed-criticality tasks. We first establish the theoretical foundations upon which to determine whether if a mixed-criticality task set is schedulable under given memory-access and computation priorities; and then based on these theoretical conclusions, we further present how to apply the well-known Audsley's algorithm to find the optimal priority assignment for both memory-access and computation phases. Extensive experiments have been conducted and the experimental results validate the effectiveness of our proposed approach.

1 Introduction

Mixed-criticality design paradigm, which tends to integrate the tasks of different criticality levels on the common hardware platform, is an increasing design trend for future real-time and embedded systems [1]. When designing complex embedded systems, in addition to the safety requirement, more and more non-functional requirements relating to production cost, power consumption and weight are enforced to the system design. Therefore, complex embedded systems such as the automotive and avionics industries are evolving into mixed criticality system design [2]. Examples involve the unmanned aerial vehicle (UAV), which integrates the HI-criticality functionalities such as flight-control tasks and LO-criticality tasks such as photo capturing tasks [3] on the same platform.

Since hardware resources are limited, LO-criticality and HI-criticality tasks on the same platform may compete for the shared resources and cause tasks to miss their deadlines. HI-criticality tasks, such as flight-control tasks in the UAV system, are crucial in the entire system and a deadline miss will result in catastrophic consequences. To ensure the safety of system and guarantee HI-criticality tasks always meet their deadlines, two worst case execution times i.e. worst case execution time by design and a more pessimistic one, worst case execution time by certification, are set for a HI-criticality task [2]. Initially, a mixed-criticality system is considered operating under the LO-mode, when any task

O. Gervasi et al. (Eds.): ICCSA 2016, Part II, LNCS 9787, pp. 140–156, 2016.
DOI: 10.1007/978-3-319-42108-7_11

executes over the designed worst case execution time, the system changes to the HI-mode immediately to signal that a situation beyond designed behaviors and needs to be taken care of. To meet the safety and efficiency requirements, a mixed-criticality system is deemed to be schedulable if both of following conditions are satisfied: (1) both LO-criticality and HI-criticality tasks are guaranteed to meet their deadlines under the LO-mode; and (2) HI-criticality tasks are also guaranteed to meet their deadlines under the HI-mode [4]. Determining whether a given mixed-criticality system is schedulable has been proven to be NP-hard [2].

To address the mixed-criticality schedulability issue, extensive research has been conducted and some well-known earliest deadline first (EDF) and fixed-priority (FP) based approaches have been published in the literature. However, to the best of our knowledge, the current research mainly focuses on computation intensive mixed-criticality systems, i.e. the tasks are assumed to be computation only. By the advent of many-core processors, the offered computing power is continuously increasing and the slow memory-access becomes the bottleneck. Though typical real-time tasks are computation intensive, their memory-access time to pre-fetch the required instructions and data may not be ignorable any more [5]. Since memory-access and computation phases are free of resource contention, i.e. different tasks' memory-access and computation phases can be executed in parallel, these existing computation intensive mixed-criticalityscheduling approaches can not be directly applied. In this paper, by taking memory-access time delay into consideration, we model each mixed-criticality task as a two-phase execution, i.e. memory-access phase and computation phase. In addition, based on the newly proposed model, we further study how to use fixed-priority based algorithm to schedule mixed-criticalitytasks without violating their deadline constraints. Our main contributions are four-fold:

- propose a new two-phase execution task model in mixed-criticality systems;
- establish the theoretical foundations for schedulability test of mixed-criticalitytasks with two-phase execution;
- propose Dual-Audsleyapproach to schedule the mixed-criticalitytasks based on the newly proposed task model; and
- experimentally validate the effectiveness of our proposed scheduling approach.

The rest of the paper is organized as follows. We first discuss related work in Sect. 2. The system models and our target problem are defined in Sect. 3. The theoretical foundations for mixed-criticalitytasks schedulability test are established in Sect. 4. In Sect. 5, we present in details how to use Dual-Audsleyalgorithm to schedule mixed-criticalitytasks with two-phase execution. Experimental results are illustrated in Sect. 6. Finally we conclude the work and point out future work in Sect. 7.

2   Related Work

The real-time community has started to address of mixed-criticality task set scheduling issues in recent few years. In [6], Baruah was the first to discuss how to apply earliest deadline first (EDF) algorithms in scheduling mixed-criticality task sets. To ensure the schedulability of a mixed-criticality task set, the EDF with virtual deadline (EDF-VD) scheduling algorithm was proposed [7]. The EDF-VD algorithm assigns HI-criticality tasks reduced deadlines to ensure that HI-criticality tasks are schedulable even if they overrun their normal worst case execution time. To further improve the schedulability over EDF-VD algorithm, Ekberg [8] proposed a greedy approach, which utilized demand-bound function analysis, to determine a task set's schedulability under EDF algorithm. Both Baruah's EDD-VD and Ekberg's greedy algorithm take the approach of terminating all LO-criticality tasks when system enters into the HI-mode, and hence the performance of LO-criticality tasks will be severely degraded. To provide a guaranteed minimum level of service to LO-criticality tasks when system enters into the HI-mode, Su [9] considered using elastic task models [10] to increase LO-criticalitytasks' period and hence reduce their competition against HI-criticality tasks but allow LO-criticalitytasks to execute when possible. By noticing that postponing HI-criticality task execution can promote earlier execution of LO-criticalitytasks, Park [11] developed a scheme called criticality based EDF which delays the execution of HI-criticality tasks as late as possible but without causing deadline violations. Dynamic resource reservation to further improve the Quality-of-Service of LO-criticality tasks were studied in [12].

In addition to the EDF based scheduling algorithms, fixed-priority (FP) based algorithms have also been studied extensively. Audsley's algorithm was proposed in [13] as an optimal fixed-priority assignment, Vestal [14] later extended the algorithm to schedule multi-criticality tasks and Baruah [15] further applied Audsley's algorithm to schedule tasks in mixed-criticality systems.

However, all the above research focuses on computation intensive tasks only. Since many-core technology is emerging, slow memory-access is becoming the bottleneck and real-time community just started to look into this issue. Considering memory-access time is not ignorable, Melani gave the response-time analysis for tasks with both memory-access and computation demand [5]. In this paper, we are to extend the research for tasks with both memory-access and computation phases in mixed-criticality systems.

3   Models and Problem Formulation

3.1   System Models

In this paper, we focus on the dual criticality system [16,17] with tasks of two different criticalities. In particular, for a given mixed-criticality task set $\Gamma = \{\tau_1, \tau_2, ..., \tau_n\}$, each task τ_k is defined by $\tau_k = (\chi_k, E_k, M_k, C_k, T_k, D_k)$, where $\chi_k \in \{LO, HI\}$ is task's criticality level. A mixed-criticality system can run in either the LO-mode or the HI-mode. E_k is a pair $(E_k(LO), E_k(HI))$ where $E_k(\chi)$ is the worst case execution time of τ_k when executing under χ mode. For HI-criticality tasks, $E_k(HI) > E_k(LO)$, while for LO-criticality tasks, $E_k(HI) = E_k(LO)$. Initially, the system runs under LO-mode, when any of task instance exceeds its $E_k(LO)$ time limitation, the system switches to the HI-mode immediately. To guarantee the entire system safety, when system runs under HI-mode, LO-criticalitytasks will be suspended and hence HI-criticality tasks are able to utilize all the available resources to catch up their deadlines [8,12,16,17].

Different with the traditional mixed-criticalitysystems that tasks are assumed to be computation only. In this paper, $\forall \tau_i \in \Gamma$, task execution includes two phases: one memory-access phase and one computation phase. Though the model itself is simple, typical real-time tasks, such as image and signal processing tasks without data written back well fits this model [5]. M_k is a pair $(M_k(LO), M_k(HI))$ and $M_k(\chi)$ is the worst case memory access time under mode χ. Similarly, C_k is a also pair $(C_k(LO), C_k(HI))$ and $C_k(\chi)$ is the worst case computation time under mode χ. It is worth pointing out that $E_k(\chi) = M_k(\chi) + C_k(\chi)$ and $M_k(HI) = M_k(LO)$. Tasks are periodic with period and deadline as T_k and D_k, respectively. We further assume that tasks have constrained deadlines, i.e., $D_k \leq T_k$.

In addition, a task set is defined to be mixed-criticality (MC) schedulable if the following conditions are both satisfied [7]:

1. Both the HI-criticality and LO-criticality tasks meet their deadlines when the system operates under LO-mode.
2. All HI-criticality tasks are guaranteed to meet their deadlines when system executes under the HI-mode.

3.2   Problem Formulation

Before formulating the problem the paper is to address, we introduce the following notations to be used later.

Q_M: task set memory access priority order.
$Q_M(\tau_i)$: the memory access priority of task τ_i in priority order Q_M.
Q_C: task set computation priority order.
$Q_C(\tau_i)$: the computation priority of task τ_i in priority order Q_C.
$hpm(\tau_i)$: the set of tasks with memory access priority higher than $Q_M(\tau_i)$.
$hpc(\tau_i)$: the set of tasks with computation priority higher than $Q_C(\tau_i)$.
$hpmc(\tau_i)$: the set of tasks with memory access priority higher than $Q_M(\tau_i)$ and computation priority higher than $Q_C(\tau_i)$.
$hpcH(\tau_i)$: the set of HI-criticality tasks with computation priority higher than $Q_C(\tau_i)$.
$R_M^L(\tau_i)$: worst case response time of task τ_i in memory access phase under LO-mode.
$R_C^L(\tau_i)$: worst case response time of task τ_i in computation phase under LO-mode.
$R_M^H(\tau_i)$: worst case response time of task τ_i in memory access phase under HI-mode.
$R_C^H(\tau_i)$: worst case response time of task τ_i in computation phase under HI-mode.
$R^L(\tau_i)$: the worst case response time of task τ_i under LO-mode including both memory access and computation phases.

$R^H(\tau_i)$: worst case response time of task τ_i under HI-mode including both memory access and computation phases.

$u_L(\tau_i)$: task τ_i LO-mode utilization, where $u_L(\tau_i) = \frac{E_i(\text{LO})}{T_i}$.

$u_H(\tau_i)$: task τ_i HI-mode utilization, where $u_H(\tau_i) = \frac{E_i(\text{HO})}{T_i}$.

$U_\chi(\Gamma_L)$: LO-criticality task set χ-mode utilization, where $U_\chi(\Gamma_L) = \sum_{\tau_i \in \Gamma_L} u_\chi(\tau_i)$ and $\chi \in \{\text{HI}, \text{LO}\}$.

$U_\chi(\Gamma_H)$: HI-criticality task set χ-mode utilization, where $U_\chi(\Gamma_H) = \sum_{\tau_i \in \Gamma_H} u_\chi(\tau_i)$ and $\chi \in \{\text{HI}, \text{LO}\}$.

With the above notations, we are ready to formulate our problem as follows:

Problem 1. Given a mixed-criticality task set $\Gamma = \{\Gamma_{LO}, \Gamma_{HI}\}$ with tasks having two-phase execution, i.e. memory access and then computation, develop a priority assignment strategy, i.e. determine Q_M and Q_C to guarantee task set Γ is MC schedulable under fixed-priority scheduling.

As mentioned above, a task set Γ is MC schedulable only if both the HI-criticality and LO-criticality tasks meet their deadlines under LO-mode, and also all HI-criticality tasks meet their deadlines under the HI-mode. Hence, to guarantee a task set Γ is MC schedulable, the worst case response time of both LO-criticality and HI-criticality tasks should not exceed their deadlines under LO-mode; while the worst case response time of HI-criticality tasks also should not exceed their deadlines under HI-mode. In other words, the following Inequations must be satisfied:

$$\forall \tau_i \in \Gamma : R^L(\tau_i) \leq D_i, \tag{1}$$

and

$$\forall \tau_i \in \Gamma_H : R^H(\tau_i) \leq D_i. \tag{2}$$

In the following section, we are to address the problem in two steps. First, we establish the theories to calculate the $R^L(\tau_i)$ and $R^H(\tau_i)$ under given task set memory access priority order Q_M and computation priority order Q_C, and then present our strategy to find the Q_M and Q_C under which the mixed-criticality task set is guaranteed to be MC schedulable.

4   Theoretical Foundation

In this section, assuming the memory access priority order Q_M and computation priority order Q_C are given as a priori, we are to establish the theoretical foundations upon which to calculate $R^L(\tau_i)$ and $R^H(\tau_i)$. The following definitions are given first to simplify the presentation.

Definition 1 (Phase-Transition Instant). *The time instant at which the task is transited from memory-access phase to computation phase.*

Definition 2 (Mode-Transition Instant). *The time instant at which the task is transited from LO-mode execution to HI-mode execution.*

Definition 3 (Memory-Interfering Task Instance). *Suppose J_k is a task instance of τ_k, if J_k's memory-access is interfered by some task instant J_i's memory-access, then J_i is said to be a Memory-Interfering Task Instance of J_k.*

Definition 4 (Computation-Interfering Task Instance). *Suppose J_k is the a task instance of τ_k, if J_k's computation is interfered by some task instant J_i's computation, then J_i is said to be a computation-interfering task instance of J_k.*

Definition 5 (Dual-Interfering Task Instance). *Suppose J_k is a task instance τ_k, if some task instant J_i is both the memory-interfering task instance and computation-interfering task instance of J_k, then J_i is said to be a dual-Interfering task instance of J_k.*

For traditional real-time systems with all the tasks are of the same criticality, assuming each task's memory-access and computation phases are at the same priority, Melani et al. [5] proposed an approach to calculate the task's worst case execution time. Apparently, their approach can not be directly applied for tasks in mixed-criticalitysystems. In the following, we present our approach to calculate mixed-criticalitytasks' worst case response times with memory-access and computation phases may be in different orders.

As mentioned above, a mixed-criticalitytask set is schedulable only if both the LO-mode and HI-mode schedulability tests, i.e. formula (1) and (2), are satisfied. We first discuss how to calculate task worst case response time under LO-mode. Since each task τ_k's execution includes memory access and computation phases, i.e. the task accesses required instructions and data from the memory and then finishes the computation, its worst case response time under LO-mode can be calculated as:

$$R^L(\tau_k) = R_M^L(\tau_k) + R_C^L(\tau_k) \tag{3}$$

where $R_M^L(\tau_k)$ is the worst case response time in memory-access phase and $R_C^L(\tau_k)$ is the one in computation phase.

Prior to presenting the calculation of $R^L(\tau_k)$, we give the following lemma first.

Lemma 1. *Assuming tasks are of the same criticality, but each task's memory-access and computation phases may set at different priority levels, the worst case response time of a task instance J_k released by task τ_k is achieved when*

1. *dual-interfering task instances complete their memory-access phases an infinitely small amount of time earlier than the phase-transition instant of J_k;*
2. *memory-interfering only task instances complete their memory-access phases an infinitely small amount of time earlier than the phase-transition instant of J_k;*

3. *computation-interfering only task instances complete their memory-access phases an infinitely small amount of time after the phase-transition instant of J_k;*
4. *all computation-interfering task instances released after the phase-transition instant of J_k are with null memory-access phases.*

Proof. The scenario having the conditions (1)–(4) is illustrated in Fig. 1, where the memory-access priority order $Q_M = \{Q_M(\tau_1), Q_M(\tau_2), ..., Q_M(\tau_k), Q_M(\tau_{k+1})\}$ and $Q_C = \{Q_C(\tau_{k+1}), Q_C(\tau_1), Q_C(\tau_2), ..., Q_C(\tau_k)\}$, i.e. tasks $\tau_1, ..., \tau_{k-1}$ have both higher memory-access and computation priority than τ_k, while task τ_{k+1} has higher computation but lower memory access priority than τ_k. We prove that under this scenario both the $R_M^L(\tau_k)$ and $R_C^L(\tau_k)$ of task instance J_k are maximized.

Conditions (1) and (2) indicate that all the higher priority memory interfering task instances are almost completed at the same time. To prove the $R_M^L(\tau_k)$ value is maximized under this scenario, we analogize the synchronous periodic release pattern which results in the maximum response time [18], i.e. shift right all higher memory-access phases until they complete an infinitely small amount of time ahead of J_k's phase-transition instant. Under this scenario, all memory-interfering task instances' phase-transition instants are aligned with J_k, which will result in the longest time inference to J_k and hence $R_M^L(\tau_k)$ is maximized.

From standard response time analysis [19], we know that $R_C^L(\tau_k)$ value is maximized when J_k and all the task instances with computation at a higher priority release their computation phases synchronously, which is stated as conditions (3) and (4). A task instance with a higher computation phase priority may be J_k's dual-interfering task instance or computation only interfering task

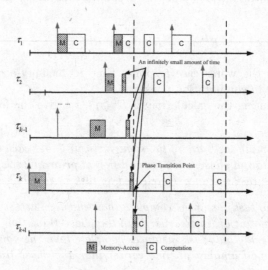

Fig. 1. Worst case response time scenario

instance. Under the scenario which results in the maximal $R_M^L(\tau_k)$, all memory-interfering task instances' phase-transition instants are aligned with J_k's, which indicates the computation phases of all dual-interfering task instances and J_k are also aligned. In addition, condition (4) ensures the computation only interfering task instances have null memory-access phases, which reveals that the computation phases of J_k and its computation only task instances are also released synchronously. All these conclude the proof.   □

When a mixed-criticalitysystem executes under LO-mode, both LO-criticalityand HI-criticalitytasks are required to meet their deadlines, hence, we can treat all LO-criticalityand HI-criticalitytasks at the same criticality level. Based on this observation, according to Lemma 1, when system runs under LO-mode, task worst case response time can be derived using the following lemma.

Lemma 2. *Given a MC task set* $\Gamma = \{\Gamma_L, \Gamma_H\}$, *the memory access priority order* Q_M *and computation priority order* Q_C, *when system executes under LO-mode,* $\forall \tau_k \in \Gamma$, *the worst case response time of memory-access phase and computation phase can be calculated by fixed-point iteration of the following equations:*

$$R_M^L(\tau_k) = M_k(LO) + \sum_{\tau_i \in hpm(\tau_k)} \left\lceil \frac{R_M^L(\tau_k)}{T_i} \right\rceil M_i(LO) \tag{4}$$

and,

$$R_C^L(\tau_k) = C_k(LO) + \sum_{\tau_i \in hpc(\tau_k)} \left\lceil \frac{R_C^L(\tau_k) + R_M^L(\tau_i)}{T_i} \right\rceil C_i(LO) \tag{5}$$

respectively.

Proof. With the standard response time analysis [19], for a task instance J_k released by task τ_k, if its memory-access response time is $R_M^L(\tau_k)$, then a task with higher memory access priority, i.e. $\tau_i \in hpm(\tau_k)$ will contribute $\left\lceil \frac{R_M^L(\tau_k)}{T_i} \right\rceil M_i(LO)$ blocking time. Hence, $R_M^L(\tau_k)$ can be calculated by fixed-point iteration of the following equation:

$$R_M^L(\tau_k) = M_k(LO) + \sum_{\tau_i \in hpm(\tau_k)} \left\lceil \frac{R_M^L(\tau_k)}{T_i} \right\rceil M_i(LO)$$

Next, we present how to calculate J_k's worst case computation response. According to Lemma 1, $R_C^L(\tau_k)$ is maximized when all higher computation phases are released synchronously with J_k. Hence, suppose J_k's computation response time is $R_C^L(\tau_k)$, under such scenario, $\tau_i \in hpc(\tau_k)$, the number of incoming task instances released from τ_i can be calculated as:

$$\left\lceil \frac{R_C^L(\tau_k) - (T_i - R_M^L(\tau_i))}{T_i} \right\rceil$$

plus the current processing task instance, the total number of interfering task instances to be completed will be:

$$\left\lceil \frac{R_C^L(\tau_k) - (T_i - R_M^L(\tau_i))}{T_i} \right\rceil + 1$$

Since the longest block time of τ_i's released instance is $C_i(\text{LO})$, the total blocking time will be:

$$\left\lceil \frac{R_C^L(\tau_k) + R_M^L(\tau_i)}{T_i} \right\rceil \cdot C_i(\text{LO})$$

These conclude the proof of Lemma 2.   □

With Lemma 2, task's worst case response time under LO-mode $(R^L(\tau_k))$ can be obtained. In the following, we present how to calculate task's worst case response time under HI-mode $(R^H(\tau_i))$.

Lemma 3. *Given a MC task set $\Gamma = \{\Gamma_L, \Gamma_H\}$, the memory access priority order Q_M and computation priority order Q_C, then $\forall \tau_k \in \Gamma_H$, its worst case response time under HI-mode, i.e. $R^H(\tau_k)$ can be expressed as:*

$$R^H(\tau_k) = \max_{x \in \{E_k(\text{LO}), R^L(\tau_k)\}} R^H(\tau_k, x) \qquad (6)$$

where $R^H(\tau_k, x)$ represents the worst case response time under HI-mode when the mode-transition instant happens at time x and it can be calculated by fixed-point iteration of the following equation:

$$R^H(\tau_k, x) = E_k(\text{HI}) - E_k(\text{LO}) + x + \sum_{\tau_i \in hpcH(\tau_k)} \left\lceil \frac{R^H(\tau_k, x) - x}{T_i} \right\rceil C_i(\text{HI}) \quad (7)$$

among which, $hpcH(\tau_k)$ indicate the set of tasks with higher computation priority than that of τ_i.

Proof. When a HI-criticality task τ_k runs over $E_k(\text{LO})$, the system will change to HI-mode. Since τ_k's worst case response time under LO-mode is $R^L(\tau_k)$, the system mode could occur at any time x, but no earlier than $E_k(\text{LO})$ and no later than $R^L(\tau_k)$, i.e. $E_k(\text{LO}) \leq x \leq R^L(\tau_k)$. With standard response time analysis, the task τ_k's worst case response time under HI-mode when system mode changes at x can be obtained as:

$$R^H(\tau_k, x) = E_k(\text{LO}) + B_k^L(x) + E_k(\text{HI}) - E_k(\text{LO}) + B_k^H(R^H(\tau_k, x) - x) \quad (8)$$

where $B^L(x)$ denotes the blocking time from tasks with higher priority under LO-mode, $E_k(\text{HI}) - E_k(\text{LO})$ is task overrun under the HI-mode, and $B_k^H(R^H(\tau_k, x) - x)$ is the blocking time from tasks with higher priority under HI-mode. Suppose system changes mode at time x, then we have

$$x = E_k(\text{LO}) + B_k^L(x) \qquad (9)$$

Since $E_k(LO) \geq M_k$, which indicates that the task has completed its memory-access phase under the LO-mode, and the whole HI-mode execution is to finish the remaining computation. Hence, it is not hard to find that, for a task $\tau_k \in \Gamma_H$ having computation priority higher than τ_k, it will introduce the longest blocking time for τ_k under HI-mode if τ_k is memory-access free and its computation time is up to $C_k(HI)$. Based these analysis, we have

$$B_k^H(R^H(\tau_k, x) - x) = \sum_{\tau_i \in hpcH(\tau_k)} \left\lceil \frac{R^H(\tau_k, x) - x}{T_i} \right\rceil C_i(HI) \qquad (10)$$

With Eqs. (9) and (10), (8) can be re-written as:

$$R^H(\tau_k, x) = E_k(HI) - E_k(LO) + x + \sum_{\tau_i \in hpcH(\tau_k)} \left\lceil \frac{R^H(\tau_k, x) - x}{T_i} \right\rceil C_i(HI) \quad (11)$$

Since x could be any time but no earlier than $E_k(LO)$ and no later than $R^L(\tau_k)$, hence, the worst case response time can be calculated as

$$R^H(\tau_k) = \max_{x \in \{E_k(LO), R^L(\tau_k)\}} R^H(\tau_k, x)$$

These conclude the proof of Lemma 3.   □

With Lemmas 2 and 3, $\forall \tau_i \in \Gamma$, we are able to calculate its worst case response time $R^L(\tau_k)$ and $R^H(\tau_k)$, and then use formula (1) and (2) to determine whether the mixed-criticalitytask set is schedulable.

5   Memory-Processor Priority Co-assignment for Mixed-Criticality Task Set

In the above section, we established the theories to decide if a mixed-criticalitytask set is schedulable by assuming the memory-access priority order Q_M and computation priority order Q_C are known as a priori. In this section, we discuss how to determine Q_M and Q_C to optimize a MC task set's schedulability under the assumption that each task is assigned a single priority, i.e. a task's memory-access and computation are assigned the same priority.

For traditional mixed-criticalitytask sets, i.e. tasks have computation phase only, a well know priority assignment algorithm, i.e. Audsley's algorithm [13] has been proven to be applicable [20]. For self-containment, we present the major steps of Audsley's priority assignment algorithm as follows:

1. priorities are assigned from lowest to highest order; if the lowest priority is assigned, the next higher one becomes the lowest available one;
2. assign the lowest available priority to an unassigned task which is schedulable under the assignment, ties are broken at random;
3. repeat the above step until all tasks are got assigned.

It is worth pointing out that, a HI-criticality task is schedulable only if it can meet its deadline under both HI-mode and LO-mode executions; while a LO-criticalitytask meets deadline under LO-mode is deemed to be schedulable. Though Audsley's algorithm is applicable for traditional mixed-criticalitytask sets, it can not be used directly for mixed-criticalitytasks with both memory-access and computation phases. The following example illustrates the issue.

Example 1. Suppose we have $\Gamma = \{\tau_1, \tau_2, \tau_3\}$ and let's try to use Audsley's algorithm, i.e. Step (1)–(3) for priority assignment. According to Step (2), initially all tasks are unassigned. We first should use formula (1) to check if τ_1 is schedulable at lowest available memory-access and computation priority, and hence $R_C^L(\tau_1)$ needs to be calculated.

When checking the schedulability of τ_1, tasks τ_2 and τ_3 are assumed to be at higher memory-access and computation priorities. Therefore, according to formula (5), $R_M^L(\tau_2)$ and $R_M^L(\tau_3)$ must be calculated first in order to get $R_C^L(\tau_1)$. However, as the memory-access priority of τ_2 and τ_3 are un-determined at this point, $R_M^L(\tau_2)$ and $R_M^L(\tau_3)$ can not be obtained.   □

Example 1 reveals the issue that in order to calculate $R_C^L(\tau_k)$, $\forall \tau_i \in hpc(\tau_i)$, $R_M^L(\tau_i)$ must be known as a priori. In the following, we discuss how to solve this issue.

According to the definition of $R_M^L(\tau_i)$, i.e. the worst case memory-access response time of task τ_i, if the task τ_i is MC schedulable, we have:

$$R_M^L(\tau_i) \leq T_i - C_i(\text{LO}). \tag{12}$$

Hence, the following equation can be used to calculated $R_C^L(\tau_k)$ instead:

$$R_C^L(\tau_k) = \sum_{\tau_i \in hpc(\tau_k)} \left(\left\lceil \frac{R_C^L(\tau_k) - C_i(\text{LO})}{T_i} \right\rceil + 1 \right) C_i(\text{LO}) + C_k(\text{LO}) \tag{13}$$

With the formula (13), $R_C^L(\tau_k)$ is able to be calculated without knowing $R_M^L(\tau_i)$ at first, and hence the issue elaborated in Example 1 has been addressed.

Suppose all tasks' memory-access and computation phases must be executed sequentially, according to the standard response time analysis, task's worst case response time under LO-mode (denoted as $R_*^L(\tau_k)$) can be obtained as:

$$R_*^L(\tau_k) = E_k(LO) + \sum_{\tau_i \in hpc(\tau_k)} \lceil R_*^L(\tau_k)/T_i \rceil E_i(LO)$$

However, in real-word scenario, due to free of resource contention, different tasks' memory-access and computation can be run in parallel, hence, we have $R^L(\tau_k) \leq R_*^L(\tau_k)$. Therefore, $R^L(\tau_k)$ can be further refined as:

$$R^L(\tau_k) = \max\{R_M^L(\tau_k) + R_C^L(\tau_k), R_*^L(\tau_k)\} \tag{14}$$

where $R_M^L(\tau_k)$ and $R_C^L(\tau_k)$ are calculated using formula (4) and (13), respectively.

In addition to $R_C^L(\tau_k)$, according on formula (2), task's worst case response time under HI-mode, i.e. $R^H(\tau_k)$, must also be obtained in order to determine whether a HI-criticality task is schedulable. As stated in Lemma 3, in order to obtain $R^H(\tau_k)$, the fixed-point iteration of equation (7) for every possible $x \in \{E_k(\text{LO}), R^L(\tau_k)\}$ must be calculated, which is time consuming. Next, we present the Lemma 4 to simplify the calculation.

Lemma 4. *Given a MC task set* $\Gamma = \{\Gamma_L, \Gamma_H\}$, *the memory access priority order* Q_M *and computation priority order* Q_C, *then* $\forall \tau_k \in \Gamma_H$, *its worst case response time under HI-mode, i.e.* $R^H(\tau_k)$ *can be calculation by fixed-iteration of the following equation:*

$$R^H(\tau_k) = E_k(HI) - E_k(LO) + \sum_{\tau_i \in hpcH(\tau_k)} \left\lceil \frac{R^H(\tau_k)}{T_i} \right\rceil C_i(HI)$$

$$+ x_0 - \sum_{\tau_i \in hpcH(\tau_k)} \left\lfloor \frac{x_0}{T_i} \right\rfloor C_i(HI) \tag{15}$$

where,

$$x_0 = \arg\max_{x \in \Omega} \left(x - \sum_{\tau_i \in hpcH(\tau_k)} \left\lfloor \frac{x}{T_i} \right\rfloor C_i(HI) \right) \tag{16}$$

and

$$\Omega = \bigcup_{\tau_i \in hpcH(\tau_k)} m \cdot T_i \ with \ m \cdot T_i \le R^L(\tau_k) \qquad \square$$

Proof. We prove this Lemma by simplifying the Eqs. (6) to (15).
    Since

$$\left\lceil \frac{R^H(\tau_k, x) - x}{T_i} \right\rceil \le \left\lceil \frac{R^H(\tau_k, x)}{T_i} \right\rceil - \left\lfloor \frac{x}{T_i} \right\rfloor$$

the Eq. (7) can be re-written as:

$$R^H(\tau_k, x) = E_k(HI) - E_k(LO) + \sum_{\tau_i \in hpcH(\tau_k)} \left\lceil \frac{R^H(\tau_k, x)}{T_i} \right\rceil C_i(HI)$$

$$+ x - \sum_{\tau_i \in hpcH(\tau_k)} \left\lfloor \frac{x}{T_i} \right\rfloor C_i(HI) \tag{17}$$

Suppose that,

$$x_0 = \arg\max_{x \in \{E_k(LO), R^L(\tau_k)\}} \left(x - \sum_{\tau_i \in hpcH(\tau_k)} \left\lfloor \frac{x}{T_i} \right\rfloor C_i(HI) \right)$$

Since $\left\lfloor \frac{x}{T_i} \right\rfloor$ only changes at multiples of T_i, the calculation of x_0 can even be simplified as

$$x_0 = \arg\max_{x \in \Omega} \left(x - \sum_{\tau_i \in hpcH(\tau_k)} \left\lfloor \frac{x}{T_i} \right\rfloor C_i(\mathrm{HI}) \right)$$

where

$$\Omega = \bigcup_{\tau_i \in hpcH(\tau_k)} m \cdot T_i \text{ with } m \cdot T_i \leq R^L(\tau_k)$$

With the above analysis, now we can re-write Eq. (6) as:

$$R^H(\tau_k) = E_k(\mathrm{HI}) - E_k(\mathrm{LO}) + \sum_{\tau_i \in hpcH(\tau_k)} \left\lceil \frac{R^H(\tau_k)}{T_i} \right\rceil C_i(\mathrm{HI})$$

$$+ x_0 - \sum_{\tau_i \in hpcH(\tau_k)} \left\lfloor \frac{x_0}{T_i} \right\rfloor C_i(\mathrm{HI})$$

All the above concludes the proof.   □

It is not hard to find that, based on Lemma 4, applying fixed-point iteration of equation (15) only once can get the $R^H(\tau_k)$, which greatly reduces the time cost.

So far, we have addressed the issues when applying Audsley's priority assignment algorithm for mixed-criticalitytasks with dual-phase execution, i.e. memory-access and then computation. Next, we set up experiments to evaluate the performance of the modified Audsley's algorithm, called dual-Audsley algorithm, regarding to mixed-criticalitytask set schedulability.

6   Evaluation

In this section, we conduct a set of experiments to evaluate the schedulability performance of proposed Dual-Audsley. For traditional mixed-criticalitytasks with single phase execution, a well-known Adaptive Mixed-Criticality (AMC) scheduling algorithm, has been proposed for priority assignment [20]. In our experiments, we set AMC as the baseline by treating task's two-phase execution as a single, i.e. executing memory-access and computation sequentially.

6.1   Experimental Setting

In the following experiments, high- and low-criticality task sets are generated using UUniFast algorithm [21]. There are total six tasks in a task set, among them, three are of HI-criticality and the other three are of LO-criticality. In particular, the following steps are used to generate a valid task set:

- The utilization of HI-criticality and LO-criticality task set are $U_H(\Gamma_H)$ and $U_L(\Gamma_L)$, respectively. The individual task utilization $u_H(\tau_i)$ and $u_L(\tau_i)$ are uniformly distributed in $[0, U_H(\Gamma_H)]$ and $[0, U_L(\Gamma_L)]$, respectively;
- Task's period T_i is randomly selected from $[50, 100]$;
- HO-criticality task's execution time $E_i(HI)$ is set as $T_i \cdot u_H(\tau_i)$ and $E_i(LO) = \lambda \cdot E_i(HI)$, where λ is a random value within the range $[0.4, 0.8]$;
- LO-criticality task's execution time $E_i(HI)$ is set as $T_i \cdot u_L(\tau_i)$ and $E_i(LO) = E_i(HI)$;
- Task's memory-access time $M_i(HI) = M_i(LO) = \gamma \cdot E_i(LO)$, where γ is the memory access ratio with range $[0.4, 0.9]$;
- Task's computation time $C_i(\chi) = E_i(\chi) - M_i(\chi)$.

6.2 Experiment Results and Discussions

The performance of compared priority assignment strategies is evaluated by the schedulability ratio, which is defined as the number of task sets passing the schedulability test over the total number of randomly generated task sets.

In our experiments, 1000 task sets are randomly generated and the schedulability ratios are evaluated by varying LO-criticality task set utilization $U_L(\Gamma_L)$, HI-criticality task set utilization $U_H(\Gamma_H)$ and memory access ratio γ, respectively.

In the first set of experiments, we fix the memory access ratio $r = 0.5$, LO-criticality task set utilization $U_L(\Gamma_L) = 0.4$ and vary HI-criticality task set utilization $U_H(\Gamma_H)$ from 0.4 to 0.8. From the experiment results depicted in Fig. 2, we can see that, the schedulability ratios of both compared approaches are dropped when the $U_H(\Gamma_H)$ is increasing. However, the drop rate of Dual-Audsleyalgorithm is much slower than that of AMC algorithm; when $U_H(\Gamma_H) = 0.7$, the schedulability ratio of Dual-Audsleyapproach is over 30 % higher than that of AMC algorithm.

In the second set of experiments, we set $U_H(\Gamma_H) = 0.4$ and vary $U_L(\Gamma_L)$ from 0.4 to 0.8. The experiment results are shown in Fig. 3. Analogous to the trend shown in Fig. 2, the proposed Dual-Audsleyapproach achieves higher schedulability ratio than AMC algorithm, especially under higher $U_L(\Gamma_L)$.

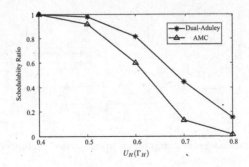

Fig. 2. Schedulability ratio comparison under varied $U_{tot}(H)$ with $U_{tot}(L) = 0.4$

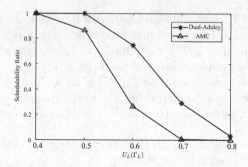

Fig. 3. Schedulability ratio comparison under varied $U_{tot}(L)$ with $U_{tot}(H) = 0.4$

The impact of memory-access ratio variation is evaluated in the last set of experiments. We set $U_H(\Gamma_H) = 0.5$, $U_L(\Gamma_L) = 0.5$ and the memory-access ratio γ is varied from 0.4 to 0.9. From the results illustrated in Fig. 4, the AMC approach is insensitive to the memory-access ratio change. This is due to the fact that, AMC treats the two-phase execution as a single phase, and hence as long as the total execution time remains the same, the schedulability ratio will not be impacted. But in contrast, the variation of memory-access ratio heavily impacts the performance of the proposed Dual-Audsleyapproach. When $\gamma < 0.7$, increasing the value of γ improves its schedulability ratio. But, in contrast, when $\gamma > 0.7$, higher value of γ results in lower schedulability ratio. This is because different tasks' memory-access and computation phases can be executed in parallel, hence tasks with "balanced" memory-access and computation phases are more tentative to be schedulable. It is interesting that the "balance point" is not at 0.5 but around $\gamma = 0.7$. This is due to the definition of γ, which is defined to be the ratio of memory-access time and LO-mode execution time. Actually, a HI-criticality task may execute up to the longer one, i.e. HI-mode execution time. Therefore, the "balance point" is be over 0.5.

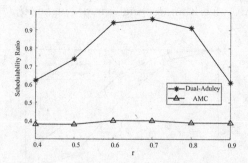

Fig. 4. Schedulability ratio comparison under varied memory-access ratio

7   Conclusion

In this paper, by taking both memory-access and computation contention into consideration, we studied how to schedule tasks in mixed-criticality systems under fixed-priority based algorithm. Different with traditional mixed-criticality tasks assumed to be computation only, we developed a new two-phase task model by incorporating the memory-access time cost. Based on the newly developed task model, we first presented the schedulability test to determine whether a mixed-criticalitytask set is schedulable under given memory-access and computation priority assignment. In addition, we further discussed how to apply the well-known Audsley's algorithm to determine the optimal priority assignment. The experiment results revealed that the proposed approach outweighed the AMC algorithm under various system configurations.

It is worth mentioning that our proposed priority assignment approach is based on the assumption that a task's memory-access and computation phases are assigned the same priority, our next step is to release this assumption by assigning memory-access and computation phases at different priority level to further improve the schedulability ratio. In addition, we are building a 100-core hardware platform and plan to evaluate the proposed models and approaches under practical scenarios.

References

1. Burns, A., Davis, R.I.: Mixed criticality systems: a review. Department of Computer Science, University of York, East Lansing, Michigan, Technical report MCC-1(b), February 2013
2. Baruah, S., Burns, A., Davis, R.: Response-time analysis for mixed criticality systems. In: 2011 IEEE 32nd Real-Time Systems Symposium (RTSS), pp. 34–43, November 2011
3. Barhorst, J., Belote, T., Binns, P., Hoffman, J., Paunicka, J., Sarathy, P., Scoredos, J., Stanfill, P., Stuart, D., Urzi, R.:A research agenda for mixed-criticality systems. In: Cyber-Physical Systems Week, April 2009
4. Ekberg, P., Yi, W.: Bounding and shaping the demand of mixed-criticality sporadic tasks. In: 2012 24th Euromicro Conference on Real-Time Systems (ECRTS), pp. 135–144, July 2012
5. Melani, A., Bertogna, M., Bonifaci, V., Marchetti-Spaccamela, A., Buttazzo, G.: Memory-processor co-scheduling in fixed priority systems. In: Proceedings of the 23rd International Conference on Real Time andNetworks Systems, ser. RTNS 2015, pp. 87–96 (2015)
6. Baruah, S., Vestal, S.: Schedulability analysis of sporadic tasks with multiple criticality specifications. In: Euromicro Conference on Real-Time Systems, 2008. ECRTS 2008, pp. 147–155, July 2008
7. Baruah, S., Bonifaci, V., D'Angelo, G., Li, H., Marchetti-Spaccamela, A., Van der Ster, S., Stougie, L.: The preemptive uniprocessor scheduling of mixed-criticality implicit-deadline sporadic task systems. In: 2012 24th Euromicro Conference on Real-Time Systems (ECRTS), pp. 145–154, July 2012
8. Ekberg, P., Yi, W.: Bounding and shaping the demand of generalized mixed-criticality sporadic task systems. Real-Time Syst. 50(1), 48–86 (2014)

9. Su, H., Zhu, D.: An elastic mixed-criticality task model and its scheduling algorithm. In: Design, Automation Test in Europe Conference Exhibition (DATE), 2013, pp. 147–152, March 2013

10. Buttazzo, G., Lipari, G., Abeni, L.: Elastic task model for adaptive rate control. In: The 19th IEEE Real-Time Systems Symposium, 1998. Proceedings, pp. 286–295, December 1998

11. Park, T., Kim, S.: Dynamic scheduling algorithm and its schedulability analysis for certifiable dual-criticality systems. In: Proceedings of the Ninth ACM International Conference on Embedded Software, ser. EMSOFT, pp. 253–262. ACM, New York (2011)

12. Li, Z., Ren, S., Quan, G.: Dynamic reservation-based mixed-criticality task set scheduling. In: 2014 IEEE International Conference on High Performance Computing and Communications, 2014 IEEE 6th International Symposium on Cyberspace Safety and Security, 2014 IEEE 11th International Conference on Embedded Software and Systems (HPCC, CSS, ICESS), pp. 603–610, August 2014

13. Audsley, N.C.: On priority asignment in fixed priority scheduling. Inf. Process. Lett. **79**(1), 39–44 (2001)

14. Vestal, S.: Preemptive scheduling of multi-criticality systems with varying degrees of execution time assurance. In: Proceedings of the 28th IEEE International Real-Time Systems Symposium, ser. RTSS 2007, pp.239–243 (2007)

15. Baruah, S., Chattopadhyay, B.: Response-time analysis of mixed criticality systems with pessimistic frequency specification. In: 2013 IEEE 19th International Conference on Embedded and Real-Time Computing Systems and Applications (RTCSA), pp. 237–246, August 2013

16. Baruah, S., Bonifaci, V., D'Angelo, G., Li, H., Marchetti-Spaccamela, A., Megow, N., Stougie, L.: Scheduling real-time mixed-criticality jobs. IEEE Trans. Comput. **61**(8), 1140–1152 (2012)

17. de Niz, D., Lakshmanan, K., Rajkumar, R.: On the scheduling of mixed-criticality real-time task sets. In: 30th IEEE Real-Time Systems Symposium, 2009, RTSS 2009, pp. 291–300, December 2009

18. Liu, C.L., Layland, J.W.: Scheduling algorithms for multiprogramming in a hard-real-time environment. J. ACM **20**(1), 46–61 (1973)

19. Joseph, M., Pandya, P.K.: Finding response times in a real-time system. Comput. J. **29**(5), 390–395 (1986)

20. Baruah, S., Burns, A., Davis, R.: Response-time analysis for mixed criticality systems. In: 2011 IEEE 32nd Real-Time Systems Symposium (RTSS), pp. 34–43 (2011)

21. Bini, E., Buttazzo, G.: Measuring the performance of schedulability tests. Real-Time Syst. **30**(1–2), 129–154 (2005)

A Generalized Ant Routing Mechanism Framework in Mobile P2P Networks

Dapeng Qu[1], Di Zhang[1], Dengyu Liang[1], Xingwei Wang[2($\boxtimes$)],
and Min Huang[3]

[1] College of Information, Liaoning University, Shenyang, China
[2] College of Software, Northeastern University, Shenyang, China
dapengqu@lnu.edu.cn
[3] College of Information Science and Engineering, Northeastern University,
Shenyang, China

Abstract. With the rapid development of mobile peer-to-peer networks (MP2P) and the diversification of users' demand, routing mechanism has become an important research focus in MP2P. Because of the characteristic of self-organization, Ant Colony Optimization (ACO) has been widely applied in designing routing mechanism in various networks. In ant routing protocols, the node routes by perceiving the pheromone in networks and deposits pheromone to direct the subsequent routing. In this paper, the relationship between ACO and routing protocols in MP2P is discussed, and some representative ant routing protocols are chosen to compare and analyze to get the general ant routing principles. Then a generalized ant routing mechanism framework is proposed, and a corresponding ant routing protocol as the generalized solution is produced and its performance is shown through simulation experiments.

Keywords: Ant routing mechanism · Ant Colony Optimization · Routing framework · Mobile P2P networks

1 Introduction

With the rapid development of wireless communication technology and huge success of peer-to-peer model in wired networks, mobile peer-to-peer networks (MP2P) emerged gradually and attracted more and more attention. The main objective of MP2P is achieving resources sharing and cooperative service through direct interaction among mobile terminals. Because there is not a unified definition, and for the convenience of study, we take the definition given by Mario Gerla, etc. in "P2P MANETs - New Research Issues". That is, MP2P takes peer-to-peer mode operation over Mobile Ad Hoc Network (MANET) [1]. Therefore, MP2P has some characteristics, such as self-organization, dynamic topology and constrained resources, etc., which make routing mechanism design difficultly [2].

Ant Colony Optimization (ACO) is inspired by the ant foraging behavior in nature [3]. Ants achieve the communication by perceiving and depositing pheromone which is a chemical substance with special scent. For the characteristics of self-organization, dynamic adjusting, etc., ACO has been successfully applied in some combination

© Springer International Publishing Switzerland 2016
O. Gervasi et al. (Eds.): ICCSA 2016, Part II, LNCS 9787, pp. 157–169, 2016.
DOI: 10.1007/978-3-319-42108-7_12

optimization problems, such as traveling salesman problem, knapsack problem, and arrangements problem, and so on [4]. In 1998, Marco Dorigo and Gianni A. Di Caro proposed AntNet [5]. As the first routing protocol based on ACO, AntNet attracts wide attention from researchers and it is found that the characteristic of ACO is suitable for computer networks, especially designing adaptive routing protocols in wireless networks [6]. Moreover, some other bio-inspired algorithms, such as bee colony algorithm [7], particle swarm algorithm [8] and fish swarm algorithm [9] and so on, have also been applied to design routing protocols [10–12].

There have been plenty of related research on ant routing protocols in mobile networks, but most of them often just apply ACO directly and lack relevant in-depth analysis. The existed surveys or reviews [13–15] just list various ant routing protocols and are short of sufficient comparison and analysis. Therefore, in this paper, we compare and analyze some representative ant routing protocols after discussed the relationship between ACO and routing protocols, and propose a generalized ant routing mechanism framework and a corresponding general routing protocol to help future related research. Without loss of generality, we cover ant routing protocols in MANET [16], Wireless Sensor Networks (WSN) [17], Delay/Disruption Tolerant Networks (DTN) [18], and Vehicular Ad hoc Network (VANET) [19], etc., for they are all self-organized, dynamic and resource constrained. The contributions of our work are summarized as follows:

- We compare and analyze eight representative ant routing protocols, and clarify their basic running mechanisms and corresponding characteristics, including the used ant types, pheromone strength and usage, and route way and so on.
- We propose a generalized ant routing mechanism framework. It is capable of producing the generalized solution which can be adapted and tailored to the actual requirements. Moreover, it offers help for the application of ACO in other domains and provides a useful reference for the applications of other SI algorithms.
- We produce a general ant routing protocol which is taken as the generalized solution to the framework, and it establishes an important foundation for future research and guides the future related routing protocol design.

The rest of this paper is organized as follows. The basic ideas of ACO and design challenges of routing protocols in MP2P are analyzed in Sect. 2. The representative ant routing protocols are reviewed and some general ant routing principles are drawn in Sect. 3. The generalized ant routing mechanism framework is proposed in Sect. 4. A generalized solution is produced from the above framework and simulation experiments are analyzed in Sect. 5. Conclusion is drawn in Sect. 6 with future research anticipated.

2   ACO and Routing Protocols

2.1   Basic Idea of ACO

As a meta-heuristic algorithm, ACO achieves optimization by ants perceiving and depositing pheromone, which is an indirect communication way and takes environment change as medium. Therefore, the basic idea of ACO is how to deposit and perceive pheromone, and that of ant routing mechanism is how to perceive pheromone to help routing and deposit pheromone to guide routing. The detailed analysis is as follows:

Pheromone Perception and Solution Construction. The ants move among nodes and each ant chooses the next node among the allowed neighborhood of the current node in probabilistic or deterministic way based on the perceived pheromone and some heuristic information. When an ant from the source node reaches the destination node, it constructs a solution between them (See details in Sect. 3.2).

Pheromone Deposition and Solution Update. When an ant reaches the destination node, its record information would be extracted and the corresponding solution constructed would be evaluated. Then the destination node generates a new ant which inherits the records of the corresponding old ant. The new ant will return to the source node in reverse direction and deposit pheromone based on the evaluation along its way to update future solutions (See details in Sect. 3.2).

To improve performance, all the existed ant routing protocols add some additional strategies, such as records more information in ants to reflect the quality of routing paths more accurately, and introduces some new kinds of ants to exploit the routing paths better, and so on.

2.2   Design Challenge of Routing Protocols in MP2P

There are some challenges for designing routing protocols in MP2P as follows:

- Constrained node energy: The nodes are battery powered, some are non-rechargeable, and have constrained energy. The node without energy has to quit the networks. Moreover, not only the individual node energy, but also the balance of overall energy should be considered. Only in this way, can the energy hole be avoided and the network lifetime be prolonged.
- Self-organized network structure: Because of the movement of nodes and the constrained energy, the nodes might incur in failures or exhaust their energy, and the network topology structure would change at any time. An effective routing protocol should be resilient to such dynamic and generally unpredictable variations and sustain a long-term availability for essential network services.
- High computational complexity: With the emergence of various new kinds of networked applications, especially multimedia based, and the diversified demands of users, the Quality of Service (QoS) should be supported. It is shown that QoS routing with two and more additive constraints is NP-Complete problem [20]. Moreover, the nodes are typically equipped with a low-end Central Processing Unit (CPU) and have constrained memory. Therefore, an effective routing protocol should have minimal processing overhead.

3 Survey of Ant Routing Protocols

The excellent performance of ACO attracts wide focus from researchers and a large amount of ant routing protocols have been proposed. However, restricted by space, in this section, we choose and compare eight representative ant routing protocols, and draw some general principles.

3.1 Discussion of Representative Ant Routing Protocols

Ant colony based Routing Algorithm (ARA) [21] consists of route discovery phase, route maintenance phase and route failure handling phase. In the route discovery phase, a Forward Ant (FANT) is broadcasted by the sender and relayed by the neighbors of the sender to their neighbors. When a FANT reaches the destination node, the record information is extracted and FANT is destroyed. Then the destination node creates a corresponding Backward Ant (BANT) which returns to the source node along the way that the corresponding FANT takes in a reverse way. The FANTs and BANTs establish the pheromone track to the source and destination node respectively. When a node relays a data packet toward the destination to a neighbor node, it increases the corresponding pheromone value by a constant value. That is, the path is strengthened by the data packets. If a node finds a route failure, it searches an alternative link, informs its neighbors, and backtracks to the source node. ARA is the first ant routing protocol proposed for wireless network.

Ant Routing Algorithm for Mobile Ad hoc networks (ARAMA) [22] is self-configured by measuring and collecting nodes' and links' parameters in the nodes' indices. Moreover, Negative Backward Ants (NBANT) are used to deemphasize an unwanted path by reducing the probability of visiting, and Destination Trail Ants (DTANT) generated by the destination node are used to reduce the connection setup time and insure the delivery of data by moving randomly and modifying the pheromone values to be biased toward the destination.

Ant Routing Algorithm for Mobile ad hoc Networks (AntHocNet) [23] is a hybrid multipath algorithm which contains reactive path discovery phase and proactive path maintain phase. In the former, reactive FANTs and BANTs are launched to find multiple paths. In the latter, while a data session is going on, the paths are probed, maintained and improved proactively by proactive FANTs and BANTs. Moreover, the path repair FANTs and BANTs are used to repair the failed paths locally.

Hybrid ant colony optimization (HOPNET) [24] is based on ACO and zone routing framework of bordercasting. The network is divided into zones which are the nodes' local neighborhood. Each node has two routing tables, namely Intrazone routing table (IntraRT) and Interzone routing table (InterRT). IntraRT is proactively maintained by internal FANTs so that a node can obtain a path to any nodes within its zone quickly and InterRT is setup reactively when a routing path outside a zone is required.

Ant-based multi-QoS routing metric algorithm (AntSensNet) [25] builds a hierarchical structure on Wireless Multimedia Sensor Networks (WMSN). It introduces routing models with four QoS metrics, namely delay, packet loss, energy and memory, and then chooses suitable paths to meet QoS requirements of an application while simultaneously reduces the consumption of constrained resources as much as possible. There are four values in a pheromone table and each value is a pheromone trail concentration for each QoS metric. The novel pheromone table contributes to the evaluation of each QoS metric made by the sink.

Simple Ant Routing Algorithm (SARA) [26] aims at offering a low overhead routing solution. In route discovery phase, a Controlled Neighbor Broadcast (CNB) mechanism is used. In route maintenance phase, only data packets are used to refresh the paths of active sessions. Finally, in route repair phase, an incremental deep search procedure is utilized as a way of restricting the number of nodes used to repair a route.

Greedy Ant (GrAnt) [27] combines ACO and a greedy transition rule which allows to select the most promising forwarder nodes or exploit previously found good paths. GrAnt takes advantage of the population-based search of ACO and the rapid adaptation of its learning framework.

Annovative ACO based Routing Algorithm (ANTALG) [28] considers a random selection of source and destination nodes and exchanges the ants between them, and creates the pheromone tables and data structures to record the trip time of nodes during ants movement. ANTALG operates using reinforcement learning to define a model of optimal routing behavior which is not merely searching shortest-hop paths, but also considers the quality of the links making up those paths.

3.2   General Ant Routing Principles

From the above subsection, we could draw some general ant routing principles:

- Almost each ant routing protocol builds the routing paths between the source node and the destination node by the FANTs and BANTs. While some generate FANTs reactively to save energy, some generate FANTs proactively to get the routing in time, and others generate FANTs hybridly.
- To improve performance, most ant routing protocols add new kinds of ants, for example, DBANT and DTANT in ARAMA are to accelerate search. Moreover, the pheromone deposited by ants implies more information, such as bandwidth, delay, packet loss, and node energy and so on. To simulate pheromone aging, some protocols fuse evaporation into deposition and the other design an independent process. They are formulated in Eqs. (1) – (2) respectively.

$$\tau_{i,j}^d = \begin{cases} (1-\gamma)\tau_{i,j}^d + g(\tau_{i,j}^d), & \text{incoming link} \\ (1-\gamma)\tau_{i,k}^d, & \text{otherwise} \end{cases} \tag{1}$$

$$\tau_{i,j}^d = (1-\gamma)\tau_{i,j}^d \tag{2}$$

where $\tau_{i,j}$ is the pheromone value corresponding to neighbor node v_j for destination node v_d at node v_i, γ is the evaporation coefficient and $0 \leq \gamma \leq 1$, and $g(\tau_{i,j})$ is a function in $\tau_{i,j}$, and it increases with the increase of $\tau_{i,j}$. In Eq. (1), the incoming link is strengthened by $g(\tau_{i,j})$ and other links are decreased by the evaporation process.

In Eq. (2), each link is decreased periodically. The evaporation process copies with the dynamic topology in wireless networks well.

- Not only pheromone, but also heuristic information is the basis of data packets route. Moreover, there are three routing ways, namely probability, determinacy and hybrid. While the former means that the node routes among several candidate paths probabilistically based on their pheromone value, and it naturally helps the load balancing, the middle one means that the node routes following the path with maximum pheromone value, and it contributes exploration of the built routing paths, and the latter combines them and takes their advantages simultaneously. They are formulated in Eqs. (3) – (5) respectively.

$$p_{i,j}^d = \begin{cases} \dfrac{f(\tau_{i,j}^d, \eta_{i,j}^d)}{\sum\limits_{k \in N_{i,allowed}^d} f(\tau_{i,k}^d, \eta_{i,k}^d)}, & j \in N_{i,allowed}^d \\ 0, & otherwise \end{cases} \tag{3}$$

$$p_{i,j}^d = \begin{cases} 1, & (f(\tau_{i,j}^d, \eta_{i,j}^d) = max(f(\tau_{i,k}^d, \eta_{i,k}^d)))\&\&(j, \forall k \in N_{i,allowed}^d) \\ 0, & otherwise \end{cases} \tag{4}$$

$$p_{i,j}^d = \begin{cases} 1, & (\alpha \geq P_{th})\&\&(f(\tau_{i,j}^d, \eta_{i,j}^d) = max(f(\tau_{i,k}^d, \eta_{i,k}^d)))\&\&(j, \forall k \in N_{i,allowed}^d) \\ \dfrac{f(\tau_{i,j}^d, \eta_{i,j}^d)}{\sum\limits_{k \in N_{i,allowed}^d} f(\tau_{i,k}^d, \eta_{i,k}^d)}, & (\alpha < P_{th})\&\&(j \in N_{i,allowed}^d) \\ 0, & otherwise \end{cases} \tag{5}$$

where $p_{i,j}^d$ is the probability of selecting neighbor v_j for the destination v_d at node v_i, $\eta_{i,j}^d$ is the local heuristic value of link $e_{i,j}$, and can represent neighbor queue delay, remaining battery energy, etc., and $0 \leq \eta_{i,j}^d \leq 1$, $f(\tau_{i,j}^d, \eta_{i,j}^d)$ is a function of $\tau_{i,j}^d$ and $\eta_{i,j}^d j$. $N_{i,allowed}^d$ is the set of all feasible neighbor nodes defined by the ant's information and the routing constraints (i.e., the guarantee of loop free, the QoS constraint, etc.). P_{th} is a threshold value and α is a random number. The choice of probabilistic or deterministic routing way is decided by the relationship between P_{th} and α.

- From the viewpoint of development of ant routing protocols, more and more factors are taken into account. In the beginning, only the hop count is considered in ARA, and heuristic information is considered by ARAMA, GrANT and ANTALG later. With the wide application of mobile networks, energy issue and various QoS parameters attract attention.

- It is noteworthy that ant routing protocols have some similarities to the common routing protocols. For example, ARA and Ad hoc On-Demand Distance Vector routing protocol (AODV) [29] are two representative and basic routing protocols in individual domain. The FANT in ARA is similar to the Route Request packets (RREQ) in AODV for their on-demand generation, broadcast moving way, and extermination by the destination node. Similarly, the BANT in ARA is similar to the Route Reply packets (RREP) in AODV.

- Finally, we can draw that each ant routing protocol has individual advantage and disadvantage based on No Free Lunch Theory [30].

The detailed comparisons of representative ant routing protocols are as shown in Table 1.

Table 1. Characteristic comparison of representative ant routing protocols

Ant routing protocol	Ants	Pheromone strengthen	Pheromone information	Route information
ARA	FANT, BANT	FANT BANT data packets	hop	pheromone
ARAMA	FANT, BANT NBANT, DTANT	BANT NBANT DTANT	hop, energy	pheromone + heuristic
AntHocNet	reactive FANT and BANT proactive FANT and BANT path repair FANT and BANT	reactive BANT proactive BANT path repair BANT	delay	pheromone
HOPNET	Internal FANT External FANT External BANT	Internal FANT External FANT External BANT	delay	pheromone
AntSensNet	CANT, FANT, BANT MANT, DANT	BANT MANT	delay, packet loss, energy, memory	pheromone
SARA	FANT, BANT C_FANT R_FANT, R_BANT	FANT, BANT R_FANT, R_BANT	delay	pheromone + hop
GrAnt	FANT, BANT	BANT	quality of paths, hop	pheromone + heuristic
ANTALG	FANT, BANT PFANT, NANT, EANT	BANT PFANT NANT EANT data packet	delay	pheromone + heuristic

4 Generalized Ant Routing Mechanism Framework

From the above section, we could draw three key schemes in ant routing protocols:

- Various kinds of ants are used to build, maintain and recover routing paths. For different objectives, each kind of ants has different characteristics, but they have some identical properties and experience similar lifetime.
- There is a pheromone table which is the heart of ant routing protocols maintained by each node to denote the quality of paths. The numeric value reflects the quality of corresponding routing paths more accurately than the binary state in common routing protocols.
- Each node routes the data packets based on pheromone and/or heuristic information. The probabilistic and deterministic routing way have their own advantages respectively.

Therefore, we propose a generalized ant routing mechanism framework, as shown in Fig. 1, which consists of three modules, Ant Module, Node Module and Data Packet Module. They implement the architecture and operation of an ant routing protocol in wireless network. Figure 1 summarizes the characteristics and functions of each module, and the detailed explanation is provided as follows, which is based on the general principles drawn in Sect. 3. It not only provides a common reference framework to describe, compare and analyze different ant routing protocols, but also defines a generalized architecture to guide the design of ant routing protocols in future.

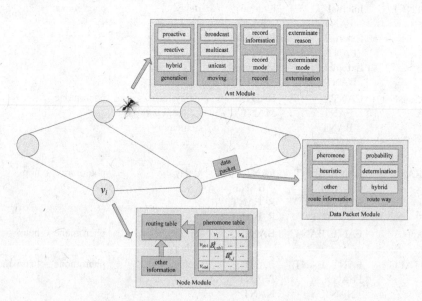

Fig. 1. Generalized ant routing mechanism framework

4.1   Ant Module

This module is the core in ant routing protocols, deals with various related issues of ants and consists of generation, moving, record and extermination. The four units cover the whole lifetime of an ant, and each ant has its own settings in each unit. There are different ants in each routing protocol. For example, there are CANT, FANT, BANT, MANT and DANT in AntSensNet. Even the ants sharing the same name in different routing protocols have own strategies for their different objectives. For example, both FANT in SARA and FANT in ARA are generated by the source node reactively, record the hop count on its way, and are exterminated by the destination node. While the former moves in a CNB way and the latter moves in a broadcast way.

4.2   Node Module

This module deals with nodes and contains routing table, pheromone table and other information which is possible additional data structure holding statistic information about node and network state to evaluate paths and make decision. The routing table which depends on pheromone table and other information serves for data packets and ants. For example, the FANT in ANTALG probabilistically moves based on the normalized sum of pheromone and heuristic value. The modification and update of the information in Node Module are realized through the ant and data packets moving and/or the regular evaporation process.

4.3   Data Packet Module

This module deals with the data packets whose forwarding is the key issue of a routing protocol. There are two main units, namely route information and route way. As already discussed in above subsections, data packets are routed based on pheromone, and/or heuristic, other information, such as energy, trust value and so on.

5   A Generalized Solution

5.1   A Generalized Solution from the Framework

A generalized solution can be produced from the above generalized framework, and the solution can be adapted and tailored to the actual requirements. Thus, we produce a generalized solution based on the analysis of different ant routing protocols in Sect. 3.1 and the general ant routing principle in Sect. 3.2. The source node generates the FANT reactively and the destination node evaluates the corresponding routing paths based on the various information recorded by the received FANTs. Then the corresponding BANTs which are generated by the destination node go to the source node and deposit the corresponding pheromone.

From Table 1, we can get the pheromone information mainly to reflect the QoS parameters and energy. We divided them into two types, namely "the bigger the better" and "the smaller the better" [31]. Obviously, bandwidth and remaining energy belong to the former, and delay, delay jitter and error rate belong to the latter. We design two evaluation functions as follows:

$$db(x) = \begin{cases} 1, & x > x_U \\ e^{-e^{(x-x_U)/(x_L-x_U)}}, & x_L \leq x \leq x_U \\ 0, & x < x_L \end{cases} \tag{6}$$

$$ds(x) = \begin{cases} 1, & x < x_L \\ 1 - e^{-e^{(x-x_U)/(x_L-x_U)}}, & x_L \leq x \leq x_U \\ 0, & x > x_U \end{cases} \tag{7}$$

In Eq. (6), x is an independent variable and represents the parameter to be evaluated; x_l and x_u represent the lower and upper bound of its value respectively. db(x) increases with x, and the bigger the value of the db(x), the higher the quality of x is. The situation of the Eq. (7) is just the opposite to that of Eq. (6). It is worthy to note that, to the QoS parameters, the value of the bounds could be determined according to the application type [32]. The bounds of remaining energy ratio are set as 0 and 1 to simplify calculation.

Now we get the evaluation value of each routing path from the source node to the destination node, and the bant deposits the value on the nodes along its paths as the pheromone information. That is, some intermediate nodes have the various pheromone information denoting each candidate routing path to the destination node. They are denoted as $\tau(en)_{i,j}^d$, $\tau(bw)_{i,j}^d$, $\tau(dl)_{i,j}^d$, $\tau(jt)_{i,j}^d$, and $\tau(er)_{i,j}^d$.

We design a generalized ant routing protocol (GANT) based on the above analysis. The nodes only maintain the delay pheromone information, and there is not any route maintenance information. More pheromone information is easy to add in GANT based on the actual requirements.

5.2   Simulation Study

We use NS2 [33] to simulate our proposed GANT and compare it with the well known AODV. There are 50 mobile nodes in a rectangular field with 1500 m*600 m. Each node has a radio propagation range of 250 m. The node mobility takes random waypoint model, and the random speed is uniformly distributed between 0 and 20 m/s. Once the destination node is reached, and another destination is randomly chosen after a random pause time. User Datagram Protocol (UDP) and Constant Bit Rate (CBR) are used as the transport protocol and the traffic respectively. The size of a packet data payload is 512 bytes, and the generation rate is 1 packet/s. The traffic is generated between a source node and a destination node randomly. A simulation executes for 600 s.

We use two metrics to do performance evaluation and comparison: Packet Delivery Ratio (PDR), i.e., the ratio of the number of packets successfully received by the destination node to the total number delivered by the source node; Average Routing Overhead (ARO), i.e., the ratio of the number of control packets to the number of data packets.

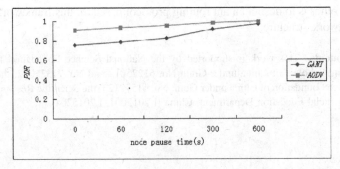

(a) Packet delivery ratio

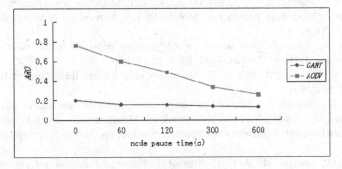

(b) Average routing overhead

Fig. 2. Performance comparison of GANT and AODV

As shown in Fig. 2(a), GANT gets lower PDR than AODV, but with the increment of node pause time, the difference gradually become smaller and smaller. The reason is that the simple mechanism of GANT weaken its ability to manage the dynamic topology. However, in the static topology, the simple and effective mechanism makes GANT get the similar performance with AODV. As shown in Fig. 2(b), GANT gets better performance than AODV in ARO. The simple and effective mechanism is the main reason.

6 Conclusions and Future Work

In this paper, a generalized ant routing mechanism framework is proposed for MP2P. Some representative ant routing protocols are chosen and discussed to compare and analyze to get the general ant routing principles for various wireless networks. A generalized ant routing mechanism framework is proposed to provide the unified reference and help future research on routing protocol. Moreover, a general ant routing protocol is designed and simulation experiments are taken to verify its performance.

In the future, we will improve our work further from two directions. One is to exploit the effectiveness of ACO in designing routing protocols from the viewpoint of theory. The other is to design an ant routing protocol based on this framework for new wireless network structure.

Acknowledgments. This work is supported by the National Science Foundation for Distinguished Young Scholars of China under Grant No. 61225012 and No. 71325002; the National Natural Science Foundation of China under Grant No. 61572123; the Scientific Research Fund of Liaoning Provincial Education Department, China (L2013001, L2015204).

References

1. Gerla, M., Lindemann, C., Rowstron, A.I.T.: P2P MANETs-new research issues. In: Perspectives Workshop: Peer-to-Peer Mobile Ad Hoc Networks – New Research Issues, pp. 9–12 (2005)
2. Malatras, A.: State-of-the-art survey on P2P overlay networks in pervasive computing environments. J. Netw. Comput. Appl. **55**, 1–23 (2015)
3. Dorigo, M., Birattari, M., Stützle, T.: Ant colony optimization. IEEE Comput. Intell. Mag. **1**(4), 28–39 (2007)
4. Dorigo, M., Stützle, T.: Ant colony optimization: overview and recent advances. In: Gendreau, M., Potvin, Y. (eds.) Handbook of Metaheuristics. International Series in Operations Research & Management Science, vol. 146, pp. 227–263. Springer, New York (2010)
5. Caro, G.D., Dorigo, M.: AntNet: distributed stigmergetic control for communications networks. J. Artif. Intell. Res. **9**, 317–365 (1998)
6. Ducatelle, F., Caro, G.A.D., Gambardella, L.M.: Principles and applications of swarm intelligence for adaptive routing in telecommunications networks. Swarm Intell. **4**(3), 173–198 (2010)
7. Bitam, S., Mellouk, A.: QoS swarm bee routing protocol for vehicular ad hoc networks. In: Proceedings of IEEE International Conference on Communications (ICC 2011), pp. 1–5. IEEE Press, Kyoto (2011)
8. Khouadjia, M.R., Jourdan, L., Talbi, E.: Adaptive particle swarm for solving the dynamic vehicle routing problem. In: Proceedings IEEE/ACS International Conference on Computer Systems and Applications (AICCSA 2010), pp. 1–8. IEEE/ACM Press, Hammamet (2010)
9. Shan, X., Jiang, M., Li, J.: The routing optimization based on improved artificial fish swarm algorithm. In: Proceedings of IEEE World Congress on Intelligent Control and Automation (WCICA 2006), pp. 3658–3662. IEEE Press, Da Lian (2006)
10. Babaoglu, O., Canright, G., Deutsch, A., Caro, G.A., Ducatelle, F., Gambardella, L.M., Ganguly, N., Jelasity, M., Montemanni, R., Montresor, A., Urnes, T.: Design patterns from biology for distributed computing. ACM Trans. Auton. Adapt. Syst. **1**(1), 1–40 (2006)
11. Meisel, M., Pappas, V., Zhang, L.: A taxonomy of biologically inspired research in computer networking. Comput. Netw. **54**(6), 901–916 (2010)
12. Saleem, M., Caro, G.A.D., Farooq, M.: Swarm intelligence based routing protocol for wireless sensor networks: survey and future directions. Inf. Sci. **181**(20), 4597–4624 (2011)
13. Shokrani, H., Jabbehdari, S.: A survey of ant-based routing algorithms for mobile ad-hoc networks. In: Proceedings of International Conference on Signal Processing Systems, pp. 323–329. IEEE Press, Singapore (2009)

14. Marwaha, S., Indulska, J., Portmann, M.: Biologically inspired ant-based routing in mobile ad hoc networks (MANET): a survey. In: Proceedings of Symposia and Workshops on Ubiquitous, Autonomic and Trusted Computing (UIC-ATC 2009), pp. 12–15. IEEE Press, Brisbane (2009)
15. Singh, G., Kumar, N., Verma, A.K.: Ant colony algorithms in MANETs: a review. J. Netw. Comput. Appl. **35**(6), 1964–1972 (2012)
16. Boukerche, A., Turgut, B., Aydin, N., Ahmad, M.Z., Boloni, L., Turgut, D.: Routing protocols in ad hoc networks: a survey. Comput. Netw. **55**(13), 3032–3080 (2011)
17. Akkaya, K., Younis, M.: A survey on routing protocols for wireless sensor networks. Ad Hoc Netw. **3**(3), 325–349 (2005)
18. Cao, Y., Sun, Z.: Routing in delay/disruption tolerant networks: a taxonomy, survey and challenges. IEEE Commun. Surv. Tutorials **15**(2), 654–677 (2012)
19. Sharef, B.T., Alsaqour, R.A., Ismail, M.: Vehicular communication ad hoc routing protocols: a survey. J. Netw. Comput. Appl. **40**, 363–396 (2014)
20. Wang, Z., Crowcroft, J.: QoS routing for supporting resource reservation. IEEE J. Sel. Areas Commun. **14**(7), 1228–1234 (1996)
21. Gunes, M., Sorges, U., Bouazizi, I.: ARA the ant-colony based routing algorithm for MANETs. In: Proceedings of International Conference on Parallel Processing Workshops (ICPPW 2002), pp. 1–7. IEEE Press, British Columbia (2002)
22. Hussein, O.H., Saadawi, T.N., Lee, M.J.: Probability routing algorithm for mobile ad hoc networks' resource management. IEEE J. Sel. Areas Commun. **23**(12), 2248–2259 (2005)
23. Caro, G.D., Ducatelle, F., Gambardella, L.M.: AntHocNet: an adaptive nature-inspired algorithm for routing in mobile ad hoc networks. Eur. Trans. Telecommun. (ETT) **16**(5), 443–455 (2005). Special Issue on Self Organization in Mobile Networking
24. Wang, J., Osagie, E., Thulasiraman, P., Thulasiram, R.K.: HOPNET: a hybrid ant colony optimization routing algorithm for mobile ad hoc network. Ad Hoc Netw. **7**(4), 690–705 (2009)
25. Cobo, L., Quintero, A., Pierre, S.: Ant-based routing for wireless multimedia sensor networks using multiple QoS metrics. Comput. Netw. **54**(17), 2991–3010 (2010)
26. Correia, F., Vazao, T.: Simple ant routing algorithm strategies for a (multipurpose) MANET model. Ad Hoc Netw. **8**(8), 810–823 (2010)
27. Vendramin, A.C.K., Munaretto, A., Delgado, M.R., Viana, A.C.: GrAnt: inferring best forwarders from complex networks' dynamics through a greedy ant colony optimization. Comput. Netw. **56**(3), 997–1015 (2012)
28. Singh, G., Kumar, N., Verma, A.K.: ANTALG: an innovative ACO based routing algorithm for MANETs. J. Netw. Comput. Appl. **45**, 151–167 (2014)
29. Perkins, C.E., Royer, E.M., Das, S.R., Marina, M.K.: Performance comparison of two on-demand routing protocols for ad hoc networks. IEEE Pers. Commun. **8**(1), 16–28 (2001)
30. Wolpert, D.H., Macready, W.G.: No free lunch theorems for optimization. IEEE Trans. Evol. Comput. **1**(1), 67–82 (1997)
31. Wang, X., Qu, D., Huang, M., Li, K., Das, S., Zhang, J., Yu, R.: Multiple many-to-many multicast routing scheme in green multi-granularity transport networks. Comput. Netw. **93**(1), 225–242 (2015)
32. End-user multimedia QoS categories ITU-T G.1010 (2001)
33. NS2. http://www.isi.edu/nsnam/ns/

WACA: WearAble Device Based Home Control Assistant Framework

Bonhyun Koo[1], Simon Kong[1], Hyejung Cho[1], and Lynn Choi[2(✉)]

[1] Convergence System Team, DMC R&D Center, Samsung Electronics,
Seoul R&D Campus, 34, Seongchon-Gil, Seocho-Gu, Seoul 06765, Korea
{bonhyun.koo,sinon.kong,cho115}@samsung.com
[2] The Department of Electronics and Computer Engineering, Korea University,
Anam-Dong, Sungbuk-Ku, Seoul 136-701, Korea
lchoi@korea.ac.kr

Abstract. In this paper, we have analyzed requirements and pain points of legacy wearable-based IoT and smart home services. In conventional system, we are required to wear additional accessory devices, and to move to specific areas to control and manage the home devices. In order to solve these restrictions, we have implemented WACA framework by using well-known general wearable devices. We also present the gesture-interaction based device control architecture using the wearable devices to provide more efficient and valuable smart home services. Based on the experiment results, we show that the proposed system has the potential as wearable device-based universal control assistance architecture.

Keywords: Wearable · Wearable device · Smart watch · Device control · Smart home · Smart building

1 Introduction

With the advent of various IoT services, technologies to connect to a single one network in conjunction with various devices have been introduced. IoT is a service technology to enable the seamless connection of network-based IT devices. Representative IoT service required area is our home space which is composed of various home appliances such as TV, lighting, and washing machine. One of representative control technology for these devices in smart home is wearable based control service. Recently a variety of IT companies like Samsung Electronics, Sony, and Google has been introduced various type of wearable devices such as Watch-phone and Glass types in MWC2014. These devices are mainly focused on the health application services or Augmented Reality (AR) services for providing incidental information by using its internal multiple application sensors.

In this paper, we talks about the previous related work on these wearable devices based home control services in Sect. 2. In Sects. 3 and 4, we propose a novel system control framework for Smart home in order to extend the role of wearable devices as the main controller and share the information among these devices.

© Springer International Publishing Switzerland 2016
O. Gervasi et al. (Eds.): ICCSA 2016, Part II, LNCS 9787, pp. 170–179, 2016.
DOI: 10.1007/978-3-319-42108-7_13

2 Motivation and Considerations

According to the increasing popularity of wearable devices, research results in the field of various applications are introduced based on these devices. However, these devices are required for additional equipment like gloves, ring, and so on. It has also many restrictions for the physical spaces to use the control service. In this section, we investigate the previous research problems in the domain of wearable based Home control and describe the necessary factors to solve or mitigate the problems.

Figure 1 shows the different types of wearable units for the representative Home control services [1]. Figure 1(A) shows the 'Gaze Tracking Device', proposed by Sang-Myung Univ., for a air-condition control system. However, users should wear the headband-type device to obtain the distance for detecting eye movements. This should be improved in terms of the user convenience and interface. Figure 1(B) is a research which is proposed by Ottawa University, with respect to the technique that can control CE device (e.g. TV) by using gesture recognition with 3D depth camera easily. Unfortunately, there is a restriction in terms of tracking area of the 3D camera so as to recognize the gesture pattern [2]. Figure 1(C) shows Ring-type controller, which is called 'Magic Ring', proposed by Aizu University. This could cause inconvenience to finger movements due to the additional H/W equipment such as RF transmission units [3]. Figure 1(D) illustrates a research 'Rubber Glove Input Interface' of Ryukoku University which can control TV by Glove-type wearable device [4, 5]. Although each finger movement can be used as a method of classifying various control commands, but it should be improved in terms of wearing of glove device essentially.

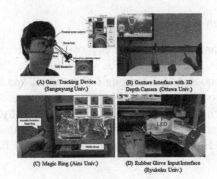

(A) Gaze Tracking Device (Sangmyung Univ.)

(B) Gesture Interface with 3D Depth Camera (Ottawa Univ.)

(C) Magic Ring (Aizu Univ.)

(D) Rubber Glove Input Interface (Ryukoku Univ.)

Fig. 1. Home control applications with various wearable devices

In order to provide more efficient Smart Home solution in our home domain, we propose a wearable-based control assistant framework-'WACA' that allows users to control CE devices by minimizing of wearing additional equipment and overcoming space constraints. We also introduce how to share information among multiple wearable devices and extend the controlling commands by applying WACA framework.

Figure 1 shows the different types of wearable units for the representative Home control services [1]. Figure 1(A) shows the 'Gaze Tracking Device', proposed by Sang-Myung Univ., for a air-condition control system. However, users should wear the headband-type device to obtain the distance for detecting eye movements. This should be improved in terms of the user convenience and interface. Figure 1(B) is a research which is proposed by Ottawa University, with respect to the technique that can control CE device (e.g. TV) by using gesture recognition with 3D depth camera easily. Unfortunately, there is a restriction in terms of tracking area of the 3D camera so as to recognize the gesture pattern [2]. Figure 1(C) shows Ring-type controller, which is called 'Magic Ring', proposed by Aizu University. This could cause inconvenience to finger movements due to the additional H/W equipment such as RF transmission units [3]. Figure 1(D) illustrates a research 'Rubber Glove Input Interface' of Ryukoku University which can control TV by Glove-type wearable device [4, 5]. Although each finger movement can be used as a method of classifying various control commands, but it should be improved in terms of wearing of glove device essentially.

In order to provide more efficient Smart Home solution in our home domain, we propose a wearable-based control assistant framework-'WACA' that allows users to control CE devices by minimizing of wearing additional equipment and overcoming space constraints. We also introduce how to share information among multiple wearable devices and extend the controlling commands by applying WACA framework.

3 Proposed System Architecture

In this section, we present the proposed WACA framework architecture, and then describe Bayesian-estimation pattern recognitions based on 'Snap' motion for CE device control.

3.1 Framework Design

The proposed WACA framework architecture is composed of wearable devices, Smartphones, gateway devices, and end devices as shown in Fig. 2. Wearable devices (A) consist of Watch-type phones, e.g. Samsung Galaxy Gear, and Fit. Smartphones (B) take care of relaying of control commands from wearable devices to gateways. Gateway device (C) is mainly in charge of controlling End devices. Finally, End devices (D) are operated by receiving and executing a control command. At this time, we adopted a de facto ZigBee-based Home Automation Profile (HAP) as the interface between

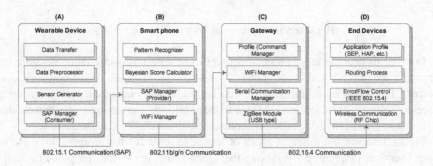

Fig. 2. A component design of the proposed architecture

Gateway and End Device uses. The interface between Smartphones and wearable devices uses Bluetooth based SAP (Samsung Accessory Profile).

3.2 Gesture Patterns

In our proposed framework, we used two watch phones (Samsung Galaxy Gear) on both wrists respectively so as to control target devices. The right Gear is used for controlling devices by two gesture patterns, Snap and Rotation, as shown in Fig. 3(A) and (C). On the other hand, the left Gear (Fig. 3(B)) is used for selecting a specific device among target devices [6]. In this paper, we have used 3-types of controlled devices such as LED Bulb, Smart Plug, and Door lock. (Smart Plug is an outlet type device which supports the function of On/Off control, as well as energy consumption.)

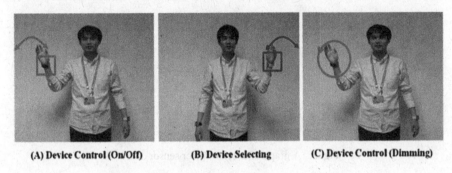

(A) Device Control (On/Off)   (B) Device Selecting   (C) Device Control (Dimming)

Fig. 3. Proposed gesture patterns with watch phones

How to send control commands by applying snap and rotation gestures is as follows. If we would like to select a specific controlled device, we can use the left snap as shown in Fig. 4(A). The right snap is used when we send simple control commands just like 'On' or 'Off' regarding Smart Plug or LED. Figure 4(B) describes the rotation gesture pattern regarding Dimming-Level control for LED Bulb.

- Pattern I: *Snap*

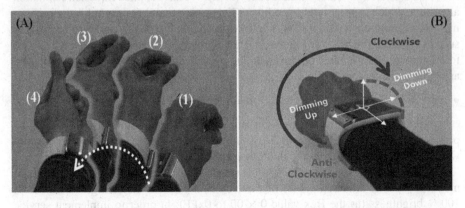

Fig. 4. Gesture control with watch phone (galaxy gear): (A) Wrist Snap and (B) Rotation

In our proposed WACA framework, we have used the raw sensor data of internal accelerometer in Gear to recognize the gesture patterns. The variation of accelerometer sensor 3-axis for the snap gesture (shown in Fig. 4(A)) is the following Fig. 5. At this time, the variation graph of 'snap' can be separated into X, Y, and Z axis respectively as illustrated in Fig. 5(B), (C), and (D).

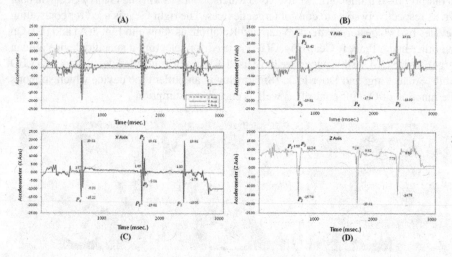

Fig. 5. A graph of accelerometer sensor data

In the case of general snap gesture, it can be visualized as the graph shape with P1, P2, and P3 edge points. However, it could not be guaranteed to draw as the same pattern. In other words, this means that the variable data, such as P4 and P5, could be generated in figure. Previous researches, which used a pre-defined simple threshold based decision-making approaches, could make the low success rate regarding the normal patterns. The other problem is that control commands are miss-transmitted to the target devices after recognized as the Normal pattern even though there are not intended. This matter is could be sometimes caused by Watch phone characteristics that almost always wear on the wrist. We call is as 'False Positive Rate' (FPR). We have applied Bayesian estimation algorithm, which is proposed by Fredman [7], based probability thresholds to decision-making information to reduce FPR as following. In order to predict the variable patterns such as P4 and P5 in Fig. 5, we can estimate P3 probability value after moving from P1 to P2 through the training processing. The overall processing flow of the proposed approach is the following Fig. 6.

- Pattern II: *Rotation (Clockwise & Anti-clockwise)*

Unlike the simple controls, e.g. 'On' and 'Off' by snap gesture as discussed in the previous section, we will talk about the Dimming-level control by using Rotation gesture which could be configured as the multiple level values.

We have applied a de facto HAP profile standard based on ZigBee stack to LED bulb modules [9]. The dimming values of these bulbs are able to be configured from 0 to 100 % brightness (as the Hex value 0×00 to 0xFF). In order to implement service

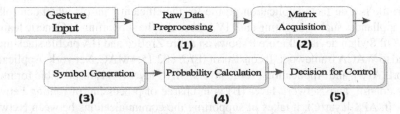

Fig. 6. Overall procedure of gesture (Snap) processing

scenarios with Watch Phone, we implemented the rotation-based gesture patterns as following; as shown in Fig. 7(A), when rotating anti-clockwise by using the right wrist wearing the watch phone, the dimming-level value is increased. On the contrary, in the case of rotating clockwise decreases the dimming-level value as illustrated Fig. 7(C).

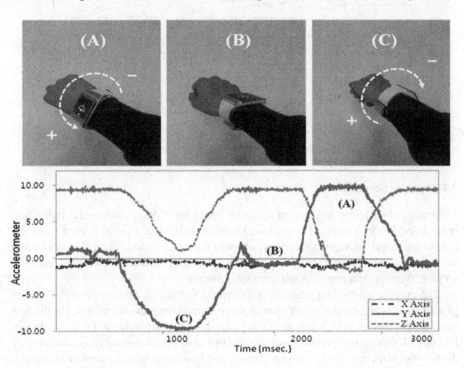

Fig. 7. Accelerometer-based rotation (clockwise/anti- clockwise) gesture for dimming control

3.3 Control Protocol Design

In this section, we present the entire design of the protocol stack and frame structure which is used to control end devices.

In our proposed WACA framework, the protocol between Gateway device and end devices has been implemented based on ZigBee. In addition, we have utilized the Home Automation Profile 1.2 on ZigBee application layer standardized by ZigBee Alliance.

HA profile is a de facto application profile for the efficient controls when controlling home appliances such as Lighting and HVAC besides the supporting of Generic features of On/Off Switch devices. Figure 8 shows an entire ZigBee and HA profile stack that is applied to WACA framework. It consists of IEEE 802.15.4 MAC, Network, Application Support (APS), and Application Profile layers. MAC layer (A) is responsible for media access control, and Network layer (B) is in charge of processing of routing between nodes. In APS layer(C), it takes of supporting the communications between Network and Application layer like the definition of Service Object. Finally, ZigBee Cluster Library (ZCL) based Profile layer processes the real command payloads.

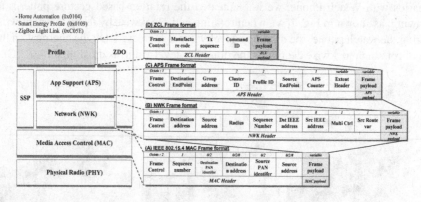

Fig. 8. A structure of ZigBee profile stack

3.4 Service Scenarios

There are two different application scenarios based on WACA framework: Individual control and Mode (Coming Home) based control. Individual control is used when we control the target devices respectively, whereas the case of Mode control is used when we want to control all devices at the same time in order to provide the convenience UX service. We now describe in detail the each scenario.

The purpose of the first scenario is to control individual devices through the user gesture such as wrist snap and rotation as we presented the previous section. As the first step, user wearing Watch phone (Gear) performs a gesture input as illustrated in Fig. 9(A). If this gesture pattern is not an abnormal case, this command is sent to Gateway device over WiFi communication interface (using Bluetooth interface between Gear and Smartphone). Gateway transmits a received command after converting to ZigBee protocol. Finally, End devices (i.e. LED or Smart plug) are executed by the received command request like 'Turn On' or 'Dimming-Level Control'.

The second scenario is utilized to control all devices concurrently. Let is consider the following actual practice. We should open the door-lock to enter the home. After entering, we also have to control a variety of devices respectively, again and again. In order to improve these kinds of user inconvenience, we considered the second scenario. Based on these considerations, we have implemented the service scenario that all of devices are operated if the door-lock is changed as unlock status. As mentioned above,

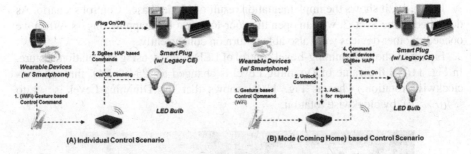

Fig. 9. A service flow of mode-based control

we can provide the user conveniences without controlling of each device respectively as providing the one-step controlling.

4   Experiment Result and Evaluation

In this section, we present an implementation of the proposed WACA framework-based gesture application services and device controls, as well as an experimental result in order to validate the proposed framework. In addition, we discuss the performance evaluation of the gesture recognition rates in order to prove the efficient of our proposed Bayesian based approach.

The following Fig. 10 is our test-bed environment composed of proposed experimental devices. We built our experimental test-bed in which there are consists of a watch phone (Galaxy Gear) and a smart phone (Note3), Gateway (Samsung HomeSync), as well as USB-type ZigBee dongle for converting protocol from WiFi to ZigBee HAP as described in Table 2. End devices composed of LED Bulbs, Smart plugs, and a door-lock.

Fig. 10. Before (A) and After (B) mode-based control scenario by snapping fingers

In order to support the Mode-based service scenario (the second one) as introduced in Sect. 3.4, we implemented that Gateway device is concurrently available to send 'On' commands to Smart plug and LED Bulb upon receiving the control command of the door-lock.

In Fig. 10, it shows the implementation result of Mode- based Control scenario. As described in the Sect. 3, we can open the door-lock just by snapping fingers. When we opened it, other devices were also able to turn on concurrently.

Figure 11 shows Dimming-Level control of LED Bulb by using the rotation gesture. In Fig. 11(A), the value of Dimming Level is changed to '234(0xEA)' through anti-clockwise rotation, whereas Fig. 11(B) shows that the Dimming-Level is set to '34(0 × 24)' by clockwise rotation.

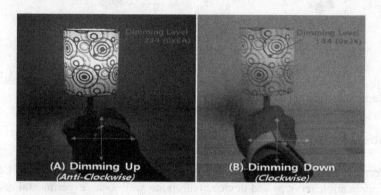

Fig. 11. Dimming-level control by using galaxy gear (watch phone)

5 Conclusion and Future Works

In this paper, we investigate the previous research problems in the domain of wearable-based home control and describe the necessary factors to solve or mitigate the problems. We also proposed a novel wearable based WACA framework for CE device control that can be applied to a variety of wearable devices. In order to validate the proposed WACA framework, we presented the 'Snap' motion based decision-making patterns for home device control. We also demonstrated three control scenarios of home control. Based on the implementation and verification results, we claimed that the proposed WACA framework can be used as a URC controller which supports multiple device control solution.

References

1. Hwan, H., et al.: Object recognition and selection method by gaze tracking and SURF algorithm. In: Proceedings of International Conference on Multimedia and Signal Processing (CMSP), pp. 261–265 (2011)
2. Dan, I., et al.: An intelligent gesture interface for controlling TV sets and set-top boxes. In: Proceedings of IEEE International Symposium on Applied Computational Intelligence and Informatics, pp. 159–164 (2011)
3. Lei, J., et al.: A Brand-independent low memory footprint universal remote control method for resource constrained wearable remote controller. In: Proceedings of IEEE 2nd Global Conference on Consumer Electronics, pp. 367–368 (2013)

4. Shohei, M., et al.: Remote control with switches on fingertips. In: Proceedings IEEE International Conference on Consumer Electronics, pp. 688–689 (2012)
5. Lei, J., et al.: A unified method for multiple home appliances control through static finger gestures. In: Proceedings of IEEE/IPSJ International Symposium on Applications and the Internet, pp. 82–90 (2011)
6. Shaowei, C., et al.: Interacting with a self-portrait camera using motion-based hand gestures. In: Proceedings of the 11th Asia Pacific Conference on Computer Human Interaction, pp. 93–101 (2013)
7. Friedman, N., et al.: Efficient Bayesian parameter estimation in large discrete domains. In: Advances in neural information processing systems 11. MIT Press, Cambridge (1999)
8. Sanhyung, C., et al.: SAD: web session anomaly detection based on parameter estimation. Comput. Secur. **23**, 312–319 (2004)
9. ZigBee Home Automation Standard. http://www.zigbee.org/Standards/ZigBeeHomeAutomation/Overview.aspx
10. Louis-Philippe, M., et al.: Head gesture recognition in intelligent interfaces: the role of context in improving recognition. In: Proceedings of International Conference on Intelligent User Interfaces, pp. 32–38 (2006)
11. Daniel, K., et al.: MAGIC summoning: towards automatic suggesting and testing of gestures with low probability of false positives during use. J. Mach. Learn. Res. **14**, 209–242 (2013)
12. Koo, B.H., et al.: R-URC: RF4CE-based universal remote control framework using smartphone. In: Proceedings of International Conference on Computa-tional Science and its Applications (ICCSA 2010), pp. 311–314 (2010)

Media, Screen, Input, and Context Sharing System for D2D Services in Smart TV 2.0

Taeho Kong, Junghyun Bum, and Hyunseung Choo[✉]

College of Information and Communication Engineering,
Sungkyunkwan University, Suwon, Korea
{kong0512, bumjh, choo}@skku.edu

Abstract. A major effort has been put in recent years on the development of Device-to-Device(D2D) communications which provides wireless connectivity for higher data rates and system capacity. Accordingly, various information sharing systems between smart devices have been developed and widely disseminated to make full use of the wireless network environment. However, due to low compatibility and performance current sharing systems hardly meet the user's needs for seamless and bidirectional sharing features. In this paper, we propose an integrated sharing system among smart devices. The proposed system provides functional features not only the media, screen, and input data but also the user context usage information of their mobile devices. Our system can fully support the role of a source device and a destination device whereas previous systems provide only one-directional data sharing. Furthermore, proposed system can support high availability and usability by providing the integrated sharing environment. Through the experimental development, we saw that our system utilized for the heterogeneous contexts and showed the validity of the future IoT multimedia system design.

Keywords: Data sharing · Context sharing · Screen mirroring · Media sharing · Input sharing · Smart TV · Smart device

1 Introduction

The number of people who owns more than two smart devices such as smartphones, tablets, smart TV has been increasing recently. According to using multiple devices, there is a rising demand for the technology which enables to link and share data among these smart devices. The importance of such technology is growing because it not only resolves problems like data fragmentation and manipulation difficulties due to multiple devices but also creates synergies that can improve usability and efficiency of each device. Sharing technologies among smart devices have advanced from simple sharing of data to share of the various things such as input and output signals, detailed user information data, and so on. Also, sharing technologies are very important to construct a future technology like Smart TV 2.0 - the future TV technology for smart living that unites of opened TV platform technology and new technology like cloud, virtualization, human interaction, device integration.

© Springer International Publishing Switzerland 2016
O. Gervasi et al. (Eds.): ICCSA 2016, Part II, LNCS 9787, pp. 180–194, 2016.
DOI: 10.1007/978-3-319-42108-7_14

The technologies related to linking and sharing among smart devices have been developed and deployed by a large number of organizations and enterprises. Most of the related systems which are currently being in service have been developed by those who operate smart device platforms. Apple's AirPlay and Google's Chromecast are the most typical case of services that developed by the platform owners. These systems provide the general functionality to share smartphones' media and screens with other smart devices. Organizations like Wi-Fi Alliance have developed Miracast or Wi-Fi Direct. Electronics manufacturers also established Digital Living Network Alliance (DLNA) and had been developing the associated technologies. The current technologies are mostly focused on sharing media or mirroring screen. It is expected that a large number of research organizations and enterprises will make a technical progress in the field of media and context sharing over multiple devices in the future.

AirPlay and Chromecast enable to share media and screen from mobile device to smart TV, but unfortunately, they don't provide the opposite direction to share. The problem is that it is not possible to share things bilaterally. Also, each device was designed to take a distinct role of a server or a client in those systems, so it leads to poor usability and relatively low performance. In this paper, it is proposed that the bidirectional sharing system could share media, screens and mobile context over multiple devices regardless of each size of devices by removing the restrictions between platforms. The implemented system also is faster and more flexible owing to integrated applications by functional.

The system, eliminating the limitations of sharing platforms can be variously utilized. For example, the contents of one device such as smart TV can be shared to entertain and work collaboratively through several smart devices. By using the integrated applications, it can be expected to take advantage of such synergies that the collected, shared data or mirrored screens are utilized in many situations.

The remainder of this paper is organized as follows. Section 2 describes the recent research status of multiple devices' sharing technologies and the techniques applied by the proposed system. Next, we propose the sharing system which can transfer media, screens, input signals, and the collected mobile contexts among smart devices and present the usage and detailed development skills in Sect. 3. Then, it follows functional features and performance evaluation compared to the conventional system in Sect. 4. Finally, Sect. 5 provides concluding remarks.

2 Related Work

Recently a variety of technologies has been developed and implemented to share data between multiple devices including input/output signals. Smart device platform holders mostly commercialize Those. For a prominent example, Apple Inc. announced AirPlay technology in 2010. AirPlay is a proprietary protocol that allows wireless streaming from Apple's small smart devices, like the iPhone, iPad to its set-top box, Apple TV of audio, video, device screens, and photos. Google develops Chromecast. It is designed as small dongles; the devices play audio/video contents on a smart TV by directly streaming it via Wi-Fi from the Internet or a local network. Media contents can be shared only from Google Cloud server, not smart devices. In addition to that, Miracast and WiDi provide smart device's screen mirroring technologies. DLNA also provides media sharing technologies. As noted above, many different companies are implementing smart sharing technologies, but their technical features are largely similar [1].

The UPnP technology is an extension of plug-and-play, a technology for dynamically attaching devices directly to a computer that enables simple and robust connectivity from many different vendors [2]. UPnP is an internet protocol and designed to run on HTTP. UPnP protocol follows six steps: Addressing, Discovery, Description, Control, Event notification, and Presentation. A device gets an IP address from a Dynamic Host Configuration Protocol (DHCP) server in the addressing step. A control point discovers other devices in the same network by sending a search request in the discovery step. The UPnP discovery protocol is known as the Simple Service Discovery Protocol (SSDP). SSDP allows that control point to search actively for devices of interest on the network, or listen passively to the SSDP alive messages of the device. A device that has just joined the network can send a multicast message to show its presence. In the description step, a control point obtains the device's description of the location (URL) by sending an HTTP request. In the control step, a control point sends a suitable control message to the control URL for the service provided in the device description. In the event notification step, a device sends event messages to a control point when its state changes. The final step in UPnP networking is a presentation. If a device has a URL for presentation, then the control point can retrieve a page from this URL, allow a user to control the device and view device status.

Digital Living Network Alliance (DLNA) is a group defining guidelines that incorporate several existing public standards, including Universal Plug and Play (UPnP) for media management and device discovery and control, and widely used digital media formats [3, 4]. DLNA uses HTTP protocol for transmitting media. The DLNA Device Classes are separated as follows: Digital Media Server (DMS), Digital Media Controller (DMC), Digital Media Renderer (DMR). DMS stores and manages media content and transmits data if requested from other devices. DMC finds content stored on DMS and instruct DMR to play the content. DMR plays content as instructed by DMC.

Virtual Network Computing (VNC) is a graphical desktop sharing system that uses the Remote Frame Buffer protocol (RFB) to control another computer remotely [5]. RFP protocol consists of two parts. Display Protocol, which displays from given output data is based on a single graphics primitive: "put a rectangle of pixel data at a given x, y position". Input protocol is simply sent to the server by the client whenever the user presses a key or pointer button, or whenever the pointing device is moved. First by sending 'ServerInit' and 'ClientInit' messages the server and the client are initialized. Once initialized, the server and the client send messages each other. There are several 'client to server message' types. 'SetPixelFormat' message sets the format in which pixel values should be sent in FramebufferUpdate messages and set display resolution and color. 'FramebufferUpdateRequest' message sends a request to the server for updating the display. 'KeyEvent' sends a signal when entering a specific key and 'PointerEvent' sends a signal when moving a mouse pointer. There also are 'server to client message' types. 'FramebufferUpdate' consists of a sequence of rectangles of pixel data which the client should put into its framebuffer. 'Bell' message rings a bell on the client if requested. RFP supports the fine image compression, so it can be shared the display screen within the low bandwidth. As the VNC system is implemented to open source, any platforms can use it for free.

Synergy is a software application for sharing a keyboard and mouse between multiple computers [6]. Synergy consists of the server and the client based on TCP/IP networks. When the client connects to the server's IP address, the server controls the clients with its connected keyboard and mouse. The multi-monitor environment can be built by setting the location of the client. Synergy can be used in a desktop environment primarily. A mobile device can be used to only the client.

SystemSens has been implemented to help analyze usage context from smart mobile devices running on Android OS by collecting data and transferring to the server [7]. SystemSens consists of the client based on Android OS and the Web-based server. The client continuously runs as a background Android service. The system is composed of sensors, service, database and the Uploader. SystemSens supports two types of sensors. Event-based sensors generate a log record whenever the corresponding state changes. Polling sensors record the corresponding information at regular intervals. The mainly collected sensing data are calling and SMS records, application-related records, battery usage records. Communications between the server and the client use HTTP post and data send by JSON format. The server consists of MySQL on Linux running Apache. It receives a post request and inserts data records in a database table.

3   System Design and Development

3.1   System Outline

Current sharing system between the smart device on the market has only one or two features. For example, Apple's Airplay, or Google's Chromecast has two features: media sharing and screen sharing. The system proposed in this paper is the 4-in-1 integrated system - not only media sharing and screen sharing features but also input sharing and context sharing. The integrated system enhances usability and compatibility and creates a whole new service by combining more than two features.

Herewith the details of providing features of the proposed system. First, media sharing system gets rid of limitations and feature gaps between smart devices rather than the current media sharing system, and it integrates all media sharing related features in one application for increasing convenience. Second, the screen sharing system also eliminates the difference of limitations like media sharing system. For example, the current system can only share the smartphone's screen to smart TV's, but the proposed system can share the opposed direction that sharing smart TV's screen to smartphone's. This system can be utilized in many situations. Third, input sharing system is the expanded system from only PC platform applied features to variant smart devices. If the user requires a multi - monitor system through smart devices, this system can be used very effectively. Fourth, context sharing system is based on SystemSens, and it can save smart devices' various contexts on the server. It can recognize and utilize from user's usage history on smart devices. For example, the system can save the user's history that usage of media, screen, and input sharing, and it can automatically prepare sharing system whenever the user usually seems to use it.

We developed under the environment using the Linux-based operating system for high stability and performance and combined the development operating system to Ubuntu 12.04 LTS for compatibility. We used Google Android for Smart device platform, and combined Android version to 4.0.3 Ice Cream Sandwich. We used Eclipse Juno 4.2.2 by Integrated Development Environment for compatibility.

3.2   Media Sharing System

Apple Airplay and Google Chromecast are current media sharing systems on the market which are DLNA-based applications. Most media sharing systems work as they transfer media from mobile device to smart TV, but on Chromecast, the Mobile-based smart device only works as a controller, so that transferring media takes place from cloud server to smart TV. However, it has no difference that system can only simple media sharing from mobile device to smart TV. It is easy to share media from smartphone to smart TV, but it is difficult to share media from smart TV to smartphone.

The proposed media sharing system in this paper developed under UPnP and DLNA protocol. The previous DLNA system has a fixed role - server, renderer, and controller, but the developed system eliminates the limitation of roles so that it can be more useful. For example, users can share media saved on the smart TV to the mobile device, or share media from a mobile device to another mobile device, or share media from smart TV to other smart TV. It is also able to combine all roles from DLNA into one application; that is DLNA Server, or DLNA Renderer can use simultaneously from one device. Figure 1 shows the concept of the proposed media sharing system.

The Media sharing system is developed by Cling, the open-sourced UPnP library for the Java platform and it developed an integrated application for Android. This system consists of lots of modules, such as account and authority module, a search module, a selection module, a server module, renderer module, a controller module and a transfer module. Account and authority module creates and manages accounts, and it can give limitation of roles to a certain device. Search module can search all devices that connected to the internal network. We developed this module used search API from Cling. Selection module can select rules from the user's input. Server module prepared the media list for controller device that can be the selection of media. Renderer module received from the server and prepared to play. Controller module receives the server's media list, and it can make the connection between server and renderer. Transfer module connects from server to renderer, then transfers media. Figure 2 shows the module of the proposed media sharing system.

Media sharing application can run on the integrated application for any device and also can take a Server role easier. If the application starts from a certain device, it will be taken the controller's role. After starting applications, system searches which devices can work server or renderer. Then, the user can select server and renderer on left side menu; right side menu shows the media stored in selected server. If the user selects a media, the device requests server and renderer for transfer media and play.

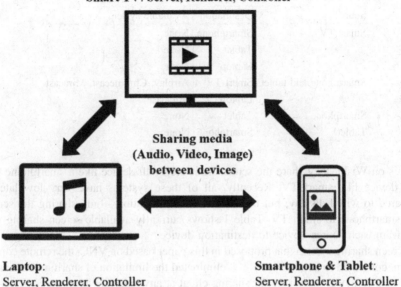

Fig. 1. The concept of the media sharing system.

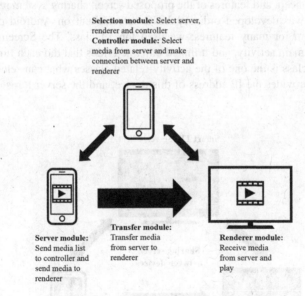

Fig. 2. The module of the media sharing system.

3.3 Screen Sharing System

Apple Airplay, Google Chromecast, Wi-Fi Alliance Miracast, Intel WiDi are currently being used as screen sharing systems. These systems have all of the same functional

Table 1. Available screen sharing systems from source to destination.

Source	Destination	Available system
Smart TV	Smartphone	None
	Tablet	None
	Laptop	None
Smartphone and tablet	Smart TV	Airplay, Chromecast, Miracast
	Laptop	Miracast
Smartphone	Tablet	None
Tablet	Smartphone	None

features on Wi-Fi, they share the screen from the small device like a smartphone to a large device like smart TV. Recently, all of these systems have very low latency compared to wired display, but the system has a limitation - only sharing the screen from smartphone to smart TV. Table 1 shows currently available screen sharing systems from each source device to destination device.

Screen sharing system that proposed in this paper based on VNC, the remote control system, compared to another system, it eliminated the limitation of sharing direction so it can take a role of the server for sharing client at any smart devices' cases. It can be sharing to two devices from the same screen. It can start sharing simply using NFC tagging, so this system provides high usability rather than another system. Figure 3 and Table 2 shows concept and features of the proposed screen sharing system respectively.

This system was developed ordinary VNC server/client on Android device. We designed a system for many features; we used lots of class. The Screenshare class connected to the main activity, and it linked other activities that did each function. The ScreenshareTab class is the one of the activity-related classes what can select from the menu screen, it provides the IP address of this device, and the server turns on/off each

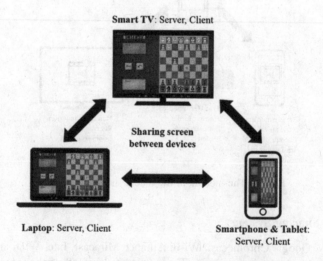

Fig. 3. The concept of the screen sharing system.

Table 2. Detailed features of the proposed screen sharing system.

Features	Description
Account management	Managing screen sharing accounts like to create, modify and delete accounts
Authority management	Managing authority of accounts that permission of features
Server settings	Managing server devices that receive input data and send Frame-based pixels
Client settings	Managing client devices that send input data andreceiveFrame-based pixels
Connect settings	Managing connects between server and clients
Input management	Managing input data using Input protocol based on RFB
Framebuffer management	Managing Framebuffer using screen protocol based on RFB
Transfer management	Managing client-server network message based on RFB

function. The SettingsTab class is also one of the activity-related classes, it provides various settings for the server, creates and manages user account. The ScreenActivities class is also one of the activity-related classes, it provides client-related works like creates an account to a server, inputs server's IP address, and tries to connect to the server. The ScreenUtils class is the substantial class that server and client communicate each other, and this class gets and updates screen from the server or sends an input signal to a server. The GLRenderer class decodes screen what receives from the server; the EventsTracking class encodes input signal what sends to the server. The VNC module, the substantial core of this system written in C language, so we have to use NDK to add a module in Java-based Android application environment. Figure 4 shows the module of the proposed screen sharing system.

The server device activates a ScreenShare application. If users press 'Turn on' button then it starts the server. The client needs an account for communication with the server, the account can be created on the server directly, or the client can request to the server to create an account for whom wants unique ID and password. After creating an account, users input ID and password according to the account and try connecting to the server. If the connection is successful, the client gets a screen from the server, and the client can send touch events or keyboard events to the server for control.

One of the features of the system can use not only the independent Android-based application but also a Web browser for screen sharing. It can connect to the server when connecting to the IP address of the server in a Web browser like Fig. 5; it can share the screen directly after submitting the account ID and password. Screen sharing client on the web can send not only touch event to the server but also keyboard event using the independent text form input window.

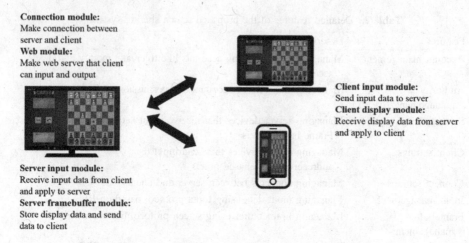

Connection module:
Make connection between
server and client
Web module:
Make web server that client
can input and output

Client input module:
Send input data to server
Client display module:
Receive display data from server
and apply to client

Server input module:
Receive input data from client
and apply to server
Server framebuffer module:
Store display data and send
data to client

Fig. 4. The module of the screen sharing system.

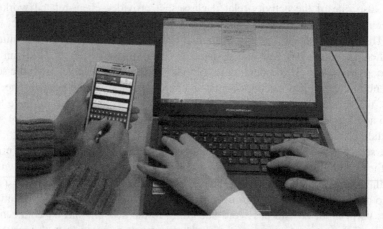

Fig. 5. Connect to the screen sharing server use smartphone and laptop.

3.4   Input Sharing System

The previous input sharing system is divided into two types. The first is wired based
sharing system using input switch; the second is wired/wireless network based input
sharing system. However, the input switch based system is very uncomfortable because
it requires extra control when the switch inputs a signal to another device. Therefore,
the network-based input sharing system becomes more popular. One of the best things
about network based input sharing system is Synergy. Synergy makes connecting to
many devices under the server-client model so that it can be shared server's input

devices according to tile-based device order. Nevertheless, Synergy considers only PC platform; it has not enough features about sharing between the various devices.

Input sharing system proposed in this paper is based on Synergy, but it developed for many smart devices available. Therefore, the user can control smart TV using the laptop's keyboard and mouse, or control the PC using smartphone's touch screen. In this system, many devices can utilize together because it can construct by controlling devices at once. Figure 6 shows the concept of the proposed input sharing system.

Because of Synergy is the system developed by C, we used NDK for the running system to the Android platform and added the front-end module for using server/client features. If the user wants to set input server, front-end helps setting server and gets input sharing connection requests from a client. We added class and activity that arrange the device position order on the front end. We also developed connection server classes for client devices. For the case of mobile device roles server, we developed 'Mobile device server mode' that always displays its keyboard and touchpad.

The server device runs the application and sets server preferences. The user can select the server connected its keyboard and mouse, or its software keyboard and touchpad. But if the user selects the server's software and touchpad, the server works only for an input device. The client device runs the application, and if entering server's IP address, then it's connected to the server. Once the client connected to the server, the server can handle it by tile-based positioning for connecting clients. For example, if the user arranges two clients by a line, and moves right the mouse pointer when the pointer is on first device's screen, the pointer moves on the second device's screen.

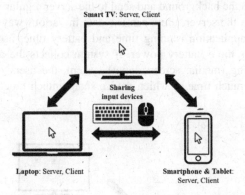

Smart TV: Server, Client

Sharing input devices

Laptop: Server, Client

Smartphone & Tablet: Server, Client

Fig. 6. The concept of the proposed input sharing system.

3.5   Context Sharing System

Context data of smart device have a lot of potentials, and they are future-oriented utilizing information [8–10]. It is one of the most important data for smoothly working intelligence contents provider system. However, the integrated context data system is hardly found. Almost context data collection system collects few types of data. If many systems are used for collecting data, data fragmentation is also concerned.

The proposed context sharing system in this paper is based on SystemSens; that is a variety of context data collection system for smart devices, but we developed system expandable by adding more features. The collectible data in this system is battery status, call-related information like call length and caller's phone number, SMS related information like arrival time and sender's phone number, Application related information like app ID, running length and running count. The system is installed on each smart device and sends information to server regularly. By analyzing the collected data, users can grasp the smart device usage pattern easily. It also can be added more data quickly if the user wants to collect from the smart device. Figure 7 shows the concept of the proposed context sharing system.

The proposed system is constituted of Android-based client and web-based server. Figure 8 shows the design of the system. The Android based client collects various data and sends to server regularly. For collecting data, the system uses many classes. PhoneStateReceiver class is developed for collecting call-related data. SmsReciever and SmsContentObserver class are developed for collecting SMS-related data. CurrentReader class is developed to collect battery-related data. ActivityLogger class is developed for application-related data. All of the collected data convert to JSON data, then SystemSens class sends data to the server. Communication with the server uses HTTP POST format; the server parses JSON after receiving data and inserts into MySQL DB. Figure 9 shows the procedure of SystemSens class that collects data from the database and sends to HTTP POST server. The user can collect additional data easily by adding code SystemSens class and creating a new class for collecting data.

First, install and run the client on the smart device to activate the system. Then, the client collects data on the background and send to the server regularly. If the system has stored lots of data on the server, users can use data in various ways. For example, by combining with an application running time and battery time, users can find which application consumes more battery power. If system collects the additional data like usage of other sharing (media, screen, input) system, the users can identify which sharing system uses much time or which device shares much time.

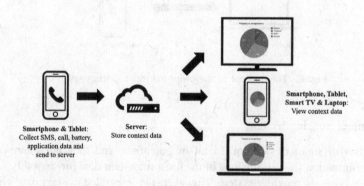

Fig. 7. The concept of the context sharing system.

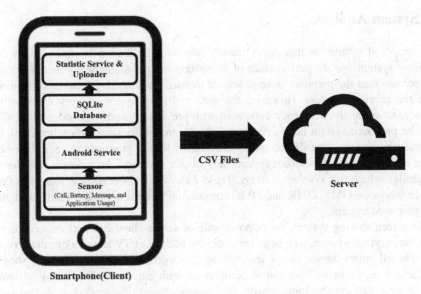

Fig. 8. System design of the context sharing system.

```
procedure TryUpload
    c←database cursor
    customUrl←address of server
    content[]←array of data
    i←0
    while c≤end of record do
        //Send select query and get data to con-
    tent[i]
        content[i]←query(select, data[c])
        c←c+1, i←i+1
    end while
    //Send content to server using HTTP POST
    sendPost(content[], customUrl)
    c←c-i
    while c≤end of record do
        //Send delete query
        query(delete, data[c])
        c←c+1
    end while
end procedure
```

Fig. 9. The procedure of sending context data to the server.

4 System Analysis

The proposed system in this paper mainly improves utilization compared with the previous system, but the performance of the system has not dramatically more increase or decrease than the previous system. So, we focused on analyzing features comparing with the previous system. However, the screen sharing system is very sensitive to latency, so we analyzed latency compared with previous systems and original VNC.

The proposed system in this paper mainly eliminates limitation of each system. In the existing systems, the device works only one role, but proposed system abolishes these notions. In media sharing system, any device can work as a Server, Renderer, Controller, which role is selected freely. Table 3 shows available devices that work that device works on DMS, DMR and DMC comparison of current media sharing systems and proposed system.

In screen sharing system, the opposite role of server-client can perform compared with the previous system, so a large smart device like smart TV can work as the server, and a small smart device like a smartphone can work as the client. Table 4 shows available screen sharing situation in comparison with current media sharing systems and the proposed system. Input sharing system and context sharing system also provide better utilization than the previous systems.

For analyzing latency of the screen sharing system between smart devices, we evaluated performance in the same Wi-Fi environment. The environment of the evaluation is following: We installed the screen sharing application on Odroid, which is the replacement of the smart TV based on Android, and smartphone. TV works as a role of the server; smartphone works as a role of the client. The non-Android laptop connects to the server using a web browser. After pre-works, we measure latency by generated

Table 3. Comparison of media sharing systems.

	Server	Renderer	Controller
Airplay	Smartphone	Smart TV	Smartphone
Chromecast	Cloud server	Smart TV	Smartphone
DLNA	Smart TV	Smart TV, Smartphone	Smart TV, Smartphone
Proposed System	Smart TV, Smartphone	Smart TV, Smartphone	Smart TV, Smartphone

Table 4. Comparison of screen sharing systems.

System	Available Screen Sharing (Source to destination)
Airplay	Smartphone to Smart TV
Chromecast	Smartphone to Smart TV
Proposed System	Smart TV to Smartphone Smartphone to Smart TV

Table 5. Comparison of average latency of screen sharing systems.

System	Average latency
Airplay	0.5 s
Chromecast	0.8 s
Miracast	0.7 s
VNC	1.2 s
Proposed System	1.2 s

touch event that refreshes the screen with proposed system and previous system. We measure latency ten times each system; then we looked for average. Finally, the latency of this system results in 1.2 s average. According to Table 5, the performance of the system has high latency compared with a commercial system, and similar to original VNC. However, the proposed system can perform screen sharing with a different type device, and provide no limitation of server/client works. Therefore, we decide that the proposed system has a competitive advantage than the previous systems.

5   Conclusion

We have lived the era that one person uses more than two smart devices. So the necessity of data or input/output signal sharing technology between smart devices is continuously increasing. However, the current sharing system between smart devices on the market has the problem of only one-sided sharing, so these systems have low utilization. The proposed system in this paper, lots of information like media, screen, input, and context can be shared regardless of smart device types. Also, we performed some tests and confirmed that proposed system's features and performance within a permissible range. The system can have contributed to future technology like smart TV 2.0. In the future, we will find the way of increasing performance of the screen sharing system and a sort of additional useful data for collecting. After gathering the information, we will start to improve the system.

Acknowledgement. This work was supported in part by MSIP, Korea, under the G-ITRC support program (IITP-2015-R6812-15-0001), Smart TV 2.0 Software Platform (No.10041244), and Priority Research Centers Program (NRF-2010-0020210).

References

1. Hsu, C., Chen, D., Huang, C., Hsu, C., Chen, K.: Screencast in the wild: performance and limitations. In: Proceedings of the 22nd ACM International Conference on Multimedia, pp. 813–816 (2014)
2. Jeronimo, M., Weast, J.: UPnP Design by Example: A Software Developer's Guide to Universal Plug and Play. Intel Press, Hillsboro (2003)
3. Digital Living Network Alliance, DLNA networked device interoperability guidelines v1.5 (2006)

4. Yu, T., Lo, S.: A remote control and media sharing system based on DLNA/UPnP technology for smart Home. In: (Jong Hyuk) Park, J.J., Kee-Yin Ng, J., Jeong, H.-Y., Waluyo, B. (eds.) Multimedia and Ubiquitous Engineering. Lecture Notes in Electrical Engineering, vol. 240, pp. 329–335. Springer, Netherlands (2013)
5. Richardson, T., Stafford-Fraser, Q., Wood, K.R., Hoper, A.: Virtual network computing. IEEE Internet Comput. 2(1), 33–38 (1998)
6. Synergy - Mouse and keyboard sharing software. https://synergy-project.org/ (2014)
7. Falaki, H., Mahajan, R., Estrin, D.: SystemSens: a tool for monitoring usage in smartphone research deployments, MobiArch. In: ACM International Workshop on Mobility in the Evolving Internet Architecture, pp. 25–30 (2011)
8. Perera, C., Zaslavsky, A., Christen, P., Georgakopoulos, D.: Context-aware computing for the internet of things: a survey. IEEE Commun. Surv. Tutorials 16(1), 414–454 (2014)
9. Tangmunarunkit, H., Kang, J., Khalapyan, Z., Ooms, J., Ramanathan, N., Estrin, D., Hsieh, C., Longstaff, B., Nolen, S., Jenkins, J., Ketcham, C., Selsky, J., Alquaddoomi, F., George, D.: Ohmage: A General and Extensible End-to-End Participatory Sensing Platform. ACM Transactions on Intelligent Systems and Technology 6(3), 1–21 (2015)
10. Li, Y., Zhou, G., Ruddy, G., Cutler, B.: A Measurement-Based Prioritization Scheme for Smartphone Applications. Wireless Pers. Commun. 78(1), 333–346 (2014)

Indoor Location: An Adaptable Platform

Mário Melo[1(✉)], Gibeon Aquino[2], and Itamir Morais[2]

[1] Federal Institute of Rio Grande do Norte, Natal, Brazil
mario.melo@ifrn.edu.br
[2] Federal University of Rio Grande do Norte, Natal, Brazil
gibeon@dimap.ufrn.br, itamir.filho@imd.ufrn.br

Abstract. Nowadays, it is clear that location systems are increasingly present in people's lives. In general people often spend 80–90 % of their time in indoor environments, which include shopping malls, libraries, airports, universities, schools, offices, factories, hospitals, among others. In these environments, GPS does not work properly, causing inaccurate positioning. Currently, when performing the location of people or objects in indoor environments, no single technology can reproduce the same results achieved by the GPS for outdoor environments. One of the main reasons for this is the high complexity of indoor environments where, unlike outdoor spaces, there is a series of obstacles such as walls, equipment and even people. Thus, it is necessary that the solutions proposed to solve the problem of location in indoor environments take into account the complexity of these environments. In this paper, we propose an adaptable platform for indoor location, which allows the use and combination of different technologies, techniques and methods in this context.

1 Introduction

Nowadays, it is clear that location systems are increasingly present in people's lives. These systems can help people solve different kinds of problems in a great variety of situations [5]. People in general often spend 80–90 % of their time in indoor environments [17], which include shopping malls, libraries, airports, universities, schools, offices, factories, hospitals, among others. Because of this, services that allow location in indoor environments have been gaining special attention. One of the reasons for the increase in popularity for this type of application is the popularization of portable devices such as cell phones, smartphones, PDAs, tablets and notebooks, which already have many built-in hardware such as WiFi, Bluetooth, GPS and inertial sensors.

Given this popularity, there is a growing need for the research and development of technologies that meet the high efficiency at low cost requirements in order to provide a variety of interesting applications for indoor spaces [12]. In addition to this, according to the OpusResearch [16], by 2018 about $ 10 billion should be directly or indirectly spent by the indoor location research area, which demonstrates a large market opportunity. Among the types of applications found in the market, those who perform targeted advertising arise as the most promising ones.

© Springer International Publishing Switzerland 2016
O. Gervasi et al. (Eds.): ICCSA 2016, Part II, LNCS 9787, pp. 195–206, 2016.
DOI: 10.1007/978-3-319-42108-7_15

In addition, given the growth in the use of mobile devices in order to obtain location, occasion in which the applications increasingly use the environments' information, location in indoor environments arises as a major ubiquitous information. The main types of applications that benefit from this feature are: augmented reality, school campus, guided museum tours, shopping malls, shop navigation, deposits, targeted advertising, among others.

There has been an increase in the demand for the location of people or objects in indoor environments to be a reliable and accurate [9]. In this sense, large technology companies such as Google [3] and Apple [1] are currently investing in this research area in order to develop solutions for location in indoor environments. This shows that this problem remains unsolved, that is, there is still no technology or combination of technologies that can solve the problem in an acceptable manner and with low costs [10]. One of the main reasons for this is the high complexity of indoor environments where, unlike outdoor environments, there are differents obstacles such as walls, equipment and even people [11].

Thus, it is necessary that the solutions proposed to solve the problem of location in indoor environments take into account the complexity of these environments. Melo and Aquino [12] showed that there is a tendency for these solutions to combine the use of different technologies, sources of information, location techniques, among other features, which will allow the adaptation of the solutions to the various complexities that can be found in these environments. In order to come up with a solution that fits all of the environments and that can be adapted to their specific characteristics, this paper proposes an adaptable platform that enables the combination and use of many techniques and technologies in order to obtain the location of people or objects in indoor environments.

The paper is organized as follows: Sect. 2 will be dedicated to the presentation of the proposed platform, detailing its requirements, architecture and components. In Sect. 3, the evalutation of the platform's behavior applied to the context of location in indoor environments using WIFI and RFID technologies is presented. Section 4 presents some related works, comparing them and pointing out strengths and weaknesses. Finally, Sect. 5 presents the final considerations, as well as future studies for this work.

2   Platform Architecture

The proposed platform defines a number of components that may have their functionalities adapted in order to enable the reception, processing and storage of many heterogeneous data that make up an indoor environment. Furthermore, the possibility of the adaptation of its components allows the platform to be adapted to different indoor environments and enables it to meet the specificities found in each one of them. Hence, new approaches, algorithms or techniques can be added without the need to adjust the base architecture.

In Fig. 1, the general architecture of the platform is presented. This platform has a predefined set of components which are called main components. The proprosed plataform has 8 main components which are: I/O Manager, Request

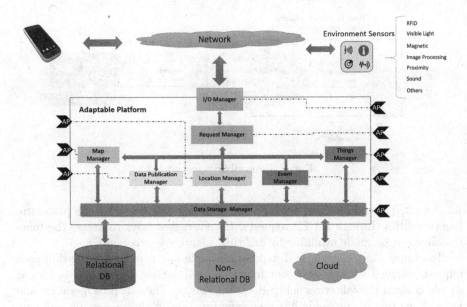

Fig. 1. Adaptable platform architecture overview

Manager, Map Manager, Data Publication Manager, Location Manager, Event Manger, Things Manager and Data Storage Manager. Each one of these components provides an adapter interface (AI) so that they may have its functionalities adapted by the adapter components. Moreover, given the separation of the architecture in different components, it is possible to adapt specific components, which making it optional to extend all of the platform's components in order to include or modify the capabilities.

The process of adapting the functionalities of the main components is shown in Fig. 2. Each component allows the adaptation of its functionalities through the AI using the Whiteboard Pattern [8]. Instead of searching in a directory for the component that implements the required functionality, the main component registers the interest in components that are able to implement the adaptation of its functionalities. So whenever an adapter component is registered in the component directory, it checks if there is any main component interested in it. If there is, it is notified. Upon receiving the notification, the main component has access to the registered component instance, which performs the adaptation of its functionality.

2.1 I/O Manager

This is the component that handles the input and output towards the platform. The main goal of this component is to ensure that the platform enables the handling of heterogeneous data, allowing the input of any information regardless the technology or unit used in order to provide many types of location services.

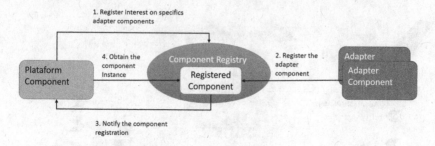

Fig. 2. Extensibility process

As default, this component provides an interface for access to the plataform functionalities through HTTP requests. However, new ways to access the functionality can be made available by creating adapters components.

To ensure the reception and making available any type of information each request received by the platform follows a well-defined format. This format should contain the following fields: Request, Version, Data Type, Operation and Data. Each one of the fields has a specific function in the processing and execution of operations within the platform. By detailing each of these fields, we have:

- **Request:** In this field, the type of processing that the user wishes to perform on the request shall be indicated. Thus, the user shall indicate the standard component of the processing layer that will be forwarded to the request.
- **Version:** Will indicate which version the functionality that is being requested is in, since the platform's functionalities may require evolution but not all of the client applications can or shall evolve in sync with the platform.
- **Data Type:** This field will indicate to the processing component, among all of the services recorded in its context, which one shall process the request.
- **Operation:** Will indicate which service shall be performed to process the request in an extended component.
- **Data:** Will contain the data to be used when processing the platform's operations.

2.2   Request Manager

This component acts as a dispatcher identifying the type of request sent to the platform, forwarding the request to the main component so it can perform its processing. This component can have its functionalities adapted. Thus, new requests for processing flow can be created, considering aspects not implemented by the main components.

2.3   Data Publication Manager

It is the component responsible for allowing the data produced on the platform to be accessed by client applications. This component will carry out the provision of information received and produced by the platform using the content providers made available by the *Data Storage Manager* component in order to provide this information.

2.4 Map Manager

If the request type is related to the maps domain, this is the component that will be responsible for receiving the request, transforming the information for this domain and invoking the service that will process the request. This component will also be responsible for providing the services for map registration, as well as for map retrieval and for registering and retrieving points of interest associated with them, among others.

2.5 Location Manager

This is the main component of the platform and is responsible for performing the processing of location information. This component's extensions must be able to execute three different behaviors. The first behavior is to turn the received information into data that can be processed by the component. The second is once the information is in a format in which the component is able to process it, the extended component shall identify the best strategy for processing this information. In this processing stage, it might to use different location algorithms, a combination of them as well as single algorithm that can perform the fusion of information in order to generate new types of information. Finally, the received and processed information must be sent to storage so that other functionalities can use them. This is the only platform main component which does not have any default functionality implemented. This component allows the inclusion of different forms to calculate the location using different types of information and techniques.

2.6 Event Manager

This component will allow the use of the Push interaction mode [2], in which it is possible to perform the registration of events. When the registered event occurs, a notification is sent to the registered clients. In order to optimize the use of resources such as network traffic and processing, these notifications are performed asynchronously. This component realize the implementation of the publish-subscribe pattern.

2.7 Things Manager

This component will be responsible for managing the platform, performing tasks such as registration, modification and removal of "things" that are necessary to platform operation. These "things" can be environments' equipment, devices, management data, among others.

2.8 Data Storage Manager

This component will serve as a data storage provider to the main components access the information persisted by the plataform. This component support different forms of data storage, i.e., relational database, non-relational database, cloud storage services and others. The platform's main components indicates the types of data that they wish to manipulate and receive, in return, they will receive the data storage provider that performs all of the persistence procedures. The Fig. 3 illustrates how this mechanism occurs showing an example of the use of repositories for data fusion. Due to the adaptable nature of the platform, each component can be attached to a different storage strategy, and use the strategy that best suits the type of data that will be consumed or stored.

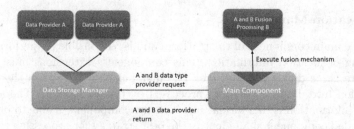

Fig. 3. Data provider storage disponibilization process

3   Evaluation

In order to evaluate the proposed platform, an Android application was used to collect data from WIFI access points present in the given environment. In addition to these data, other data from RFID tags found in specific points of the environment were also collected. In every data collection, the data was sent to the platform, which processe it and calculate the estimated position of the device. In order to carry out this evaluation, the scenario used to for this is presented in the Fig. 4. To execute this evaluation it was needed to create four adapter componentes which are:

– **RFID Service.** This component implements the functionality that processes data obtained from the RFID tags by the client applications. This component requires the following information in order to carry out the correct processing: tag identifier and device identifier. Upon receiving this information, it is stored and a new position is assigned for the device.
– **WIFI Fusion Service.** This component performs the adaptation of the location functionality using WIFI and RFID data. This component requires to use as input the map identifier information, device identifier and the list of access points information. When starting the device positioning calculation process,

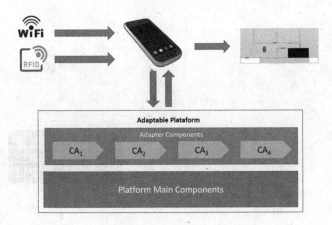

Fig. 4. Evaluation scenario

it is checked if there is any position obtained by the RFID at the last second. If there is, the calculation is not performed and the current position will be the last one obtained. If not, the Weighted Centroid algorithm [4, 7] is performed, in which the estimated position using coordinates (x, y) is obtained by calculating the geometric centroid formed by the coordinates (x, y) of all access points. With these initial coordinates, a new position is assigned for the device.

– **WIFI Location Storage and RFID Location Storage.** These components are responsible for performing the storage of the WIFI and RFID data received by the platform. The chosen mode of persistence for the storage of information was a non-relational database called MongoDB [13].

The environment used for this assessment is a conventional office in a commercial building measuring $115\,\mathrm{m^2}$. In this environment, there are four access points, so the existing infrastructure was used to perform this assessment. In addition, there is one RFID tag attached to a wall in one of the rooms in the environment. Thus, a walking test was performed within the office and its result is shown in Fig. 5. In the Figure, the environment map shows two routes. The dashed line represents the actual route, or the route that was taken by the user. The solid line represents the route calculated by the platform using the strategies defined by the adaptable components.

The evaluation was performed under two facets: main components processing time compared to the total plataform processing time and the easiness of the creation and incorporation of adapters components. In order to assess the processing time of the designed platform, the measurement of the time required for a request to be fully processed within it was performed. Thus, the time spent from the moment the request was received until the response was sent to the requestor was measured. The requests send to the platform was to calculate the indoor location of the device. This request was chosen because is the most costly to the platform. The execution context in which the platform was implemented

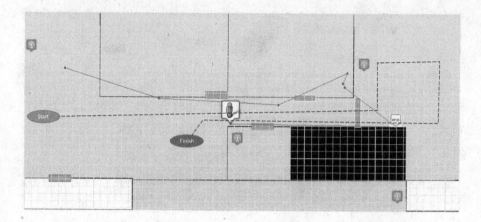

Fig. 5. Android app using the proprosed platform for indoor location

is an Intel Core i7 1.8 GHz machine with 8 GB of ram. However, for this assessment, the memory usage was limited to 512 MB. Accordingly, the average time that it takes a certain amount of requests sent at a time to be processed by the platform was calculated. The results are shown in Fig. 6, in which the X axis represents the amount of requests sent and the Y axis represents the average processing time in milliseconds. In addition, for each amount of requests sent, ten samples were collected and their average time was calculated.

It is possible to note that the main component processing time does not exceed 0,11 ms which represents 11 % of the total time to processing the requests sent to the platform. This demonstrates that the greatest amount of average processing time is spent with the logical associated with the adapters components. Another fact that was possible to note is that the plataform was able to keep the average processing time in stable values even increasing load 100 times higher.

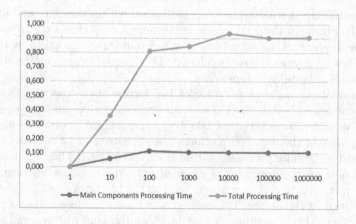

Fig. 6. Requests processing time

Table 1. Number of lines of code for the evaluation

Component	Lines of code	Lines for adaptations
WIFI fusion service	157	16
WIFI location storage	127	9
RFID service	160	16
RFID location storage	143	9
Total	587	50

Regarding the easiness to create and incoporate adaptations, we collected the amount of lines of code implemented in each of the adapter component, as shown in Table 1. On this count, all of the lines of code in the source codes file were accounted for, including imports, statements, etc. In addition, we also counted the lines of code that are directly related to tasks that is necessary to incorporate these adapter components to the platform. Therefore, were required total 64 lines of code for the components to be coupled to the platform. Thus, it was necessary less than 9 % of total lines of code to incorporate the adaptabilities the platform.

4   Related Work

In this section, related works directly associated with what was proposed in this work will be presented. We will highlight the main contributions of each of the listed works and will try to show the main differences between them and the proposed approach.

The Location Stack Architectural model introduced by Hightower et al. [6] shows a standardized model for developing location systems using a layer architectural model which refers to standard ways to perform the combination of data from different sources using a software architecture based on the Open System Interconnect (OSI) network communication model.

As the main difference between the Location Stack and the proposed platform's architecture, it is possible to note that the fact that it defines rigid layers - just like the OSI model in which a particular layer only receives data from the layer right below it and only exports the processed data to the layer right above it - makes the architecture only a little flexible. This little flexibility makes it hard to extend the functionalities, which differs greatly from the proposed architecture, whose main feature is to allow the adaptability of all components and not set all of the responsibilities or input and output information that they must obtain/produce. Furthermore, although its architecture can also be organized in layers, there is an exchange of information between layers in both directions, as well as among the components of the processing and the persistence layers.

In Najib et al. [14], a middleware for location in indoor environments called MapUme was presented. The MapUme allows the fusion of data from various types of sensors. This process occurs through the implementation of a fusion engine that is built using the abstract factory pattern. It also has a distributed processing capability, which allows scalability and high performance. The architecture used is based in an architecture proposed by Hightower et al., in which it is organized in layers. Also, it uses the service-oriented architectural model. This middleware has, as main requirements, scalability, usability and adaptability.

Despite being based on a modular architecture and divided into layers, this middleware allows the adaptability of its components. However, the adaptability is only allowed for some of the architecture's components. Even when the adaptability is allowed, it must occur with a high degree of coupling, since the new implementation must be built and coupled to the middleware. This feature makes it very different from the proposed platform, which allows the inclusion of new features without the need for interrupting the platform implementation. Associated with this, there is the fact that the proposed platform uses the Publish/Subscribe architectural model, which ensures that once instantiated and registered, the new services will be available for use within the platform.

The LOC8 proposed by Stevenson et al. [18], in its turn, is a framework developed in JAVA for location in indoor and outdoor environments, which supports a specific context and space model. It has a layer-based architecture with a robust separation of concerns. In addition, it supports different types of information from the sensors, as well as the extension and customization of the fusion algorithms. It offers an API that allows the developers to use queries in order to obtain positioning information without having to know how the sensor data is obtained. As in Hightower et al., the proposed architecture is based on layers following the OSI model. Its main focus is to enable queries on the data found in the proposed location model.

The LOC8 framework is quite different from the proposed platform since it focuses on creating a model for the environment and the positions based on ontologies, as well as providing a query mechanism to support the use of location in pervasive systems. The proposed platform has the pervasive feature since it enables the use of the information found in the environment, however it does not create a specific model for the environments and does not define how the queries will be carried out in order to obtain the location. It features a fully flexible model in which each application can set its own specificities.

In Ranganathan et al. [15], a middleware sensitive to location called "MiddleWhere" is proposed. This platform integrates multiple location technologies as well as presents the consolidated location data to customer applications. In order to do this, it uses a layered architecture to collect information from sensors and persist the information in a special database; an intelligence engine to perform data fusion in order to obtain the location information; and a service layer to make such information available. It also allows the pull and push interaction mode for communication with applications. Unlike the approach proposed in this paper, this middleware only allows a small degree of adaptability, making it only possible to vary the runtime of the insertion and removal of new sources of information. The proposed platform's architecture allows the use of new and different sources of information at runtime, as well as allows the modification of any of the behaviors found in the platform.

5 Conclusion and Future Works

This work presented the design, implementation and assessment of an adaptable platform for location of people and objects in indoor environments, which allows the use of different technologies, information sources, techniques, among other characteristics, in order to adapt itself to different and complex indoor environments. Furthermore, it was shown that the proposed platform is divided into main and adaptaer components. These main components are provided by the platforms architecture and can have their functionalities adapted by the adapter components. It was shown that all main components of the platform may have their functionality adapted, allowing its use in many different environments or technologies.

The proposed platform was evaluate using an implementation in which WIFI data obtained from access points and RFID data obtained using tags were used. To collect such data an Android application was built. In addition to get the data from the environment, the application sent it to the platform in order to calculate its current position, which was presented on a map. Furthermore, the platform were evaluated under two aspects: main components processing time compared to the total plataform processing time and the easiness of the creation and incorporation of adapters components. The main components processing time cause minor impact on the total process time, even with increasing amounts of requests sent. Regarding the easiness of creating and incorporating adaptability's, indicates to be an easy task. We realized that the amount of necessary lines of code on the specific parts of the platform up less than 10 % of the total.

As future works, we intend to assess platform operation using other technologies and techniques in order to identify potential improvements in the proposed architecture. Furthermore, there is the intent to evaluate other characteristics related to platform implementation, such as processing and distributed scalability. Besides, we also intend to assess the use of cloud computing concepts and big data.

Acknowledgment. This work was partially supported by the National Institute of Science and Technology for Software Engineering (INES) funded by CNPq under grant 573964/2008–4.

References

1. Apple. Footprint: Indoor positioning with core location (2015). http://developer. apple.com/library/ios/samplecode/footprint/
2. Bhide, M., Deolasee, P., Katkar, A., Panchbudhe, A., Ramamritham, K., Shenoy, P.: Adaptive push-pull: disseminating dynamic web data. IEEE Trans. Comput. **51**(6), 652–668 (2002)
3. Google. Indoor maps (2014). http://www.google.com/intl/pt-br/maps/about/ partners/indoormaps/
4. Han, D., Andersen, D.G., Kaminsky, M., Papagiannaki, K., Seshan, S.: Access point localization using local signal strength gradient. In: Moon, S.B., Teixeira, R., Uhlig, S. (eds.) PAM 2009. LNCS, vol. 5448, pp. 99–108. Springer, Heidelberg (2009)

5. He, J., Wang, Q., Zhan, Q., Liu, B., Yu, Y.: A practical indoor toa ranging error model for localization algorithm. In: 2011 IEEE 22nd International Symposium on Personal Indoor and Mobile Radio Communications (PIMRC), pp. 1182–1186. IEEE (2011)

6. Hightower, J., Brumitt, B., Borriello, G.: The location stack: a layered model for location in ubiquitous computing. In: Proceedings Fourth IEEE Workshop on Mobile Computing Systems and Applications 2002, pp. 22–28 (2002)

7. Thomas Konrad, P.W.: Wifi compass wifi access point localization with android devices. Master's thesis, Information Security program at St. Polten University of Applied Sciences (2012)

8. Kriens, P., Hargrave, B.: Listeners considered harmful: The whiteboard pattern. Technical whitepaper, OSGi Alliance (2004)

9. Lemic, F., Handziski, V., Wirstrom, N., Van Haute, T., De Poorter, E., Voigt, T., Wolisz, A.: Web-based platform for evaluation of RF-based indoor localization algorithms. In: 2015 IEEE International Conference on Communication Workshop (ICCW), pp. 834–840 (2015)

10. Lymberopoulos, D., Giustiniano, D., Lenders, V., Rea, M., Andreas Marcaletti, A., et al.: A realistic evaluation, comparison of indoor location technologies: experiences and lessons learned. In: ACM/IEEE International Conference on Information Processing in Sensor Networks, pp. 178–189 (2015)

11. Melo, M., Aquino, G.: A taxonomy of technologies for fingerprint-based indoor localization. In: 7o Simpósio Brasileiro de Computao Ubíqua e Pervasiva (SBCUP 2015), pp. 111–120. SBC (2015)

12. Melo, M., Aquino, G.: Categorization of technologies used for fingerprint-based indoor localization. In: Eleventh International Conference on Systems and Networks Communications 2015. ICSNC 2015, pp. 25–29. IARIA (2015)

13. MongoDB (2015). https://www.mongodb.org

14. Najib, W., Klepal, M., Wibowo, S.B.: Mapume: scalable middleware for location aware computing applications. In: 2011 International Conference on Indoor Positioning and Indoor Navigation (IPIN), pp. 1–6, September 2011

15. Ranganathan, A., Al-Muhtadi, J., Chetan, S.K., Campbell, R., Mickunas, M.D.: MiddleWhere: a middleware for location awareness in ubiquitous computing applications. In: Jacobsen, H.-A. (ed.) Middleware 2004. LNCS, vol. 3231, pp. 397–416. Springer, Heidelberg (2004)

16. Opus Research. Mapping the indoor marketing opportunity (2015)

17. Simoni, M., Jaakkola, M.S., Carrozzi, L., Baldacci, S., Di Pede, F., Viegi, G.: Indoor air pollution and respiratory health in the elderly. Eur. Respir. J. **21**(40 suppl), 15s–20s (2003)

18. Stevenson, G., Ye, J., Dobson, S., Nixon, P.: LOC8: a location model and extensible framework for programming with location. IEEE Pervasive Comput. **9**(1), 28–37 (2010)

Sequential and Parallel Hybrid Approaches of Augmented Neural Networks and GRASP for the 0-1 Multidimensional Knapsack Problem

Bianca de Almeida Dantas[(⊠)] and Edson Norberto Cáceres

Faculty of Computing, Federal University of Mato Grosso do Sul,
Campo Grande, MS, Brazil
{bianca,edson}@facom.ufms.br

Abstract. There are a lot of problems, whose solutions are based on 0-1 multidimensional problem. Since this combinatorial optimization problem is $\mathcal{NP}$-hard, one of the approaches to solve it is the use of metaheuristics. For this problem, even heuristics consume a lot of time to find a solution, which motivates the search for alternatives capable of making the use of such techniques less time-consuming. Among these alternatives, the use of parallelization strategies deserves to be highlighted, once it may lead to reduced execution times and/or better quality results. In this work we propose a hybrid approach of augmented neural networks and GRASP for the 0-1 multidimensional knapsack problem, we describe a sequential and a GPGPU implementation and measure the achieved speedups. We also compare our results with the ones obtained by other metaheuristics. The obtained results show that the proposed approach can achieve better quality solutions than some of the other algorithms found in the literature. These solutions can lead to better solutions to real problems that can be modeled with the 0-1 MKP.

Keywords: Metaheuristics · Multidimensional knapsack problem · Augmented neural networks · GRASP · GPGPU

1 Introduction

A wide range of real-world problems can be modeled and solved using the 0-1 multidimensional knapsack problem (0-1 MKP), such as capital budgeting problem, cutting-stock problem and resources allocation in distributed systems. The capital budgeting problem is the process of evaluating, comparing and selecting the needed projects given limited resources, aiming to maximize the profits and minimize the costs [3]. Several works [23,33] presented alternatives for the solution of this problem using 0-1 MKP considering that a maximum value of projects has to be selected subject to limitations on budgets [26]. The cutting-stock problem consists of minimizing the cost of filling an order for specified numbers of lengths of material to be cut from given stock lengths of given cost [15].

Partially supported by CNPq and CAPES.

O. Gervasi et al. (Eds.): ICCSA 2016, Part II, LNCS 9787, pp. 207–222, 2016.
DOI: 10.1007/978-3-319-42108-7_16

According to Haessler and Sweeney [18], Gilmore and Gomory [15,16] made the first advances in solving cutting-stock problems using linear programming by solving an associated knapsack problem. After these initial works, different authors [7,17] also used the association established by Gilmore and Gomory. Efficient alternatives to allocate available resources are fundamental in the context of distributed systems; the problem of deciding how the resources should be shared among the processors is known as the resource allocation problem and it also can be modeled as a knapsack problem. A large number of works have been using this approach, Gavish and Pirkul [14], for instance, addressed the allocation of processors and databases in distributed systems and Vanderster [31] described an alternative to allocate resources in computational grids, as well as proposed a novel approach to fault-tolerant task scheduling.

Since 0-1 MKP is a $\mathcal{NP}$-hard problem, there is a growing demand for techniques capable of obtaining good quality solutions in feasible time, among such techniques metaheuristics have been extensively studied. A metaheuristic is a set of concepts used for the definition of heuristic methods which may be applied to different optimization problems. Generally, few modifications are needed to apply metaheuristics to obtain good quality solutions for different problems [6]. Several works address the use of metaheuristics to solve combinatorial optimization problems and their results show the effectiveness of such methods, motivating the study of their different variants to solve problems for which obtaining an exact solution is difficult, in particular to the $\mathcal{NP}$-hard class of problems, such as the 0-1 MKP, which have been used as object of the implementations described in this work.

Some of the most commonly used metaheuristics to solve the 0-1 MKP are: genetic algorithms, simulated annealing, tabu search and neural networks [32]. Niar and Freville [24] proposed a parallel metaheuristic based on tabu search and showed that the application of parallelization techniques made it possible the dynamic and automatic configuration of some search parameters, which allowed to reduce the time needed to achieve a solution. Genetic algorithms (GAs) have been explored by many authors, such as Khuri et al. [20], Hoff et al. [19] and, in particular, Chu and Beasley [8] whose work deserves to be highlighted. Chu and Beasley considered problem specific information in order to improve the quality of the obtained solutions compared to the ones of previous works. Besides they were responsible for generating a set of test instances, larger than the ones previously known, with which they showed their algorithm was effective for large instances. A parallel implementation of GAs was proposed by Posadas et al. [25] to solve middle-sized instances; the authors used the libraries GALib [1] and OOMPI [2] to implement the features of GAs and communication, respectively. The ant-colony optimization (ACO) was used by Liu and Lv [22] and by Fingler et al. [13]; the first authors described a parallel algorithm using the map-reduce programming model which was run with large test instances using cloud computing. The authors of the second work implemented a parallel version using GPGPU with CUDA of the strategy proposed by Solnon and Bridge [30]; the proposed algorithm achieved good quality results with reduced execution times.

Deane and Agarwal [11] described an augmented neural network with two possible heuristics to construct a solution and then, based on this work, proposed a neurogenetic algorithm [12], a hybrid approach of augmented neural networks and genetic algorithms, which showed that interleaving iterations from both techniques can lead to better results than each technique used separately. Dantas and Cáceres [9,10] presented sequential and parallel implementations of genetic algorithms and augmented neural networks. The parallel strategy using GPGU with CUDA implemented by the authors showed that good quality solutions can be achieved in significantly smaller execution times compared to sequential alternatives.

Considering the good results that were obtained by the metaheuristics and by their parallel approaches, we propose a new hybrid metaheuristic associating GRASP to an augmented neural network for the solution of the 0-1 multidimensional knapsack problem. We present a sequential and a GPGPU version of the proposed approach. For the GPGPU implementation, we have chosen strategies that have got good parallel performance and measured the achieved speedups. We also compared the obtained results to the ones from previous works and applied statistical tests to validate our results that are competitive. The remainder of this work is organized as follows. In Sects. 2, 3 and 4, we describe the fundamentals of the 0-1 multidimensional knapsack problem, augmented neural networks and GRASP, respectively. The proposed solution and its implementation details are presented in Sect. 5 and the obtained results are shown in Sect. 6. Finally, in Sect. 7 we present our conclusions and ideas for future work.

2   0-1 Multidimensional Knapsack Problem

The 0-1 multidimensional knapsack problem is a variation of the classical 0-1 knapsack problem in which, given a set of n different items with their associated values and m different resources with their capacities, we have to decide which items must be put in the knapsack aiming to maximize its value without extrapolating the capacity of any resource. Formally the problem can be described by (1) [8]:

$$\max\{\sum_{j=1}^{n} v_j x_j\}$$

$$\text{subject to:} \sum_{j=1}^{n} r_{ij} x_j \leq b_i, i = 1, \ldots, m, \tag{1}$$

$$x_j \in \{0, 1\}, j = 1, \ldots, n.$$

where n is the number of items, m is the number of resources, $V = \{v_1, v_2, \ldots, v_n\}$ is an array with the values of the items and $B = \{b_1, b_2, \ldots, b_m\}$ stores the capacities of the resources. The matrix $R = \{r_{11}, \ldots, r_{1n}, \ldots, r_{m1}, \ldots, r_{mn}\}$ stores the demand for each resource by each item.

3   Augmented Neural Networks

An artificial neural network is a set of *artificial neurons*, with directed *connections*, whose operation is determined by its topology and by the properties of its neurons [28]. A pair of neurons i and j are connected in order to propagate the *activation* a_i from i to j; each connection also contains an associated numerical *weight* w_{ij} which determines the strength and the signal of the connection. A neuron computes the weighted sum of its input signals according to (2) [28]:

$$in_j = \sum_{i=0}^{n} w_{ij}a_i. \tag{2}$$

The neuron applies an *activation function* g to the result of this sum to produce the output, according to (3) [28]:

$$a_j = g(in_j) = g(\sum_{i=0}^{n} w_{ij}a_i). \tag{3}$$

An *augmented neural network* is a metaheuristic which associates a heuristic to the usual operation of a neural network, such heuristic guides the decision process through the network layers [4]. The topology of the neural network is defined considering the heuristic characteristics which determines the operation of the input, activation and output functions of each network layer. The variability in the search for solutions is achieved with changes in the weights associated to the connections between the neurons.

In this work we revisit the augmented neural network approach for 0-1 MKP previously implemented by Dantas and Cáceres [10] and initially proposed by Deane and Agarwal [11] which associates the heuristics proposed by Senju and Toyoda [29] and by Kochenberger et al. [21] to the neural network. The first one, which we refer to as ST, considers all the items are initially inside the knapsack and then the items are removed one by one until the solution becomes feasible. The second heuristic acts inversely, the knapsack is initially empty and the items are added to the knapsack until the attempt to add an item makes the solution unfeasible. Both approaches need to establish a policy in which the items are removed or added to the knapsack, this is done using the concept of *pseudoutility*. The pseudoutility of an item is computed considering its value and how much it consumes of the capacity of each resource. Due to the better results achieved by KMW heuristic in [10], hereafter we focus on the concepts and use of KMW. The pseudoutility in KMW is calculated using (4) [11]:

$$psUt_i = v_i/\{\sum_{j=1}^{m} r_{ji}/rc_j\}, i = 1, \ldots, n \tag{4}$$

where $RC = \{rc_1, rc_2, \ldots, rc_m\}$ is an m-position array which stores the remaining capacity of each resource. The choice of the item to be added to the knapsack is made in non-ascending order of pseudoutility; the solution is considered complete when the attempt to add the chosen item extrapolates the capacity of any resource. In this work we implement a variant which tries to add other items with smaller pseudoutilities, if the initially chosen item can not be added.

The augmented neural network executes a predefined number of *epoches* composed of *iterations*. Each epoch is responsible for assembling a complete feasible solution whose items were added individually in different iterations. The neural network returns the best solution achieved throughout its epoches. Since the items are added considering the items pseudoutilities, a *weight vector* is used in order to avoid the insertions to be made in the same order in different epoches; instead of using only the pseudoutilities to establish the order of the items, their values are multiplied by the correspondent weights from the weight vector. In the first epoch the weight vector $W[1\ldots n]$ is initialized with 1 in each position and, at each epoch $k > 1$, it is updated using (5) [10]:

$$w_i^k = w_i^{k-1} \pm \alpha * \epsilon * \beta * v_i, i = 1, \ldots, n \tag{5}$$

where α is the learning rate of the network, β is a random value between 0 and 1 which determines if a sum or a subtraction will be held and allows the new weights to be different from each other, otherwise the items would be added in the same order at every epoch – consequently leading to the same solution –, v_i is the value of item i and ϵ is the *gap* between the best known solution and the one obtained in previous epoch calculated according to (6).

$$\epsilon = (optSol - curSol)/optSol * 100 \tag{6}$$

where *optSol* is the value of the best known solution and *curSol* is the value of the solution obtained.

The neural network we implemented, initially proposed by Deane and Agarwal [11], has three layers:

- Input layer: one node which determines the activity to be executed;
- Hidden layer (items layer): n nodes representing the n items of the problem instance;
- Output layer (knapsack layer): decides which item will be added to the knapsack and deactivated in the subsequent iterations of current epoch.

The input layer is activated by one of the following values:

- Zero: maximum number of epoches has been reached and the network must terminate its execution;
- One: a new iteration of current epoch must be performed;
- Two: a new epoch must be initialized.

Once the number of epoches has not been reached, the execution continues with hidden layer receiving a signal for each node which indicates if the corresponding item is a candidate for addition to the knapsack, i.e. if the item is still active. In the first iteration of each epoch, every node is active; from second iteration on, each node receives a signal from the output layer which indicates if the corresponding item was added in the previous iteration, and therefore must be deactivated in the subsequent iterations, or if it remains active. The output layer is responsible for choosing an item to be added to the knapsack based on the heuristic and on the calculated pseudoutilities. If a suitable item is found, the layer sends a signal to the input layer to execute a new iteration of current epoch and a signal to each node of hidden layer indicating if the nodes continue active or must be deactivated. If no suitable item is found, the layer sends a signal to the input layer to start a new epoch or to terminate the execution if the maximum number of epoches has been reached.

The network can perform backtracking if there is a relatively long sequence of epoches without improvement in the global solution, as proposed by Deane and Agarwal [11]; the size limit of this sequence is provided as a configuration parameter of the network. When this limit is reached, the network is restored to the state of the epoch in which the current best solution was achieved and then the execution continues to search for solutions in a new neighborhood.

4 GRASP

Greedy Randomized Adaptive Search Procedure (GRASP) [27] is a multi-start metaheuristic that performs a sequence of iterations, each one consisting of two phases: construction and local search. The construction phase is responsible for building an initial feasible solution; the local search explores the neighborhood of this solution until a local maximum (or minimum) is found. The result of the whole process is the best overall solution. Algorithm 1 illustrates the pseudocode of GRASP main program [27].

```
GRASP( maxIterations ,  seed )
{
    Read_Input () .
    for  (k = 1;  k < maxIterations;  k++)
    {
        Solution  ←  BuildRandomSolution ( seed )
        Solution  ←  LocalSearch ( Solution )
        UpdateSolution ( Solution ,  BestSolution )
    }
    return  BestSolution
}
```

Algorithm 1. GRASP main program.

The BuildRandomSolution() function is responsible for building an initial solution using a greedy strategy. A key concept in this procedure is the Restricted Candidate List (RCL) which consists of the best elements that can be added to the partial solution without making it unfeasible. The choice of these elements is made using a greedy evaluation function that assures the RCL contains the elements which contribute the most to maximize (or minimize) the incremental costs to the solution according to the problem to be solved.

The number of elements in the RCL is limited by one of two manners [27]: cardinality-based or quality-based. In the cardinality-based option, the list is built with a predetermined number p of the best available elements. The second option considers a threshold parameter $\alpha \in [0, 1]$ which guides the choice of the elements. Every available element e which does not destroy the feasibility of the solution and whose quality $c(e) \in [c^{max} - \alpha(c^{max} - c^{min}), c^{max}]$ is added to the RCL, considering we are dealing with a maximization problem and that c^{max} and c^{min} are the maximum and the minimum costs of the available items, respectively. The elements to be added to the solution are randomly selected from the RCL, thus parameter α controls the greediness and the randomness levels of the algorithm; when $\alpha = 0$ the algorithm is purely greedy, similarly, when $\alpha = 1$, it is purely random. Algorithm 2 [27] shows the pseudocode of BuildRandomSolution() procedure which receives the seed to the pseudorandom number generator as parameter.

```
BuildRandomSolution(seed)
{
    Solution ← ∅
    Evaluate incremental costs of candidate elements
    while (Solution is not complete)
    {
        Build RCL
        Randomly select an element s from RCL
        Solution ← Solution ∪ {s}
        Re-evaluate Solution cost
    }
    return Solution
}
```

Algorithm 2. Procedure for constructing random initial solution.

The local search phase aims to improve the quality of initial solution by iteratively searching the solution neighborhood and eventually replacing it with a better solution. If no such solution is found, the procedure is terminated. LocalSearch() procedure, illustrated in Algorithm 3 [27], implements the steps of this phase and considers $N(solution)$ as the solution neighborhood and that we are working with a maximization problem.

```
LocalSearch(Solution)
{
        while (Solution is not locally optimal)
        {
            Find s ∈ N(Solution) with f(s) > f(Solution)
            Solution ← s
        }
        return Solution
}
```

Algorithm 3. Procedure for performing local search.

5 Proposed Solution

Dantas and Cáceres [10] presented a sequential implementation, based on Deane and Agarwal [11], and two parallel implementations of an augmented neural network for 0-1 MKP, one using message passing with MPI and other using GPGPU with CUDA. The most promising results were achieved using the KMW heuristic, in which the item with the highest pseudoutility which does not make the solution unfeasible is added at each iteration. In order of facilitate the search for such an item, in the approach used by the authors, the elements were sorted in non-ascending pseudoutility order at each iteration, this operation made the process expensive. Although good results were achieved, these previous implementations were limited by the fact that the initial solution was always the same due to the deterministic greedy process of its construction, which always adds the item with larger pseudoutility at each step.

In this work, we propose a new approach to the previous implementation of augmented neural networks using KMW heuristic, aiming to expand the explored solutions space and improve the performance of the algorithms. To achieve these objectives, we propose a hybrid approach which associates the assembly of a restricted candidate list (RCL) from GRASP to the augmented neural network. The pseudocode of the sequential implementation of the proposed approach is illustrated in Algorithm 4, which receives as parameters the threshold parameter $factor$, the maximum number of epochs max and the network learning rate α.

```
GRASP_AugNN(factor, max, α)
{
    Read_Input()
    while (input signal is not zero)
    {
        if (signal is one)
            Begins next iteration of the epoch
        else
        {
            Remove all the knapsack items
            Compute error of previous epoch and update weights
            Increment epoch counter and start a new epoch
        }
        Compute pseudoutilities of active nodes
        Build RCL and randomly pick a viable item s from RCL
        if (s was found)
            Add s to the solution and send 1 to input layer
        else
            if (number of epoches equals max)
                Send 0 to input layer
            else
                Send 2 to input layer
    }
}
```

Algorithm 4. Sequential hybrid pseudocode.

In order to accomplish this association, we implemented a novel method to choose the item to be added to the solution: instead of using the greedy strategy of choosing the best item in every iteration, we build a RCL composed of the items with the highest pseudoutilities and, then, randomly select a viable item from it, i.e., an item whose addition to the solution does not make it unfeasible. If there is such an item, the solution and its respective value is updated and the signal 1 is sent to the input layer in order to continue the construction of current solution, otherwise, the execution is finished, if the maximum number of epoches has been achieved, or a new epoch is initialized. We chose to limit the size of the RCL using the quality policy with a threshold parameter that defines the level of greediness of the process.

Due to the significant execution time reduction achieved by the GPGPU version described in [10], we also implemented our proposed hybrid approach using GPGPU with CUDA. In this version, CPU reads the input file and allocates memory for the storage of the necessary structures used in the algorithm. The input data are copied to GPU's memory and then the launched threads execute the same steps of sequential algorithm, each one separately constructing a feasible solution to the problem instance, without any communication with the other threads. CPU is also responsible for managing the epoches, assuring each GPU thread executes $maxEpoch/(nBlocks * nThreads)$ epoches, where $maxEpoch$ is the number of epoches provided as input, $nBlocks$ and $nThreads$ are, respectively, the number of blocks and threads per block of the GPU grid.

At each epoch, CPU launches the auxiliary kernel `Init_Epoch()`, which performs operations to initiate the construction of a solution: remove the items from the knapsack, reset the remaining capacities array to the total capacities of the resources and update the items weights according to the gap of previous epoch (or set them to 1, if it is the first epoch). Procedure `My_Kernel()` is then responsible for performing the operations of a single epoch of the hybrid augmented neural network in each one of the $nBlocks * nThreads$ threads. At each time this kernel is executed, $nBlocks * nThreads$ feasible solutions are constructed in parallel and their corresponding gaps are computed. After `My_Kernel()` finishes its execution, CPU finds and stores the best solution. Algorithms 5, 6 and 7 show the steps of our GPGPU implementation of the hybrid augmented neural network.

```
GPGPU_GRASP_AugNN( factor , maxEpoch, nBlocks , nThreads , α)
{
    Read_Input_File ()
    for (epoch ← 1 , maxEpoch/(nBlocks * nThreads))
    {
        Init_Epoch(Solution , RC, W)
        My_Kernel()
        CPU finds and stores best solution found
    }
}
```

Algorithm 5. GPGPU GRASP and augmented neural network hybrid pseudocode.

```
Init_Epoch(Solution , RC, W)
{
    Solution  ←  ∅
    Reset the remaining capacities array RC
    Update the weights array W
}
```

Algorithm 6. Pseudocode of the Init_Epoch() kernel.

```
My_Kernel(Solution , RC, W)
{
    do
    {
        Compute pseudoutilities of the active items
        Build RCL
        Randomly pick a viable item s from RCL
        if (s was found)
        {
            Solution  ←  Solution ∪ {s}
            Update remaining capacities array RC
        }
    } while (Some item can be added to Solution)
    Compute the gap of Solution
}
```

Algorithm 7. Pseudocode of the My_Kernel() kernel.

The assembly of the RCL does not require the items to be sorted, we only traverse the list three times: the first one to find the maximum pseudoutility, the second one to find the minimum pseudoutility and the latter to effectively assemble the RCL. Since the size of the RCL is usually significantly smaller than the total number of items, the search for the item to be added is not expensive, even if all the items need to be tested.

In this new approach, we also adapted (5) used for updating weights in [10]. In the new equation, we removed the factor representing the item value, since it is already considered in the computation of the pseudoutility. Using original equation the item value was considered twice and this situation led the items with highest associated values to be favored in relation to those with smaller values, consequently, it reduced the variability of the network. Equation (7) shows our new alternative to update the weight vector.

$$w_i^k = w_i^{k-1} + \alpha * \epsilon * \beta, i = 1, \ldots, n \tag{7}$$

where α is the network learning rate, ϵ is the gap between the best known solution and the one achieved in previous epoch and β is a random value between 0 and 1. One can notice we also modified the equation to always perform an addition.

6   Comparative Results

The implemented programs were executed in a computer with an Intel I7-3770 processor of 3.4 GHz, 24 GB RAM, 256 GB of SSD storage, 1 TB hard disk and

NVIDIA Geforce GT 640 video card with 384 processing cores and 1 GB of memory. We used the instances of ORLIB library [5] to test our implemented programs. The library is composed of a set of 270 test instances considering 5, 10 or 30 resources and 100, 250 or 500 items, generated according to a "tightness factor" δ which assumes the values 0.25, 0.50 or 0.75. To measure the quality of the obtained solutions, we used the gap between the values of the obtained solution and the best known solution, computed according to (6).

For each instance the programs were executed 30 times, considering 10000 epochs, and the average gaps and execution times were computed, as well as the standard deviations from the obtained values. The GPGPU version was executed using 100 threads aiming to make each thread execute 100 epochs; according to our experiments when each thread execute less than 100 epochs, the quality of the achieved solution decreased, this characteristic limited the number of used threads to up to 100. When we tested configurations with fewer threads, the number of epochs to be executed by each one was not sufficiently reduced and the execution time increased, mainly because the processing capacity of each thread is significantly lower when compared to the CPU. We tested different grid configurations with 100 threads and the one which led to smaller execution times contained 25 blocks with 4 threads per block. Since the demand for simultaneous accesses to the device global memory is a bottleneck in GPGPU programs, we used texture memory to store the data of the knapsack, such as the item values, the resources capacities and the demands for the resources; other constant values were stored in constant memory. Both texture and constant memories are smaller than global memory but allow faster accesses, leading to a significant reduction in execution time. Table 1 shows the comparison between the gaps and execution times of our proposed implementations, as well as the speedups achieved grouped by instances configurations.

Analyzing the values in the table, we notice that using the same number of epochs, the GPGPU implementation showed better quality solutions than the ones obtained by the sequential version. This improvement is due to the fact that with the association of RCL, each thread was capable of exploring a different neighborhood, allowing to expand the explored solution area instead of exhaustively exploring only the neighborhood of a single initial solution. Additionally, the GPGPU version also achieved speedup in most configurations up to 30 % in the one with 30 resources and 500 items, which is a fair improvement considering each thread is basically executing the sequential algorithm and there are numerous operations that are inherently sequential.

The speedup values achieved make sense because larger configurations take better advantage of the subdivision of the epochs among the GPU threads, because in these configurations the sequential version has to manipulate a large amount of data for a number of epochs 100 times greater than the parallel version. When the amount of data is smaller, the overhead introduced by thread management and the smaller computing power of each thread compared to the CPU does not allow the same level of speedup. Because of this, searching for more effective alternatives for parallelization using GPGPU is still of great interest.

Table 1. Average gaps and execution times grouped by configurations.

m	n	δ	Seq. hybrid		GPGPU hybrid		
			Gap	Time (s)	*Gap*	Time (s)	Speedup
5	100	0.25	0.77	1.59	0.69	1.85	0.86
		0.50	0.47	2.98	0.31	2.97	1.00
		0.75	0.25	4.31	0.22	3.88	1.11
	250	0.25	0.51	9.58	0.20	10.59	0.90
		0.50	0.21	18.01	0.12	17.15	1.05
		0.75	0.15	26.17	0.05	22.45	1.17
	500	0.25	0.20	38.36	0.07	40.96	0.94
		0.50	0.10	73.02	0.04	66.63	1.10
		0.75	0.06	106.46	0.02	87.04	1.22
10	100	0.25	1.37	2.12	1.05	2.36	0.90
		0.50	0.84	4.17	0.74	3.97	1.05
		0.75	0.41	6.14	0.41	5.16	1.19
	250	0.25	0.81	13.12	0.45	13.87	0.95
		0.50	0.53	25.67	0.29	23.25	1.10
		0.75	0.25	37.85	0.15	30.16	1.26
	500	0.25	0.34	52.51	0.17	54.12	0.97
		0.50	0.20	103.48	0.10	91.34	1.13
		0.75	0.11	153.38	0.06	118.68	1.29
30	100	0.25	2.03	4.27	1.88	4.82	0.89
		0.50	1.20	8.80	1.11	8.44	1.04
		0.75	0.59	13.41	0.89	10.94	1.23
	250	0.25	1.93	27.49	1.28	29.64	0.93
		0.50	1.00	55.62	0.73	51.34	1.08
		0.75	0.47	84.06	0.36	66.30	1.27
	500	0.25	1.22	112.46	0.87	118.69	0.95
		0.50	0.62	226.19	0.48	203.79	1.11
		0.75	0.35	339.80	0.25	261.41	1.30

The promising performance of the GPGPU version of the proposed algorithm can be clearly noticed in the graph illustrated in Fig. 1, which presents the comparison of the gaps, grouped by number of items, of the algorithms proposed in this work compared to the ones achieved by Deane and Agarwal's [11] original implementation of an augmented neural network and Dantas and Cáceres [10] implementations, Deane and Agarwal's [12] neurogenetical approach, the ACO algorithm proposed by Fingler et al. [13] and the GA of Beasley [5].

As illustrated by the graph, the GPGPU implementation of the proposed hybrid metaheuristic achieved good quality results in every configuration,

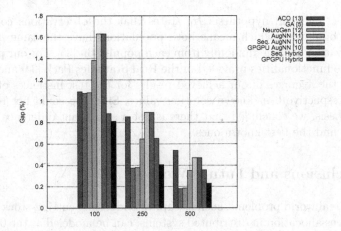

Fig. 1. Gaps grouped by number of items. (Color figure online)

specially in the configurations with 100 items; with 250 or 500 items the gaps are slightly larger than the ones obtained by the genetic algorithm and the neurogenetical approach.

Figure 2 shows the compared execution times of our implemented programs, the information related to the other implementations are not shown because they were not executed using the same hardware configuration. We must also highlight that the GPGPU implementation of the augmented neural network from [10] was executed using only 100 epoches which justifies its significantly reduced execution times.

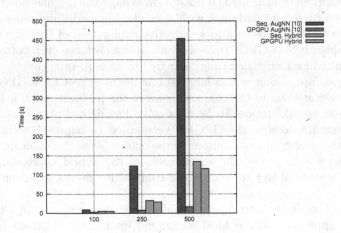

Fig. 2. Times grouped by number of items. (Color figure online)

Aiming to validate the behavior of our algorithms we used the Wilcoxon rank sum test [35] which is applied to independent samples from different populations. The null hypothesis H_0 states that the observations come from the same population, which means both populations have the same median if each observation is made randomly from each population [35]. For our purposes, we used the functionalities provided in the Real Statistics Pack [34] and applied the test to the averages of our achieved results for the test instances of ORLIB and their respective best known solution values. Since we could not reject the null hypothesis, we concluded that there is not a significant difference between our results and the best known ones.

7   Conclusions and Future Work

Important real world problems, such as capital budgeting, cutting-stock problem and resources allocation in distributed systems, can be modeled as the 0-1 MKP. Good solutions of this optimization combinatorial problem can be used to solve them. Since the 0-1 MKP is a $\mathcal{NP}$-hard problem, in the last years, the use of metaheuristics to solve it has been subject of a large number of researches. Nevertheless, there are cases in which even heuristics demand very long time to execute and the association of parallelization strategies to the heuristics have been proved effective.

In this work we proposed a hybrid metaheuristic approach which associates the RCL, used in GRASP for building its initial solution, to the execution of an augmented neural network. We implemented a sequential and a parallel version using GPGPU with CUDA. The obtained results showed that the association of the RCL to the basic operation of the augmented neural network added variability to the search process and improved the quality of the solutions. Additionally, the use of multiple threads made it possible to explore the neighborhood of different initial solutions and, although with a smaller number of iterations executed by each thread, this allowed a more effective exploration of the solution space and, consequently, our GPGPU version was able to achieve even better quality solutions in reduced execution time compared to its sequential counterpart. The achieved speedups, albeit not as large as one might expect for a GPGPU program, are relevant due to the characteristics of the problem which is known to be of difficult parallelization. It deserves to be highlighted that, despite of the relatively small speedups, the GPGPU version led to improved solutions. We compared the quality of our solutions to the ones achieved by other metaheuristics and the results proved the effectiveness of the hybrid approach. We also applied an statistical test to confirm our competitive results were not achieved by chance.

As future work, we intend to associate other characteristics of GRASP to the hybrid approach, such as local search and intensification phases aiming to improve the quality of the solutions. Furthermore, we intend to study alternative strategies of parallelization using GPGPU in order to take better advantage of the resources available by modern GPUs.

References

1. GAlib - A C++ Library of Genetic Algorithm Components. http://lancet.mit.edu/ga/
2. OOMPI - Object Oriented Message Passing Interface. http://www.osl.iu.edu/research/oompi/
3. Akpan, N.P., Etuk, E.H., Essi, I.D.: A deterministic approach to a capital budgeting problem. Am. J. Sci. Ind. Res. 2(3), 456–460 (2011)
4. Almeida, R., Steiner, M.T.A.: Aplicação de uma rede neural aumentada para resolução de problemas de corte e empacotamento utilizando novas estratégias de aprendizagem. Simpóio Brasileiro de Pesquisa Operacional (2013)
5. Beasley, J.E.: OR-Library: distributing test problems by electronic mail. J. Oper. Res. Soc. 41(11), 1069–1072 (1990)
6. Blum, C., Roli, A.: Metaheuristics in combinatorial optimization: overview and conceptual comparison. ACM Comput. Surv. 35(3), 268–308 (2003)
7. Carvalho, J.M.V.: Exact solution of cutting stock problems using column generation and branch-and-bound. Int. Trans. Oper. Res. 5(1), 35–44 (1998)
8. Chu, P.C., Beasley, J.E.: A genetic algorithm for the multidimensional knapsack problem. J. Heuristics 4(1), 63–86 (1998)
9. de Almeida Dantas, B., Cáceres, E.N.: Implementações paralelas para o problema da mochila multidimensional usando algoritmos genéticos. In: Anais do XLVI Simpósio Brasileiro de Pesquisa Operacional, pp. 1984–1994 (2014)
10. de Almeida Dantas, B., Cáceres, E.N.: A parallel implementation to the multidimensional knapsack problem using augmented neural networks. In: Proceedings of the 2014 Latin American Computing Conference (CLEI), pp. 570–578 (2014)
11. Deane, J., Agarwal, A.: Neural metaheuristics for the multidimensional knapsack problem. Technical report (2012)
12. Deane, J., Agarwal, A.: Neural, genetic, and neurogenetic approaches for solving the 0-1 multidimensional knapsack problem. Int. J. Manag. Inf. Syst. 17(1), 43–54 (2013). First Quarter 2013
13. Fingler, H., Cáceres, E.N., Mongelli, H., Song, S.W.: A CUDA based solution to the multidimensional knapsack problem using the ant colony optimization. Procedia Comput. Sci. 29, 84–94 (2014). 2014 International Conference on Computational Science
14. Gavish, B., Pirkul, H.: Allocation of databases and processors in a distributed computing system. Manag. Distrib. Data Process. 31, 215–231 (1982)
15. Gilmore, P.C., Gomory, R.E.: A linear programming approach to the cutting-stock problem. Oper. Res. 9(6), 849–859 (1961)
16. Gilmore, P.C., Gomory, R.E.: A linear programming approach to the cutting-stock problem - Part II. Oper. Res. 11(6), 863–888 (1963)
17. Gramani, M.C.N., França, P.M.: The combined cutting stock and lot-sizing problem in industrial processes. Eur. J. Oper. Res. 174, 509–521 (2006)
18. Haessler, R.W., Sweeney, P.E.: Cutting stock problems and solution procedures. Eur. J. Oper. Res. 54, 141–150 (1991)
19. Hoff, A., Løkkentangen, A., Mittet, I.: Genetic algorithms for 0/1 multidimensional knapsack problems. In: Proceedings Norsk Informatikk Konferanse, pp. 291–301 (1996)
20. Khuri, S., Bäck, T., Heitkötter, J.: The zero/one multiple knapsack problem and genetic algorithms. In: Proceedings of the 1994 ACM Symposium on Applied Computing, pp. 188–193 (1994)

21. Kochenberger, G., McCarl, B., Wyman, F.: A heuristic for general integer programming. Dec. Sci. **5**, 36–44 (1974)
22. Liu, R.T., Lv, X.J.: Mapreduce-based ant colony optimization algorithm for multi-dimensional knapsack problem. Appl. Mech. Mater. **380–384**, 1877–1880 (2013)
23. McMillan, C., Plaine, D.R.: Resource allocation via 0-1 programming. Dec. Sci. **4**, 119–132 (1973)
24. Niar, S., Freville, A.: A parallel tabu search algorithm for the 0-1 multidimensional knapsack problem. In: Proceedings of the 11th International Symposium on Parallel Processing (IPPS 1997), pp. 512–516 (1997)
25. Posadas, C.B., Ayala, D.V., Garcia, J.S., Silverio, S.L., Vidal, M.T.: A solution to multidimensional knapsack problem using a parallel genetic algorithm. Int. J. Intell. Inf. Process. **1**(2), 47–54 (2010)
26. Rardin, R.L.: Optimization in Operations Research. Prentice-Hall Inc., New Jersey (1998)
27. Resende, M.G.C., Ribeiro, C.C.: Greedy randomized adaptive search procedures. In: Glover, F., Kochenberger, G. (eds.) Handbook of Metaheuristics, pp. 219–249. Kluwer, Boston (2002)
28. Russel, S., Norvig, P.: Artificial Intelligence - A Modern Approach, 3rd edn. Prentice Hall, Upper Saddle River (2010)
29. Senju, S., Toyoda, Y.: An approach to linear programming with 0-1 variables. Manag. Sci. **15**, B196–B207 (1968)
30. Solnon, C., Bridge, D.: An ant colony optimization meta-heuristic for subset selection problems. In: Nedjah, N., Mourelle, L. (eds.) Systems Engineering Using Swarm Particle Optimization, pp. 7–29. Nova Science Publishers, New York (2006)
31. Vanderster, D.C.: Resource allocation and scheduling strategies using utility and the knapsack problem on computational grids. Ph.D. thesis, University of Victoria (2003)
32. Varnamkhasti, M.: Overview of the algorithms for solving the multidimensional knapsack problems. Adv. Stud. Biol. **4**(1), 37–47 (2012)
33. Weingartner, H.M.: Mathematical Programming and the Analysis of Capital Budgeting Problems. Prentice-Hall, Englewoods Cliffs (1963)
34. Zaiontz, C.: Real statistics using Excel - free download. http://www.real-statistics.com/free-download/
35. Zaiontz, C.: Real statistics using Excel - Wilcoxon rank sum test for independent samples. http://www.real-statistics.com/non-parametric-tests/wilcoxon-rank-sum-test/

Computational Verification of Network Programs for Several OpenFlow Switches in Coq

Hiroaki Date and Noriaki Yoshiura(✉)

Department of Information and Computer Sciences, Saitama University,
255, Shimo-ookubo, Sakura-ku, Saitama, Japan
yoshiura@fmx.ics.saitama-u.ac.jp

Abstract. OpenFlow is a network technology that enables to control network equipment centrally, to realize complicated forwarding of packets and to change network topologies flexibly. In OpenFlow networks, network equipment is separated into OpenFlow switches and OpenFlow controllers. OpenFlow switches do not have controllers that usual network equipment has. OpenFlow controllers control OpenFlow switches. OpenFlow controllers are configured by programs. Therefore, network configurations are realized by software. This kind of software can be created by several kinds of programming languages. NetCore is one of them. The verification method of NetCore programs has been introduced. This method uses Coq, which is a formal proof management system. This method, however, deals with only networks that consist of one Open-Flow switch. This paper proposes a methodology that verifies networks that consist of several OpenFlow switches.

1 Introduction

Large scale networks consist of many Layer 2 or Layer 3 network switches. Each of the network switches requires to be configured by network operators. Modification of network topologies or configurations requires modification of configurations of all network equipment. This modification takes much cost for network operations. Moreover, operations of different kinds of network equipment are harder than operations of the same kind of network equipment. One of the aims of OpenFlow is to simplify operations of much network equipment. OpenFlow is a network technology that enables to control network equipment centrally, to realize complicated forwarding of packets and to change network topologies flexibly [5]. In OpenFlow networks, network equipment is separated into OpenFlow switches and OpenFlow controllers. Exactly, OpenFlow is a protocol between OpenFlow switches and OpenFlow controllers. OpenFlow switches do not have controllers that usual network equipment has. OpenFlow controllers control OpenFlow switches. OpenFlow controllers are configured by programs. Therefore, network configurations are realized by software. OpenFlow is one of software defined networks (SDNs).

Network switches in OpenFlow forward packets like Layer 2 or Layer 3 switches, but the way of forwarding packets is not configured in network switches.

O. Gervasi et al. (Eds.): ICCSA 2016, Part II, LNCS 9787, pp. 223–238, 2016.
DOI: 10.1007/978-3-319-42108-7_17

The way of forwarding packets is decided by OpenFlow controllers as follows; OpenFlow switches send messages to OpenFlow controllers after receiving packets. The messages include the information of the packets that are received by the OpenFlow switches. The information includes source and destination IP addresses, source and destination port numbers, protocol number and so on After receiving the messages, OpenFlow controllers send messages to OpenFlow switches. The messages include the instructions of forwarding the packets. OpenFlow switches forward the packets according to the instructions in the messages. In OpenFlow, the instructions are called flow entries. After receiving the flow entries, OpenFlow switches keep the flow entries in itself. OpenFlow switches receive the same kind of packets (for example, the packets that have the same destination IP addresses) and forward the packets according to the flow entries without sending the messages to OpenFlow controllers. Therefore, OpenFlow switches do not have controllers in itself and the way of forwarding packets is not configured in OpenFlow switches. Flow entries are kept in a predefined time or as long as possible. After the time is expired or when cache memory for flow entries is full, the flow entries are erased. Modification of network configuration requires modification of the configuration of OpenFlow controllers without modification of OpenFlow switches. Even if a network consists of network switches that are manufactured in different companies, modification of configurations of OpenFlow controllers is enough to modify the network configuration in the case that the network switches can deal with OpenFlow protocol.

There is several hardware of OpenFlow controllers, but some software enables usual PCs to be used as OpenFlow controllers. The examples of such software are Trema [18], NetCore [6] and so on. NetCore is a programming language that can be used in functional programming language Haskell [17]. This feature is preferable for program verification and some researches use Coq for NetCore program verification. Coq is an interactive theorem prover and a formal proof management system [16]. If Coq proves that a program satisfies the properties that should be satisfied by the program, Coq reduces the cost of checking programs and guarantees the correctness of programs better than software testing. There are several researches that use Coq to verify software [2,10]; one is to use Coq to guarantee programs by typing [12]. In the classical program verification approach, a programs and their specifications are described separately and the verification procedure proves that the programs satisfy their specifications. Coq enables to create and verify programs simultaneously. In this case, specifications are expressed as types [9].

There are also several kinds of researches that verify OpenFlow programs: testing method [4,14,15], model checking [1,7,8], and proof system [11,13]. There are also several researches that use Coq to verify the programs in NetCore. Suppose that pg is a NetCore program, P is the precondition of pg and Q is the postcondition of pg. To verify the program pg, Coq proves a Hoare logic formula $\{P\}pg\{Q\}$, which represents that if P is satisfied before the program pg is executed, Q is satisfied after the program pg is executed. The previous research [11] verifies Hoare logic formulas by two steps; the first step is to calculate the minimum precondition that satisfies the postcondition Q after executing the program *pg*. The second step is to check whether P implies the minimum precondition. The previous research verified NAT (Network Address Translation) in NetCore programs but dealt with only network topologies that consist of one network switch.

This paper proposes a methodology that verifies programs for network topologies that consist of several network switches. Especially, this paper verifies that NetCore programs do not generate looping packets in the networks. First, to realize this verification, this paper tries verifying a whole of a program for several network switches. However, this paper cannot verify a whole of a program because of shortage of memory and execution time. Therefore, this paper verifies a program for each network switch by Coq and manually checks that a whole program for a network does not generate looping packets by using the result of verification in Coq. As a result, this paper proposes the methodology of verification of programs and shows that the methodology is efficient.

This paper is organized as follows; Sect. 2 explains NetCore and Coq. Section 3 explains the methodology of verification. Section 4 shows the example of using the methodology that is proposed in this paper. Section 5 discusses the proposed methodology. Section 6 concludes this paper.

2 Preliminary

2.1 NetCore

This section explains NetCore programs. NetCore is a declarative programming language and the programmers describe only the way of forwarding packets. For example, let me set up the program for an OpenFlow controller so that an OpenFlow switch sends packets that are received from network interface "Port 2", to network interface "Port 1" and drops all other packets. First, the following program of NetCore is created.

```
Definition pg1 := WILD /=> FWD 1.
```

This program sends all packets to "Port 1". The following explains the syntax of Netcore briefly; "Definition pg1 :=" defines the program "pg1", the left side of "/=>" is a condition and the right side of "/=>" is an action. "WILD" represents true, that is to say "WILD" represents all packets. "FWD 1" represents sending packets to "Port 1". This program is not enough because the program does not specify the condition of forwarding packets. The following program restricts the packets that are forwarded.

```
Definition pg2 := RESTRICT pg1 BY PORT=2.
```

This program sends only packets that are received by "Port 2", to "Port 1". "RESTRICT x BY y" represents that the packets that satisfy the condition "y" is applied to program "x". "PORT=2" represents the condition that the packets are received by "Port 2". The program "pg2" realizes that an OpenFlow switch sends packets that are received by network interface "Port 2", to network interface "Port 1" and drops all other packets. NetCore can represent many kinds of conditions and processes for packets. For example, NetCore can describe static NAT, firewalls and so on.

2.2   Coq

This subsection explains how NetCore programs are verified by Coq. This verification is based on Hoare logic [3]. A Hoare logic formula is the following description

$$\{P\}pg\{Q\}$$

where P is a precondition, pg is a NetCore program and Q is a postcondition. The Hoare logic formula represents that if the packets that arrive at an OpenFlow switch satisfy a precondition P, a post condition Q is satisfied after a program pg is applied to the packets. Verification of programs is to check whether Hoare logic formulas are satisfied. The previous researches proposed the method of checking a Hoare logic formula $\{P\}pg\{Q\}$. The method first deduces the weakest precondition so that a postcondition is satisfied after a program pg is applied to packets; in the following, $wp(pg, Q)$ is defined to be the weakest precondition. Next, the method checks whether P satisfies $wp(pg, Q)$ to check whether $\{P\}pg\{Q\}$. The following shows the verification of pg2 by using Coq.

```
Lemma verification: |-[PORT=2] pg2 [PORT=1].
Proof. checker. Qed.
```

In this description, "Lemma verification" means that Coq checks whether a proof "verification" holds or not. |-[PORT=2] pg2 [PORT=1] represents a Hoare logic formula {a packet arrives at Port 2}pg2{the packet departs from Port 1}. Concretely, "checker", which is a function for verification, verifies that this Hoare logic formula holds. The function "checker" calculates $wp(pg, Q)$ and checks whether P implies $wp(pg, Q)$. In this case, "Qed" shows that the Hoare logic formula holds. This paper uses the same method that is proposed by the previous research.

3   Methodology of Verification

This section explains the methodology that is proposed in this paper. For explanation, this section uses a network topology in Fig. 1. This section creates programs for the network topology and verifies that the programs do not generate looping packets. This section generates two programs: one program that does not generate looping packets and the other program that generates looping packets. Appendix A is a program that does not generate looping packets and Appendix B is a program that generates looping packets.

First, this paper tries to prove that the program in Appendix A does not generate looping packets; that is to say, this paper tries to prove a Hoare logic

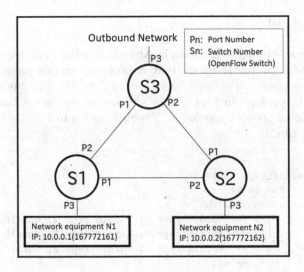

Fig. 1. Network topology

formula {P}*pg*{Q} where P is a precondition, Q is a postcondition and *pg* is the program in Appendix A. However, the program in Appendix A is too large to prove the Hoare logic formula. In fact, this paper cannot prove the Hoare logic formula by Coq because it lacks memories and takes much time.

Nowadays, CPUs become powerful and memory size of PCs become large, but the program in Appendix A is still large for verification by Coq. Those who have the ability of using Coq may prove the Hoare logic formula in the current PCs. However, many people cannot obtain this ability and it is important to prove the Hoare logic formula without high abilities of Coq and Hoare logic. Therefore, this paper proposes the methodology that enables those who are not familiar with Coq or Hoare logic to verify programs by Coq.

The overview of the methodology is as follows; suppose that the methodology checks that a program satisfies a property *Prop* at a network that consists of several network switches.

1. To create the properties that each switch should satisfy so that the program satisfies the property *Prop*.
2. To check that each switch satisfies the properties by Coq.
3. To prove that the program satisfies the property *Prop* by using the results that each switch satisfies the properties.

The point of this methodology is the first step, which is to create the properties for each switch. The first step separates the property *Prop* into several small properties for each switch. However, this separation cannot be accomplished automatically. In this paper, this separation is accomplished manually.

4   Verification

The section proves that the program in Appendix A does not generate looping packets by using the methodology that is proposed in this paper. The proof requires the description of network topologies in Coq. The Fig. 2 is a description of the network topology in Coq. There are several researches that deal with the descriptions of network topologies. The description in Fig. 2 is based on the previous research [11].

```
Definition def_topo: topo :=
{| ports := fun sw =>
            match sw with
            | 1 => [Word16.repr 1; Word16.repr 2; Word16.repr 3]
            | 2 => [Word16.repr 1; Word16.repr 2; Word16.repr 3]
            | 3 => [Word16.repr 1; Word16.repr 2; Word16.repr 3]
            | _ => nil
            end;
num_links := 3;
switch_topo := fun loc =>
            match loc with
            |Build_location 1 x =>
             if Word16.eq x (Word16.repr 1)
             then Some (Build_location 2 (Word16.repr 2))
             else if Word16.eq x (Word16.repr 2)
                  then Some (Build_location 3 (Word16.repr 1))
                  else None
            |Build_location 2 x =>
             if Word16.eq x (Word16.repr 1)
             then Some (Build_location 3 (Word16.repr 2))
             else None
            | _ => None
            end;
switch_topo' := fun loc =>
              match loc with
              |Build_location 2 x =>
               if Word16.eq x (Word16.repr 2)
               then Some (Build_location 1 (Word16.repr 1))
               else None
              |Build_location 3 x =>
               if Word16.eq x (Word16.repr 1)
               then Some (Build_location 1 (Word16.repr 2))
               else if Word16.eq x (Word16.repr 2)
                    then Some (Build_location 2 (Word16.repr 1))
                    else None
              | _ => None
              end |}.
```

Fig. 2. Network topology

```
Definition s_to_s' (ver_topo: topo)(loc_from loc_to: location) : bool :=
  match (ver_topo.(switch_topo) loc_from) with
    | Some loc' => loc_eq loc' loc_to
    | _ =>
      match (ver_topo.(switch_topo') loc_from) with
        |Some loc' => loc_eq loc' loc_to
        |_ => false
      end
  end.
end.
```

Fig. 3. Function

In "Def_topo" in Fig. 2, "ports" represents the network interfaces for each network switch. "num_links" represents the number of connections. Each of the network switches in Fig. 1 has three network interfaces and the network topology has three connections. "switch_topo" and "switch_topo'" shows the connections among switches. Concretely, "switch_topo" and "switch_topo'" are functions that map a switch number and a port number to the other switch number and the other port that are connected with the switch number and the port number. The function "s_to_s'" in Fig. 3 shows connection between switches.

The function "s_to_s'" receives a network topology, two pairs of a switch number and a port number as the first, second and third arguments. If the two pairs of the switch number and the port number are connected in the network topology, the function outputs true. Otherwise, the function outputs false.

By using this description, this paper checks whether both programs of Appendices A and B satisfy the property, which is loop free. This paper uses a PC that has Ubuntu 14.04 LTS as OS, Core i5-6600K as CPU and 8 GB memory. This paper uses Coq several times for several proofs and to execute each proof takes less than five seconds.

4.1 Overview of Verification

This paper verifies the programs according to the methodology that is proposed in this paper. First, this paper tries to verify that the program in Appendix A does not generate looping packets. According to the methodology, the property that looping packets are not generated is separated into several properties that are satisfied by each switch. This paper focuses on three kinds of packets whose destinations are 10.0.0.1, 10.0.0.2 and the others. This paper uses Coq to check how each switch deals with these kinds of packets. The results of using Coq are used to verify that looping packets are not generated in the program in Appendix A.

Figure 4 shows the proofs of Coq for the program in Appendix A. Figure 4 includes the proof results of thirteen lemmas. Each lemma represents a property of each switch. For example, Lemma tst1 represents that the packets that are received by Port 3 of Switch 1 and whose destination is 10.0.0.2 are forwarded to

```
Lemma tst1 : triple (SWITCH=1 AND PORT=3 AND NWDST =167772162)
                pg100 (PORT=1).
Proof. checker. Qed.
Eval compute in s_to_s' def_topo ( sw_pt 1 1) ( sw_pt 2 2).
Lemma tst1': triple (SWITCH=2 AND NOT PORT=3 AND (PORT=1 OR PORT =2) AND
                NWDST =167772162) pg200 (PORT =3).
Proof. checker. Qed.


Lemma tst2 : triple (SWITCH=1 AND PORT=3 AND NOT NWDST=167772162 AND
                NOT NWDST =167772161) pg100 (PORT =2).
Proof. checker. Qed.
Eval compute in s_to_s' def_topo ( sw_pt 1 2) ( sw_pt 3 1).
Lemma tst2': triple (SWITCH =3 AND (PORT =1 OR PORT =2) AND NOT
                NWDST =167772161 AND NOT NWDST =167772162) pg300 ( PORT =3).
Proof. checker. Qed.


Lemma tst3 : triple (SWITCH =2 AND PORT =3 AND NOT NWDST =167772161 AND
                NOT NWDST =167772162) pg200 ( PORT =1).
Proof. checker. Qed.
Eval compute in s_to_s' def_topo ( sw_pt 2 1) ( sw_pt 3 2).
Lemma tst3': triple (SWITCH=2 AND NOT PORT=3 AND (PORT=1 OR PORT=2) AND
                NWDST =167772162) pg200 ( PORT =3).
Proof. checker. Qed.


Lemma tst4 : triple (SWITCH=2 AND PORT=3 AND NDST=167772161) pg200
                (PORT=2).
Proof. checker. Qed.
Eval compute in s_to_s' def_topo ( sw_pt 2 2) ( sw_pt 1 1).
Lemma tst4': triple (SWITCH =1 AND NOT PORT =3 AND (PORT=1 OR  PORT=2)
                AND NWDST =167772161) pg100 ( PORT =3).
Proof. checker. Qed.


Lemma tst5 : triple (SWITCH=3 AND PORT=3 AND NWDST =167772161) pg300
                (PORT=1).
Proof. checker. Qed.
Eval compute in s_to_s' def_topo ( sw_pt 3 1) ( sw_pt 1 2).
Lemma tst5': triple (SWITCH =1 AND NOT PORT =3 AND (PORT =1 OR PORT =2)
                AND NWDST =167772162) pg100 (PORT =3).
Proof. checker. Qed.


Lemma tst6 : triple (SWITCH =3 AND PORT=3 AND NWDST =167772162) pg300
                ( PORT =2).
Proof. checker. Qed.
Eval compute in s_to_s' def_topo ( sw_pt 3 2) ( sw_pt 2 1).
Lemma tst6': triple ( SWITCH =2 AND NOT PORT =3 AND (PORT=1 OR  PORT=2)
                AND NWDST =167772162) pg200 ( PORT =3).
Proof. checker. Qed.


Lemma tst7 : triple (SWITCH =3 AND PORT =3 AND NOT NWDST =167772161 AND
                NOT NWDST =167772162) pg300 ( NOT WILD ).
Proof. checker. Qed.
```

Fig. 4. Proof of the program in Appendix A

Port 1 of Switch 1. Lemma tst1' represents that the packets that are received by Port 1 or Port 2 of Switch 2 and whose destination is 10.0.0.2 are forwarded to Port 3 of Switch 2, and Lemma tst6 represents that the packets that are received by Port 3 of Switch 3 and whose destination is 10.0.0.2 are forwarded to Port 2 of Switch 3. These three lemmas, which are proved by Coq, are related with the packets whose destination is 10.0.0.2. These lemmas imply that the packets arrive at the network N2. As a result, the packets whose destinations are 10.0.0.2 do not loop in the network topology. Figure 4 includes several lemmas, which are proved by Coq, for the packets whose destination are 10.0.0.1 and the others. As the case of the packets whose destinations are 10.0.0.2, these lemmas imply that the packets whose destination are 10.0.0.1 and the others do not loop in the network topology.

Figure 5 shows the proofs of Coq for the program in Appendix B. Since this program generates looping packets, the lemmas in Fig. 5 show that the packet whose destination is 10.0.0.2 loop in the network topology.

```
Lemma tst1 : triple (SWITCH =3 AND PORT=3 AND NWSRC=167772164 AND
               NWDST=167772162) pg300 (PORT=1).
Proof. checker. Qed.

Eval compute in s_to_s' def_topo (sw_pt 3 1) ( sw_pt 1 2).

Lemma tst2 : triple (SWITCH =1 AND PORT=2 AND NWSRC =167772164 AND
               NWDST =167772162) pg100 ( PORT =1).
Proof. checker. Qed.

Eval compute in s_to_s' def_topo ( sw_pt 1 1) ( sw_pt 2 2).

Lemma tst3 : triple (SWITCH=2 AND PORT=2 AND NWSRC=167772164 AND
               NWDST=167772162) pg200 (PORT=1).
Proof. checker. Qed.

Eval compute in s_to_s' def_topo ( sw_pt 2 1) ( sw_pt 3 2).

Lemma tst4 : triple (SWITCH=3 AND PORT=2 AND NWSRC=167772164 AND
               NWDST =167772162) pg300 (PORT =1).
Proof. checker. Qed.
```

Fig. 5. Proof of the program in Appendix B

5   Discussion

The verification that is described in the previous section shows that one program creates looping packets and the other does not create looping packets. Actually this paper would like to verify a whole of all programs. However, the function "checker" does not terminate to verify the whole of all programs because the verification is short of memory of the PC and takes much time. Therefore, this paper verifies a behavior of each switch and those who try to verify programs use the results of the verification for each switch to prove manually that the a program does not generate looping packets.

The network topology has three network switches and there are three destinations of packets; the first is N1, the second is N2 and the third is outbound. All switches may deal with all destination packets. Therefore, this paper requires three proofs for each of three switches and totally requires nine proofs to verify that a program does not generate looping packets. Since this paper deals with the simple network topology, nine proofs are enough for the verification. However, verifications in complicated network topologies require many proofs for many network switches. Moreover, those who try to verify programs must deal with many proofs. Therefore, the methodology that is proposed in this paper is hard to use for complicated network topologies.

6   Conclusion

This paper proposed the methodology that verifies the programs that are described in NetCore. In this methodology, manual proof and Coq verify the programs. The future works are to develop the method to verify the programs automatically and to improve the proposed methodology to deal with complicated network topologies.

A   Appendix (Program that Does Not Generate Looping Packets)

```
(* s1 *)

(* s1_p1 *)

Definition pg101 := WILD /= > FWD 1.

Definition pg102 := RESTRICT pg101 BY (PORT=3 AND NWDST=167772162).

(* s1_p2 *)
Definition pg103 := WILD /= > FWD 2.

Definition pg104 := RESTRICT pg103 BY (PORT=3 AND (NOT NWDST=167772162)).

(* s1_p3 *)
Definition pg105 := PORT =1 /= > FWD 3.

Definition pg106 := PORT =2 /= > FWD 3.
```

```
Definition pg107 := RESTRICT (pg105 PAR pg106) BY
                    (NWDST=167772161 AND NOT PORT =3).

(* s1_behave *)
Definition pg100 := RESTRICT ( pg102 PAR pg104 PAR pg107 ) BY SWITCH =1.

(* s2_p1 *)
Definition pg201 := WILD /= > FWD 1.

Definition pg202 := RESTRICT pg201 BY (PORT=3 AND (NOT NWDST=167772161)).

(* s2_p2 *)

Definition pg203 := WILD /= > FWD 2.

Definition pg204 := RESTRICT pg203 BY ( PORT =3 AND NWDST=167772161).

(* s2_p3 *)
Definition pg205 := PORT =1 /= > FWD 3.

Definition pg206 := PORT =2 /= > FWD 3.

Definition pg207 := RESTRICT ( pg205 PAR pg206 ) BY
                    ( NWDST =167772162 AND NOT PORT =3).

(* s2_drop *)

(* s2_behave *)
Definition pg200 := RESTRICT (pg202 PAR pg204 PAR pg207 ) BY SWITCH =2.

(* s3_p1 *)
Definition pg301 := WILD /= > FWD 1.

Definition pg302 := RESTRICT pg301 BY (PORT =3 AND NWDST=167772161).

(* s3_p2 *)
Definition pg303 := WILD /= > FWD 2.

Definition pg304 := RESTRICT pg303 BY ( PORT =3 AND NWDST=167772162).

(* s3_p3 *)
Definition pg305 := WILD /= > FWD 3.

Definition pg306 := RESTRICT pg305 BY ((PORT =1 OR PORT=2) AND
                    NOT PORT =3 AND NOT NWDST =167772161 AND
                    NOT NWDST =167772162).

(* s3_behave *)
Definition pg300 := RESTRICT ( pg302 PAR pg304 PAR pg306 ) BY SWITCH =3.
```

```
Require    Import   List.
Require    Import   Bool.
Require    Import   Arith.
Require    Import   WP.

Record topo : Type :=
  {
    ports : nat -> list Word16.t ;
    num_links : nat ;
    switch_topo : location -> option location ;
    switch_topo ': location -> option location
  }.

Require Import List.
Import ListNotations.
Close Scope Z_scope.
Definition def_topo : topo :=
  {| ports := fun sw = >
          match sw with
            | 1 = > [ Word16.repr 1; Word16.repr 2; Word16.repr 3]
            | 2 = > [ Word16.repr 1; Word16.repr 2; Word16.repr 3]
            | 3 = > [ Word16.repr 1; Word16.repr 2; Word16.repr 3]
            | _ = > nil
          end ;
      num_links := 3;
      switch_topo := fun loc = >
                match loc with
                  | Build_locatio n 1 x = >
                    if Word16.eq x ( Word16.repr 1)
                    then Some ( Build_location 2 ( Word16.repr 2))
                    else if Word16.eq x ( Word16.repr 2)
                        then Some ( Build_location 3 ( Word16.repr 1))
                        else None
                  | Build_location 2 x = >
                    if Word16.eq x ( Word16.repr 1)
                    then Some ( Build_location 3 ( Word16.repr 2))
                    else None
                  | _ = > None
                end ;
      switch_topo ' := fun loc = >
                match loc with
                  | Build_lo cation 2 x = >
                    if Word16.eq x ( Word16.repr 2)
                    then Some ( Build_location 1 ( Word16.repr 1))
                    else None
                  | Build_lo cation 3 x = >
                    if Word16.eq x ( Word16.repr 1)
                    then Some ( Build_location 1 ( Word16.repr 2))
                    else if Word16. eq x ( Word16.repr 2)
                        then Some (Build_location 2 (Word16.repr 1))
```

```
                              else None
                |  _ = > None
              end |}.
```

```
Definition loc_eq ( loc1 loc2 : location ): bool :=
  match loc1 , loc2 with
    | Bu ild_location sw1 pt1 ,
      Build_location sw2 pt2 = > beq_nat sw1 sw2 && Word16.eq pt1 pt2
  end.
```

```
Definition s_to_s' (ver_topo:topo)(loc_from loc_to:location):bool:=
  match ( ver_topo.( switch_topo ) loc_from ) with
    | Some loc' => loc_eq loc' loc_to
    | _ =>
        match ( ver_topo.( switch_topo') loc_from) with
          | Some loc' = > loc_eq loc' loc_to
          | _ = > false
        end
  end.
```

```
Definition sw_pt ( n : nat )( m : BinNums.Z ): location :=
  Build_location n ( Word16.repr m ).
```

B   Appendix (Program that Generates Looping Packets)

```
(* s1_p1 *)
Definition pg101 := WILD /= > FWD 1.
```

```
Definition pg102 := RESTRICT pg101 BY ((PORT=2 OR PORT=3) AND
                NOT PORT=1 AND NWDST=167772162).
```

```
(* s1_p2 *)
Definition pg103 := WILD /= > FWD 2.
```

```
Definition pg104 := RESTRICT pg103 BY ( PORT =3 AND NOT NWDST=167772162).
```

```
(* s1_p3 *)
Definition pg105 := WILD /= > FWD 3.
```

```
Definition pg106 := RESTRICT pg105 BY ((PORT=1 OR PORT=2) AND NOT PORT=3
                AND NWDST=167772161).
```

```
(* s1_behave *)
Definition pg100 := RESTRICT ( pg102 PAR pg104 PAR pg106 ) BY SWITCH =1.
```

```
(* s2_p1 *)
Definition pg201 := WILD /= > FWD 1.
```

```
Definition pg202 := RESTRICT pg201 BY ((PORT=2 OR PORT =3) AND
                NOT PORT=1).


(* s2_behave *)
Definition pg200 := RESTRICT pg202 BY SWITCH =2.


(* s3_p1 *)
Definition pg301 := WILD /= > FWD 1.


Definition pg302 := RESTRICT pg301 BY ((PORT =2 OR PORT=3) AND NOT PORT=1
                    AND (NWDST=167772161 OR NWDST=167772162)).


(* s3_p3 *)
Definition pg303 := WILD /= > FWD 3.


Definition pg304 := RESTRICT pg303 BY PORT =1.


(* s3_behave *)
Definition pg300 := RESTRICT (pg302 PAR pg304 ) BY SWITCH =3.


Record topo : Type :=
  {
    ports : nat -> list Word16.t ;
    num_links : nat ;
    switch_topo : location -> option location ;
    switch_topo ': location -> option location
  }.


Definition def_topo : topo :=
  {| ports := fun sw = >
            match sw with
              | 1 = > [ Word16.repr 1; Word16.repr 2; Word16.repr 3]
              | 2 = > [ Word16.repr 1; Word16.repr 2; Word16.repr 3]
              | 3 = > [ Word16.repr 1; Word16.repr 2; Word16.repr 3]
              | _ = > nil
            end ;
    num_links := 3;
    switch_topo := fun loc = >
                match loc with
                  | Build_locatio n 1 x = >
                    if Word16.eq x ( Word16.repr 1)
                    then Some ( Build_location 2 ( Word16.repr 2))
                    else if Word16.eq x ( Word16.repr 2)
                        then Some ( Build_locati on 3 ( Word16.repr 1))
                        else None
                  | Build_locatio n 2 x = >
                    if Word16.eq x ( Word16.repr 1)
                    then Some ( Build_location 3 ( Word16.repr 2))
                    else None
                  | _ = > None
```

```
                   end ;
     switch_topo ' := fun loc = >
                   match loc with
                   | Build_lo cation 2 x = >
                     if Word16.eq x ( Word16.repr 2)
                     then Some ( Bui ld_location 1 ( Word16.repr 1))
                     else None
                   | Build_lo cation 3 x = >
                     if Word16.eq x ( Word16.repr 1)
                     then Some ( Bui ld_location 1 ( Word16.repr 2))
                     else if Word16.eq x ( Word16.repr 2)
                          then Some (Build_location 2 ( Word16.repr 1))
                          else None
                          | _ = > None
                   end |}.

Definition loc_eq ( loc1 loc2 : location ): bool :=
  match loc1 , loc2 with
  | Bu ild_location sw1 pt1 ,
    Build_location sw2 pt2 = > beq_nat sw1 sw2 && Word16.eq pt1 pt2
  end.

Definition s_to_s' (ver_topo:topo)(loc_from loc_to:location ):bool :=
  match ( ver_topo.( switch_topo ) loc_from ) with
  | Some loc' = > loc_eq loc' loc_to
  | _ =>
  match (ver_topo.( switch_topo') loc_from ) with
               | Some loc' = > loc_eq loc' loc_to
               | _ = > false
          end
    end.

Open Scope Z_scope.
Definition sw_pt (n : nat)(m : BinNums.Z): location :=
  Build_location n ( Word16.repr m).
```

References

1. Canini, M., Venzano, D., Perešíni, P., Kostić, D., Rexford, J.: A NICE way to test openflow applications. In: Proceedings of the 9th USENIX Conference on Networked Systems Design and Implementation (2012)
2. Garcia, R., Tanter, É., Wolff, R., Aldrich, J.: Foundations of typestate-oriented programming. ACM Trans. Program. Lang. Syst. 36(4), 1–44 (2014)
3. Hoare, C.A.R.: An axiomatic basis for computer programming. Commun. ACM 12(19), 576–580 (1969)
4. Kuzniar, M., Peresini, P., Canini, M., Venzano, D., Kostic, D.: A SOFT way for openflow switch interoperability testing. In: Proceedings of the 8th International Conference on Emerging Networking Experiments and Technologies, pp. 265–276 (2012)

5. McKeown, N., Anderson, T., Balakrishnan, H., Parulker, G., Peterson, L., Rexford, J., Shenker, S., Turner, J.: OpenFlow: enabling innovation in campus networks. SIGCOMM Comput. Commun. Rev. **38**(2), 69–74 (2008)

6. Monsanto, C., Foster, N., Harrison, R., Walker, D.: A compiler and runtime system for network programming languages. In: Proceedings of the 39th Annual ACM SIGPLAN-SIGACT Symposium on Principles of Programming Languages, pp. 217–230 (2012)

7. Majumdar, R., Tetali, S.D., Wang, Z.: Kuai: a model checker for software-defined networks. In: Proceedings of the 14th Conference on Formal Methods in Computer-Aided Design, pp. 163–170 (2014)

8. Sethi, D., Narayana, S., Malik, S.: Abstractions for model checking SDN controllers. In: Formal Methods in Computer-Aided Design, pp. 145–148 (2013)

9. Sheard, T., Stump, A., Weirich, S.: Language-based verification will change the world. In: Proceedings of the FSE/SDP Workshop on the Future of Sofware Engineering Research, pp. 343–348 (2010)

10. Siek, J., Taha, W.: Gradual typing for functional languages. In: Proceedings of the Scheme and Functional Programming Workshop, pp. 81–92, September 2006

11. Stewart, G.: Computational verification of network programs in Coq. In: Gonthier, G., Norrish, M. (eds.) CPP 2013. LNCS, vol. 8307, pp. 33–49. Springer, Heidelberg (2013)

12. Tanter, É., Tabareau, N.: Gradual certified programming in Coq. In: Proceedings of the 11th Symposium on Dynamic Languages, pp. 26–40 (2015)

13. Ball, T., Bjørner, N., Gember, A., Itzhaky, S., Karbyshev, A., Sagiv, M., Schapira, M., Valadarsky, A.: VeriCon: towards verifying controller programs in software-defined networks. SIGPLAN Not. **49**(6), 282–293 (2014). PLDI 2014

14. Wu, Y., Haeberlen, A., Zhou, W., Loo, B.T.: Answering why-not queries in software-defined networks with negative provenance. In: Proceedings of the Twelfth ACM Workshop on Hot Topics in Networks, pp. 1–7 (2013)

15. Wundsam, A., Levin, D., Seetharaman, S., Feldmann, A.: OFRewind: enabling record and replay troubleshooting for networks. In: Proceedings of the USENIX Annual Technical Conference (2011)

16. The Coq Proof Assistant. https://coq.inria.fr

17. Haskell. https://www.haskell.org

18. NEC: "Trema Openflow Controller". http://trema.github.com/trema/

Parallelizing Simulated Annealing Algorithm in Many Integrated Core Architecture

Junhao Zhou[1], Hong Xiao[1], Hao Wang[2(✉)], and Hong-Ning Dai[3]

[1] Faculty of Computer, Guangdong University of Technology,
Guangzhou 510006, China
[2] Big Data Lab, Faculty of Engineering and Natural Sciences,
Norwegian University of Science and Technology, 6009 Ålesund, Norway
hawa@ntnu.no
[3] Faculty of Information Technology,
Macau University of Science and Technology, Macau, China

Abstract. The simulated annealing algorithm (SAA) is a well-established approach to the approximate solution of *combinatorial optimisation* problems. SAA allows for occasional *uphill moves* in an attempt to reduce the probability of becoming stuck in a poor but locally optimal solution. Previous work showed that SAA can find better solutions, but it takes much longer time. In this paper, in order to harness the power of the very recent hybrid *Many Integrated Core Architecture* (MIC), we propose a new parallel simulated annealing algorithm customised for MIC. Our experiments with the *Travelling Salesman Problem* (TSP) show that our parallel SAA gains significant speedup.

Keywords: Parallel computing · Simulated annealing · MIC optimization

1 Introduction

The simulated annealing algorithm (SAA) [8] is a well-established approach to the approximate solution of *combinatorial optimisation* problems. SAA is based on an analogy to the behaviour of physical systems in the presence of a heat bath. It improves the accuracy of traditional *local optimisation* technique, in which an initial solution is iteratively improved with small local changes until no such change produces a better solution. SAA allows for occasional *uphill moves* in an attempt to reduce the probability of becoming stuck in a poor but locally optimal solution. In this way, SAA can produce more accurate solutions.

Travelling Salesman Problem (TSP) [6] is a NP-hard combinatorial optimisation problem. We are given n cities and distances between all the cities, the goal is to find a shortest path in which the salesman visits each city exactly once. TSP has many applications and is established as the prototypical example of an NP-hard combinatorial optimisation problem. Previous work [7] showed that, compared to local optimisation techniques, SAA can find better solutions for TSP, but it takes much longer time.

© Springer International Publishing Switzerland 2016
O. Gervasi et al. (Eds.): ICCSA 2016, Part II, LNCS 9787, pp. 239–250, 2016.
DOI: 10.1007/978-3-319-42108-7_18

As commonly agreed, SAA is computationally intensive [10], researchers in the high performance computing community have developed various parallel and distributed SAA in different paradigms, including shared memory [9], message-passing [4], hybrid-parallel [2], and the recent MapReduce [10]. Very recently, the hybrid architecture with CPUs and coprocessors, such as the *Many Integrated Core Architecture* (MIC) by Intel®, has become one of the most promising platforms and initiated many highly parallel applications [3]. The existing parallel versions of SAA are not customised for such a new hybrid architecture, so they cannot harness fully the power of MIC.

In this paper, we propose a new parallel simulated annealing algorithm customised for MIC. Our experiments with the *Travelling Salesman Problem* (TSP) show that our parallel SAA gains significant speedup.

The remainder of this paper is organised as follows: Sect. 2 presents our improved simulated annealing algorithm. Section 3 describes our parallel simulated annealing algorithm. Experimental results with TSP are presented in Sect. 4. We conclude and point to our future work in Sect. 5.

2 Simulated Annealing Algorithm

The analogy in physics of simulated annealing algorithm is that initially the solid is preheated to a sufficient temperature, i.e., all particles in the solid is disordered (the state of the highest entropy), then the temperature decreased slowly and particles gradually return to order, i.e., entropy decreases. If the initial temperature is high enough and the cooling process is slow enough, eventually all particles will stay in the lowest energy state (lowest entropy).

2.1 The Basic Idea

Simulated annealing is an approach that attempts to avoid entrapment in poor local optima by allowing an occasional uphill move. As typically implemented, the simulated annealing approach involves a pair of nested loops and two additional parameters, a cooling ratio r, $0 < r < 1$, and an integer temperature length L. In Step 3 of the algorithm, the term frozen refers to a state in which no further improvement in $cost(S)$ seems likely. Note that $e^{-\Delta/T}$ will be a number in the interval $(0, 1)$ when $\Delta t'$ and T are positive, and rightfully can be interpreted as a probability that depends on $\Delta t'$ and T. The probability that an uphill move of size will be accepted diminishes as the temperature declines, and, for a fixed temperature T, small uphill moves have higher probabilities of acceptance than large ones.

The algorithm, within Markov chain lengths, iteratively execute the *"to generate new solutions - judge - accept/discard"* process, corresponding to the solid tends to thermal equilibrium at a certain constant temperature process. When the algorithm terminates, it returns an *approximately-optimal* solution.

Algorithm 1. Simulated Annealing Algorithm [7]

    Data: Cooling ratio r and length L
    Result: approximate solution S
1 Initialize solution S;
2 Initialize temperature $T > 0$;
3 **while** *not yet* frozen **do**
4   **for** $i \leftarrow 1$ **to** L **do**
5   Pick a random neighbor S' of S;
6         $\Delta \leftarrow (cost(S') - cost(S))$;
7   **if** $\Delta \leqslant 0$ // `downhill move`
8   **then**
9 | $S \leftarrow S'$;
10   **end**
11   **if** $\Delta \geq 0$ // `uphill move`
12   **then**
13 | $S \leftarrow S'$ with probability $e^{-\Delta/T}$;
14   **end**
15   **end**
16     $T \leftarrow rT$ (reduce temperature);
17 **end**

2.2   Generation of New Solutions

The generation of new solutions to the simulated annealing algorithm can be divided into the following four steps:

The first step is to generate new solutions in the solution space. In order to facilitate subsequent computation and to reduce execution time, new solutions are generated through simple transformations, such as replacing or swapping all or part of the elements in the previous solution. Note that the transformations to generate new solutions determines the neighbourhood structure of the new solution, thus has an impact on the cooling schedule.

The second step is to calculate the new solution to the difference between the corresponding objective function. The best way to calculate the difference between the objective function is calculating increments. The fact that, for most applications, this is the fastest method to calculate the difference between the objective function.

The third step is to determine whether the new solution is accepted, the judgement is based on an acceptance criterion, the most commonly accepted criteria are Metropolis criterion: If $t' < 0$ is accepted S' as the new current solution S, otherwise the probability $exp(-t'/T)$ to accepts S' as the new current solution S.

The fourth step is to determine when a new solution be accepted. With a new solution to replace the current solution, it simply corresponds to the current solution in converting part to generate new solutions to be implemented in time, an amendment to the objective function value. At this time, the current solution to achieve the first iteration. On this basis, we can begin the next round of testing.

And when a new solution is judged to be discarded, then the current solution on the basis of the original continue to the next round of testing.

Simulated annealing algorithm is independent of the initial value, the algorithm obtained the solution and the initial solution state S (is the starting point iterative algorithm) irrelevant. Simulated annealing algorithm has asymptotic convergence and has been shown theoretically to a probability 1 converge to the global optimal solution.

2.3   To Solve TSP Using Simulated Annealing Algorithm

In the Travelling Salesman Problem, there are n cities, with the number $(1, ..., n)$ representative. Distance between city i and city j is $d(i, j)$ $i, j = 1, ..., n$. TSP problem is to find the total length of the shortest path, which visited each city exactly once the domain of a circuit (see Fig. 1).

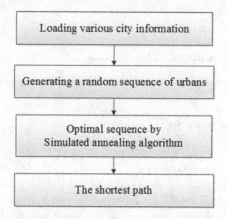

Fig. 1. Program flow chart

A solution S corresponds to a path in which the salesman visits each city exactly once, the visiting sequence is numbered by $1, ..., n$. That is, the cities that the salesman visits in sequence are denoted $(w_1, w_2, ..., w_n)$, and noted $w_{m+1} = w_1$. The cost function of a path is the total length, and the goal is to find a path with minimum cost:

$$f(w_1, w_2, ..., w_n) = \sum_{j=1}^{m}(w_j, w_{j+1})$$

Assume the current solution $(w_1, w_2, ..., w_n)$ is transformed to a new solution $(u_1, u_2, ..., u_n)$, then the cost function for the difference:

$$\Delta f = f(u_1, u_2, ..., u_n) - f(w_1, w_2, ..., w_n) = \sum_{j=1}^{n} d(u_j, u_{j+1}) + \sum_{j=1}^{n} d(w_j, w_{j+1})$$

Temperature management is also one of the difficult problems of simulated annealing algorithm. In the practical applications, it is necessary to consider the computational complexity of practical problems, therefore the following cooling method is often used:

$$T(t + 1) = k/timesT(t)$$

Where k is a constant positive and it is slightly less than 1.00. t is the number of times of cooling.

Depicted in Fig. 2 is the process to solve TSP based on simulated annealing algorithm.

2.4 To Improve Simulated Annealing Algorithm for TSP

In the original SAA, a large enough number M is set for the iteration of the outer loop to ensure that the temperature can drop to the stable state. However, due to the characteristics of simulated annealing, the temperature will drop faster in the beginning. When the temperature drop to a certain level, it starts to stabilise and decreases more slowly. The number of cycles to reach stability is unknown

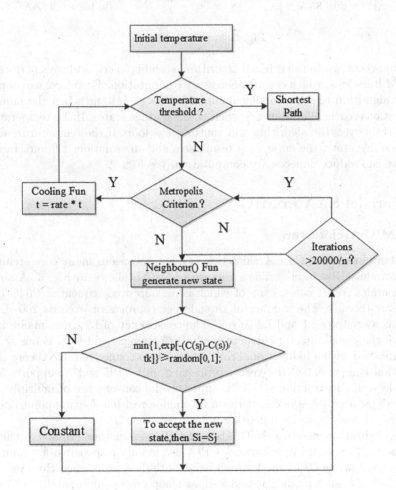

Fig. 2. Simulated annealing algorithm flowchart

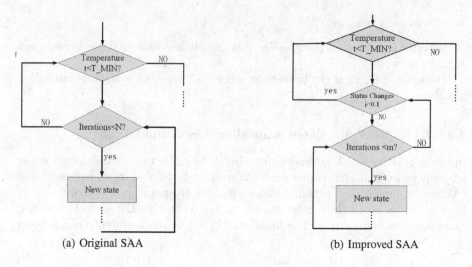

(a) Original SAA   (b) Improved SAA

Fig. 3. Improved SAA

prior to execution. In the original algorithm, stability check is always performed after M iterations, which cause unnecessary computation. Therefore, we propose in our algorithm to set a smaller number $N(N << M)$ and when the number of iterations reaches N, it checks whether the state is stable. If the temperature is not stable yet, the algorithm will continue the loop. If the temperature drop to the stable state, the outer loop terminates and the solution is found. In this way, we can reduce unnecessary computation (see Fig. 3).

3   Parallel SAA in MIC

3.1   MIC Architecture

The Many Integrated Core Architecture is a heterogeneous many-core architecture introduced by Intel®. The structure of MIC is shown in Fig. 4. A single chip contains 57–61 cores, each of which is an improved sequential P54C core with four threads. The number of threads per coprocessor exceeds 200. Each core has a separate L1 and L2 caches. The consistency of L2 cache maintained through the global directory structure. GDDR5 memory and PCI, E bus access are connected with a bidirectional ring network interconnection. KNC core is an important component of the vector processing unit VPU and it supports 512-bit wide vector instruction set. VPU supports the convergence of multiply-add instructions, each cycle it can perform 32 single-precision floating-point operations, or 16 double-precision floating-point operations [1].

The hybrid architecture of GPUs and CPUs is another popular architecture for highly parallel applications. CPUs are mainly responsible for complex control flows, while GPUs are for high-bandwidth data processing. However, the GPUs lack of branch prediction and feedback loop processing capabilities [5] and the coordination between CPUs and GPUs cause complexity with an increased

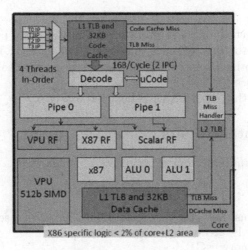

Fig. 4. MIC architecture

burden on programmers. In contrast, MIC architecture supports highly parallel applications and it provides facilities for traditional parallel programmers to adapt their programs to the new platform. The features such as high computing density, high performance/power ratio and ease of programming makes MIC be an appealing platform. MIC offers good flexibility for programming. A coprocessor can also be seen as an independent node. MIC can be configured in three modes: (1) NATIVE mode, the program runs only on the MIC; (2) OFFLOAD mode, that CPU will need to calculate the portion of the MIC card OFFLOAD to perform parallel computing; (3) peer mode, procedures applicable to more complicated process, such as MPI program in the CPU and main function MIC terminal simultaneously initiated. Our parallel SAA is implemented in NATIVE mode.

Nowadays, major high-performance architectures use *Performance Monitoring Unit* (PMU), which is a specific type of register to detect and count events from various micro-architectural components. These events provide details of the interaction between programs and the architecture. In our implementation, we have used the PMU-based Vtune® software by Intel® to measure the performance of our algorithm (see Fig. 4).

3.2 A Parallel Simulated Annealing Algorithm

The parallelization of the SAA is as follows:

(1) Initialisation: Generating k problem, which is the size M. The function value is calculated in parallel for each initial solution. And retain the sub-problems local optima.
(2) Compute and collect all k sub-optimal solutions to the problem and identify the global optimal solution.

The computation is divided and distributed to multiple processes. The No.0 Process performs scheduling and summation. Each of other processes will produce the optimal solution to each subset as local optima. The No.0 process collects all local optimal solutions and identifies an optimal solution as a global optimal solution (see Fig. 5) [12].

The process of the parallel SAA is depicted in Fig. 6.

4   Experimental Results

The server configuration for our experiments is as follows:

– Processor: Intel® Xeon® CPU E5-2620 v2 @ 2.10 GHz;
– Memory (RAM): 63 GB;
– Operating systems: Red Hat Enterprise Linux Server release 6.3 (Santiago);
– Coprocessor: Intel® Xeon Phi® coprocessor 5110P;
– Coprocessor memory: 8 GB.

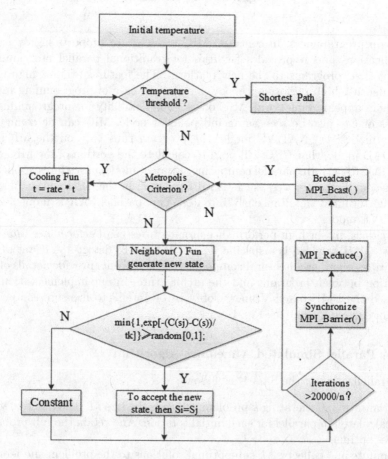

Fig. 5. Distribution of computational tasks

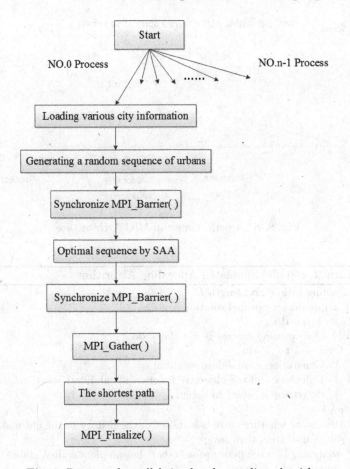

Fig. 6. Process of parallel simulated annealing algorithm

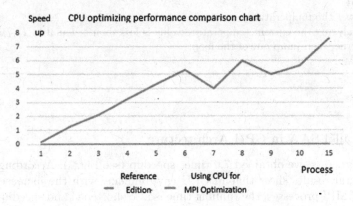

Fig. 7. SAA performance in CPU Architecture

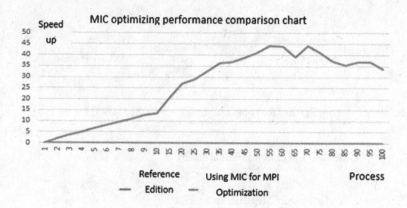

Fig. 8. SAA performance in MIC Architecture

Algorithm 2. Parallel Simulated Annealing Algorithm

Data: Cooling ratio r and length L
Result: approximate optimal solution S

1 **while** $T > T_{MIN}$ **do**
2   **while** *The sampling process is not stable* **do**
3   **for** $i \leftarrow 1$ **to** M **do**
4   Generating a candidate solution;
5   Calculation of the objective function of candidate solutions, and determine whether to accept;
6   **end**
7   Determine whether there is a subset of the history in the optimal solution, if true, then save;
8   Analyzing this temperature, whether the sample stability, stable then jump out of the loop;
9   **end**
10   Lower the temperature;
11   Analyzing the current temperature value is less than or equal to T_{MIN}, if it is true, then jump out of the loop;
12 **end**

4.1   Parallel SAA in CPU Architecture

With 15 processs, we obtained 7.6 times speedup (see Fig. 7). According to the experimental results show that in CPU environment, with the increase in the number of MPI processes, the running time is also shortened and speedup significantly increased. In the figure there have been some turning point, because the scheduling process appeared to be unstable. The overall trend goes up. However, due to the fact that the number of processes is rather limited, maximum speedup is not achieved.

4.2   Parallel SAA in MIC Architecture

Figure 8 shows the experimental results of the parallel SAA in MIC architecture. The horizontal axis represents the number of processes, the ordinate represents speedup. On the super computer named 'yuan', compared with CPU in the case of 15 processes, the algorithm in MIC architecture is 20 times faster. Using the MIC architecture with 70 processes, our algorithm achieved the highest speedup of 44.2 times.

With the many-core capability of MIC, our algorithm gets more number of processes. In this experiment, the algorithm harnesses 100 processes. When the number of processes is in the range of 50–60, the speedup reached its peak, to about 45 times. When the number of processes is in the range of 60–70, the speedup appeared to float because large number of processes got initiated so the process scheduling became unstable. With more than 70 processes, the speedup began to decline. This is because a large number of processes send the results, they caused increased burden on I/O, which decrease the overall efficiency.

5   Conclusions and Future Work

In this paper, we propose a new parallel simulated annealing algorithm customised for the recent MIC architecture. Our experiments with the Travelling Salesman Problem show very promising results with significant speedup and great potential for scaling-up.

For the next step, we intend to further improve our parallel simulated annealing algorithm and look for opportunities to parallelize other algorithms using the MIC architecture. In addition, we are exploring the applicability of the simulated annealing algorithm in real industrial applications [11].

Acknowledgement. The work described in this paper was partially supported by Macao Science and Technology Development Fund under Grant No. 096/2013/A3 and the NSFC-Guangdong Joint Fund under Grant No. U1401251 and Guangdong Science and Technology Program under Grant No.2015B090923004.

References

1. Bo, S., Yong, Z.G., Shao-hua, W., Xiao-wei, L., Qing, Z.: Research of offload parallel method based on MIC platform. Comput. Sci. **41**(6), 477–480 (2014)
2. Choong, A., Beidas, R., Zhu, J.: Parallelizing simulated annealing-based placement using GPGPU. In: International Conference on Field Programmable Logic and Applications, FPL 2010, pp. 31–34. IEEE (2010)
3. Guo, S., Dou, Y., Lei, Y.: GPU parallel optimization of the oceanic general circulation model pop. Comput. Eng. Sci. **34**(8), 147–153 (2012)
4. Hansen, P.B.: Studies in Computational Science: Parallel Programming Paradigms, 1st edn. Prentice Hall PTR, Upper Saddle River (1995)
5. Jie, F., Guohua, Z.: Parallel ant colony optimization algorithm with GPU-acceleration based on all-in-roulette selection. Comput. Digital Eng. **39**(5), 23–26 (2011)

6. Johnson, D.S.: Local optimization and the traveling salesman problem. In: Paterson, M.S. (ed.) Automata, Languages and Programming. LNCS, vol. 443, pp. 446–461. Springer, Heidelberg (1990)

7. Johnson, D.S., Aragon, C.R., McGeoch, L.A., Schevon, C.: Optimization by simulated annealing: an experimental evaluation; part i, graph partitioning. Oper. Res. **37**(6), 865–892 (1989)

8. Kirkpatrick, S., Gelatt Jr., C., Vecchi, M.: Optimization by simulated annealing. Science **220**(4598), 671–680 (1983)

9. Ma, J., Li, K.p., Zhang, L.Y.: The adaptive parallel simulated annealing algorithm based on TBB. In: 2nd International Conference on Advanced Computer Control, ICACC 2010, vol. 4, pp. 611–615. IEEE (2010)

10. Radenski, A.: Distributed simulated annealing with MapReduce. In: Di Chio, C., et al. (eds.) EvoApplications 2012. LNCS, vol. 7248, pp. 466–476. Springer, Heidelberg (2012)

11. Wang, H., Osen, O., Li, G., Li, W., Dai, H.N., Zeng, W.: Big data and industrial internet of things for the maritime industry in northwestern norway. In: IEEE Region 10 Conference, TENCON 2015 (2015)

12. Wei, W.: The research on parallel algorithm of simulation annealing. Comput. Knowl. Technol. **3**(7), 1523–1524 (2008)

On Efficient SC-Based Soft Handoff Scheme in Proxy Mobile IPv6 Networks

Byunghun Song[1], Youngmin Kwon[1], Hana Jang[2], Jongpil Jeong[3], and Jun-Dong Cho[3(✉)]

[1] IoT Convergence Research Center, Korea Electronics Technology Institute (KETI), Seongnam 463-816, Korea
{bhsong, kwon.youngmin}@gmail.com
[2] Graduate School of Information and Communications,
Sungkyunkwan University, Sungkyunkwan-Ro, Jongno-Gu, Seoul, Korea
hnsh77@naver.com
[3] College of Information and Communication Engineering,
Sungkyunkwan University, 2066 Seobu-ro,
Jangan-gu, Suwon, Kyunggi-do, Korea
{jpjeong, jdcho}@skku.edu

Abstract. This study examines mobility support functions in soft handoff and IP-based mobile network, and distinguishes the control areas in the cell's range areas. Several important characteristics of cell configurations for soft handoff are used to propose new structures for mobile network's efficient session control (SC). A fixed-point strategy is proposed in order to not only determine the handoff traffic's arrival speed, but stably calculate the loss probabilities or set the optimum guard channel numbers. We suggest a False Handoff Sessions (FHS) for improving the channel use efficiency based on mobility information. Numerical analyses indicate the efficiency of the presented Markov chain model and the advantage of proposed soft handoff method.

Keywords: Continuous Time Markov Chain (CTMC) · Soft handoff · Session control (SC) · Proxy mobile IPv6 · False Handoff Sessions (FHS)

1 Introduction

Soft handoff is a critical function in IP-based mobile networks. Mobile Node (MN) receives the signals from the multiple Local Mobility Anchor (LMA) via their soft handover region [1, 2]. A combination of a soft handoff [3], Markov chain [4, 5], and the compensation network probability model [3, 6], is introduced to evaluate the performance of a system incorporating a soft handoff. However, for IP-based mobile systems the loss probability of the closure in the form of a solution is being studied. IP-based application of the various cellular mechanisms for characterization of the soft handoff in mobile networks is presented to demonstrate the distribution of the LMA and session [7–14]. Since the handoff region accounts for 50 % of the total cell area

J-D. Cho—Distinguished Visiting Professor, North University of China (Shanxi 100 people plan)

from the IP-based mobile networks, in general about 30 % may occur out of the channel, because use efficiency of traffic channels requires multiple channels simultaneously to reduce the soft handoff request. This queue priority session control portion, such as handoff (SC) technique [1], the guard channel reservation [13], parameter optimization [8] and the like have been proposed to address this problem.

New modeling techniques for assessing the performance index of a soft handoff method for soft handoff include the performance analysis of the IP-based mobile network. First, network-based mobility management [15] based on separating the handoff region introduces the expansion and improvement of mobility management schemes [3, 16] with various handoff, then develops an algorithm to predict that the relative mobility of the MN in the system. In addition, the Markov chain model based on various performance analysis models to derive a closed solution for the measurement and calculation of the loss formulas and reliable solution was used in an algorithm to determine the optimum number of the guard channels.

The proposed model is a direct extension of that proposed for the performance of hard handoff channel assignment [17] in a mobile network. Analysis model of the soft handoff in a mobile network [17] has been proposed in a variety of situations [18] to investigate the key factors that determine the performance of the soft handoff, and compare the performance of hard handoff, and soft handoff. Therefore, this channel guard, priority handoff atmosphere, and the doctor did not consider the concept of handoff request. An analysis model is proposed to integrate more functions associated with the soft handoff. In addition, a stable algorithm is provided.

This paper is structured as follows. We give an overview of the IP-based mobility management schemes in Sect. 2, point out our proposed scheme in Sect. 3, and discuss the performance evaluation for our proposed scheme in Sect. 4. Finally, Sect. 5 provides our conclusions and future research plans.

2 Related Work

Mobile IPv6 (MIPv6) [19] is a protocol proposed to manage IPv6-based mobility. The MN moves to a new link in the Internet outside the home link; to a host-based protocol that allows a network attachment point to communicate even in the duration before change. Mobility means that there is no change before and after the movement of the MN in an upper layer of the network layer. The Hierarchical MIPv6 (HMIPv6) [20] is a host-based protocol proposed as a way to reduce the handoff delay the MN in MIPv6 network. This handoff delay is increased by performing the Binding Update (BU) to the Home Agent (HA) and Correspondent Node (CN), movement of the other subnet unnecessary overhead and the delay of the signal. HMIPv6 [21, 22] is proposed to solve this problem by adding a node called Mobility Anchor Point (MAP) to locally manage the movement of the MN. MAP is defined as the movement of the local HA in the visited network and decrease in the update delay limits the number of signals in the MIPv6 domain around the network. The MN in the MAP domain of the 2 temporary configurations to the IP including Regional CoA (RCoA) and On-Link CoA (LCoA) is MN/CN BU; exchange of messages between requires simply changing the LCoA carried out while the MN is in the domain or Access Network (AN) on the MAP.

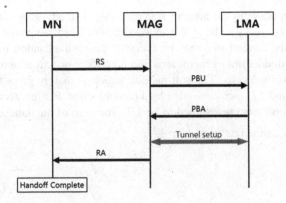

Fig. 1. Handover operation of the PMIPv6.

Finally, the PMIPv6 [23, 24] was adopted in this paper. MIPv6 is often limited in mobility such as in increased amount of movement of the resources due to signaling on the radio link between the terminal and the access router, the performance and resources are limited. This problem has been emerging as a network-based protocol, PMIPv6 (Proxy MIPv6) [19, 25, 26]. Figure 1 shows the operating procedures of the PMIPv6. A PMIPv6 LMA and Mobile Access Gateway (MAG), and Authentication, Authorization and Accounting (AAA) server are required in the new component to be made. LMA is a kind of HA role for the mobile node in the domain.

MAG is mainly the first hop to the mobile node that is directly connected and located in the AR, and performs signaling on behalf of the mobile node. The mobile node performs the L2 access authentication, and the MN to the MAG informs the process. MAG performs the authentication procedure with the AAA, and updates the current location for sending a PBU message to the LMA, it must charge the mobile node the terminal. LMA that receives the PBU is sent a PBA message to MAG by the service of the terminal. LMA is in service and ready to create a bi-directional tunnel between the LMA and MAG using the address of the MAG. MAG connected to the MN sends a message to allocate to their HNP (Home Network Prefix) who's LMA and the IP address is assigned to the RA. Once the settings for connection to MAG is sent to the LMA associated with the tunnel, all traffic coming from the mobile node and the LMA, LMA is to be sent to the MN the MAG manages all traffic coming from the outside.

3 Proposed Scheme

In IP-based mobile networks, MAG calculates a received signal strength of the MN with a number of multi-channel and wide bandwidth measurement. If the current power received from the MAG of the MN is in excess of any handoff start time T_{add} predefined, the MN sends an RS in MAG carries information of the MN to the current target MAG. If the MN receives the handoff command from the MAG, it detects the MN moves to the execution phase in the handoff preparation stage. When the current is reduced below the power received by the MN's MAG any cell handoff completion time T_{drop}, MN moves to the completion phase in the handoff step executed. The advantage of handoff in an IP-based mobile networks is the ability to use different reception paths. That is at least one MAG from the handoff step may be executed during the handoff time.

Cells can be divided into general area and handoff region. It is assumed that each cell is surrounded by 6 cells. The soft handoff region is mainly gives a signal by the MAG. The T_{add} and T_{drop} are controlled by a random value. For the overlapping region in a cell, the handoff ratio is assumed to be 1/3. The ratio of the handoff region to the entire cell area is defined as $a^2 \frac{\sqrt{3}}{2} \Big/ 3a^2 \frac{\sqrt{3}}{2}$.

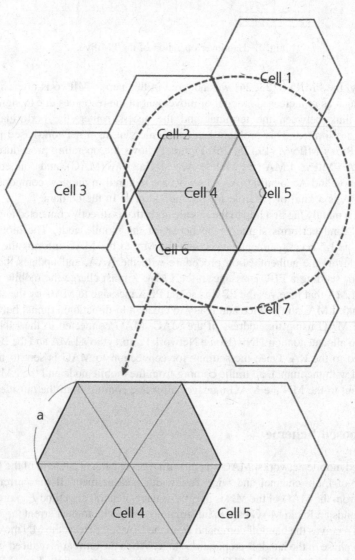

Fig. 2. Soft handoff of the System.

The cross section of the cell is considered a soft handoff region. All the MN has 2 handoff signal in the step execution. (Figure 2) the soft handoff process may have more than 1 MAG from the handoff execution step. MN can be received the multiple signals from the MAGs. If one of the signal strength in the handoff execution step is less than T_{drop} that the MN moves in the general area of the target cell from the handover region. The exact structure of soft handoff in IP-based mobile networks are irregular cell boundaries, such as signaling traffic conditions and the MN's movement, due to various factors are difficult to explain. To simplify the problem, we assume the following:

1. IP-based mobile network has the same number of all the size, shape, and the cell, the area of each cell is substantially in the form of exact circle.
2. MN initiated the communication is uniformly distributed throughout each cell and, MN can be carried out up to a communication time.
3. Cells in the system are properly distributed and symmetrically, and each cell is surrounded by 6 other cells.
4. In the handoff region, the MN has 2 signals up to the execution phase in the handoff.

Figure 3 is based on the assumption of the above shows an example of area and perimeter. The general area and the handoff region can be divided into a range of cells. The soft handoff region is shown by the intersection of the target cell and neighboring cells. Each MN reserves 2 channels for different transmission. Handoff region may be further divided into 2 regions of the target control area and the surrounding area control. MAG control of the target cells in the target area has a stronger signal than the surrounding cells in the handoff execution phase of MN.

By default, the link from the parent to select the MAG in the IP-based network, the received signal of the cell has a higher reception intensity in the handoff execution step. When the signal strength received from the MAG decrease away from the MN and the MAG, it is assumed that the MN is closer to increases toward MAG. Handoff area at the MN can detect the signal strength coming from the MAG. And can be spread at the same time to the channel of the current time t. This is a forward link channel used to transmit some system parameters. The mobile devices are used to build the MAG and the forward traffic channel. This is important for time synchronization. $ps(t, i)$ is the signal strength from the services you receive MAG measured when time t work by MN_i. And $cr\_ps(t, i)$ is the ratio of the change to the $ps(t, i)$.

$$cr\_ps(t, i) = \frac{ps(t + \Delta t, i) - ps(t, i)}{\Delta t} \tag{1}$$

Δt is the update time information in a mobile network. Rate of change in signal strength and, (in the handoff area) of the MN such as mobility relative position, direction of movement and can be estimated by the speed. This estimate is MAG is a strong signal strength, signal strength $ps(t, i)$ and $ps(t, i)$ handoff execution phase of MN_i in the process of handoff should be close to the MN than MAG. In addition, for sensing the rate of change ratio $cr\_ps(t, i)$ from MAG, a handoff in-progress in the MN has to move toward the MAG. $cr\_ps(t, i)$ ratio change value is greater, the faster the speed. If when $|cr\_ps(t, i)| < \varepsilon$, ε is appropriate to select a value, MN is assumed

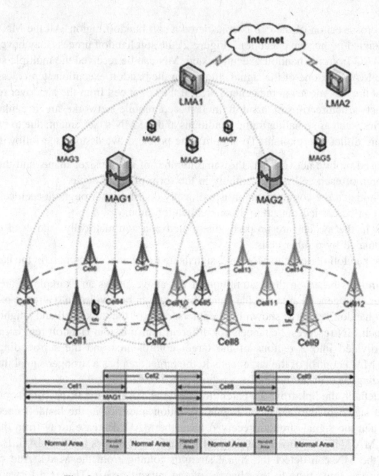

Fig. 3. Soft handoff of the IP based mobile network.

to be fixed. Therefore, the area of the defined cell is possible to calculate the measured value of $ps(t,i)$ and $cr\_ps(t,i)$.

The area of the range of cells is determined that the signal intensity of the MN in the area to determine if greater than T_{add}. Soft handoff region of the 2 cells is determined by the 2 signal strength in both the target and the surrounding MAG MAG verify larger than T_{add}. Also, $cr\_ps(t,i)$ of MAG in the general area of the target cell is higher than the T_{add}, all the signal strength is in the vicinity of the MAG is determined by checking the case that within a T_{drop}. The target cell and the neighboring cell region in the handoff region is indistinguishable by comparing the measured signal strengths. $ps(t,i)$ is defined as the power of the transmission signal received at around the MAG by the electric power, and $ps_N(t,i)$ is received in the MAG of the target cells in the transmission signal by the MN in the soft handoff region MN. If $ps_T(t,i)$ is greater than the $ps_N(t,i)$, MN must be in the target cell area. If $ps_T(t,i)$ is less than the $ps_N(t,i)$, MN must be in the area surrounding the cell. The overall structure of the algorithm for estimating the relative mobility running on an IP-based network is as follows.

- Step 1: Determine the location of the handoff session

(assuming that more than one can be connected)
for (m: number of the MN in the handoff region of the target MAG)
for (q: a desired value of the time t)
Time t is obtained by the MN_i when the value of the intensity of the $ps_T(t,i)$
       signal to be measured from the target MAG.
Time t is obtained by the MN_i when the value of the intensity of the $ps_N(t,i)$
       signal to be measured from the target MAG.
if $ps_T(t,i) > ps_N(t,i)$
MN_i belongs to the target cell area.
else
MN_i belongs to the near cell area.

- Step 2: Evaluation of the handoff session mobility

Mobility evaluation of the handoff session
Calculate the value of time t by the MN_i of the $ps_T(t,i)$ signal strength is
      measured from when the MAG.
Calculate the value of the variation ratio $cr\_ps(t,i)$ of the $ps_T(t,i)$.
if $|cr\_ps_T(t,i)| < \varepsilon$
The MN does not move.
if $|cr\_ps(t,i)| > 0$
The MN moves to the target MAG.
else
MN is moved to peripheral MAG get out of the target MAG.

When implementing an actual mobile network, due to the fading characteristics of the wireless channel, it is necessary to propose a mobility assessment for important details. Several samples of the received signal strength may be used to retrieve the value of one of the $ps(t,i)$. Measurement results can be shared between different MAG in order to improve the accuracy of the predictive. Although the proposed new soft handoff scheme is based on user mobility information, effective and reliable evaluation does not consider the mobility.

4 Performance Evaluation

We develop a Continuous Time Markov Chain (CTMC) for the soft handoff method. Performance indices defined in a closed form expression is derived in Sect. 2. Section 3 addresses the fixed-point iteration to determine a handoff arrival probability.

4.1   Markov Chain Model for Soft Handoff and the Handoff Queue

In the same all around the cell is a statistic and can be considered for model performance of a single cell in an IP-based network in a state that operates independently. Type of session is a new session and cell handoff the session, there are 2 types. The arrival of these signals are assumed to be Poisson λ_n and λ_h. For arrival of a new session, it can be seen that it may occur in the control area (hexagon) of the target cell. N is the number of channels in the channel pool of cells and each cell is assumed to be limited and preferred method than that after deleting the session handoff, to reserve a g channels of all the channels available for handoff to a new session.

All handoff requirements are considered to be found in the fully proposed model and proceeds rapidly if the allocation of the channel in a moment, if possible. Handoff queue is used in situations in which the absence of air channels available handoff session arrives. The maximum length of a handoff queue is l_e. In addition, the exponential distribution index of the cell that covers the entire area to be regarded as a u_c^{-1} respectively with u_{dc}^{-1} about grabbing the channel time T_c, average downtime T_{dc}. Session is a cell which is a handoff request, in the middle of air is terminated by force when moving out of the radio coverage area of the neighboring cell. The stay time of the exponential distribution is considered to be a factor u_l. In view of the proposed home, and the life cycle of the base model follows the CTMC. In the state of the CTMC, $C(t)$ means a crowded channel of the target cell can be added in the mobile in the handoff queue for the time t.

The life-cycle ratio is dependent on the state. If the number of channels is greater than the $N - g$, the production rate of the CTMC, because a new session is blocked the formula (2).

$$\Lambda(n) = \begin{cases} \lambda_{1c} + \lambda_h, & \text{if } n < N - g \\ \lambda_h & \text{if } N - g \leq n < N + l_e \end{cases} \qquad (2)$$

It can be obtained where the $\lambda_{1c} = \lambda(1 - \beta/2) + \lambda_n(\beta/2)\lambda_b$. β is a new session to block the possibility of the cell to be calculated later and the ratio of the peripheral cell area of the target cell. The new session will be seen that the target cell is divided into two parts. One is the target area and the other area of the cell is added from the general surrounding area that the request is denied by the surrounding cells of the new cell. $M(n)$ of CTMC is

$$M(n) = \begin{cases} n(\mu_c + \mu_{dc}), & \text{if } n \leq N \\ N(\mu_c + \mu_{dc}) + (n - N)(\mu_c + \mu_l), & \text{if } N \leq n < N + l_e \end{cases} \qquad (3)$$

The number of channels is less than $C(t)$ or equal to N, cell is waiting for the handoff queue. Channel is primarily due to the completion and start off the session on the target cell. If the number of channel $C(t)$ can be use is larger than N or smaller than the $N + l_e$, the N channel assignment and some handoff call will wait in the queue. In this case, some of sessions in the queue, the session is deleted from the target cell cannot get the channel until they are moved out of the radio range of the neighboring cell. The steady-state probability is defined as $p_n = \lim_{t \to \infty} \Pr(C(t) = n), n = 0, 1, \cdots, N + l_e$.

In addition, this is defined as $A = (\lambda_{1c} + \lambda_h)/(\mu_c + \mu_{dc})$, $A_1 = \lambda_h/(\lambda_{1c} + \lambda_h)$ and $A_2 = (\mu_c + \mu_l)/(\mu_c + \mu_{dc})$.

So you can see from the pool is the Eq. (4) of the CTMC.

$$p_n = p_0 \begin{cases} A^n, & \text{if } n < N - g \\ \frac{A^n}{n!} A_1^{n-(N-g)}, & \text{if } N - g \leq n \leq N \\ \frac{A^n}{(N! \prod\limits_{j=1}^{n-N} (N+jA_2))} A_1^{n-(N-g)}, & \text{if } N < n \leq N + l_e \end{cases} \tag{4}$$

Obviously, $\sum\limits_{n=0}^{N+l_e} p_n = 1$. Accordingly, it is possible to obtain an expression for the p_0, as shown in Eq. (5).

$$p_0 = \left[\sum_{n=0}^{N-g-1} \frac{A^n}{n!} + \sum_{n=N-g}^{N} \frac{A^n}{n!} A_1^{n-(N-g)} + \sum_{n=N+1}^{N+l_e} \frac{A^n}{N! \prod\limits_{j=1}^{n-N} (N+jA_2)} A_1^{n-(N-g)} \right]^{-1} \tag{5}$$

Blocking probability means that the probability of blocking in this paper as seen from the point of view of the cell such as $P_b(N, g, l_e)$ and viewing from the point of view of the system such as P_{BS}. Blocking probability of a new session is the formula (6) and Eq. (7).

$$P_b(N, g, l_e) = \sum_{n=N-g}^{N+l_e} p_n \tag{6}$$

$$P_{BS} = \beta P_b P_b + (1 - \beta) P_b \tag{7}$$

If a new session is in a soft handoff region is blocked in a cell, the opportunity arises to other cells. That is, the reason for using more than two of the above formula for the P_b to block the possibility of the system.

P_{ds} is the formula of the existing handoff scheme, 8 are given in the points of view in the system represents the probability of missing handoff FHS scheme.

$$P_{ds}(N, g, l_e) = p_{N+l_e} + \sum_{n=N-g}^{N+l_e} (n - N) \mu_l \frac{p_n}{\lambda_h} \tag{8}$$

The incoming handoff sessions are missing in the two situations. First, a handoff queue is to be filled (and do not know whether to use any channel) is sometimes expressed as a formula in the first handoff request to the device (8). Second, before the channel is assigned to the target cell is represented as the second with the queued handoff request to handoff at the time of moving out of the radio range of the neighboring cells, as is represented by the formula (8).

4.2 Numerical Results and Discussion

By applying a certain fixed value $A = 20$, $A_1 = 0.5$, $\mu_1 = 0.024$, $\beta = 0.3$, $l_e = 4$, $\mu = 0.04$ in the channel number N, P_{ds} and P_{bs} composed the loss probability for the different value g. N is not assign to the 1000 as a number of large. And we assign the value of parameters $N = 24$, $g = 2$, $\beta = 0.3$, $\lambda_n = 0.01$, $\mu_c = 0.01$, $\mu_l = 0.024$, $\mu_{dc} = (T_{dc})^{-1} = 0.03$ and $l_e = 4$ in consideration of the system to obtain the results shown in Figs. 4 and 5.

Figures 4 and 5 shows a conventional soft handoff approach. New handoff request for the blocking probability and λ_h shows the blocking probability of a new handoff request about the number of channels N and g of the proposed schemes for FHS. It is slightly better than the conventional soft handoff technique, because the new call blocking probability of the FHS plan saves some channel resources for the new session requests from the FHS techniques.

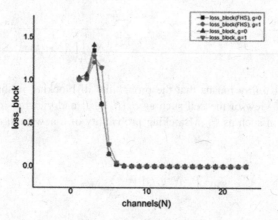

Fig. 4. Handoff request block for the FHS and existing handoff techniques about N and g.

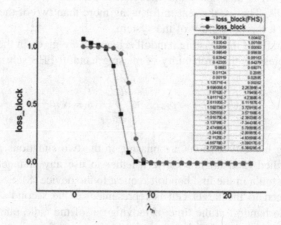

Fig. 5. Handoff request block probability for the FHS and existing handoff techniques about λ_h

5 Conclusion

In this paper, we propose a new view of the relative movement of the cellular shape and soft handoff algorithm in IP-based mobile networks. Then, to relatively increase the channel utilization of the mobility prediction system and displayed based on handoff by reducing the probability of missing the new soft handoff method of the FHS (False Handoff Sessions). The Markov chain model is proposed to obtain a formula for the system loss in order to analyze the performance mode mi of the soft handoff system in an IP-based network. In soft handoff with the IP-based mobile network is explained by adding a practical concept than the falling session. It reliably calculates the loss probability of formula using an efficient algorithm and proposes to solve the problem of determining the optimal number of the guard channels. Therefore, fixed-point strategy was developed to determine the arrival rate of the cell handoff. Handoff scheme of the developed formula of loss FHS techniques effectively shows improved performance over conventional handoff method in numerical results for both the new session blocking probability and the probability of missing handoff.

Future research will continue in consideration of the analysis according to the model, and the associated method for a soft handoff than in the normal case is set to the activity of each session to have more than one channel, instead of one.

Acknowledgement. This work was supported by Institute for Information & communications Technology Promotion(IITP) grant funded by the Korea government(MSIP) (No. 10047049, Development of smart service infrastructure for vulnerable group safety based on IoT).

This research was supported by the Ministry of Trade, Industry and Energy (MOTIE), Korea, through the Education Support program for Creative and Industrial Convergence (Grant Number S-2016-0117-000).

References

1. Chang, J.W., Sung, D.K.: Adaptive channel reservation scheme for soft handoff in DS-CDMA cellular systems. IEEE Trans. Veh. Technol. **50**(2), 341–353 (2001)
2. Wong, D., Lim, T.J.: Soft handoff in CDMA mobile system. IEEE Pers. Commun. **4**(6), 6–17 (1997)
3. Prakash, R., Veeravalli, V.V.: Locally optimal soft handoff algorithms. IEEE Trans. Veh. Technol. **52**(2), 347–356 (2003)
4. Lachlan, L.H., Payne, J.B.: Queueing model for soft-blocking CDMA systems. In: Vehicular Technology Conference, VTC 1999-Fall, IEEE VTS 50th, vol.1, pp. 436–440, September 1999
5. Leu, A.E., Mark, B.L.: Discrete-time analysis of soft handoff in CDMA cellular networks. In: IEEE International Conference on Communications, ICC 2002, vol.5, pp. 3222–3226, April 2002
6. Ma, Y., Han, J.J., Trivedi, K.S.: Call admission control for reducing dropped calls in CDMA cellular systems. Comput. Commun. **25**(7), 689–699 (2002)
7. Asaeda, H., Seite, P., Xia, J.: PMIPv6 Extensions for Multicast. IETF Internet Draft, draft-asaeda-multimobpmip6-extension-01 (2009)

8. Bargh, M.S., Hulsebosch, B., Eertink, H., Heijenk, G., Idserda, J., Laganier, J., Prasad, A.R., Zugenmaier, A.: Reducing handover latency in future IP-based wireless networks: proxy mobile IPv6 with simultaneous bindings. In: WoWMoM 2008, pp. 1–10, June 2008

9. Hui, M., Chen, G., Deng, H.: Fast Handover for Multicast in Proxy Mobile IPv6. IETF Internet Draft, draft-huimultimob-fast- handover-00 (2009)

10. Jeon, S., Kim, Y., Lee, J.: Mobile Multicasting Support in Proxy Mobile IPv6. IETF Internet Draft, draft-sijeon-multimob- mms-pmip6-02, March 2010

11. Lee, J.H., Chung, T.M., Pack, S., Gundavelli, S.: Shall we apply paging technologies to proxy mobile IPv6. In: Proceedings of the 3rd International Workshop on Mobility in the Evolving Internet Architecture, pp. 37–42, August 2008

12. Li, Y., Su, H., Su, L., Jin, D., Zeng, L.: A comprehensive performance evaluation of PMIPv6 over IP-based cellular networks. In: IEEE 69th Vehicular Technology Conference, VTC Spring 2009, pp. 1–6, April 2009

13. Magagula, L., Falowo, O., Chan, H.: Enhancing PMIPv6 for better handover performance among heterogeneous wireless networks in a micro-mobility domain. EURASIP J. Wirel. Commun. Netw. **2010**(24), 1–13 (2010)

14. Melia, T., Bernardos, C.J., De la Oliva, A., Giust, F., Calderon, M.: IP flow mobility in PMIPv6 based networks: solution design and experimental evaluation. Wirel. Pers. Commun. **61**(4), 603–627 (2011)

15. Kong, K.S., Lee, W., Han, Y.H., Shin, M.K.: Handover latency analysis of a network-based localized mobility management protocol. In: IEEE International Conference on Communications, ICC 2008, pp. 5838–5843, May 2008

16. Banerjee, N., Wu, W., Das, S.K.: Mobility support in wireless Internet. IEEE Wirel. Commun. **10**(5), 54–61 (2003)

17. Harine, G., Marie, R., Puigjaner, R., Trivedi, K.: Loss formulas and their application to optimization for cellular networks. IEEE Trans. Veh. Technol. **50**(3), 664–673 (2001)

18. Lin, Y.B., Pang, A.C.: Comparing soft and hard handoffs. IEEE Trans. Veh. Technol. **49**(3), 792–798 (2000)

19. Ritika, D.K.V.: A reviwes on handover processes in IPv6 mobility management protocols. Int. J. Adv. Res. Comput. Commun. Eng. **4**(5), 482–487 (2015)

20. Mathi, S., Lavanya, M., Priyanka, R.: Integrating dynamic architecture with distributed mobility management to optimize route in next generation internet protocol mobility. Indian J. Sci. Technol. **8**(10), 963–974 (2015)

21. Han, Y.H., Min, S.G.: Performance analysis of Hierarchical MobileIPv6: Dose it improve Mobile IPv6 in terms of handover speed? Wirel. Pers. Commun. **48**(4), 463–483 (2009)

22. Lee, J.H., Han, Y.H., Gundavelli, S., Chung, T.M.: A comparative performance analysis on hierarchical mobile IPv6 and proxy mobile IPv6. Telecommun. Syst. **41**(4), 279–292 (2009)

23. Giust, F.: Distributed Mobility management for a flat architecture in 5G mobile networks: solutions, analysis and experimental validation. Doctoral dissertation, Universidad Carlos III de Madrid, Spain, pp. 1–195 (2015)

24. Lee, J.H., Chung, T.M.: How much do we gain by introducing route optimization in Proxy Mobile IPv6 networks? Ann. Telecommun. Ann. des Teleommun. **65**(5-6), 233–246 (2010)

25. Guan, J., Zhou, H., Yan, Z., Qin, Y., Zhang, H.: Implementation and analysis of proxy MIPv6. Wirel. Commun. Mobile Comput. **11**(4), 477–490 (2011)

26. Gundavelli, S., Leung, K., Devarapalli, V., Chowdhury, K., Patil, B.: Proxy Mobile IPv6. IETF RFC 5213 (2008)

Distributed Computing Infrastructure Based on Dynamic Container Clusters

Vladimir Korkhov[✉], Sergey Kobyshev, Artem Krosheninnikov,
Alexander Degtyarev, and Alexander Bogdanov

St. Petersburg State University,
7/9 Universitetskaya Nab., St. Petersburg 199034, Russia
v.korkhov@spbu.ru

Abstract. Modern scientific and business applications often require fast provisioning of an infrastructure tailored to particular application needs. In turn, actual physical infrastructure contains resources that might be underutilized by applications if allocated in dedicated mode (e.g., a process does not utilize provided CPU or network connection fully). Traditional virtualization technologies can solve the problem partially, however, overheads on bootstrapping a virtual infrastructure for each application and sharing physical resources might be significant. In this paper we propose and evaluate an approach to create and configure dedicated computing environment tailored to the needs of particular applications, which is based on light-weight virtualization also known as containers. We investigate available capabilities to model and create dynamic container-based virtual infrastructures sharing a common set of physical resources, and evaluate their performance on a set of test applications with different requirements.

Keywords: Virtualization · Containers · Virtual cluster

1 Introduction

Constant development of computer hardware and software together with the development of computing methods and algorithms stimulates new ways of bringing together software and hardware, matching application requirements and resources, thus mapping programs to computing infrastructures. Virtualization technologies started a new era of tailoring computing environment to the needs of users and applications. However, flexibility of full- and para-virtualization approaches is hold back by some limitations causing extra overheads, resource consumption and lack of dynamics. Container-based virtualization, a new generation of virtualization techniques, can give better answers to create a flexible and dynamic distributed computing infrastructure with small overhead.

Containers as a way to create a dedicated environment for running applications have been around for years. However, the boost of new interest to

© Springer International Publishing Switzerland 2016
O. Gervasi et al. (Eds.): ICCSA 2016, Part II, LNCS 9787, pp. 263–275, 2016.
DOI: 10.1007/978-3-319-42108-7_20

them started when new technologies and tools to orchestrate their operations appeared. One of the most commonly used tool to manage container infrastructures is Docker [8].

Traditional hypervisor-based virtualization is still widely used to deploy and run applications on a wide range of platforms, however, it suffers from a number of restrictions:

- Significant overheads while running fully-virtualized guest operating systems, in particular, overheads to boot up virtual machine instances
- Lack of flexibility to allocate resources to particular processes
- Downfalls of application performance due to virtualization overheads, hypervisor mediation etc.

While hypervisor-based virtualization provides flexibility in building variety of environments, e.g. allowing to simulate and run completely different architectures and operating systems on top of each other, container-based or operating system-level virtualization is restricted by using the same core components within host and containers, in particular operating system kernel. Nevertheless, the variety of supported platforms is often not required, but low overheads and dynamics are needed.

Container-based virtualization, also referred as operating-system level or light-weight virtualization, follows a different paradigm compared to hypervisor virtualization. Containers are based on the host operating systems itself rather than on hypervisor. Containers do not virtualize hardware, which would require virtualized operating system images on each guest OS. Instead, containers virtualize OS by sharing the host OS kernel and other resources between original host environment and environments run in containers. Thus, containers provide an isolated and controlled environment that provides everything an application might need for execution without extra overheads caused by virtualizing hardware.

In this paper, we evaluate the capabilities obtained while using the OS-level virtualization technology to build a computational environment with configurable computation (CPU, memory) and network (latency, bandwidth) characteristics. Such configuration enables flexible partitioning of available physical resources between a number of concurrent applications utilizing a single physical infrastructure. Depending on application requirements and priorities of execution each application can get a customized virtual environment with as much resources as it needs or is allowed to use.

Our main interest is to use container-based computing infrastructures for parallel high-performance computing applications: parallel programs that consist of a number of processes running on computing nodes and communicating during the execution. Using containers as computing nodes can help us to control and share available computing and networking resources between concurrent parallel applications. Thus, applications with complementary requirements (e.g. fast CPU-slow network + slow CPU-fast network) can co-exist on a single physical node or a VM without affecting each other much.

The approach that we propose is complementary to the traditional queue-based batch processing used in HPC systems. Applications would not have to wait in the queue until worker nodes become fully free and available. Instead, the scheduler can control fraction of resources allocated for each application thus enabling immediate execution for applications with requirements fitting still available fraction of resources. In addition, flexible quality of service (QoS) and service-level agreement (SLA) policies can be built on top of such infrastructure: applications might be ready to get smaller amount of resources right away rather than wait in line to acquire more resources.

The paper is structured as follows: Sect. 2 gives an overview of related work in the area of container management software. Section 3 takes a closer look at comparison or containers and virtual machines for building distributed computing infrastructures. Section 4 presents and approach to simulate and predict actual application requirements that can be used for configuring container-based DCI created for particular application. Section 5 presents an experimental evaluation of building container-based computing environment for a number of test applications. Section 6 discusses the results and Sect. 7 concludes the paper.

2   Related Work

Containers are an easy way to generate large amounts of compute units, and robust monitoring, management, and orchestration are needed to cope with container crowds, where containers can be mislocated or left running forgotten.

There are a number of available tools and technologies that provide means to manage containers, maintain their lifecycle, orchestrate and monitor their execution.

One of the most popular tools for managing containers is Docker [8]. Docker introduced the concept of the container image. The Docker container image is a straightforward way to package an application and all its dependencies so that it can be executed on any modern Linux servers supporting Docker. Such portability is very important for distributed infrastructures that can be based on various platforms and versions of operating systems. In addition, Docker has tools for container deployment and orchestration, including Docker Machine, Docker Compose, and Docker Swarm. Docker Machine provides means to easily deploy Docker Engines local computer, on cloud providers, and in data centers. Docker Swarm is a native clustering solution for Docker containers. It pools together several Docker Engines into a single virtual host. Docker Compose is a way of defining and running multi-container distributed applications with Docker.

Kubernetes (originally by Google, now is a part of the Cloud Native Computing Foundation) is an open-source platform for automating deployment, scaling, and operations of application containers across clusters of hosts, providing container-centric infrastructure [11]. Kubernetes defines a set of building blocks ("primitives") which collectively provide mechanisms for deploying, maintaining, and scaling applications. These primitives are designed to be loosely coupled and extensible so that the infrastructure can meet a wide variety of different workloads. The extensibility is provided in large part by the Kubernetes API, which is used by internal components as well as extensions and containers running on Kubernetes.

Apache Mesos can be used to deploy and manage application containers in large-scale clustered environments. It abstracts CPU, memory, storage, and other compute resources away from machines (physical or virtual), enabling fault-tolerant and elastic distributed systems to easily be built and run effectively [9]. At a high level Mesos is a cluster management platform that combines servers into a shared pool from which applications or frameworks like Hadoop, Jenkins, Cassandra, ElasticSearch, and others can draw. Mesos allows developers to conceptualize their applications as jobs and tasks. In combination with a job system like Marathon, it takes care of scheduling and running jobs and tasks. Marathon is a Mesos framework for long-running services such as web applications, long computations and so on [10].

CoreOS is a Linux distribution designed to make large multiple-machine deployments secure, consistent, and reliable. Instead of installing packages via yum or apt, CoreOS uses Linux containers to manage services at a higher level of abstraction. A single service's code and all dependencies are packaged within a container that can be run on one or many CoreOS machines [13]. It uses "fleet" for cluster management and "etcd" for service discovery and keeping configuration up to date across the cluster.

3   Deploying and Running Applications in DCIs: Containers vs Virtual Machines

Container virtualization allows to virtualize physical servers at the level of operating system kernel. OS virtualization layer provides insulation and security of resources between different containers. Virtualization layer makes each container similar to a physical server. Each container maintains therein an application and workload. The main advantages of container virtualization are the following:

- Containers are maintained on the level of physical servers. Lack of virtualized hardware, the use of real equipment and direct access to drivers allows to achieve high performance.
- Each container can be scaled to the resources of a physical server.
- Virtualization on the OS level allows to achieve the highest density among the other available virtualization solutions. You can create and launch hundreds of containers on a single physical server.
- Containers use a single OS, making their support and update very simple. Applications may also be deployed in a separate environment.

In addition to the light-weight virtualization benefits, containers provide flexible and convenient ways to package and distribute software. The older ways to package, deploy and distribute applications were installing them directly on

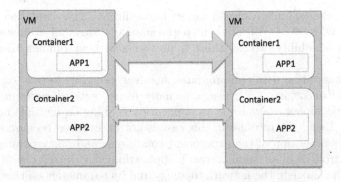

Fig. 1. Concurrent applications on container cluster

every machine with the help of operating system package managers, or creating a separate virtual machine for a particular software deployment that could be rather heavy-weight and non-portable.

Containers are based on operating-system-level virtualization rather than hardware virtualization, nevertheless they are still well isolated from each other and from the host. Containters have their own filesystems, they cannot see and influence each others processes, and their computational and network resource usage can be bounded. The latter brings us the possibility to create fully defined and controlled container-based clusters, configured to the needs of particular users and applications (Fig. 1).

Containers are not coupled to particular underlying infrastructure of filesystem; they are easy to build and portable across different types of operating systems in data centers or in clouds.

Containers are light-weight and fast; normally there is one-to-one relation between an application and a container image which enables composing a computing environment in a loosely coupled manner, built from individual blocks that can be easily created at build/release time rather than deployment time.

An important feature of containers for our research is the application-centric management that raises the level of abstraction from running an OS on virtual hardware to running an application on an OS using logical resources configured according to application requirements.

4 Simulating Container-Based Distributed Computing Infrastructure

Conducting offline simulations is often needed to preliminary assess the correctness and adequacy of the applications that interact with the network, in particular to evaluate performance of network-related algorithms. Moreover, the ability to perform reproducible experiments is necessary in the development of software. This is especially important when large-scale network applications are concerned since real network environment is dynamic, and application behavior needs to be checked in advance in variety of conditions. The simulation tools like ns [7] and OMNet++ [14] allow to perform simulations and evaluate the efficiency and scalability of algorithms or protocols without running them in real networks.

Experimentation with real programs, however, is mainly focused on the measurement of makespan, CPU usage, memory usage, network performance, etc. One of the complex examples is an application running on multiple nodes over a network. Using the Internet in this case is not encouraged because there will be no exact repetition of the experiment conditions, and many parameters can not be controlled. In addition, network applications may interact with different hosts, but a change in the network topology and its parameters on this real testbed is time consuming and can be error prone. On the other hand, the topology of choice in accordance with the specific tasks may speed up the task and show a great performance. However, the transition to a different type of tasks would require changing the entire network topology.

Here we consider an approach to create a virtual testbed, having virtual links between nodes, that allows us to create a variety of network topologies. The created nodes may have different restrictions, e.g. the limitation on the memory or the data transfer speed between nodes. Furthermore, it is possible to simulate the conditions of poor communication between two or more nodes, to limit the bandwidth to add delay when transmitting packets, to change the error ratio, i.e. the ratio of the number of incorrectly received bits (1, instead of 0 and vice versa) to the total number of bits transmitted in transmission data between nodes.

When we talk about the emulation of the network we usually refer to the ability to control artificially created environment for effective interaction with real computer networks and real-time traffic. We should note that already in 1996 David Tennenhouse et al. proposed an idea of software-centric networks, called Active Networks, but then it was not widely acknowledged [2]. Recently similar projects such as OpenVSwitch [16], Mininet [1] and others appeared; together with the container virtualization technologies they can perform various tasks, like simulating many nodes with connections of controlled quality, simulated latency, packet loss etc., even within a single node.

While reviewing the existing software, we paid attention to the following points:

1. Virtualization of nodes. The ability to simulate multiple independent sites that are running on a single physical machine can be achieved through a variety of virtualization techniques, e.g., QEMU, VirtualBox, VMware, that deliver maximum isolation between nodes, but achieve this due to the intensive use of memory, hard drive, and processor capacity. Paravirtualization reduces the load on the I/O subsystem. Various container virtualization technologies, along with the namespaces, creates the illusion of running many nodes with different network characteristics between each of them on the same machine with a totally negligible overhead.

2. Network emulation. In addition to simulating a connection, the ability to model deterioration of network quality is needed, for example to emulate packet loss, latency, BER (the number of bits inverted with respect to the original) and so on.
3. In addition, a graphical interface is preferable, at least in the form of a web-client, because it allows us to not keep in mind the whole network topology, it is easy to change network settings and to visualize the results.

We have checked a number of tools that can be used for the purposes of simulating distributed computing infrastructure:

– User-mode Linux (UML) [15]: virtualization solution that allows you to run processes isolated from the rest of the operating system in the user space. In addition, it is possible to run network services. Important difference of UML from other solutions is that it allows you to run Linux kernel version different from the version on the host. Currently, this technology is integrated into the Linux kernel, but the development is slow, in addition, the performance is lower than other available solutions.
– Manage Large Networks (MLN) [12]: means of virtual machine management, working with Xen, VMware and User-mode Linux. In addition, the MLN was able to work with Amazon EC instances, but the development of the software was frozen in 2009.
– Marionnet [18]: software for network virtualization. It allows users to define, configure and run complex simulated networks. It is also possible to combine real and virtual networks. Unfortunately, Marionnet is also based on the User-mode Linux, so we cannot consider this technology new and evolving. Moreover, for the management of network properties it uses VDE (Virtual Distributed Ethernet), which does not support bandwidth limitation mechanisms, delays and other things.
– Integrated Multiprotocol Network Emulator / Simulator - IMUNES [3]: framework on the basis of today's popular products like Docker and Open VSwitch, which allows you to create nodes, connections between them, and fine-tune the network properties. It is possible to emulate the delay in data transmission, damage to transport packets, limit the bandwidth, etc. This product also has a graphical user interface that simplifies the creation of topology and access to running nodes.

5   Building and Evaluating Container-Based Distributed Computing Environment

We have implemented a prototype of container-based distributed computing infrastructure on the cloud resources provided by Microsoft Azure. To build the infrastructure we used a set of 8 virtual machines residing in different regions (5 machines in East US; 3 machines in North Europe) with the following characteristics (see Fig. 2):

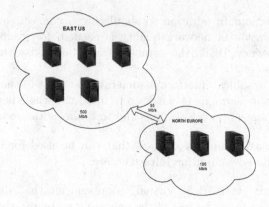

Fig. 2. Resources of experimental testbed in MS Azure

- Instance type: A1
- Cores: 1
- Memory: 1.75 GB

Current prototype implementation does not use any specific container management tools and relies only on Docker and a custom python-based toolkit developed to configure and execute containers for parallel applications (with OpenMPI deployed and configured) and control resource usage with help of a separately maintained database (see Fig. 3).

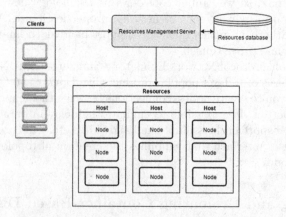

Fig. 3. Schematic view of the system

Figure 4 shows sequence diagrams illustrating the functionality of the system to manage resources and created container clusters.

We used several programs from NAS Parallel Benchmarks (NPB) suite as the applications with various requirements to the underlying infrastructure. NPB is

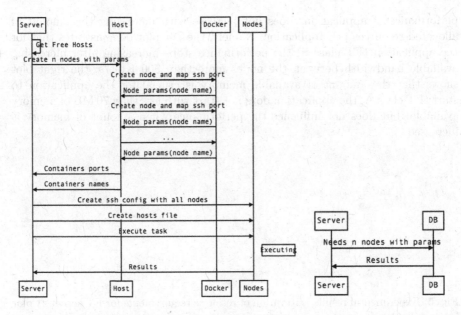

Fig. 4. Sequence diagrams of system's functionality (left: server-resources communication; right: server-database communication)

a small set of programs designed to help evaluate the performance of parallel computers and clusters. The benchmarks are derived from computational fluid dynamics (CFD) applications and consist of five kernels and three pseudo-applications. Moreover, the benchmark suite contains benchmarks for unstructured adaptive mesh, parallel I/O, multi-zone applications, and computational grids [17]. In our experiments we used MG, FT, and CG kernels:

- MG - Multi-Grid on a sequence of meshes, long- and short-distance communication, memory intensive
- FT - discrete 3D fast Fourier Transform, all-to-all communication
- CG - Conjugate Gradient, irregular memory access and communication

The aims of the experiments were the following:

- Investigate performance of parallel applications on the prototype of distributed container-based infrastructure
- Check how performance of applications with different requirements on cpu/memory/network varies depending on infrastructure configuration
- Evaluate possibilities of concurrent execution of parallel applications, minimizing their influence on each other

The first set of experiments was performed to evaluate the resource saturation point for an application: the point when adding more memory (or available cpu, or network bandwith) would not increase application performance anymore. Sample results are presented in Fig. 5. We can see that after some point the performance of applications does not increase with increasing the amount of allocated resources per application. Namely, the left plot demonstrates that for the application (FT class S) the performance stops increasing after increasing available bandwidth between the nodes more than 900 Kbit/s; the right plot shows that the amount of available memory is crucial for the application to start (FT class A, the application does not start with less than 70 MB of memory available) but does not influence the performance when amount of memory is increased.

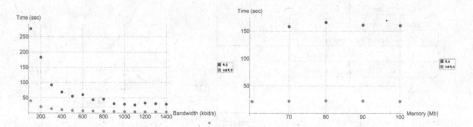

Fig. 5. Experimental results: saturation of resource requirements for FT kernel. (Color figure online)

The results presented in Fig. 6 illustrate details of application performance for a particular configuration of computing infrastructure.

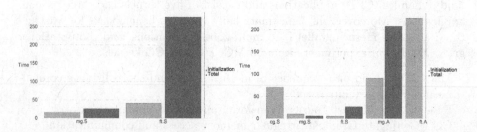

Fig. 6. Experimental results. Left: mem 256 MB, net 100 Kb/s; Right: mem 512 MB, net 1024 Kb/s. (Color figure online)

Next, we have evaluated sequential and concurrent execution of two different application kernels, MG and FT, to ensure that concurrent execution of both applications will not affect their performance in case container clusters are configured to meet the individual requirement of the applications. Figure 7 illustrates observed differences in initialization and benchmark time of MG and FT kernels in a particular virtual hardware configuration (512 MB memory, 120 Kbit/s network) for sequential and concurrent execution. Here sequential execution means allocation of the whole set of resources to each of the applications and executing them one by one; concurrent execution means execution of both kernels simultaneously in separate containers with given limitations. Figure 8 shows experimental comparison of shared and concurrent execution, where shared execution means running both MG and FT kernels in a single container simultaneously. We can observe that in this case kernels can compete for shared resources which results in overall performance degradation.

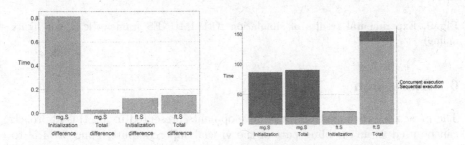

Fig. 7. Experimental evaluation of separate and concurrent execution: difference in init and benchmark time; mem 512 MB, net 120 Kb/s. (Color figure online)

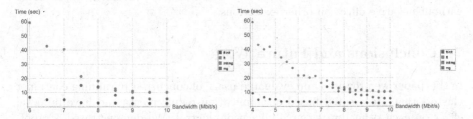

Fig. 8. Experimental comparison of shared (left) and concurrent (right) execution: mem 100 MB, lan 120 Kb/s. (Color figure online)

In order to confirm results gained on Amazon cluster, a bunch of tests were made on a local machine with IMUNES software. A star topology was generated with 8 nodes, all nodes were created from the same Dockerfile so they were completely identical. As IMUNES uses Docker and OpenVSwitch, network capacity was easy to configure. FT class A was used with different memory limitations but it wasn't possible to make test fail due to memory constraint as it was on a cluster. However the amount of memory available indeed doesn't influence the time when we allocate more than 80 mb per node. Overall, we cannot completely rely on results gained in a simulated testbed but we can at least understand how applications can scale (see Fig. 9).

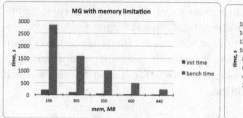 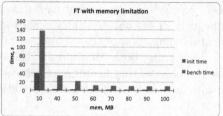

Fig. 9. Experimental results of simulation with IMUNES framework (Color figure online)

6 Discussion

The presented work continues the developments presented in [6]; this approach can be used as an enabling part of the virtual supercomputer concept [4,5] to ensure proper and efficient distribution of resources between several applications. Knowing the application demands in advance we can create appropriate infrastructure configuration giving just as much resources as needed to each particular instance of a virtual supercomputer running a particular application. Here we use containers as an enabling part of the computing infrastructure. In such a way, free resources can be controlled and granted to concurrent applications without negative effect on other executions.

7 Conclusions and Future Work

In this paper we proposed and evaluated usage of concurrently running container clusters that are created based on application requirements and have minimal effect on each other by resource allocation control. We demonstrated a proof-of-concept prototype running on Microsoft Azure cloud resources and showed experimental evaluation of its performance on a set of NAS benchmarks based on real application kernels.

Our future work will be to look more closely into container cluster management and orchestration software (e.g. Docker Swarm, Kubernetes, or Mesos) to delegate the functionality of maintaining the cluster to these tools so that we could concentrate on mechanisms of application requirements evaluation and concurrent execution of applications on distributed container resources.

Acknowledgments. The research was supported by Russian Foundation for Basic Research (projects N 16-07-01111, 16-07-00886, 16-07-01113) and St. Petersburg State University (project N 0.37.155.2014).

References

1. Lantz, B., Heller, B., McKeown, N.: A network in a laptop: rapid prototyping for software-defined networks. In: Proceedings of the 9th ACM SIGCOMM Workshop on Hot Topics in Networks, Hotnets-IX, NY, USA, 2010, pp. 19: 1–19: 6
2. Tennenhouse, D.L., Wetherall, D.J.: Towards an active network architecture. Comput. Commun. Rev. **26**(2), 5–18 (1996)
3. Salopek, D., Vasic, V., Zec, M., Mikuc, M., Vasarevic, M., Koncar, V.: A network testbed for commercial telecommunications product testing. In: Proceedings of the Softcom 2014 22th International Conference on Sotware, Telecommunications and Computer Networks, Split, September 2014
4. Gankevich, I., Gaiduchok, V., Gushchanskiy, D., Tipikin, Y., Korkhov, V., Degtyarev, A., Bogdanov, A., Zolotarev, V.: Virtual private supercomputer: Design and evaluation. CSIT 2013–9th International Conference on Computer Science and Information Technologies (CSIT), Revised Selected Papers. pp. 1–6 (2013). doi:10.1109/CSITechnol.2013.6710358
5. Gankevich, I., Korkhov, V., Balyan, S., Gaiduchok, V., Gushchanskiy, D., Tipikin, Y., Degtyarev, A., Bogdanov, A.: Constructing virtual private supercomputer using virtualization and cloud technologies. In: Murgante, B., Misra, S., Rocha, A.M.A.C., Torre, C., Rocha, J.G., Falcão, M.I., Taniar, D., Apduhan, B.O., Gervasi, O. (eds.) ICCSA 2014, Part VI. LNCS, vol. 8584, pp. 341–354. Springer, Heidelberg (2014)
6. Korkhov, V., Kobyshev, S., Krosheninnikov, A.: Flexible configuration of application-centric virtualized computing infrastructure. In: Gervasi, O., Murgante, B., Misra, S., Gavrilova, M.L., Rocha, A.M.A.C., Torre, C., Taniar, D., Apduhan, B.O. (eds.) ICCSA 2015. LNCS, vol. 9158, pp. 342–353. Springer, Heidelberg (2015)
7. NS-3 Project Homepage. http://www.nsnam.org
8. The Docker platform. https://www.docker.com
9. Apache Mesos. http://mesos.apache.org/
10. Mesosphere Marathon. https://mesosphere.github.io/marathon/
11. Kubernetes. http://kubernetes.io/
12. Manage Large Networks. http://mln.sourceforge.net/
13. CoreOS. https://coreos.com/
14. OMNeT++. https://omnetpp.org/
15. Dike, J.: User Mode Linux, p. 352. Prentice Hall, Englewood Cliffs (2006)
16. Open vSwitch. http://openvswitch.org/
17. NAS Parallel Benchmarks. http://www.nas.nasa.gov/publications/npb.html
18. Marionnet: a virtual network laboratory. http://www.marionnet.org

Building a Virtual Cluster for 3D Graphics Applications

Alexander Bogdanov, Andrei Ivashchenko[✉], Alexey Belezeko,
Vladimir Korkhov, Nataliia Kulabukhova, Dmitry Khmel, Sofya Suslova,
Evgeniya Milova, and Konstantin Smirnov

Saint Petersburg State University, 7/9 Universitetskaya nab.,
Saint Petersburg 199034, Russia
aiivashchenko@cc.spbu.ru
http://spbu.ru/

Abstract. This paper discusses a possible approach to distributed visualization and rendering system infrastructure organization, based on Linux environment with the usage of virtualization technologies. Particular attention is paid to the minutiae, which may be encountered due to the environment setup and exploitation processes, and may affect system performance and usability. Some applications and development tools are studied, as they can provide a rapid onset of computing resources exploration.

Keywords: Computer graphics · Virtualization · Distributed rendering · Remote workstation

1 Introduction

Modern scientific studies not only actively exploit information technologies, but already rely on them. Due to the significant growth of domains involved in a single simulation in general, and detailed description of each involved process in particular, proportionally to the capabilities of the newest computational resources, the amount of data that should be processed and generated has greatly increased. Such complex data sets are very difficult to consider on their own, so various visualization techniques are applied to represent them in a more understandable form to the researcher. This task could become even more complicated if real-time data processing and exploration is expected. And if enormous calculations in most cases are only a matter of time, visualization of extra-large data volumes is still unresolved.

This paper discusses one of the ways to organize an environment for various tasks connected with real-time and batch rendering, based on usage of virtualization technologies. Also a set of tools and various applications that could help to scale out an existing application, or to develop a new one, is presented here.

© Springer International Publishing Switzerland 2016
O. Gervasi et al. (Eds.): ICCSA 2016, Part II, LNCS 9787, pp. 276–291, 2016.
DOI: 10.1007/978-3-319-42108-7_21

2   GPU Delivery to the Virtual Environment

Of course, usage of virtualization approach for organizing computational environment will bring some overhead in resource consumption, but sometimes there is no other choice. This might be necessary if you are planning to share resources with other users, or this might be essential if you are getting the resources as a service this way. Since you have to use such a solution, you can try to extract some benefits out of there. For example, different virtual machine images and templates could be prepared for rapid bootstrapping of specific task-related environments. Some research, analysis and evaluation on organizing computing and networking resources as a virtual supercomputer tuned to particular application requirements have been addressed in [25, 28, 29]. Another case is also trivial: during working hours computing resources could be occupied by users' virtual machines providing remote access to the workspace and being used interactively, but at other times dormant computing capacities could be exploited to execute some heavy batch routines, like rendering with usage of ray tracing techniques, if we are talking about computer graphics related tasks.

Whereas the question about common virtual resources delivery is someway trivial due to amount of both open source and commercial virtualization solutions, the problem of supplying a guest environment with powerful graphics acceleration, especially for Linux-type operating systems, remains an open question.

The main idea of GPU virtualization is to be able to schedule GPU calls coming from a virtual machine the same way CPU calls are scheduled, and there are three main approaches presented in Fig. 1 which could be provided by virtualization platform to achieve the goal.

Virtual Shared Graphics Acceleration (vSGA) is probably the earliest technology introduced in the field of graphic resources sharing. This method requires an installation of vendor's graphics driver at the hypervisor layer, which should communicate with special VGA driver on virtual machine, as it shown in Fig. 1a. The weakest point of this solution is the presence of substantial constraints with regards to the graphics standards and platform support. Essentially, it means that vSGA could be used only with those operating systems that are supported by virtualization provider, and applications are able to run only with particular graphics API implemented for custom driver (e.g. VMWare implementation support is limited up to DirectX 9.0c and OpenGL 2.1) [36].

Virtual GPU (vGPU) is a more advanced technology available right now in that area. Right now there are only two virtualization platforms that are providing support for fully functioning GPU virtualization: XenServer by Citrix and vSphere by VMWare. VMWare ships it's products only on a commercial basis completely packed. XenServer itself is a free to use and open source virtualization server, but features of our interest are available only in the Enterprise Edition. Without a proper license GPU will be working only in a passthrough mode. Moreover, GPU virtualization should be backed by the manufacturer and physical card itself.

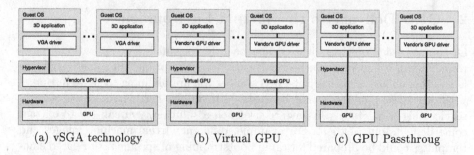

(a) vSGA technology   (b) Virtual GPU   (c) GPU Passthroug

Fig. 1. Ways of obtaining a hardware graphics acceleration on virtual machine

The main idea of vGPU is represented in Fig. 1b. The appropriate graphics driver installation here required on both hypervisor and guest operating systems. The hypervisor operates a virtual GPU manager software, which is responsible for providing a direct access to the virtualized part of hardware. The user's virtual machine should be running an original graphics driver, so mostly all features become available for the applications [14].

Still, there are some things that are working in a passthrough mode, but not available here due to technology limitations with virtual GPU or GPU sharing:

– OpenCL expression evaluation
– Physics computation with GPU acceleration
– Hardware accelerated tessellation

These restrictions are applied because graphics computing resources are not delegated straightly to the virtual GPU and could be reserved at task allocation stage. Instead, users have to use GPU computational resources in a time share manner.

If the available graphics hardware does not support virtual GPU technology, or if there is just no need to use one, generally, any virtualization solution combinations could be involved, since they support direct hardware passing to the guest virtual machine. The appropriate technology for that type of graphics acceleration delivery could be called as Virtual Dedicated Graphics Acceleration (vDGA), or just GPU passthrough [14]. The main positive aspect offered by this solution is that there is no need of installing any additional drivers or managing software on the host machine's operating system, because in that case the hypervisor is just connecting a native PCI bus device, as it shown in Fig. 1c. Since the original graphics driver could be installed to the guest operating system, the end user can be satisfied with the entire spectrum of functionality that has been incorporated by the graphics card manufacturer.

The approaches described above could be involved not only for the discrete graphics cards, but also for embedded system-on-chip solutions. Intel introduces its own Graphics Virtualization Technology standard (Intel GVT) for Intel HD Graphics coupled with processor [5]. The KVMGT and XenGT implementations are available for respective hypervisors. All three resource delivery technologies described above are available and defined with suffixes, where Intel GVT-d stands for vDGA, -s relates to vSGA and -g is for vGPU.

3 Hardware Configuration and Infrastructure Organization

To perform the installation and deployment of the graphics cluster we used three powerful Dell PowerEdge R720 workstations with the following configuration:

- 2 Intel Xeon E5-2695 v2, 12 cores/24 threads @ 2.4 GHz
- 256 GB DDR3 RAM
- 2 NVIDIA GRID K2 GPU
- Two-way 10 Gbit Ethernet for intranet organization
- External 1 Gbit Ethernet interface

PowerEdge unit itself is a certified solution for virtual desktop infrastructure organization [13]. The NVIDIA GRID K2 graphics card equipped with the server is also positioned at the market as a specialized hardware for the mentioned purposes [15]. Basically, each of NVIDIA GRID K2 modules holds on board two dedicated Kepler GPU's. Each of them has:

- GK104 745 MHz main GPU chip
- 4 GB of 2.5 GHz 256-bit GDDR5 memory
- 1536 CUDA cores
- Up to 4 displays support with 2560x1600 dimension

If the capabilities of each chip were considered individually, it could be said that it comes very close to the desktop GeForce GTX 680 solution. There are two main differences that could be noted between these two cards: each chip of GRID GPU has twice as much memory, but the core and the memory work at a lower clock rate (745 MHz against 1058 MHz for the core, and 1250 MHz against 1502 MHz for the memory) [3,15]. As for the virtualization capabilities, the whole board can handle the following virtual GPU profiles:

- 2 GRID K280Q with 4 GB of memory (passthrough mode)
- 4 GRID K260Q with 2 GB of memory
- 8 GRID K240Q with 1 GB of memory
- 16 GRID K220Q with 512 MB of memory

The good news is that virtualization profiles could be mixed; there are no constraints on using the same, if one is already activated.

VMWare vSphere platform supplied with ESXi hypervisor was used to execute virtualization tasks, since the proper license with GPU virtualization support was available for that package. All three server machines were consolidated into a single unified resource pool that allowed us to redistribute computing facilities on the fly. One of the nodes was involved to create a 3D accelerated gateway to the cluster cycling with the K220Q profile. In general, the four basic templates for future virtual machines have been created, where each one stands for its own type of vGPU profile. Each of these machines is expected to receive a proportional part to the amount of node's computational resources by default: 16th part for K220Q profile, 8th part for K240Q profile, and so on. Finally, eight computational nodes could be obtained with a passthrough mode, that enabled a whole functional stack with the following configuration:

- 6 CPU cores on a single socket 12
- 64 GB RAM
- GK104 based GPU with 1536 CUDA cores and 4 GB of memory

Such amount of RAM could be explained by a relatively small GPU memory capacity, and, therefore, the necessity of active read-back usage. The bigger amount of data should be placed on the scene, the larger fast memory area should be allocated. In a case of the need for slow storage volumes treatment, the user will experience a massive slowdown of application execution, and also a frame drop if it is a real-time application. Proper caching must be considered carefully while developing such applications.

While the platform is expected to be used for computational purposes, a high performance mode should be enabled for the hypervisor. Another tweak that should be applied is platform specific and implies power management delegation to operating system. These changes provide a correct way to tune the wattage by hypervisor on corresponding virtual machine request. Even those two small and simple attunements allow us to improve performance of synthetic single GPU benchmarks, like FurMark, up to 30 percent [17]. These improvements are true for both Citrix and VMWare virtualization platforms [4,18].

4   Preparation of Computing Environment

This section discusses the preparation stages that should be carried out on the computing and gateway nodes, which are represented with virtual machines in our case, since we are working with a virtual cluster. The initial steps include installing and configuring the video card driver, the X server and the desktop environment. Since we work with NVIDIA hardware, the description includes some context-related stages.

Before beginning the GPU driver installation process, it is necessary to ensure that the open-source implementation called nouveau is not installed, since it can bring some issues [6]. In most cases, this package comes by default with the operating system distribution if it ships with graphical subsystem preconfigured. A problem that can arise is that the operating system kernel can boot correctly only one load module responsible for the graphics hardware usage, and the second one will be just ignored. Additionally, nouveau could be put to the kernel's modules blacklist to ensure it will not be loaded.

Driver installation is trivial, and it could be useful to install some packages that are included into the CUDA Toolkit [2]. The performance analysis tools shipped with this bundle help to show the consumption of GPU resources by the applications using hardware 3D acceleration. After the installation a proper X server configuration for the headless usage should be obtained. It could be done by nvidia-xconfig utility. Sometimes you would need to set the bus id of your GPU explicitly in Xorg configuration file, because there always will be two VGA controllers shown in lspci. One will stand for the real graphical unit, and another will show the virtual device provided by the hypervisor.

There is only one condition that should be noted while choosing a desktop environment: it should not require 3D acceleration because the remote session can not provide proper functioning in that mode due to several reasons, which are described below.

While preparing the computing environment it is necessary to pay extra attention to the several major issues, which have the greatest impact on system's performance and usage experience.

The first issue that should be solved arises due to the nature of the X Window System used in Linux. X server itself is network transparent and allows to execute render routines remotely, even over the SSH channel. But, actually, all graphics computations will be processed on client machine, so it is just makes no sense. In a case of the X proxy server usage, a dedicated X server is started for the remote session without hardware acceleration available. One would think that everything is fine, because X server supports off-screen rendering, but the rendered frames will not be displayed on the remote desktop. So, an appropriate solution should be found, which will allow to achieve a server-side hardware accelerated rendering combined with the remote graphical access.

With the X proxy session initiated the problem itself is expressed in the inability to load the GLX extension, which is responsible for X-delegated render-ing command execution and graphics context initialization for the application. The first and the easiest solution that could be applied is to force the application to use an implementation of OpenGL specification called Mesa 3D, which could be used in software mode with GLX emulation enabled [12]. So, it means that all graphics operations will be performed by CPU. In a normal case Mesa could be used just as a regular OpenGL API implementation backed by GPU, but if it were possible there were no problems here. However, it does not solve the issue in full, because software rendering is not an option if we are looking forward to heavy graphics.

A more appropriate solution here would be the VirtualGL library that allows to display the framebuffer produced on another X server in the active session [23]. Functioning mechanism is presented in Fig. 2. The bootstrapping tool called vglrun injects the library on the application startup and passes the control of GL commands to it. After the frame rendering is done, it is received by the client part of VirtualGL service and appears in a local framebuffer, which could be displayed already. So, the X server responsible for the remote session is performing only 2D manipulations connected with desktop workflow.

For optimization reasons VirtualGL adds all resulting frames into processing queue and transmits them only if the client is able to accept them. To prevent the backlogging unprocessed frames are destroyed and replaced with new ones.

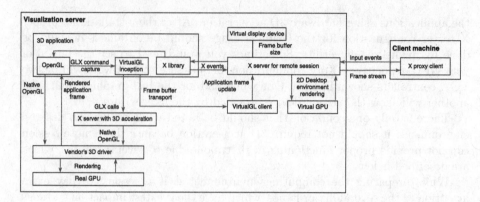

Fig. 2. VirtualGL rendering pipeline

This improves the responsiveness of interactive applications. But it is should be taken in mind that for benchmarking a frame spoiling must be switched off, as the maximum frame rate is expected as a result.

VirtualGL's toolset has several features that could be useful in such type of environment. An access to the 3D accelerated X server could be restricted to the particular user group on configuration setup. The client part of VirtualGL could be launched not only on the rendering machine, but also on the client side, so the rendered frame would be available in a local buffer immediately, avoiding an X proxy usage, but requires an extra network throughput.

It is worth noting that Windows operating system avoids all these problems, since it includes RemoteFX, which stands for a toolset that implements remote desktop access protocol specially prepared and actively exploited by vGPU [30].

The second problem follows the first and refers to the matter of proper technology choosing for virtual desktop delivery. This task could be mentioned as essential for Virtual Desktop Infrastructure (VDI) organization, because the resulting convenience of use depends exactly upon it. Since we are in need to operate with 3D accelerated applications, the chosen system should provide a sufficient frame rate for real-time interaction. There are, basically, variations of two most common solutions available, based on VNC and NX.

In addition, it is possible to refer to proprietary virtual desktop infrastructure solutions, provided by VMWare and Citrix. The latest versions of toolkit are also supporting Linux environment. However, both of them require license acquisition, therefore, they will not be discussed in depth in the article.

The core part of NX protocol and software was developed and published by NoMachine [21]. This technology is based on the usage of special NX proxy agent for the X server command sequence capturing and transmitting. That means that the X terminal should also run at the client side. Remote desktop application set, based on that solution, have some advantages, such as tolerance to the lack of good and stable bandwidth and secure tunneling. However, the usage of the X terminal on the client combined with the poor connection quality could lead to the performance drop while running the application showing intensively changing frames, e.g. video playback, or OpenGL applications. Each time the thin client will require the corresponding X hook should be sent, thus the resulting amount of frames per second will depend on quantity of requests processed through the network. In some cases, where interactive demeanor or smooth image is not urgent, that will not be vital, but if there will be a need to capture the window frames, even on the workstation's side, the desirable frame rate will not be achieved.

VNC, which is standing for Virtual Network Computing, is another standard protocol with a huge amount of open-source implementations available [32]. VNC toolset is also consists of client and server parts, however the thin client is only responsible for input capturing and frame displaying. The main difference between VNC and NX system is the fact that all computing operations associated with obtaining a frame are happening at the server side. That is why this solution could be noted as more dependent of network connection quality due to the necessity of constant framebuffer steaming large amounts of image. In other hand VNC could show the true workstations' performance because it is not influenced by third party factors. The connection could be directed through the ssh tunnel, but in most cases it should be done manually.

Thus, it is clear that VNC is a more promising technology, when the remote access for interactive content is needed. In other cases NX could be used as a free and reliable alternative.

As for the particular implementation of VNC protocol, the TurboVNC had been chosen. This program optimizes the process of frame transmission and composition through the use of algorithms for analysis and comparison of the current and previous images, and eventually transmits only those parts of the image that differ [22]. As a result, the network load is reduced. Moreover, for the frame data compression that is going to be transmitted an accelerated a version of the libjpeg library, called jibjpeg-turbo, is used. By the usage of SIMD instruction set for image encoding the same result could be achieved from two to six times faster on assurances of developers [19]. Another nice and useful feature presented is a web-based desktop delivery support using Java applet technology. The only really significant weakness on which attention can be drawn is that TurboVNC can not be integrated with authentication services used in the infrastructure. The only two methods supported right now are password authentication and PAM-based authentication, which is recommended. The first option also has two possible ways of usage: user can set a permanent passphrase or generate a one-time password with vncpasswd utility. The password hash will be stored in .vnc folder at users home location along with the desktop environment initialization script.

It should be also understood that construction of any infrastructure should start from the list of those tasks, which are intended to be solved with the help of resulting system. Therefore, the last question should be addressed to methods of preparation and use of computing resources to tasks that are typical for the area of computer graphics. At least two main categories could be assigned here that should be treated in a different manner: real-time applications, which require an urgent resource allocation in most cases, and batch tasks, like image or animation rendering, which could be queued and scheduled.

For more efficient use of computing resources, high-performance clusters are exploiting a resource management systems. These allow to handle the massive flow of jobs from different users. A manual task management comes as a problem of uncontrolled distribution and node interaction time, with presence of competition for computing resources showing up a negative impact on the entire system performance, brought up by parallel jobs. As a consequence, modern computing clusters should use a queuing system to operate with machine resources.

SLURM (Simple Linux Utility for Resource Management) is one of the systems featuring such a solution [31]. This software is shipped as free to use opensource project and its mentions could be found in the TOP500-ranked systems, including the most powerful today - Tianhe-2 cluster [16]. SLURM is able to run, manage, monitor and complete tasks on the nodes. It allows you to manage user access to computing power, according to the allocated and joint rights to perform the tasks. Also, it has a number of advantages with respect to the virtual graphics cluster.

In SLURM, as in other similar resource managers, tasks are getting carried out by scripts. The task's description could be divided into main body and socalled prologue and epilogue scripts. Prologue scripts are designed to prepare computing environments for a specific task, and an epilogue script could help to return the environment to the initial state. In a case of rendering tasks, prologue script can start up the X server on each computing node allocated by the user, and the epilogue should be responsible for killing X session on completion of the task.

Queues have a number of options, including the queue priority ordering, which makes sense in case there is a need to give a preference to the real-time rendering applications against batch tasks.

The group of researchers from Computer Engineering Department, Bogazici University have developed a plugin for SLURM profiler called AUCSCHED, designed for heterogeneous CPU-GPU clusters. It allows to reduce the usage of the computing resources up to 25 % compared with a bare SLURM system, because its current profiler's implementation could operate only with a node range, rather than computation device range [33]. AUCSCHED solves this problem, allowing to pack tasks on nodes more tightly and closely to each other, also allowing to change the number of used GPU (or other accelerating devices) at runtime [34]. Whereas the first feature is very useful for allocating a big amount of real-time tasks, both of them suit perfectly for the batch rendering jobs. As a result, AUCSCHED plugin increases efficiency while operating with several graphics cards.

5   Applications and Development Tools

Whereas all the necessary preparatory operations for computing infrastructure have been carried out, the environment evaluation should be performed and the solution of the raised visualization objectives should be designed. Thus, in this section the most popular scalable applications and development tools will be considered to get an idea on how to start using obtained resources.

The first, and, probably, the most prominent system for distributed rendering that should be noted is the Chromium, which also could be mentioned as Cr [1]. Its basic principle lies in substitution of system's graphics libraries for selected running OpenGL applications. That means, that it is able to run the graphics applications in parallel, even if there is no source code access available. Onward, the Chromium application loader will use fake library to intercept into OpenGL instruction pipeline and redirect selected command streams to computational nodes that will perform the real rendering job. The resulting rendered frame could be presented not only on single monitor, but also on multiple physical display devices combined into grid, and this is shown as one of the Chromium features.

But in spite of all the positive Chromium's aspects, it has one essential flaw, which is revealed rather quickly when the amount of graphics data and OpenGL is getting larger. Since the Chromium is the streaming system and it does not require to distribute the target applications among all active rendering nodes, certain chunks of split graphics data and corresponding OpenGL command sequences should be transferred to the working servers. When communication costs become so large that the network is no longer able to deal with them, the application gets a slowdown instead of expected frame rate or quality improvement at final.

If the development of a new application or a porting of existing one is planned, the look at the Equalizer framework should be taken, which, probably, becomes one of the most powerful development toolset in the area of distributed rendering.

Equalizer is a system for development of distributed parallel OpenGL-based applications, which provides a rich API for management of 3D graphics object allocation, and also flexible graphics pipeline and node configuration [27].

The main feature of Equalizer, which can highlight it amid other similar frameworks and libraries, is the presence of load balancing mechanisms. Also, a well-designed architecture simplifies introduction of new functionality.

Apart of Chromium, the executed application should be available at the node that is going to be used. That solves the problem, which Chromium is affected by, due to the fact that the nodes need to transmit only particular parts of the data that is required to render the area or the object for which the node is responsible. Thus, a client node that is connected to the output device, may receive just a ready frame, or the set of parts thereof that simply need to be composed.

Once written with Equalizer library, an application acquires the ability to be used with various output devices, like display wall, immersive display, virtual reality glasses and a classic single display setup of course. This makes it easy to build systems that expose benefits of augmented reality and of course virtual reality, which broke through the world of computer graphics as a buzzword in the last years. The rendering workflow for the application done in the right way could be managed just by making changes in the configuration file, that is getting passed to the both, server and client parts. The configuration itself can be made sufficiently flexible due to amount of parameters that could be involved. They include but not limited to algorithms of data sorting and distribution among nodes, data compression algorithms, the quality of resulted frames, displayable volumes of virtual space. The availability of all these things and taking them into mind while developing software itself and parameters needed allows to perform assigned tasks in a most efficient way.

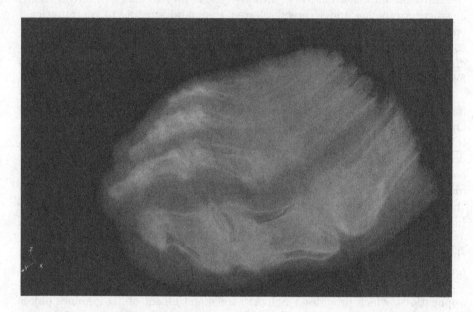

Fig. 3. Volume rendering of tomography results made with ParaView

Using Equalizer and VirtualGL in one sheaf is supported, but requires to make some specific changes to the configuration. First, the frame compression method should be explicitly set for VirtualGL at the application launch. Next, if several graphical units should be used at single node, each application's rendering pipe should be directed to the dedicated GPU manually, except the first one on each node, because it will be managed by VirtualGL itself [26].

In addition, Equalizer developers recommend to use their framework in combination with graphics engines that could provide a scene representation with the usage of scene graph data structure. In this case the whole collection of objects, which represent a virtual world, could be decomposed into a tree structure, thus making it easy to clip undesirable scene parts and distribute the load for remaining evenly. One of such libraries, which is commonly used for development of simulation systems and already has all necessary bindings available, is OpenSceneGraph [11].

Scientific visualization problem should be singled out as a separate topic of conversation, since it has its own peculiar properties. In most cases, the calculation results are represented as a discrete set of values, and the actual meaning is also depending on particular interpretation rules. The biggest difference lies in the way the display of volumetric objects. Here, the body is described not by an external polygonal mesh, but with the points set individually, which are filling the volume, so it comes for volume rendering. Visualization Toolkit (VTK in a short) could be mentioned as one of the most common solutions in this area [9]. The main value of the program is to have a programmable standardization of interfaces and data formats that not only allows, if necessary, easy to repeat current experiment, but also affects the amount of available solutions and extensions. or there could be found a specific exporter, like in case of OpenFOAM application [8].

ParaView is standing out among the rest of VTK application, and it should be referred as the main subject for the following discussion [24]. That application allows to process, render and visualize data coming from many popular scientific packages and data sources. User is able to interact with the image in real-time, or render an animation, video and image files with the selected frame set. Initial data could be supplied with additional materials, like textures, polygonal models, shaders, etc., which could be involved into graphics scene compositing. Powerful API combined with Python scripting support allows to automate actions and tap advanced features. The ability to distribute rendering tasks is implemented with the usage of ICE-T library that automatically enables the feature of displaying the image on multiple screens simultaneously [7]. Figure 3 illustrates the ParaView rendering possibilities for volume data set representing a part of the foot [35].

Two more VTK-based visualization systems that could be noted here as the applications for distributed and parallel rendering are VisBox and VisIt. The first one could help to create a virtual or augmented reality showcase, while the second one will also provide an opportunity to process data in parallel.

6   Evaluating Results

As mentioned above, the most rapid onset of system usage could be achieved with Chromium, since it provides a possibility to launch an existing software without making any changes on it. For the system performance evaluation we used some tools from GpuTest benchmarking suite that are able to show distributed rendering peculiarity [17]. All tests use the same configuration, which means FullHD resolution for the application running in a fullscreen mode and tight jpeg compression with frame spoiling enabled for the VirtualGL proxy.

The first benchmark case, called FurMark, appears as single object of a torus shape covered by fur, which is produced with a shader program, and filling the most of the display area, as shown in Fig. 4a. The object being always in transition causes fur effect to be recalculated all the time the movement is animated. This makes FurMark a great scenario to determine GPU calculating capabilities and also to scale it in a sort-first manner.

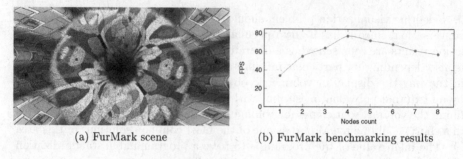

(a) FurMark scene   (b) FurMark benchmarking results

Fig. 4. Running FurMark with Chromium in sort-first mode

So, for this case the screen is split in a number of tiles equal to the amount of calculating nodes. The priority is given to horizontal division, except cases with an even number of nodes bigger than two: the screen will be also divided vertically here.

Results of evaluation are shown in Fig. 4b. As it was expected from a streaming system, performance was able to grow only until it has met the bandwidth limit, which happened after adding a fifth node. Dips on the odd node counts are indicating unequal distribution of workload. Nevertheless, the peak frame rate was twice as high as the initial one. '

The second case uses the same torus figure but in a different manner. It fills exactly the same volume as in the previous test with toruses of a smaller size, as it shown in Fig. 5a. This test covers completely different area, its main aim is to show the transition effectiveness of a multiple objects. As for distributed rendering, it is a great opportunity to show a sort-last compositing technique, based on z-buffer relation. It means that workload will be split not by screen area ownership, but by the groups of dedicated objects. At the final stage the frame will be produced using only those objects that are not overlapped by another ones, however, it leads to some overheads.

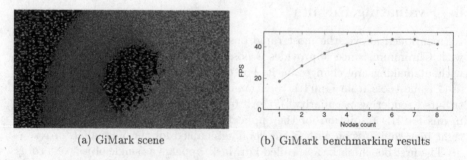

(a) GiMark scene   (b) GiMark benchmarking results

Fig. 5. Running GiMark with Chromium in sort-last mode

As it could be seen in Fig. 5b, the workload is distributed much more evenly, considering that each computing system is getting almost the same amount of data to process. Consequently, attaching additional nodes to the system, including the sixth one, brings positive impact on the result. Apparently, seventh processing unit has no influence at all, and extension up to eight nodes even leads to performance drop, which could be treated as a bandwidth limitation.

7 Conclusion

The technology stack and software environment described above reveal the possibility of Linux operating system usage as a platform for intensive computer graphics applications. The chosen solution allows not only to run applications with 3D acceleration remotely, but also to control a launch process with the usage of queue management system. The number of accelerated gateway nodes and computational nodes could be varied on demand, as well as their performance characteristics. A set of development tools and ready-to-use applications is given, which have the ability to work in parallel rendering mode in the following environment, allowing the most efficient use of available resources.

Further scope of work for virtual infrastructure improvement could be designated with effort of traditional X graphics server replacement to the alternative implementation, like Wayland or Mir, which are getting shipped as default at the latest Fedora and Ubuntu distributions respectively. Noted display management software have some advantages, such as simplified display server protocol and ability of direct framebuffer manipulation, thereby promising some experience and performance improvements [10, 20].

Acknowledgements. Research was carried out using computational resources provided by Resource Center "Computer Center of SPbU" (http://cc.spbu.ru/) and supported by grants of Russian Foundation for Basic Research (projects no. 16-07-01111, 16-07-00886, 16-07-01113) and Saint Petersburg State University (project no. 0.37.155.2014).

References

1. Chromium Documentation. http://chromium.sourceforge.net/doc/index.html. Accessed 17 Jan 2016
2. CUDA Toolkit | NVIDIA Developer. https://developer.nvidia.com/cuda-toolkit. Accessed 15 Sept 2015
3. GeForce GTX 680 | Specifications | GeForce. http://www.geforce.com/hardware/desktop-gpus/geforce-gtx-680/specifications. Accessed 03 Oct 2015
4. How to investigate and use Turbo mode, C-States, P-States in XenServer. http://xenserver.org/partners/developing-products-for-xenserver/19-dev-help/138-xs-dev-perf-turbo.html. Accessed 24 Oct 2015
5. Intel Graphics Virtualization Technology (Intel GVT). https://01.org/igvt-g. Accessed 18 Oct 2015
6. Part I. Installation and Configuration Instructions, Chap. 8. Common Problems. ftp://download.nvidia.com/XFree86/Linux-x86/256.44/README/common problems.html. Accessed 15 Sept 2015

7. Sandia National Laboratories: IceT. http://icet.sandia.gov/. Accessed 19 Jan 2016
8. The ParaView Post-processor. http://www.openfoam.org/features/paraview.php. Accessed 19 Jan 2016
9. VTK - The Visualization Toolkit. http://www.vtk.org/. Accessed 19 Jan 2016
10. Wayland FAQ. https://wayland.freedesktop.org/faq.html. Accessed 24 Mar 2016
11. Openscenegraph and equalizer. Technical report, Eyescale Software GmbH, Neuchâtel, Switzerland, April 2010
12. Mesa FAQ, 9 October 2012. http://www.mesa3d.org/faq.html. Accessed 21 Nov 2015
13. PowerEdge R720, R720xd Technical Guide, April 2012. http://partnerdirect.dell.com/sites/channel/Documents/PowerEdge-Rack-Server-R720-R720xd-Technical-Guide-April2012.pdf. Accessed 11 Sept 2015
14. Extending slurm with support for gpu ranges. Technical report, VMWare, Palo Alto, CA, United States (2013)
15. NVIDIA GRID K2 Graphics Board: Board Specification, January 2013. http://partnerdirect.dell.com/sites/channel/Documents/PowerEdge-Rack-Server-R720-R720xd-Technical-Guide-April2012.pdf. Accessed 03 Oct 2015
16. Simple Linux Utility for Resource Management, 24 November 2013. http://slurm.schedmd.com/. Accessed 29 Jan 2016
17. GpuTest - Cross-Platform GPU Stress Test, OpenGL Benchmark for Windows, Linux and OS X — Geeks3D.com. 4 March 2014. http://www.geeks3d.com/gputest/. Accessed 16 Nov 2015
18. VMware KB: Poor virtual machine application performance may be caused by processor power management settings. 28 August 2014. https://kb.vmware.com/selfservice/microsites/search.do?language=en_US&cmd=displayKC&externalId=1018206. Accessed 24 Sept 2015
19. libjpeg-turbo | About , Performance, 13 August 2015. http://www.libjpeg-turbo.org/About/Performance. Accessed 24 Sept 2015
20. Mir: Welcome to Mir, 24 March 2015. https://unity.ubuntu.com/mir/. Accessed 24 Mar 2016
21. NoMachine - A brief description of the NX protocol in version 4 or later, 21 December 2015. https://www.nomachine.com/AR11K00745. Accessed 24 Sept 2015
22. TurboVNC | About / A Brief Introduction to TurboVNC, 13 August 2015. http://www.turbovnc.org/About/Introduction. Accessed 24 Sept 2015
23. VirtualGL | Main / The VirtualGL Project, 13 August 2015. http://www.virtualgl.org/. Accessed 24 Sept 2015
24. Utkarsh Ayachit. The ParaView Guide: A Parallel Visualization Application. Kitware (2015)
25. Bogdanov, A., Degtyarev, A., Korkhov, V., Gaiduchok, V., Gankevich, I.: Virtual supercomputer as basis of scientific computing. In: Clary, T.S. (ed.) Horizons in Computer Science Research, pp. 159–198. Nova Science Publishers, New York (2015). ISBN: 978-1-63482-499-6
26. Eilemann, S.: VirtualGL Support. 22 December 2011. https://github.com/Eyescale/Equalizer/blob/master/doc/README.VirtualGL. Accessed 16 Nov 2015
27. Eilemann, S.: Equalizer Programming and User Guide. Eyescale Software GmbH, 26 July 2013
28. Gankevich, I., Gaiduchok, V., Gushchanskiy, D., Tipikin, Y., Korkhov, V., Degtyarev, A., Bogdanov, A., Zolotarev, V.: Virtual private supercomputer: Design and evaluation. In: 9th International Conference on Computer Science and Information Technologies, CSIT 2013, Revised Selected Papers, pp. 1–6 (2013)

29. Gankevich, I., Korkhov, V., Balyan, S., Gaiduchok, V., Gushchanskiy, D., Tipikin, Y., Degtyarev, A., Bogdanov, A.: Constructing virtual private supercomputer using virtualization and cloud technologies. In: Murgante, B., Misra, S., Rocha, A.M.A.C., Torre, C., Rocha, J.G., Falcão, M.I., Taniar, D., Apduhan, B.O., Gervasi, O. (eds.) ICCSA 2014, Part VI. LNCS, vol. 8584, pp. 341–354. Springer, Heidelberg (2014)

30. Isoka, D.: Understanding, Evaluating RemoteFX vGPU on Windows Server 2012 R2 — Remote Desktop Services Blog, 6 June 2014. https://blogs.msdn.micro soft.com/rds/2014/06/06/understanding-and-evaluating-remotefx-vgpu-on-windo ws-server-2012-r2/. Accessed 21 Nov 2015

31. Jette, M., Grondona, M.: Slurm: Simple linux utility for resource management. Technical report UCRL-MA-147996 REV 3, Lawrence Livermore National Laboratory, Livermore, CA, United States (2003)

32. Richardson, T., Stafford-Fraser, Q., Wood, K.R., Hopper, A.: Virtual network computing. IEEE Internet Comput. 2(1), 33–38 (1998)

33. Soner, S., Özturan, C.: An auction based slurm scheduler for heterogeneous supercomputers and its comparative performance study. Technical report, Computer Engineering Department, Bogazici University, Istanbul, Turkey (2013)

34. Soner, S., Özturan, C., Karac, I.: Extending slurm with support for gpu ranges. Technical report, Computer Engineering Department, Bogazici University, Istanbul, Turkey (2013)

35. Tierny, J.: Visualization Tutorial Exercise - Visualization with ParaView. 26 July 2014. http://www-pequan.lip6.fr/~tierny/visualizationExerciseParaView. html. Accessed 28 Feb 2016

36. Jain, M.: Under the hood of GPU Sharing technologies: vSGA and vGPU (8 January 2014).https://www.citrix.com/blogs/2014/01/08/under-the-hood-of-gpu-shar ing-technologies/. Accessed 18 Oct 2015

Great Deluge and Extended Great Deluge Based Job Scheduling in Grid Computing Using GridSim

Omid Seifaddini[✉], Azizol Abdullah, Abdullah Muhammed,
and Masnida Hussin

Universiti Putra Malaysia, Serdang, Malaysia
omid.seifaddini@gmail.com, {azizol,abdullah,masnida}@upm.edu.my

Abstract. Scheduling of jobs is one of the most important research areas of Grid computing as it has attracted so much attention since its beginning. Job scheduling in Grid computing is a NP-Complete problem due to Grid characteristics such as heterogeneity and dynamicity. Many heuristic algorithms have been proposed for Grid scheduling to avail Grid computing. However, these heuristic methods are limited by time constraints required for remapping of jobs to Grid resources in such elastic and dynamic environments. Great Deluge (GD) is a practical solution for such a problem. Therefore, this paper presents Great Deluge and Extended Great Deluge (EGD) based scheduling algorithm for Grid computing. We also present the detailed implementation of GD and EGD in a reliable simulation platform, GridSim. This has two advantages. First, it will ease the reimplementation process for future contributors since there are lots of complexity and ambiguity to develop such scheduling algorithms. Second, most of the research and experimental results, especially in the area of Grid scheduling, have used their own developed infrastructure to simulate the performance of their algorithms, thus the question remains on how well they will perform in a real world environment. We also, investigate the computation time and the number of soft constraints violations of EGD against its conventional GD algorithm. The GD scheduling algorithm is able to provide qualitative solution in shorter time for small Grid size while EGD could produce schedule in shorter time for all cases.

Keywords: Grid scheduling · Great Deluge · Heuristics · GridSim

1 Introduction

Distributed systems like Grid Computing have attracted lots of attention since the development of the Internet that enabled Grid platform to provide vital services such as data storage and processing to the users. Grid computing is currently a very successful computing platform for distributed applications [1] but it arises issues and challenges like allocation of jobs to the Grid resources

© Springer International Publishing Switzerland 2016
O. Gervasi et al. (Eds.): ICCSA 2016, Part II, LNCS 9787, pp. 292–302, 2016.
DOI: 10.1007/978-3-319-42108-7_22

with their unique characteristics. Job allocation or job scheduling is important because of the need for Grid enabled applications in optimization, e Science and computing [2]. Grid Computing with geographically distributed resources can give these applications many benefits. However, this is possible only if the resources are scheduled well.

Grid scheduling is defined as the process of making scheduling decisions involving allocating jobs to resources over multiple administrative domains. This can include searching multiple administrative domains to use a single machine or scheduling a single job to use multiple resources at a single site or multiple sites. Scheduling an application in Grid computing is significantly more complicated than scheduling an application in traditional supercomputers. A proper and good Grid scheduling should take Grid characteristics into consideration to get the true and promising potential of Grid computing. Grid characteristics to be considered are as follows:

- Resource heterogeneity: Grid resources are heterogeneous in terms of computational and network. Heterogeneity in computational resources exists due to the different architectures, different number of processors, different processor speeds. A network is heterogeneous in terms of bandwidth and communication protocols. Therefore, Grid resources having heterogeneity cannot be addressed uniformly. A good scheduling algorithm is needed to consider these heterogeneities to leverage different computing powers of diverse resources.
- Autonomy: Grid computing comprises many administrative domains and each of them has it's own policies. Grid resources are autonomous and the Grid scheduler does not have full control over the resources. The Grid scheduler should obey the resources local policies and it can not violate them thereby making Grid scheduling difficult and challenging. Autonomy also brings diversity in resource management and access control policy which necessitates the scheduler to be adaptive to different local policies.
- Application diversity: It refers to the variety of Grid applications of many users, each of them having its own particular feature. Dependent and independent jobs of application are examples of application diversity.
- Dynamic environment: The environment in the traditional distributed computing is assumed to be fixed and stable but in Grid environment both network and resources are dynamic. Grid network is dynamic as it can be shared by many parties and there will be no guaranteed bandwidth. Also the available bandwidth in network connecting grid resources might be affected by internet traffic flows, as it results in performance fluctuation. Resources are dynamic because of resource autonomy since usually, a Grid resource might not be dedicated to one application job. Besides, on one hand, a resource might not be available at a time due to network failure, and on the other hand, some new resource might join the Grid. The capability of a resource may change overtime because of contention between parities who share the resource
- Resource Selection: In the traditional distributed system such as cluster, the computation part of the application and input/output data are at the same site or it is known before application submission. This characteristic of

traditional systems makes the cost of data staging negligible. But unlike the traditional distributed systems, the computation sites of an application is chosen by the Grid scheduler based on certain known criteria since the Grid consists of many heterogeneous computing sites with storage sites connected through Wide Area Networks (WANs).

Allocation of jobs to the resource in Grid computing with mentioned characteristics is NP-complete problem and such problems are often solved using heuristic methods. Heuristic methods have demonstrated to be the right approach for such problems as using them is de facto choice [3]. Many heuristic methodologies have been applied in Grid computing such as Genetic Algorithms [4] and Tabu Search [3]. However, to the best of our knowledge, and based on literature, all of heuristic based scheduling algorithms are limited by the time constraints required for remapping the resources in such elastic and dynamic environments [5]. Great Deluge (GD) [6] is one of the simple and practical solution to such problem.

The extension of Great deluge proposed by McMullan and McCollum [5] and applied in Grid scheduling. Their aims of extension were to speed up the process while avoiding the algorithm to entrap in local optima as the algorithm reaches to optimal solution. However, there is always difficulty to re implement such a scheduling algorithm as there might be lots of assumption, no standard in system and application models and no reliable platform for evaluation.

Therefore, this study presents the detailed implementation of Great Deluge and Extended Great Deluge algorithm. Furthermore, performance of Extended Great Deluge will be compared with the conventional Great Deluge in a reliable and verified evaluation tool called GridSim [7]. Using reliable platform provides us many benefits such as reliable results, easier reimplementation for future researchers and better evaluation in more realistic environment [8]. A standard system model which represents real Grid environment based on [3] will be used for evaluation purpose.

The rest of the study is presented in the following order: Grid scheduling under the study will be discussed in the next Sect. 2. The Great Deluge and Extended Great Deluge algorithms are presented in Sects. 3 and 4 respectively. The detailed implementation steps using GridSim are discussed in Sect. 5. In Sect. 6, the evaluation tool and experimental configurations are discussed. The experimental results are presented in Sect. 7 while we conclude the paper in Sect. 8.

2   Grid Scheduling

Grid Scheduling is about the fulfilment of several requirements and conditions as constraints. Constraints are the instruments which are available for restricting the search space in the search process. The nature of the Grid scheduling problem requires considering not only validity and feasibility of solution but also quality of solutions. Therefore, the following constraints (hard constraints) that must be satisfied in order to achieve a feasible solution are presented:

- A resource cannot execute more than one job at a time. Thus two jobs could not assign to one processor at the same time, no overlapping jobs.
- A Job must not appear more than once in valid schedule.
- A resource must have enough capability to run the job (The resource must have enough timeslot to run).
- The job must not starts before the job earliest time (Arrival time).

Soft constraints should be identified to address the quality of solution. Generally, it is difficult to fulfill all soft constraints. The following soft constraints are used as measure of the overall penalty for solutions:

- Job should be scheduled to the fastest resource possible.
- The job should be assigned to the cheapest resource possible.

3   Great Deluge

Great Deluge is local search method proposed by Dueck in 1993 [6] as alternative to simulated annealing. It is one of the successful heuristic method which its effectiveness is proved not only in Grid scheduling [5] but course time tabling [9], exam time tabling [10] and etc. It is very simple yet effective algorithm as it needs only to set up the water level as its basic parameter, which makes it naturally very attractive for solving optimisation problems specially scheduling [11].

Great Deluge is threshold acceptance strategy which uses a threshold that decrease linearly to define an acceptance range for the quality of solution [11]. In other words, GD use a simpler acceptance policy for dealing with moves that lead to decrease in the quality of solution [12]. Using threshold accepting strategy instead of using probability, guides the algorithm to accept the worse solution. The algorithm accept new candidates if it is better or equal to current solution. It accepts the worse solution if and only if the penalty of the solution is less than or equal to the pre defined value water level [9].

The water level is initially set to slightly higher than the initial solution and reduced gradually to improve the quality of solution. One of the main advantages of GD over other local search methods is that better solutions could be obtained with the prolongation of the search time, although this may not be a true fact as the search space cannot be controlled. GD algorithm is used by many researchers in various area of NP complete problem such as [11,13–15]. The pseudo code of the Great Deluge based Grid scheduling is presented in Algorithm 1.

4   Extended Great Deluge

The standard Great Deluge algorithm has been extended by McMullan and McCollum [5] to improve the conventional GD performance. The Extended Great Deluge (EGD) allows the algorithm to reheat by an approach similar to Simulated Annealing [16]. Their main concern was to improve the speed at which a good solution can be found, and avoiding the algorithm not being trapped in the

---

Algorithm 1. GD Pseudo code

Data: Number of jobs and number of resources
Result: Assigning jobs to the resources
Initialize number for resources;
initialize number of jobs;
/* Generating initial random feasible solution   */
initialEvaluation= Calculate initial penalty;
f_s=initialEvaluation;
$f_{s\star}$=initialEvaluation;
b_0=initialEvaluation;
while *Temination condition not met* **do**
  Apply neighbourhood moves;
  Calculate $f_{s\star}$;
  if $f_{s\star} \leq f_s \parallel f_{s\star} \leq b_0$ **then**
    $f_s = f_{s\star}$;
    Calculate Penalty;
    $b_0 = b_0-$decayRate
  end
end
randomScheduler.getNewSchedulingTable();
Update $f_{s\star}$;

---

local optima. What distinguishes the EGD from the standard GD is the reheat mechanism, as the standard GD stops when a lack of improvement is observed for specific time while EGD employ reheat mechanism to widen the boundary. This will allow the worse solution to be applied to the current solution. The pseudo code of EGD based Grid scheduling is illustrated in Algorithm 2.

5   Implementation

In this section, we discuss the scheduling algorithm implementation in GridSim [7]. GridSim is an open source java based discrete event simulator that simulates the behaviour of Grid environment. It supports modelling and simulation of heterogeneous Grid resources, users and application models. It also provides primitives for creation of application tasks, mapping of tasks to resources, and their management.

In order to implement scheduling policy based on our own defined policy as presented in GridSim documentation[1] and [7], the developers and researchers need to implement the body of methods in *AllocPollicy* regarding to their defined scheduling policy. On the other hand, there is a class called GridSim in GridSim toolkit which is responsible for creating the resources and submitting the jobs to the Grid. The *ResourceCharacteristics* class provides support for each resource in way that each resource can have it's own characteristics such as Processing

---

[1] http://www.cloudbus.org/gridsim/doc/api/.

Algorithm 2. EGD Pseudo code

Data: Number of jobs and number of resources
Result: Assigning jobs to the resources
Initialize number for resources;
initialize number of jobs;
```
/* Generating initial random feasible solution                */
```
initialEvaluation= Calculate initial penalty;
f_s=initialEvaluation;
f_{s*}=initialEvaluation;
b_0=initialEvaluation;
while *Termination condition not met* **do**
  | Apply neighbourhood moves;
  | Calculate f_{s*};
  | **if** $f_{s*} \leq f_s \parallel f_{s*} \leq b_0$ **then**
  | | $f_s = f_{s*}$;
  | | Calculate Penalty;
  | | $b_0 = b_0 - \text{decayRate}$
  | **end**
end
randomScheduler.getNewSchedulingTable();
Update f_{s*};
if *reheat is activated* **then**
  | **while** *reheat activation is true* **do**
  | | b_0 =penalty;
  | | **if** $b_0 > f_s$ **then**
  | | | Break;
  | | **end**
  | | randomScheduler.getNewSchedulingTable();
  | | update Penalty value;
  | **end**
  | set the decay rate
end

Elements (PE) and PE ratings. The Unified Modelling Language (UML) diagram presented in Fig. 1 shows the UML class design of our overall simulation in abstract level.

The initial solution for our scheduling policy is handled using simple random algorithm. This utilises a random list of the events to be scheduled based on the individual penalties incurred during each iteration of construction. The simple random algorithm does not intend to enhance the solution itself, but simply proceed until a feasible solution is found. The first parameter used within the Great Deluge and Extended Great Deluge is the initial decay rate, which will determine how fast the search space is reduced and ultimately the condition for accepting worse moves is narrowed. The approach outlined in this paper uses a decay rate 6 which was obtained experimentally. This decay rate would push the algorithm to attempt to reach the optimal solution, but, generally, a

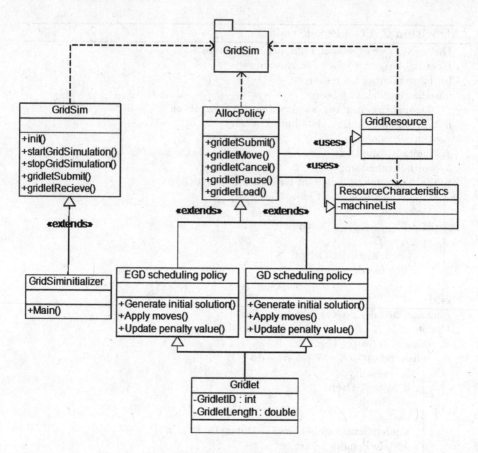

Fig. 1. UML diagram

continuous lack of improvement will happen before this is reached, at which time the reheat mechanism is activated. The reheat mechanism activation because of lack of improvement could be identified in terms of percentage or number of total moves in the process. After the reheat activation, the boundary is set to greater the current best evaluation by a similar percentage to that applied in the initial boundary setting. The decay rate is set to the quicker rate than the initial setting to speed up the search process. In this paper, the general setting chosen for the algorithm outlined is set to 25 percent of the remaining time, with the improvement wait time remaining unchanged. The heuristic moves utilized in this paper are add, drop, move and swap. Selected heuristic moves would maximise the performance of the algorithm.

6   Evaluation

The evaluation of the proposed approach is discussed and described in this section. The evaluation is a very important stage as it verifies how beneficial the proposed approach is. Simulation configuration along with the parameters are described and discussed in this section.

6.1   Configuration

For experimental purpose, a suitable workload from workload archive[2] based on the Lublin model [17] is derived. Lublin model is a detailed model of rigid jobs which include the arrival pattern, number of CPU required and execution time of each job. We have defined heterogeneous Grids of different sizes grouped into four different sets called small (32 resource/512 gridlets), medium (64 resources/1024 gridlets), large (128 resources/2048 gridlets), and very large (256 resources/4096 gridlets) for the evaluation purpose. Gridlet is a package that contains information regarding the jobs. In other words, gridlet represents the job in GridSim. In this paper, each resource is having one machine and each machine could have 1 to 16 Processing Elements (PEs) with each PE having 377 to 515 Million Instructions Per Second (MIPS). Cost of using each resources are varied from 1 to 8 Grid dollars.

6.2   Performance Metric

In computer terms performance metrics are measures used to evaluate and assess given algorithm, model or framework such as throughput, delays. The performance metrics that have been used in this study to investigate the performance of developed algorithms are computation time and number of soft constraints violations. Computation time is the amount of time consumed by the algorithm to schedule jobs to resources in this paper. Computation time is an important criterion when comparing scheduling algorithms. Depending on scheduling context, it can have crucial impact on the applicability of an algorithm [18]. Soft constraints are determined in a way to address makespan and user's costs since minimizing the soft constraints violations means minimizing the makespan and user's cost on using the Grid services.

7   Results

To investigate the performance of developed scheduling algorithms, four different case studies were considered. The case studies reflect different real Grid computing environment. An average of 30 runs is used to ensure statistical correctness. The best results (average and minimum) are highlighted with bold type. The simulation results are presented using Lublin model and GridSim discrete event simulation. Comparison of computation time and number of soft constraints violations for the developed scheduling algorithms have been presented in Tables 1, 2, 3 and 4.

Tables 1 and 3 present computation time achieved by Great Deluge and Extended Great Deluge algorithm to schedule jobs to resource for each case

---

[2] http://www.cs.huji.ac.il/labs/parallel/workload/swf.html.

Table 1. Computation time(s)

Algorithms	Small			Medium		
	Ave	Min	Max	Ave	Min	Max
GD scheduler	**217**	210	231	442	**418**	468
EGD scheduler	218	**206**	235	**433**	419	460

Table 2. Number of soft constraints violations

Algorithms	Small			Medium		
	Ave	Min	Max	Ave	Min	Max
GD scheduler	**483.50**	**474.00**	490.00	**1074.70**	**1052.00**	1087.00
EGD scheduler	487.68	475.00	495.00	1081.36	1060.00	1093.00

Table 3. Computation time(s)

Algorithms	Large			Very large		
	Ave	Min	Max	Ave	Min	Max
GD scheduler	906	869	943	1972	1903	2198
EGD scheduler	**904**	**807**	968	**1960**	**1883**	2142

Table 4. Number of soft constraints violations

Algorithms	Large			Very large		
	Ave	Min	Max	Ave	Min	Max
GD scheduler	**2319.40**	**2304.00**	2332.00	**4891.13**	4864.00	4907.00
EGD scheduler	2330.37	2312.00	2345.00	4912.00	**4859.00**	4933.00

study. As can be seen, the Great Deluge performs better over Extended Great Deluge only in small case. The EGD was able to schedule jobs to resources in less amount of time for medium, large and very large Grid sizes. The most obvious occurrence of this is in very large Grid size.

But however, Great Deluge could produce better solution quality compared to EGD since the number of soft constraints violations are less (Tables 2 and 4). Having less soft constraints violations will strongly reduce makespan and cost as the constraints are defined in such a way to address these two metrics.

By comparing the overall four Tables, we can see that Great Deluge was able to produce good quality solution in short time only for small Grid sizes. EGD was able to produce solution in short time in cost of sacrificing the solution quality. Great Deluge also scarifies time to obtain a good solution quality.

8   Conclusion and Future Work

In this work, we developed and analyse two practical solutions to time constrain of Grid scheduling based meta heuristic algorithm in a reliable simulation platform called GridSim. The detailed steps of implementation are outlined for future contributor.

Although Extended Great Deluge is an effective algorithm for Grid scheduling especially in terms of computation time, the standard Great Deluge algorithms is also successful given the limited time in which an effective scheduling solution must be achieved in small Grid size. The scheduling algorithms were not able to produce good solution quality in condition of not sacrificing time. This is a big drawback for current scheduling algorithms. It is worth mentioning that the developed scheduling algorithms were able to provide better computation time compared to the current scheduling algorithms reported in literature by [19] as they mentioned that the Tabu Search take 3.5 h to schedule jobs. Also, the scheduling algorithms provide different results on different Grid sizes which will raise the question on the applicability and ability of different scheduling algorithms in different instances of problem.

Investigation will continue to directly compare the results from the scheduling algorithm against other reported benchmark results in the current literature. We are currently working on developing scheduling algorithms to simultaneously produce good solution quality in short time. We also investigate on parameters involved in Great Deluge algorithm such as reheat and decay rate. The analysis will be carried out on the limits of these parameters and how they affect solution quality.

References

1. Talbi, E.-G., Zomaya, A.Y.: Grid Computing for Bioinformatics and Computational Biology, vol. 1. Wiley, New York (2007)
2. Xhafa, F., Duran, B., Abraham, A., Dahal, K.P.: Tuning struggle strategy in genetic algorithms for scheduling in computational grids. In: 2008 7th Computer Information Systems and Industrial Management Applications, pp. 275–280. IEEE, June 2008
3. Xhafa, F., Carretero, J., Dorronsoro, B., Alba, E.: A tabusearch algorithm for scheduling independent jobs in computational grids. Comput. Inform. **28**, 1001–1014 (2009)
4. Carretero, J., Xhafa, F.: Using genetic algorithms for scheduling jobs in large scale grid applications. J. Technol. Econ. Dev. **7**(1), 11–17 (2006)
5. McMullan, P., McCollum, B.: Dynamic job scheduling on the grid environment using the great deluge algorithm. In: Malyshkin, V.E. (ed.) PaCT 2007. LNCS, vol. 4671, pp. 283–292. Springer, Heidelberg (2007)
6. Dueck, G.: New optimization heuristics: the great deluge algorithm and the record-to-record travel. J. Comput. Phys. **104**(1), 86–92 (1993)
7. Buyya, R., Murshed, M.: Gridsim: a toolkit for the modeling and simulation of distributed resource management and scheduling for grid computing. Concurr. Comput. Pract. Exper. **14**(13–15), 1175–1220 (2002)

8. Moallem, A.: Using swarm intelligence for distributed job scheduling on the grid. Ph.D. thesis, University of Saskatchewan, Saskatoon (2009)

9. McMullan, P.: An extended implementation of the great deluge algorithm for course timetabling. In: Shi, Y., Albada, G.D., Dongarra, J., Sloot, P.M.A. (eds.) ICCS 2007, Part I. LNCS, vol. 4487, pp. 538–545. Springer, Heidelberg (2007)

10. Landa-Silva, D., Obit, J.H.: Great deluge with non-linear decay rate for solving course timetabling problems. In: 2008 4th International IEEE Conference Intelligent Systems, pp. 8-11–8-18. IEEE, September 2008

11. Kifah, S., Abdullah, S.: An adaptive non-linear great deluge algorithm for the patient-admission problem. Inf. Sci. **295**, 573–585 (2015)

12. Ozcan, E., Burke, E.K.: A reinforcement learning great-deluge hyper-heuristic for examination timetabling. Int. J. Appl. Metaheuristic Comput. **1**(1), 39–59 (2010)

13. Yang, Y., Petrovic, S.: A novel similarity measure for heuristic selection in examination timetabling. In: Burke, E.K., Trick, M.A. (eds.) PATAT 2004. LNCS, vol. 3616, pp. 247–269. Springer, Heidelberg (2005)

14. Bilgin, B., Özcan, E., Korkmaz, E.E.: An experimental study on hyper-heuristics and exam timetabling. In: Burke, E.K., Rudová, H. (eds.) PATAT 2007. LNCS, vol. 3867, pp. 394–412. Springer, Heidelberg (2007)

15. Aron, R., Chana, I., Abraham, A.: Hyper-heuristic based resource scheduling in grid environment. In: 2013 IEEE International Conference on Systems, Man, and Cybernetics, pp. 1075–1080, October 2013

16. Kendall, G., Mohamad, M.: Channel assignment in cellular communication using a great deluge hyper-heuristic. In: Proceedings. 2004 12th IEEE International Conference on Networks (ICON 2004) (IEEE Cat. No. 04EX955), pp. 2:769–2:773 (2004)

17. Lublin, U., Feitelson, D.G.: The workload on parallel supercomputers: modelling the characteristics of rigid jobs. J. Parallel Distrib. Comput. **63**, 1105–1122 (2003)

18. Dragiev, S.: Grid workow recovery as a dynamic constraint satisfaction problem. Technische Universit at Berlin Group Communication and Operating Systems. Ph.D. thesis (2008)

19. Nesmáchnow, S., Cancela, H., Alba, E.: Heterogeneous computing scheduling with evolutionary algorithms. Soft Comput. **15**(4), 685–701 (2010)

Employing Docker Swarm on OpenStack for Biomedical Analysis

Christoph Jansen, Michael Witt, and Dagmar Krefting$^{(\boxtimes)}$

University of Applied Sciences Berlin (HTW Berlin), Berlin, Germany
{christoph.jansen,m.witt,dagmar.krefting}@htw-berlin.de

Abstract. Biomedical analysis, in particular image and biosignal analysis, often requires several methods applied to the same data. The data is typically of large volume, so data transfer can become a bottleneck in remote analysis. Furthermore, biomedical data may contain patient data, raising data protection issues. We propose a highly virtualized infrastructure, employing *Docker Swarm* technology as the computing infrastructure. An underlying *Openstack* based IaaS cloud provides additional security features for a flexible and efficient multi-tenant analysis platform. We introduce the prototype infrastructure along a sample use-case of multiple versions of a machine-learning method applied to feature sets extracted from multidimensional biosignal recordings from Sleep Apnea patients and healthy controls.

Keywords: Biomedical analysis · SaaS · IaaS · OpenStack · Docker swarm

1 Introduction

Biomedical analysis is a common show-case for distributed compute infrastructures, as it typically involves large datasets and compute-intensive analysis (e.g. [1]). Parallelization on different scales has be shown to reduce overall processing-time significantly [2]. Established methods have been implemented to offer a remote analysis service, e.g. the BLAST algorithm in bioinformatics, or Freesurfer in brain imaging [3,4]. In clinical research, however, analysis may require application of different tools and methods provided by several developing groups within the community [5]. Cloud- or Gridbased biomedical data analysis is typically laid out as a predefined workflow that starts with data selection from a central data repository and ends with storage of the analysis results within this repository. In between several analysis steps can be processed [6]. Here, we see the need of a more dynamic selection of methods to be applied to a dataset. In earlier work, a system has been developed, where new MATLAB[1]-based algorithms can be uploaded and are dynamically deployed to a virtual machine or container. They are automatically recognized and integrated by the processing

---

[1] http://www.mathworks.com.

© Springer International Publishing Switzerland 2016
O. Gervasi et al. (Eds.): ICCSA 2016, Part II, LNCS 9787, pp. 303–318, 2016.
DOI: 10.1007/978-3-319-42108-7_23

pipeline [7]. However, this approach was limited to methods that could be executed in a standardized environment and code compiled for a specific MATLAB Compiler Runtime version. Nowadays, Docker offers a light-weight method to provide a full execution environment for specific methods, based on Linux containers[2]. While containers are often implemented as a stand-alone applications, they could serve as atomic building blocks for the execution of single workflow steps within the analysis. However, this requires communication between dynamically scheduled Linux containers. New developments towards container clusters, as provided by *Docker Swarm*, are promising approaches. Docker Swarm allows for dynamic provisioning of Docker containers on a distributed computing infrastructure, typically on bare metal servers. While scaling and virtualization of the execution environment is supported, virtualization of the operating system - including authentication management and network - cannot be reached. Multi-tenant use is to date barely supported. These are typically features of IaaS clouds, such as OpenStack based infrastructures[3]. In this manuscript, the implementation and employment of an infrastructure using Docker within an Openstack IaaS along a use-case of biosignal analysis is presented. Advantages and disadvantages for scientific computing of such a double-virtualized system are discussed.

2   Use Case Requirements

The goal of the infrastructure is efficient biomedical data analysis in clinical research. In particular, medical image and biosignal processing applications are addressed. An input set typically spans a few Megabytes up to Terabytes of data. Data-intensive applications require high data storage capacity and in most cases performance is strongly influenced by the amount of memory available. For such data, transfer in distributed compute infrastructures quickly becomes a bottleneck. Apart from performance issues due to the data transfer time, data transfer is often a major error source [8]. For data-intensive applications, the big-data paradigm to *bring the application to the data* is a good approach to avoid such problems. But clinical data often raise data protection issues, that require certain security measures before data transfer is allowed. If this is not the case, e.g. for anonymized data, remote storage close to the compute facilities can be employed [9,10].

Applications in this field can be compute-intensive or run a few minutes. Typically, there are a lot of different methods available from various developing groups, that need to be combined for a specific biomedical analysis. The heterogeneity of frameworks and programming languages makes it difficult for users to get all required code run on one local system. This is particularly true in clinical environments, where centrally administered workstations are used and installation of arbitrary code may conflict with local IT security policies. SaaS in

---

[2] https://www.docker.com. If not individually referenced, details about specific Docker technologies within this paper can also be found on this website.

[3] https://www.openstack.org.

Table 1. Characteristics of biomedical analysis and resource demands.

Characteristics	Compute power	Memory	Storage	Bandwidth	Security
Data intensive		high	high	high	
Compute intensive	high	medium			
Patient data					high

combination with a scientific workflow system is the typical approach, typically provided by so-called science gateways [6]. However, this approache requires data transfer and - due to predefined processing pipelines - certain expert effort to change the workflow description. If different applications shall be applied to the same data set, data transfer to all software services becomes necessary. Table 1 summarizes the complementing resource requirements due to the main characteristics of the addressed biomedical applications.

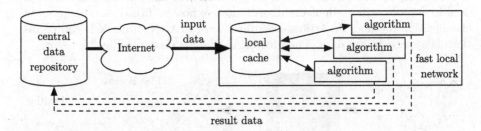

Fig. 1. Concept of the proposed infrastructure. Large input data is cached on the compute site to serve multiple methods, results are sent directly to the final storage.

To offer a reasonable compromise between these requirements, an infrastructure as depicted in Fig. 1 is proposed: the algorithms are executed remotely, but a data cache on the compute site avoids multiple data transfer through the internet, if data is required by different algorithms. However, the cache needs to be deleted as soon as it is no longer used, due to data security concerns. Result data is transferred directly to the final data storage.

3   Methods

3.1   Sample Use Case

As a use case, a current problem from a project on sleep research has been chosen. In particular, the influence of Sleep Apnea on the physiological network in different sleep stages is investigated. Sleep Apnea is a common sleep disorder, that is characterized by repetitive interruptions in breathings during the night. It is associated with excessive daytime sleepiness, as well as several other health issues, such as hypertension [11]. The term *physiological network* refers to the modelling of the human body as an organ system network [12]. It allows to apply methods from network sciences to clinical research topics, and help to reveal biomarkers and give insight into physiological synchronisation. Network nodes are represented by biosignals related to the respective organ system; for example the electrocardiogram represents the heart, or nasal airflow represents the breathing system. Overnight multidimensional biosignal recordings, so-called polysomnographies (PSG), are excellent data sources for network physiology, as up to 30 different biosignals are measured simultaneously. Crosscorrelations of all signal segment combinations are used to determine the binary link state between the organ systems. The results are matrices for each sleep stage (deep sleep, light sleep, REM and wake stage), containing the link strengths for all signal combinations. As it is for example known, that Sleep Apnea may influence cardiac oscillations [13], the influence of Sleep Apnea on the organ network is of high interest. Link strength matrices of 391 PSG from healthy volunteers and 97 PSG from patients suffering from Sleep Apnea have been determined. In a first step, classification of the Apnea patients, based on all four link strengths matrices, is used to get a first idea about the discrimintative power of the link strengths. As classification method, an artificial neural network (ANN) in various configurations is applied to the data. The processing chain is shown in Fig. 2.

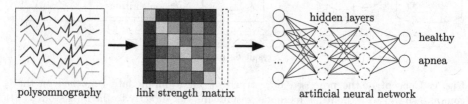

Fig. 2. Processing chain of the use case. PSGs are preprocessed. The entries resulting link strength matrices are used as features in the training of a ANN to classify Apnea patients and healthy subjects.

Classification with artificial neural networks. The data file, used in our experiments, contains link strength values of polysomnographic measures in a CSV format, where every data row describes the results for one subject. Each data row consists of 2812 link strength features. These features are used as input data for an ANN, that is trained to automatically classify healthy subjects and Sleep Apnea patients in a supervised manner. A feed-forward ANN is built from a layer-wise architecture, where every layer is specified by a certain number of neurons. There is always one input layer, one output layer and an arbitrary number of hidden layers in between. The layers are fully connected by weights controlling the strength of each connection between two neurons. During the network's training, the weights are adjusted to produce a better classification in the output layer. The code is developed using the python framework *Keras*[4]. It allows to create artificial neural networks from ready-to-use building blocks.

---

[4] http://keras.io/.

Artificial neural networks typically contain a large amount of configuration parameters. The optimal values for these parameters might be estimated by a data scientist to fit the needs of the given problem. Most of the parameter settings behave similarly, but some tweaks might improve the classification accuracy and reduce overfitting. As the robustness of the classification is of high interest in our case, a parameter scan is realized. As the chosen parameters are independent, all parameter permutations are applied to the neural network.

Data preparation: The data set contains floating point values in a range [0, 1] but also contains some missing values, that need to be replaced. Typical choices for replacement values are *0, −1* or *the mean* of the corresponding feature.

Training and testing is performed with subsets of the full dataset. A fixed splitting ratio of 80 % for training and 20 % for testing has been chosen. The subsets are stratified, so that the ratio of Apnea and healthy subjects is equal in both sets. 10 % of the training set is used as crossvalidation set to serve as reference during the training process.

Training parameters: For performance reasons the training data is processed in small batches, set to not given to *32 samples per batch*. In one training epoch every batch is given to the neural network once. The total number of epochs is set to *200* for every experiment. The number is relatively large for the chosen optimization function *rmsprop*, a quickly converging alternative to *stochastic gradient descent*.

Network parameters: The input layer is populated with the input data, and has therefore a fixed size of the feature number (2812 neurons). The output layer produces the binary classification result, consequently consists of two neurons. The number of hidden layers has been set to *0, 1 or 2*; and *128, 256, 512 or 1024* neurons for each of the hidden layers. The choice for each layer is independent. Since the number of samples in the data set is limited, overfitting might be a problem. This problem is addressed by the random reset of a fixed portion of weights to zero during training (dropout). The dropout rate after each layer can be specified as parameter settings. After the input layer the dropout has been set to *0 % or 10 %*. After each of the hidden layers the dropout was set to *0 %, 25 % or 50 %*. These parameters have shown to be reasonable choices in neural network research [14].

Tasks and output files: For some of the described parameters a fixed value has been chosen, while other parameters have three or more values. All parameter settings result in 942 permutations, that are all tested for the ANN training. The total size of the input file is about 16 MB. The algorithm produces three result files. A HDF5-formatted file holds the weights. In addition the loss/error and accuracy values for the test set are computed and stored in a CSV file, allowing to search through the 942 results and pick the best performing parameter set. The third output file contains the parameter set in JSON format, for further evaluation.

3.2 OpenStack

OpenStack is a free and open-source Infrastructure-as-a-Service (Iaas) Cloud framework. There exist OpenStack-based public clouds, but it can also be deployed in local environments. Due to the complexity of a cloud-platform, OpenStack devides the different tasks into separate services that are coupled by a service registry. This registry service also handles authentication and authorization of users and services inside OpenStack. Core functionality is provided by the Nova service that handles running the virtual machines on the connected hardware hosts. Different virtualization technologies are supported like KVM, XEN or even Microsofts Hyper-V. For network connectivity between the virtual machines and to external resources the Neutron service is used. This service provides encapsulated tenant networks, that are separated using the VXLAN-technology. This prevents virtual machines from capturing network traffic that is not associated with their client network even if the ethernet packages are sent through the same hardware interface.

3.3 Docker

Docker uses container technology to encapsulate processes on Linux hosts. *Container virtualization*, in contrast to *full virtualization* of virtual machines, allows for running arbitrary Linux distributions, but employs the Linux kernel of the underlying host. This kind of virtualization is considered more efficient, since it removes the overhead of booting an operating system. The counterpart to the hypervisor in full virtualization is the *docker-engine*, managing the execution of multiple containers on the host. It assigns a unique name to each container and connects it to a dedicated local network. In addition, Docker is able to restrict available compute resources like CPU time and RAM for each container. The deployment tool *Docker Machine* supports automatic provisioning of virtual hosts with docker-engine. Container image management is comparable to that of VM images: base images, containing the directories and binaries of a certain Linux distribution, are available in the public registry *Dockerhub*[5]. Subsequent self-configured images can be derived. Docker supports provisioning of costumized containers by so-called *Dockerfiles*. New images can be uploaded to a repository at *Dockerhub* by registered users. In the case that the selected repository is public (private repositories are available upon payment), images can be accessed without authorization. It is also possible to set up a registry in a local network, which allows for free private repositories.

Docker Cluster. With the rising use of Docker containers as lightweight alternative for full virtualization of Linux machines, the use cases shift from process isolation to container networks - each providing an atomic service. These new demands regarding orchestration and communication of multiple distributed docker-engines on different hosts are addressed by new developments within

---

[5] https://hub.docker.com/.

the Docker ecosystem. *Docker Swarm* enables cluster computing with Docker. It serves as abstraction layer, providing central access to distributed Docker containers through a single interface. It is achieved by so-called *Swarm agents* - themselves containerized services - running on each host within the cluster. An additional *Swarm manager* on one of the hosts serves as central service and access point for Docker clients. The swarm is complemented by a key-value store like Docker's *consul*. It is required that the consul container runs on a separate host. The cluster is transparent to the client. The Swarm manager selects an appropriate host and delegates the commands to the agent on this host and assigns unique container names. Currently two scheduling strategies, *binpack* and *spread*, are supported for spawning new containers within the Swarm: *binpack* fills the hosts subsequently, while *spread* selects the host with the most free resources. However, as the standard Docker network is managed by the local docker-engine, inter-host communication is not supported by *Docker Swarm* alone. Virtual Docker networks are realized using Linux network namespaces. Each virtual network is encapsulated in such a namespace and a single bridging interface in the system namespace handles sending and receiving the packages using the underlying network infrastructure. This setup is comparably simple for the Docker host- and bridge-networks, where traffic is either send through a iptables-NAT-firewall or simply bridged through the host network interface. Inter-host communication has recently been addressed by the development of *Docker Overlay Networks*[6], extending the docker-engine. The overlay network establishes an IP network using VXLAN-technology. Therefore inside the network-namespace on each host a VXLAN interface with the unique LAN-ID is created and package transmission is tunneled through this device before leaving the host. The whole process of network namespace and virtual device creation is done by the Docker service and transparent to the user. Overlay network name resolution and IP address management is handled via a key-value-storage service. Services running inside a Docker container can be accessed from arbitrary containers within the overlay network by the container name and service port.

3.4   Related Work

Container cluster. As described in Sect. 3.3, *Swarm* provides support for multi-host Docker clusters and uses simple strategies to schedule containers across hosts. Advanced scheduling systems, like Torque[7] and SLURM[8], are well known in the domain of high performance computing. Torque already added Docker support and it might as well be possible to run a Docker container from a job script with SLURM. However, it seems that currently inter-container dependencies are not supported. On the other hand specialized container orchestration technologies have been developed lately. First of all, Docker *Compose*, another official tool from the Docker ecosystem, is meant to complement *Swarm* to manage containerized applications. For example a web application might consist of

---

[6] https://docs.docker.com/engine/userguide/networking/get-started-overlay/.

[7] http://www.adaptivecomputing.com/products/open-source/torque/.

[8] http://slurm.schedmd.com/.

a database service and a web service with the application code. Both services can run in individual containers and *Compose* is used to model the dependency between both containers. Apart from the tools provided by the Docker company, the third-party projects *Kubernetes*[9] and *Mesos*[10] also provide Docker cluster and orchestration technologies. All these technologies have in common, that they are designed to support applications with a long life-cycle and to provide fault tolerance by setting up redundant containers or replacing faulty containers.

Container cluster on OpenStack. A Docker cluster on bare-metal servers does not provide any access control to the compute resources and does not have network isolation. Such security features are on the other hand provided by OpenStack, as it is specifically laid out for different tenants. Docker support is today provided by all public cloud infrastructures. However, the support of container clusters is rarely described, and seems not to be fully supported yet[11]. As public clouds are not of interest for the sensitive data employed in biomedical analysis, we focus on container cluster integration on Openstack. In earlier work we have employed a native OpenStack Docker driver[12], which replaces the hypervisor by docker-engine [7]. Probably due to the different virtualization approach, the Docker driver was not fully integrated into OpenStack, leading to stability issues affecting the whole cloud system. The fact, that docker-engine needs to run with root privileges, left questions about the backend security of the worker nodes. So it was not a fully satisfying solution. The OpenStack Docker driver is still experimental and not part of the mainline OpenStack distribution. *Docker Machine* contains an Openstack plugin, that enables the tool to provision Docker-equipped VMs within OpenStack. However, it does not support a fully automated way to set up a *Swarm* cluster. This topic has been addressed by the official OpenStack project *Magnum*[13]. It aims at enabling Docker cluster technology, like Kubernetes, Mesos and Swarm, in OpenStack Virtual Machines. It provides its own client tools for Linux Shell and Python, that are used to automatically provision a Docker cluster and to run basic container commands. In order to run advanced commands it is designated to use the native tools of the employed technology stack. Magnum is in active development. At the time, where the experiments have been carried out, the available Magnum release for the employed OpenStack version did not support inter-host communication of Docker clusters. Newer versions are announced to provide this feature, though it has not been tested by us.

Docker-based workflows in life sciences. In the area of biomedical analysis, only few container-based solutions are published yet. The Agave Platform [15] is a web service platform for reproducible research in plant biology, building

---

[9] http://kubernetes.io/.

[10] http://mesos.apache.org/.

[11] https://www.pandastrike.com/posts/20160307-docker-swarm-aws-vpc.

[12] https://wiki.openstack.org/wiki/Docker.

[13] https://wiki.openstack.org/wiki/Magnum.

on Docker and loosely coupled compute resources from external cloud services. Its scope is similar to our envisioned infrastructure, but does not provide a light-weight service for scheduling a single cluster in a private network. In the area of neuroimaging, the *Boutiques* framework is developed to allow a consistent description of container-based applications, to be easily deployed within different community infrastructures [16].

4 Implementation

To support the given use-case and the resulting concept, implementation has been based on a Docker cluster on OpenStack VMs. Docker provides lightweight dynamic provisioning of data and application services, the cluster component with the overlay network enables inter-host communication required for data transfer from the data cache. The underlying OpenStack infrastructure allows for resource isolation and restriction within the multi-tenant distributed compute infrastructure. Figure 3 shows the architecture of the employed infrastructure.

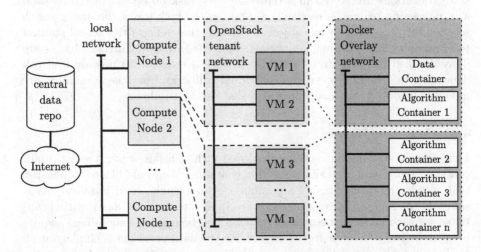

Fig. 3. Architecture of the Docker on OpenStack platform.

The core of the platform is a dedicated Application Management Service (AMS). It handles incoming tasks, schedules them for processing as soon as there are free resources in the cluster and starts the required containers. Inside the container additional custom software components handle data exchange and the execution of algorithms. These components perform callbacks to the AMS in order to inform it about the current state and potential errors.

4.1 Application Management Service

The AMS is a web service, providing a REST interface. It accepts processing requests including the task description in JSON format. The task description itself contains information about the input files, the application and the produced output files. Scheduling of incoming tasks is currently performed with a first-in-first-out strategy. A more sophisticated method is envisioned for more heterogeneous jobs. As described in the concept, data caching is supported. The AMS analyses the task description and checks if all input data is already available within existing data containers. If any data is missing, a new data container is created. It downloads the required files from the remote data source and provides them to the cluster via a web service running inside the container. As soon as all input data is available on the cluster, the algorithm container is started. It contains an executable algorithm bundled with all its dependencies. It fetches the input files from the data containers, executes the algorithm and produces output files. Output file handling is performed by the algorithm container and doesn't involve the data container, as data caching of the different results would not reduce data transfer. As the infrastructure is complex, and many remote services are involved during the task processing, extensive tracking and persistent logging is implemented. Different stateful objects are defined and their state transitions are stored in a database. The task object is created as soon as a processing request is received. As soon as the task leaves the queue and is selected for processing, a job object is derived, that reflects the actual state of the processing job (waiting, processing, succeeded, failed). Further objects are derived, when the algorithm or data container is created. To provide in-depth debugging information, even the individual callbacks from the containers are modelled as objects with certain states that can be validated.

4.2   Container Images

Two different container images are employed within this use-case: A generic data container image and a specific algorithm container. We would like to emphasize, that only in the special case of a parameter scan a single algorithm container is sufficient. As soon as different methods are applied to the data, an algorithm container for each method is required. As described before, data containers acquire input data and provide it to the cluster. As the data provision is similar for all applications, the data container is - with limitations in the transfer protocols - generic. The link to the container image is predefined in the AMS, as well as the RAM size of 512 MB. As mentioned before, the location and access information of the required files are given in the task description. The AMS passes it to the data management program in the container as command-line parameter during container creation. The data management program validates the information and creates a security token for every file. Currently two different data sources, *sftp* and *http*, are supported. In our experiments *sftp* has been employed for data transfer. It requires five parameters per file: remote hostname or IP address, username, password, as well as the path and file name on the remote host. After validating the parameters, the security tokens are sent back to the AMS with a callback. Afterwards the data management program downloads the input files. It sends a second callback to tell the AMS that all files have been downloaded and starts a web server, providing the files on request. The files can only be obtained by an algorithm container if the correct security token is provided. An algorithm container obtains the security tokens from the AMS, which will only reveal them, if the container is supposed to access the files.

In the AMS task description, the algorithm is specified by the container image URL in a Docker registry. Before an algorithm is executed, the corresponding container image is pulled to the Swarm cluster. Docker registries provide a version control and tagging system, which can be used to access a specific version of the algorithm container. However, the algorithm container needs to implement certain functionality to be able to work within the infrastructure. It does not only need the algorithm itself, but also a data management program, which downloads input files from data containers, invokes the algorithm execution, handles uploading result files and communicates with the AMS via callbacks. Therefore a base container with the necessary management program has been created. Algorithm containers should be derived from this base image adding the specific algorithm executables to work. Besides the executables, an application-specific configuration file in JSON format is required. This file contains the path to the algorithm executable to be invoked by the management program. It also contains ordered lists of the input and result files. These lists define the local file paths inside the container image. These lists must match their counterparts in the task description, where the remote input file sources and result file destinations are given. The management program downloads the remote files from a data container and rename them according to the local file names in the configuration and uploads the local result files according to their destination in the task description. This matching is necessary in order to inject different input files without the need to modify the algorithm container and to change the destination of the results. The algorithm can therefore work with fixed local file names.

4.3   Security Considerations

In Sect. 4.2 the usage of security tokens restricting file access in the Docker overlay network has been described. These tokens are only secure, as long as their transmission is encrypted. In the current implementation TLS encryption is enabled for callbacks from containers to the AMS, but not for file access from an algorithm container to a data container. This is not a problem as long as trusted code is executed in the cluster, but enabling TLS for all web servers running in data containers is considered. Generating private and public key pairs, required for encryption, could be realized inside a newly created data container, that will send its public key to the AMS. Algorithm containers can access this public key and enable a secure communication to the data container.

4.4   Environment

The OpenStack cluster consists of one management node and eight compute nodes. Four compute nodes are each equipped with four 6-Core AMD Opteron 2.4 GHz processors and 32 GB RAM, the remaining four with two 4-Core Intel

Xeon 2.67 GHz CPUs and 24 GB RAM. OpenStack Liberty Release is installed. VXLAN for network and KVM for machine virtualization, employing the built-in drivers, are used. A total number of 160 VCPUs and 220 GB RAM are available on the nodes. For our experiments Fedora 23 Cloud as host operating system is used for the VMs, as this distribution is explicity supported by Docker Machine. A small VM with 1 VCPU, 2 GB RAM and 20 GB of disk storage is reserved for the consul service. For the cluster itself, 12 VMs with each 8 VCPUs, 8 GB of RAM and 40 GB of disk storage are running, provisioned with Docker Machine. Docker Version 1.10.3 is used. Docker Swarm is in Version 1.1.3, Docker Machine in Version 0.7.0-rc1.

5   Experimental Results

Scientific goal of the experiment is optimal parameter set for the ANN to classify Apnea patients and healthy controls by the link strength features of the physiological network. Each parameter set is described as an individual task and is scheduled by the AMS.

5.1   Performance

The experiment has been carried out with different RAM size requirements for the algorithm containers. The maximum number of algorithm containers is limited by the amount of 8 GB on 11 VMs and 7.5 GB on the VM where the data container is spawned. The settings including the total runtime of the experiments are summarized in Table 2.

Table 2. RAM size settings for the algorithm containers (AC RAM), resulting maximum number of possible concurrent AC (max. sim. AC) and total runtimes of the ANN training parameter scan. One data container with 512 MB RAM is also running.

Measure	Exp. 1	Exp. 2	Exp. 3	Exp. 4
AC RAM	512 MB	1 GB	2 GB	3 GB
max. sim. ACs	191	95	47	24
Total runtime [min]	153	152	188	341

The overall performance of the cluster is basically constant for 512 MB and 1 GB RAM and increases for larger RAM sizes. This is due to the compute-intensive nature of the neural network training method. The data transfer and processing times, shown in Fig. 4, are indeed smaller for larger RAM sizes, but the higher number of simultaneous tasks at the experiments with lower RAM size outperform regarding total runtime. Larger processing times for low RAM sizes can be explained by the higher number of tasks sharing the CPUs. Larger transfer times with low RAM size can be explained by more tasks simultaneously

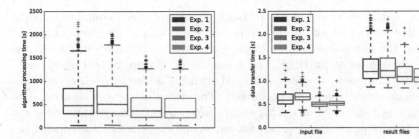

Fig. 4. *Left:* Algorithm execution times. *Right:* Data transfer times for downloading input files from a data container to algorithm containers and for uploading result files from algorithm containers to a remote data storage. (Color figure online)

trying to transfer data. Local transfer times from the data container are significantly lower as the result file transfers to a remote machine. However, direct comparison is difficult, as the result files vary with size, and each task performs three result file uploads, in contrast to one download of the input file with fixed size. Furthermore, individual input file transfer times up to 81.48 s have been measured (not shown). This indicates, that a single data container is not be sufficient to avoid data transfer to be a bottleneck.

The fact, that processing times of 512 MB and 1 GB RAM size are nearly constant, can be explained by a similar load, shown in Fig. 5. Due to the non-vanishing time of about 3 s required for task scheduling, the number of parallel containers does not reach it's maximum in Exp. 1, before the first containers are finished. The load is fluctuating strongly for low RAM, as the termination of

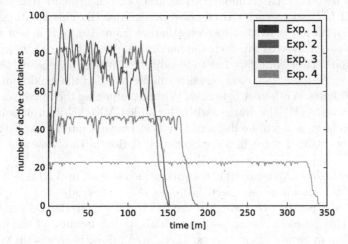

Fig. 5. Number of running algorithm containers on the cluster. At low RAM size, the number of containers is fluctuating strongly, due to a spawning overhead of about 3 s for each container. (Color figure online)

containers cannot be compensated instantly by spawning new containers. As the mean value for Exp. 1 and 2 are similar, they result in similar total runtimes. The lower the maximum number of parallel tasks is, the more constant is the load, reaching most of the time a the maximum of 47 resp. 24 containers.

In order to measure the computational overhead added by docker, two additional experiments have been performed. The described neural network algorithm has been executed in a VM, encapsulated in a docker container for the first experiment and without docker for the second experiment. With docker the mean time of twenty runs is 265.05 s and without docker it is only 199.15 s. There were no resource restrictions for the docker containers.

5.2 Use Case Evaluation

All experiments produced the same optimal parameter set, with a classification accuracy of 97.96 %. For the preprocessing step, setting all missing values to 0 is of advantage. The number of hidden layers should be set to the maximum of 2 hidden layers, with 256 neurons per layer. The optimal dropout values are 0.1 for the input layer, 0 for the first hidden layer and 0.25 for the second hidden layer. The worst parameter sets give an accuracy of 80 %. As they all have low dropout values and a high number of neurons in the hidden layers, overfitting seems to be the problem.

5.3 Infrastructure Networking Issues

Due to the complexity of the employed double-virtualized networking technologies, special attention need to be given to the network. As described in Sect. 3.2, OpenStack uses VXLAN-technology to separate and encapsulate tenant network traffic. VXLAN assigns each virtual network a unique ID. This additional transmission information is written to each ethernet frame before it is sent over the physical network. Due to this fact, the maximum ethernet frame size in the virtual machine network interface has to be adjusted. We use a value of 1450 bytes. This ensures that the frame size remains smaller or equal the maximum allowed size of 1500 bytes in ethernet networks. Without adjusting the Maximum Transmission Unit (MTU) the frame (with the added VXLAN-information) could become too large and will be discarded by host network interfaces. This adjustment can be realized through a configuration option in the Nova and Neutron services.

When using the Docker overlay network, which also set up VXLAN-Interfaces in the virtual machines (see Sect. 3.3), the MTU isn't adjusted properly. To date, it is not possible to configure the MTU values for the Docker Overlay Network. This problem was in particular hard to find because of the interface-encapsulation in network namespaces. Those aren't directly accessible with standard system tools. As a work-around the data and algorithm containers are started in *privileged*-mode. The management programs manually set the MTU values of the *eth0* and *eth1* interfaces within the containers to 1450 after startup.

Apart from these network configuration problems, the overlay network became instable during our experiments. Due to a bug in the docker-engine, the overlay network support of the Swarm agents broke at random occasions. The error was not detected by the docker-engine and therefore not by Docker Swarm, resulting in failures during container creation. The problem seems to be fixed upstream[14] for the next Docker release and will hopefully improve stability.

6   Conclusion and Outlook

The prototype of the distributed algorithm processing infrastructure has been set up and could be successfully employed for the sample use-case. Besides some stability issues in its current version, Docker provides many advantages for Biomedical Analysis use cases. The overhead for spinning up multiple tasks is low and packaging algorithms on top of the algorithm base container does not require much effort for developers. The AMS has proven to be stable and is able to track timings, as well as potential errors. However, for these relatively short-running applications, spawning times and input data transfer can become bottlenecks when many containers are scheduled. Scalability could probably be increased by additional data containers and multi-threaded spawning of the algorithm containers rather than the current sequential scheduling. The prototype use-case focused on multiple algorithms using the same input data, but was not data-intensive and did not include sensible data. The next steps are the execution of the whole signal processing on the PSGs itself and a descent security inspection including the security measures mentioned in Sect. 4.3. We currently use Dockerhub[15], the official public docker registry, but plan to employ our own private registry with restricted user access a well. Possible compatibility with other Docker-based solutions like Boutiques or Agave will be investigated, to extend the capabilities of the infrastructure.

Acknowledgements. The work is supported by the German Ministry of Education and Research (Project BB-IT-Boost, 03FH0061X5) and the German Ministry of Economic Affairs and Energy (ZIM Project BeCRF, Grant number KF3470401BZ4).

References

1. Solomonides, T. (ed.): Healthgrid Applications and Core Technologies. Proceedings of HealthGrid 2010. Studies in Health Technology and Informatics, vol. 159. IOS Press, Amsterdam (2010)
2. Glatard, T., et al.: Large-scale functional MRI study on a production grid. Future Gener. Comput. Syst. **26**(4), 685–692 (2010)
3. Gu, Y., Huang, Z.: Robinia-BLAST: an extensible parallel BLAST based on data-intensive distributed computing. In: 2014 IEEE 12th International Conference on Dependable, Autonomic and Secure Computing (DASC) (2014)

---

[14] https://github.com/docker/libnetwork/pull/1065.

[15] https://hub.docker.com/.

4. Korkhov, V., et al.: Exploring workflow interoperability for neuroimage analysis on the SHIWA platform. J. Grid Comput. **11**, 1–18 (2013)
5. Goldberger, A.L., et al.: PhysioBank, PhysioToolkit, and PhysioNet: components of a new research resource for complex physiologic signals. Circulation **101**, e215–e220 (2000)
6. Shahand, S., et al.: A data-centric neuroscience gateway: design, implementation, and experiences. Concur. Comput. Pract. Exper. **27**, 489–506 (2015)
7. Jansen, C., et al.: Extending XNAT towards a cloud-based quality assessment platform for retinal optical coherence tomographies. Scalable Comput. Pract. Exper. **16**(1), 85–102 (2015)
8. Krefting, D., et al.: Performance analysis of diffusion tensor imaging an academic production grid. In: Parashar, M., Buyya, R. (eds.) 10th IEEE/ACM International Conference on Cluster, Cloud and Grid Computing. CPS Conference Publishing Service (2010)
9. Sherif, T., et al.: CBRAIN: a web-based, distributed computing platform for collaborative neuroimaging research. Front. Neuroinformatics **8**, 54 (2014)
10. Krefting, D., et al.: Grid based sleep research analysis of polysomnographies using a grid infrastructure. Future Gener. Comput. Syst. **29**(7), 1671–1679 (2013)
11. Young, T., et al.: Population-based study of sleep-disordered breathing as a risk factor for hypertension. Arch. Intern. Med. **157**, 1746–1752 (1997)
12. Bashan, A., et al.: Network physiology reveals relations between network topology and physiological function. Nat. Commun. **3**, 702 (2012)
13. Dickhaus, H., Maier, C.: Detection of sleep apnea episodes from multi-lead ECGs considering different physiological influences. Methods Inf. Med. **46**, 216–221 (2007)
14. Srivastava, N., et al.: Dropout: a simple way to prevent neural networks from overfitting. J. Mach. Learn. Res. **15**(1), 1929–1958 (2014)
15. Dooley, R., Stubbs, J.: Dynamically provisioning portable gateway infrastructure using Docker and Agave. In: Proceedings of the 2014 Annual Conference on Extreme Science and Engineering Discovery Environment, XSEDE 2014, pp. 55:1–55:2. ACM, New York (2014)
16. Tristan, G., et al.: Boutiques: an application-sharing system based on Linux containers. Front. Neurosci. **9**

Untangling the Edits: User Attribution in Collaborative Report Writing for Emergency Management

Adrian Shatte[✉], Jason Holdsworth, and Ickjai Lee

Division of Tropical Environments and Societies, College of Business,
Law and Governance (Information Technology),
James Cook University, Cairns, Australia
adrian.shatte@my.jcu.edu.au,
{jason.holdsworth, ickjai.lee}@jcu.edu.au

Abstract. This paper describes the progress of our ongoing research into collaborative report writing for emergency management using web technologies to improve communication and shared knowledge in emergency situations. Specifically, the work presented in this paper focuses on the development of a user attribution framework that enhances the Differential Synchronisation (diffsync) technique by exploiting its diff operation. This technique improves real-time collaborative editing for emergency management by combining the benefits of user attribution with diffsync features such as convergence, scalability, and robustness to poor network environments. As a proof of concept, we implement a prototype collaborative system and report results of simulations to test scalability, efficiency, and correctness. Further, we consider the potential benefits of this framework for web-based collaborative report writing in the context of emergency management.

Keywords: Collaboration · Differential synchronisation · Edit history · Emergency management · Text editing · User attribution · Workspace awareness

1 Introduction

Emergency management is a complex field requiring collaboration between many separate entities, such as emergency response personnel, law enforcement, government, and the public [1, 2]. Collaborative report writing is an important task conducted by emergency service agencies during a disaster to collate and distribute the facts of a disaster situation to stakeholders [3]. Documents such as *Incident Reports* and *Situation Reports* (sitreps) require input from several experts and organisations, with the content of these reports constantly changing based on the availability and accuracy of information as the disaster situation progresses. This, coupled with the uncertainty of a disaster situation, contribute to the challenge of effective collaboration during an emergency.

© Springer International Publishing Switzerland 2016
O. Gervasi et al. (Eds.): ICCSA 2016, Part II, LNCS 9787, pp. 319–330, 2016.
DOI: 10.1007/978-3-319-42108-7_24

Emerging technologies, including web-based systems and mobile applications, have been demonstrated to improve communication during a disaster [2, 4, 5]. Our ongoing research into collaboration for emergency management has focused primarily on developing a framework to support real-time, collaborative report writing for emergency management practitioners [6, 7]. Based on a survey of emergency management practitioners in regional Australia, the following list of requirements was derived: real-time synchronisation, flexible locking, user attribution (e.g. maintaining a history, attributing authorship), support for desktop and mobile users, and automated summarisation techniques. This paper focuses specifically on the development of the user attribution features for the report writing framework.

User attribution is a technique for increasing awareness in a shared workspace by attributing changes to a user. Primary user attribution features include visual representation [8], displaying relative contribution of authors [9], and maintaining a history of edits [10]. These features improve awareness by indicating who is sharing the workspace, where they are working, and what they are doing [11]. Visual information increases awareness and improves the overall collaboration experience for users [12].

Our technique is built on Differential Synchronisation [13] (diffsync) by exploiting the *diff* operation to find differences between versions of a document. While other synchronisation techniques exist, such as Operational Transformation [14], diffsync has several inherent benefits for collaborative emergency management report writing. Diffsync is scalable, robust to poor network environments that can exist during disaster situations, and is naturally convergent which ensures that collaborators are viewing consistent and up-to-date information [13]. In its current specification [13], diffsync does not provide support for user attribution features such as workspace awareness, relative contribution of authors and maintaining a history of edits. Thus, this paper focuses on developing techniques within diffsync to achieve these features so that they can be used in the larger collaborative report writing framework.

To demonstrate the applicability of our technique, we developed a prototype web-based collaborative text-editing prototype that supports visual attribution, relative contribution of authors, and maintaining a history of edits. To determine the cost of our user attribution techniques on the performance of diffsync, we report the results of benchmarking simulations. Further, we consider the usefulness of this feature within the scope of report writing for emergency management.

This paper begins with a brief background on user attribution techniques in collaborative editing, followed by a description of our methods for achieving user attribution in diffsync. Next, we describe the implementation of our web-based collaborative text-editing prototype and outline how our technique is used to support user attribution. Following this, results of benchmarking experiments are presented that demonstrate that the new technique maintains efficiency and achieves correctness. Finally we conclude with a discussion on the benefits of this technique in the scope of collaborative report writing for emergency management.

2 Preliminaries

2.1 User Attribution in Collaborative Editing

Maintaining awareness of user actions and intention are crucial for successful collaboration [15]. Workspace awareness embodies knowledge of user identity, user edits (the type, position and time), and user intention [10, 11]. Mechanisms for communicating this knowledge to users helps coordinate activity, simplify communication, provide relevant assistance, and manage the dynamism between individual work and shared work [11]. We focus on three methods: visual user attribution, relative contribution of authors, and history of edits.

Visual User Attribution. Visual user attribution mitigates the risk of misunderstandings during remote collaboration [8]. A review by Koren, Guth and Klamma [16] on support libraries for shared editing on the web concluded that there are several key features needed to support workspace awareness. These features are: (1) user-specific colours; (2) mouse over effects (e.g. tooltips); (3) a list of participants; and (4) edit history. Additionally, we believe that these features should be tied to user profiles to include additional cues such as user name.

Relative Contribution of Authors. *Relative contribution* measures the edits in a document in terms of total number of edits per user and the quality of those edits. Arazy and Stroulia note several techniques for calculating user contributions on wiki systems including summing all revisions over time, using a reputation system, and weighting edits to measure their importance [9]. We believe that real-time editing systems might benefit from visual relative user contribution features by providing additional awareness to collaborators engaging in a shared activity. For example, academic workspaces could track student engagement during a collaborative editing task. Collaborative programming environments could determine which developer is responsible for a particular file in a complex program based on total number of developer edits. Therefore, it is important for modern synchronisation techniques to support relative user contribution.

History of Edits. Awareness in a collaborative workspace can be further supported by maintaining a history of revisions and edits that are displayed to the user [10]. *Change awareness* refers to the mechanisms used for detecting and tracking changes made to a document over time [10]. Change awareness can support users in planning future actions based on their own past actions in the workspace, and those of other collaborators. Techniques and tools for supporting change awareness include *diff* functions, version control systems, and algorithms with support for *undo* and *redo* methods. We believe that synchronisation techniques benefit from maintaining a history of past edits. Then users might better understand the evolution of a document during collaboration and plan future edits by reviewing existing versions. For example, a user could view changes made to sections of a document in which they did not contribute. A single user could also revisit earlier edits that may have since been deleted.

In its existing form, diffsync does not support user attribution features as all changes are merged on the server [13]. This poses a challenge to developers and users alike in untangling and attributing edits made by users in collaborative workspaces. There has been no reported work on developing techniques for diffsync to support visual attribution, relative contribution of authors, and maintaining document history. In the next section, we present methods for enabling these features in diffsync and implement them in a prototype web-based collaborative text-editing application as a proof of concept.

3   Incorporating User Attribution into Diffsync

3.1   Differential Synchronisation

Diffsync is a technique used for document synchronisation that runs a continuous cycle of *diff* and *patch* operations between client and server (see Fig. 1). The *diff* operation determines the differences between two versions of a document. The *patch* operation attempts to append those changes to another document (similar to a three-way merge [13]). The current version of diffsync is known as the "Guaranteed Delivery" implementation due to its fault tolerance on unreliable networks.

A number of characteristics make diffsync appropriate for real-time collaborative applications (including collaborative report writing for emergency management) such as: (1) nearly identical code on both the client-side and the server-side; (2) a state-based approach that can assist in providing an edit history; (3) asynchronous updates and scalability; (4) fault tolerance on unreliable or high-latency networks, such as those present in a disaster; (5) all clients are guaranteed to converge to the same document; and (6) compatibility with various document types. Moreover, diffsync is designed so that it can be appended to existing applications, thus introducing real-time collaboration to existing non-collaborative systems [13].

User attribution is not currently supported by diffsync, and edits from all users are blended together on the server [13]. This can pose challenges for both small and large collaboration tasks. For example, a single user in a document may not require visual attribution to determine ownership of edits, but might benefit from viewing a document history and also "rolling back" edits to undo changes [13]. On the other hand, multiple

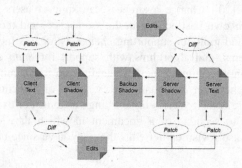

Fig. 1. The guaranteed delivery implementation of diffsync (adapted from [13]).

users (three or more) can easily become confused if there are no mechanisms for visually attributing changes to individual users [12].

The remainder of this section of the paper explains methods we determined were necessary to achieve user attribution in diffsync.

3.2   Overview of Our Technique (Attributed Diffsync)

Four additional methods are utilised to achieve user attribution in diffsync. The details of these methods are provided in this section of the paper.

Exploiting Diffs to Track Authorship. A *diff* operation returns the differences found between two versions of text, which we refer to as the *diff result*. The *diff result* contains the *type* of change detected (represented as an integer) and the *affected text* (represented as a string). For example, consider a *diff* operation between the following two strings:

```
The dog runs fast      The dog ran fast
```

The *diff result* returns the following information:

```
 0, The dog r    (text identical)
-1, u            (text removed)
 1, a            (text added)
 0, n            (text identical)
-1, s            (text removed)
 0,  fast        (text identical)
```

The user added one character and removed two characters. If we group this data with a user ID and calculate the position of the affected text in terms of index and range, then it is possible to provide a simple structure to begin attributing contributions to each user. We call this object the *attribution object*. For example:

```
attribution { userID: 1, range: 10->11, text: a }
```

The underlying structure of shared text is represented with an index-based data structure (e.g. $0 \rightarrow n - 1$) which is used to track the position of changes in a document. During each synchronisation loop the client propagates a stack of edits to the server. Each edit contains a series of *diff result* objects. Our user attribution technique works primarily on the server-side: when a stack of edits is received from a client, the server loops through each object and handles it according to the following rules:

1. If the *diff result* indicates text was added by the client, a new attribution object is stored on the server containing the user ID, start index, end index, and the text contained in the edit;
2. If the *diff result* indicates text was removed by the client, another method runs to decrease the size of the affected attribution objects. Additionally, the start and end index of other attribution objects are adjusted;
3. If the *diff result* states that a region of text remained unchanged, the server does nothing with that information.

Merging Adjacent Attribution Items. It is inefficient to store multiple attribution objects that cover adjacent characters, so we use an additional method to merge attribution objects where possible. For example, the following adjacent attributions (A) and (B) are merged to become (C):

```
(A)    { userID: 1, range: 31 -> 34, text: app }
(B)    { userID: 1, range: 34 -> 36, text: le  }

(C)    { userID: 1, range: 31 -> 36, text: apple }
```

Adjusting Range of Attribution. Since a shared document is constantly changing due to edits made by many users, attribution objects need to adjust based on those edits. Attribution objects need to be adjusted whenever text is added or delete before an existing attribution object, or text is added or deleted within an existing attribution object. For example, if User A owns an attribution on the text "Cat", but User B edits that area of the text to become "Chat", there are a few possible approaches. For example, the attribution of the entire word could be transferred to the new user.

Our approach retains the original user attribution, but splits it into two objects so that User A now owns "C" and "at". A third attribution is added for User B to contain the "h". For example:

```
{ userID: 1, range: 0->1, text: C  }
{ userID: 2, range: 1->2, text: h  }
{ userID: 1, range: 2->4, text: at }
```

Maintaining Document History. The final feature supported by our technique is maintaining edit history. Diffsync utilises version numbers on both the client and server-side to track changes. Thus, saving the state of the shared document is a simple matter of storing a *history object* on the server to store all relevant information. A history object includes the state of the server text at the time, the current version number of the server text, a record of all attributions currently owned by each user, and a timestamp. This information can be propagated to the clients for rendering on request.

4 Web-Based Collaborative Editor with User Attribution

To demonstrate the applicability and flexibility of our technique we developed a web-based prototype. This section describes the prototype application with respect to visual user attribution, relative contribution, and history of edits.

4.1 Tools

We used several tools to develop our web-based prototype, including Node.js [17], Ace editor, and the *google-diff-patch* library. These tools were selected for their interoperability, scalability, and concurrency features which are beneficial for real-time collaborative editing.

4.2   User Attribution Features

Highlighting User Contributions. Utilising the stack of attribution items, the client can render each area based on its region. In our implementation, each user is generated a random colour upon connecting to the server. This colour is stored with an attribution object, alongside the ID and username of the user. At the end of each synchronisation loop (e.g. when the client receives a server response), the client runs a function to highlight each attribution with its colour. The username and colour for each attribution area are also added to a visible panel (see Fig. 2). As a secondary user attribution characteristic, we also display author username in a tooltip (see Fig. 2).

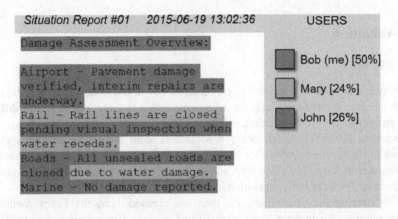

Fig. 2. Attributions owned by a user can be colour highlighted and associated with a username tooltip; relative contributions are shown as a percentage beside the username. (Color figure online)

Calculating Relative User Contribution. Utilising the same data that our technique stores on the server, our application can determine the percentage of contribution for each author. The client side inspects the attribution objects and finds the total number of characters owned by each user, and then calculates the percentage of contribution compared to the length of the shared text. This is updated during synchronisation loop, and is displayed on the screen next to the user's name (see Fig. 2).

History. Maintaining a history of edits is achieved by storing a collection of *history objects* on the server-side. A *history object* contains a version number, server text, attribution stack, and a timestamp. On each synchronisation loop a new *history object* is added to the collection if there have been changes since the last update. Using this information the client can view a history page. The client renders each version in a similar interface to the collaborative environment with editing disabled (see Fig. 3). Using a slider, the user can browse through each version in a chronological order to view how the document has changed over time, including visual user attribution.

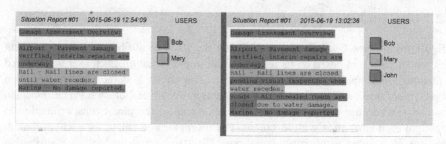

Fig. 3. Two versions of a collaborative report saved by our history technique. (Color figure online)

5   Evaluation

5.1   Efficiency

To determine the effect of our technique on efficiency of diffsync, we ran a simulation. Two implementations of our system were compared: (1) a version including diffsync with attribution, and (2) a standard implementation of diffsync. Ten (10) clients were connected to the server. Each client generated random input (using *chance.js*) and deletion to simulate a collaboration session. The average time taken to complete a full synchronisation loop was calculated across 600 synchronisation loops. This process was repeated for both implementations.

After running the simulations, the data was cleaned using the Tukey method to remove outliers. An independent-samples t-test was used to compare the synchronisation times. No significant difference was found between the synchronisation time for both groups ($t = 8.81$, $p > 0.05$ *n.s.*). This indicates that there was no significant effect on synchronisation speed in our test conditions.

5.2   Scalability

Previous studies have determined that diffsync is highly scalable [13]. To verify that scalability was maintained in our user attribution implementation as the number of collaborators increased, we conducted the following benchmarking test using the same script to generate random input as described in Sect. 5.1. All clients were provided with a random identification number upon connecting to the server which was used in place of a user name.

While smaller collaborative scenarios such as pair-programming require support for only a small number of users, other contexts require larger groups of users, e.g. an undergraduate course with interactive workshops. Thus, we simulated both small and large numbers of users. Five conditions were tested: one user (1), two users (2), five users (5), ten users (10), and twenty-five users (25).

Each condition was run over 250 synchronisation loops and the average synchronisation time was recorded. The results of this scalability experiment are shown in Fig. 4. These results indicate that our implementation follows a linear trend in scalability within the parameters tested.

5.3   Correctness

To evaluate the accuracy of our technique at correctly tracking user contributions, we conducted another simulation with three (3) clients. Using a modified version of the custom script from the previous benchmarking studies, each client was given a probability of adding text. Client 1, 2 and 3 had 50 %, 30 %, and 20 % probability of adding text on a synchronisation loop respectively. The simulation was run over one-thousand (1000) synchronisation loops. At the end of the simulation the percentage calculated for each user was recorded. We expected that the total contributions for all three users would be 100 %, and that individual contribution should be close to the expected values due to the random nature of the script. The results are shown in Fig. 5 and indicate that our technique is accurate.

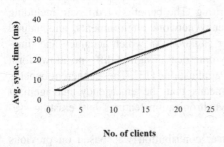 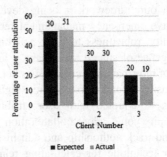

Fig. 4. Average synchronisation time shows linear trend up to 25 clients.

Fig. 5. Relative contribution of three clients at the end of the simulation.

5.4   Benefits for Collaborative Report Writing in Emergency Management

The user attribution features of our prototype ensure that users are responsible for their own contributions. This can have effects on the quality and accuracy of contributions in an emergency situation as organisations and individuals are held accountable for their work. Awareness is also increased as users will know who is working in the document, what changes they are making, and where those changes are being made. This is important in providing context to the content, as users can identify the accuracy and reliability of information based on its source. Further, the history feature of our technique may improve situation awareness. In an emergency situation, the status of an emergency and incoming information can change quickly. The sitrep will be continuously updated to reflect the current situation. Users can review past changes and who made those changes to better assess how the emergency situation and sitrep has changed over time. It can also enforce a higher level of accountability for changes that have been made in the document, as other users can review the actions of any user.

6 Discussion

Based on the results of our benchmarking studies it is evident that our technique did not significantly impact the average synchronisation time between clients. This means that the inclusion of user attribution in our system should not inhibit the consistency of shared data between clients. Additionally, we found that our technique has a linear trend of scalability within the range tested. As mentioned in the introduction and background, features like visual attribution, relative contribution of authors, and version history have been demonstrated to increase awareness and improve outcomes in collaborative workspaces. Thus, our technique provides these features while still performing synchronisation efficiently for several users.

While other synchronisation techniques exist that support user attribution features [9, 18], our framework is being developed specifically for the requirements of collaborative emergency management report writing. The benefits of diffsync including scalability, robustness to poor and high latency network environments, and natural convergence of shared text, make it ideal for situations in which accuracy and timeliness of information is critical. Further, the symmetrical design of diffsync allows for it to be easily appended to existing, single-user applications [13]. These inherent benefits of diffsync coupled with user attribution means that the emergency management community could adopt collaboration and synchronisation features into existing systems and track authorship and edit history.

Our technique for calculating relative user contribution was based on previous research [19], however this technique has limitations. For example, a user could maliciously erase other authors' content to increase their own relative contribution. Whether this is a problem depends on the domain of the application and how importantly individual contributions are considered.

Another potential research direction that has been suggested is supporting edit rollbacks in diffsync [13]. The history feature we demonstrated in this paper is a powerful tool for rolling back the entire document to a specific version (including the state of user attributions) and could easily be extended to provide redo mechanisms in addition to support for rollbacks of individual user edits.

As the work presented in this paper is part of ongoing research into developing a framework for collaborative report writing, there is also room for further development. Additional features for future studies include flexible locking, support for multi-synchronous editing (e.g. mobile users and desktop users), crowdsourcing, and automated summarisation techniques.

7 Conclusion

We developed a technique for user attribution using diffsync document synchronisation. Our technique exploited the behaviour of the underlying *diff* operation, while also introducing additional methods to achieve consistency of user attribution across clients. Using our technique, we incorporated three user attribution features into a web-based collaborative text editor: visual attribution, relative user contribution, and storing a history of edits. While our technique has yielded positive results in its ability to provide user attribution features for a collaborative application, there are still many improvements that can be made. Our future work will focus on further optimisations and usability testing to determine the effectiveness and ease-of-use of our techniques. We will also seek feedback on how well our technique can be used to support workspace awareness in real collaborative report writing tasks for emergency management.

References

1. Waugh, W.L., Streib, G.: Collaboration and leadership for effective emergency management. Public Adm. Rev. **66**, 131–140 (2006)
2. Ludwig, T., Reuter, C., Pipek, V.: What you see is what i need: mobile reporting practices in emergencies. In: Bertelsen, O.W., Ciolfi, L., Grasso, M.A., Papadopoulos, G.A. (eds.) ECSCW 2013: Proceedings of the 13th European Conference on Computer Supported Cooperative Work, 21–25 September 2013, Paphos, Cyprus, pp. 181–206. Springer, London (2013)
3. Schulz, A., Paulheim, H., Probst, F.: Crisis information management in the web 3.0 age. In: Proceedings of ISCRAM (2012)
4. Yin, J., Lampert, A., Cameron, M., Robinson, B., Power, R.: Using social media to enhance emergency situation awareness. IEEE Intell. Syst. **27**(6), 52–50 (2012)
5. Yates, D., Paquette, S.: Emergency knowledge management and social media technologies: a case study of the 2010 Haitian earthquake. Int. J. Inf. Manage. **31**(1), 6–13 (2011)
6. Shatte, A., Holdsworth, J., Lee, I.: Web-based collaborative document writing for emergency management. In: 2016 49th Hawaii International Conference on System Sciences (HICSS), pp. 217–226 (2016)
7. Shatte, A., Holdsworth, J., Lee, I.: Multi-synchronous collaboration between desktop and mobile users: a case study of report writing for emergency management. In: 26th Australasian Conference on Information Systems (ACIS) (2015). acis2015.unisa.edu.au
8. Birnholtz, J., Ibara, S.: Tracking changes in collaborative writing: edits, visibility and group maintenance. In: Proceedings of the ACM 2012 Conference on Computer Supported Cooperative Work, pp. 809–818 (2012)
9. Arazy, O., Stroulia, E.: A utility for estimating the relative contributions of wiki authors. In: Third International AAAI Conference on Weblogs and Social Media (2009)
10. Tam, J., Greenberg, S.: A framework for asynchronous change awareness in collaborative documents and workspaces. Int. J. Hum. Comput. Stud. **64**(7), 583–598 (2006)
11. Gutwin, C., Greenberg, S.: Effects of awareness support on groupware usability. In: Proceedings of the SIGCHI Conference on Human Factors in Computing Systems, pp. 511–518 (1998)
12. Kraut, R.E., Fussell, S.R., Siegel, J.: Visual information as a conversational resource in collaborative physical tasks. Hum. Comput. Interact. **18**(1–2), 13–49 (2003)
13. Fraser, N.: Differential synchronization. In: Proceedings of the 9th ACM Symposium on Document Engineering, pp. 13–20 (2009)
14. Sun, D., Xia, S., Sun, C., Chen, D.: Operational transformation for collaborative word processing. In: Proceedings of the 2004 ACM Conference on Computer Supported Cooperative Work, pp. 437–446 (2004)
15. Carroll, J.M., Neale, D.C., Isenhour, P.L., Rosson, M.B., McCrickard, D.S.: Notification and awareness: synchronizing task-oriented collaborative activity. Int. J. Hum. Comput. Stud. **58**(5), 605–632 (2003)

16. Koren, I., Guth, A., Klamma, R.: Shared editing on the web: a classification of developer support libraries. In: 9th International Conference Conference on Collaborative Computing: Networking, Applications and Worksharing (Collaboratecom), pp. 468–477 (2013)
17. Tilkov, S., Vinoski, S.: Node. js: using javascript to build high-performance network programs. IEEE Internet Comput. **14**(6), 80–83 (2010)
18. Shen, H., Sun, C.: Improving real-time collaboration with highlighting. Future Gener. Comput. Syst. **20**(4), 605–625 (2004)
19. Ding, X., Danis, C., Erickson, T., Kellogg, W.A.: Visualizing an enterprise wiki. In: CHI 2007 Extended Abstracts on Human Factors in Computing Systems, pp. 2189–2194 (2007)

An Improved Reconfiguration Algorithm for VLSI Arrays with A-Star

Junyan Qian, Zhide Zhou, Lingzhong Zhao$^{(\boxtimes)}$, and Tianlong Gu

Guangxi Key Laboratory of Trusted Software,
Guilin University of Electronic Technology, Guilin 541004, China
`cszide@gmail.com, qjy2000@gmail.com`

Abstract. This paper describes a novel technique to speed up the reconfiguration for the VLSI arrays. We propose an efficient algorithm based on A-star algorithm for accelerating reconfiguration of the power efficient VLSI processor subarrays to meet the real-time constraints and lower power consumption of the embedded systems. The proposed algorithm treats the problem of constructing a local optimal logical column that has the minimum number of long interconnects as a shortest path problem. Then the local optimal column can be constructed by utilizing A-star algorithm with appropriate heuristic strategy. The proposed algorithm greatly reduces the number of visits to the fault-free PEs for constructing a local optimal logical column and effectively decreases the reconfiguration running time. Experimental results show that the computation time can be improved by more than 38.64 % for a 128 × 128 host array with fault density of 20 %, without loss of harvest.

Keywords: Reconfiguration · VLSI array · Shortest path · A-star algorithm

1 Introduction

Most signal and image processing algorithms can be high-speed implemented by using Mesh-connected VLSI processor arrays due to its simplicity, scalability, and structural regularity. However, as the density of the VLSI arrays increases, the probability of the occurrence of faults in the arrays also increases. Thus, large scale VLSI processor arrays must provide fault-tolerant techniques to reconfigure the system to improve the operability and dependability. One of the approaches to achieve the dependability of VLSI array is degradation approach, in which, all PEs are treated in a uniform way and the fault tolerance is obtained by constructing a logical subarray using as many fault-free PEs as possible. The most significant degradation approaches before 1990 are summarized in [1]. Kuo and Chen [2] studied the problem of reconfiguring two-dimensional degradable VLSI arrays under three different switching and routing constraints, namely, (1) *row and column bypass*, (2) *row bypass and column rerouting*, and (3) *row and column rerouting*. They have shown that most reconfiguration problems

© Springer International Publishing Switzerland 2016
O. Gervasi et al. (Eds.): ICCSA 2016, Part II, LNCS 9787, pp. 331–343, 2016.
DOI: 10.1007/978-3-319-42108-7_25

under different rerouting constraints are *NP-complete* [2]. An optimal algorithm based on greedy strategy was proposed by Low and Leong [3], named GCR, to construct a maximum target array (MTA) in linear time that contains the selected rows. This optimal algorithm was employed in [4], resulting in an efficient heuristic reconfiguration algorithm under the row and column rerouting constraint. Jigang and Srikanthan [5] have simplified the row-selection scheme for the rows to be excluded and proposed a partial rerouting scheme in [6] to accelerate the reconfiguration of the target array. A more efficient algorithm was sequentially reported to further increase the harvest in [7], based on an integrated row and column rerouting constraint. In addition, a novel preprocessing and partial rerouting technique to accelerate the reconfiguration of degradable VLSI arrays was proposed in [8].

Minimizing the total interconnection length (*inter-length* for short) of a target array is known to lead to lesser routing cost, capacitance and dynamic power dissipation. To this end, many studies have addressed the efficient algorithms to reduce the total *inter-length* of a target array, such as [9–11]. A dynamic programming approach (denoted as ALG06 in this paper) was introduced for reducing power dissipation of a logical array in [9] by reducing the number of long-interconnects. However, ALG06 does not address minimizing the total interconnection length of the target array. Therefore, a divide-and-conquer algorithm (denoted as ALG14 in this paper) was proposed for a tightly-coupled MTA in [10], resulting in significant improvements over ALG06 in terms of the total interlength. However, these two algorithms all depend on a procedure LDP (local optimal algorithm by dynamic programming) to construct the local optimal logical column, and it is expensive to visit all fault-free PEs in the local area for constructing a local optimal logical column. Thus, it is significant to reduce the number of visits to the fault-free PEs in the local area to improve the runtime of reconfiguration.

In this paper, we propose an improved algorithm based on A* (A-star) algorithm to accelerate the reconfiguration of VLSI array. We treat the problem of constructing a local column as a shortest path problem. Then the local optimal column that has the minimum number of long interconnects can be constructed by utilizing A* algorithm with appropriate heuristic strategy. Experimental results show that the runtime of the proposed algorithm is reduced very well compared with previous works.

2   Preliminaries

2.1   Architecture and Rerouting Schemes

A *host array H* (*degradable array*) is defined as the original processor array after manufacturing, which may contain faulty PEs. A fault-free subarray of the host array after reconfiguration is called a *target array* (or *logical array*). The rows (columns) in the host array and logical array are called *physical rows (columns)* and *logical rows (columns)*, respectively.

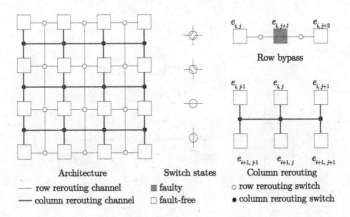

Fig. 1. Architecture and routing manners of a 4×4 array linked by switches. (Color figure online)

Figure 1 shows an example of the host array connected by four-port switches. Generally, two basic reroute schemes are used to reconfiguration, named, *row bypass* and *column reroute*. As shown in Fig. 1, in row bypass scheme, $e_{i,j}$ can directly communicate with $e_{i,j+2}$ and the data (include both the data and control signals) will bypass $e_{i,j+1}$ through the internal bypass. In column reroute scheme, $e_{i,j}$ can connect directly to $e_{i,j'}$ through the external switches while $e_{i+1,j}$ is faulty, where $|j - j'| \leq d$, d is named compensation distance. As same as in [9,10], the compensation distance d is limited to 1. Based on this limitation of *compensation distance*, let $row(u)$ $(col(u))$ denote the physical row (column) index of the PE u, the lower adjacent set $Adj^+(u)$ and the upper adjacent set $Adj^-(u)$ of each fault-free PE u in the row R_i is defined as follows:

Definition 1. *For each fault-free PE u in R_i:*

(1) $Adj^+(u)=\{v : v \in R_{i+1}, v$ is fault-free and $|col(u) - col(v)| \leq 1\}$ for $1 \leq i \leq m-1$.

(2) $Adj^-(u)=\{v : v \in R_{i-1}, v$ is fault-free and $|col(u) - col(v)| \leq 1\}$ for $2 \leq i \leq m$.

(3) For arbitrary $v \in adj^+(u)(Adj^-(u))$, v is called the lower (upper) left adjacent PE, the lower (upper) middle adjacent PE, or the lower (upper) right adjacent PE of u if $col(v) - col(u) = -1$, $col(v) - col(u) = 0$, $col(v) - col(u) = 1$, respectively.

As shown in Fig. 2, there are six possible types of link-ways for a target array [9], which can be classified into two classes based on the number of the switches used. One is called the *short interconnect*, which uses one switch to connect neighboring PEs; the other is called the *long interconnect*, which uses two switches. In Fig. 2, (a) and (d) are short interconnects, while the others are long interconnects.

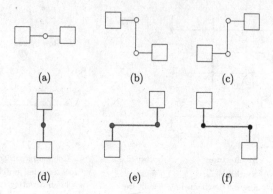

Fig. 2. Short and long interconnects.

2.2   A-Star Algorithm

A-star search algorithm is a widely used graphic searching algorithm, which combines the merits of both depth-first search algorithm and breadth-first algorithm. It uses the evaluation function(usually denoted $f(n)$) to guide and determine the order in which the search visits nodes in the tree. The evaluation function is given as $f(n) = g(n) + h(n)$, where $g(n)$ is the actual cost from the initial node (start node) to node n (i.e. the cost finding of optimal path), $h(n)$ is the estimated cost of the optimal path from node n to the target node (destination node), which depends on the heuristic information of the problem area [12].

3   The Proposed Algorithms

In this section, we propose an efficient algorithm to accelerate ALG14. Without loss of generality, we assume that the target array contains the selected rows $R_1, R_2, \cdots, R_m$. Assume B_l , B_r are two logical columns passing through each physical row of the $m \times n$ host array. We say that $B_l \leq B_r$ if the ith PE in B_l lies to the left of, or is identical to, the ith PE in B_r , for $1 \leq i \leq m$. In this paper, $A[B_l, B_r]$ indicates the area that consists of the PEs bounded by B_l and B_r(including B_l and B_r). The logical columns B_l and B_r are called the left boundary and the right boundary of the area, respectively.

Definition 2. *A logical column is called the local optimal logical column related to B_l and B_r if and only if it is the logical column of the minimal number of the long interconnects in $A[B_l, B_r]$* [9].

Suppose that B_l is the ith logical column generated by GCR [3] with the left-to-right manner and B_r is the $(k - i + 1)$th logical column generated by GCR with the right-to-left manner (where k is the total number of the logical columns), the literature [9] indicates that B_l and B_r is not independent, i.e., $B_l \leq B_r$, there exists at least one common PE between B_l and B_r, and the area $A[B_l, B_r]$ is the largest area available to produce the ith local optimal logical column.

3.1   Motivations

In algorithm LDP, the local optimal column is constructed by dynamic programming. For the PEs u and v, where $v \in Adj^+(u)$, the cost of the interconnections between u and v, denoted as $c(u, v)$, is set to $|col(u) - col(v)|$. The cost of a logical column is defined as the sum of the costs of all interconnections in the logical column. From the definition of $Adj^+(u)$, we obtain that the cost on each interconnection is 0 or 1, for $|col(u) - col(v)| \leq 1$.

Let $c(u, R_i)$ be the cost of the local optimal column from the PE u to the physical row R_i , and let $\mathcal{L} = Adj^+(u) \bigcap A[B_l, B_r]$. For arbitrary $u \in A[B_l, B_r]$, $c(u, R_i)$ can be calculated recursively in the following manner, for $i \leq m$ [9].

$$c(u, R_i) = \begin{cases} 0, \text{if } row(u) = i, \\ \min_{v \in \mathcal{L}} \{c(u, v) + c(v, R_i)\}, \text{if } row(u) < i. \end{cases} \quad (1)$$

$c(u, R_i)$ is set to ∞ if $\mathcal{L}$ is empty. The cost of the shortest path in $A[B_l, B_r]$ from R_1 to R_m is calculated by

$$c(R_1, R_m) = \min_{u \in R_1} \{c(u, R_m)\}. \quad (2)$$

The calculation of the local optimal logical column is based on bottom-up technique. Initially, for each $v \in R_m \bigcap A[B_l, B_r]$, the cost of v is set to be 0, i.e., $c(v, v) = 0$. Then, the cost $c(u, R_i)$ can be calculated by using the formula (1) for each $u \in R_{i-1} \bigcap A[B_l, B_r]$, for $i = m - 1, m - 2, \cdots, 1$. After obtaining the cost $c(u, R_m)$ for each $u \in A[B_l, B_r]$, LDP selects the minimum cost of the PE $u \in R_1 \bigcap A[B_l, B_r]$ according to the formula (2) and forms the local optimal logical column in the trace-back process with the start of u. For a detailed description, see [9].

However, it is expensive, obviously, to visit all fault-free PEs in the local area for constructing a local optimal column, because LDP needs to calculate the shortest logical column of all fault-free PE in the local area to the destination row, and selects the PE with minimum cost to construct the local optimal column. Thus, it is significant to reduce the number of visiting fault-free PEs as constructing a local optimal column.

Assume PE p is a common PE between B_l and B_r, and let r be the row index of p, thus the logical optimal column can be divided into two parts, one is the upper part that passes the first row to the rth row, and the other is the lower part that passes the $(r+1)$th row to mth row. Therefore, if we treat the first row and the last row in the local area bounded by B_l and B_r as two virtual PE T_1 and T_m respectively, i.e., $T_1 = \{v, v \in R_1 \bigcap A[B_l, B_r]\}$, $T_m = \{v, v \in R_m \bigcap A[B_l, B_r]\}$, then we can transform the problem of constructing logical optimal column into the shortest path problem of p to T_1 and p to T_m.

Algorithm 1. LOC_A(H, B_l, B_r, P)

Input: an $m \times n$ host array.
Output: P-the local optimal column.
list $itsc := \{u, u \in B_l \cap B_r\}$;
list $T_1 := \{v, v \in R_1 \cap A[B_l, B_r]\}$;
list $T_m := \{w, w \in R_m \cap A[B_l, B_r]\}$;
$\eta := $ the size of $itsc$;
for $i := 1$ **to** η **do**
  if $row(itsc[i+1]) - row(itsc[i]) \neq 1$ **then**
    | Shtest_link($B_l, B_r, itsc[i], itsc[i+1], P$);
  else
    | Add $itsc[i]$ to the path P;

if $itsc[0] \notin T\_1$ **then**
  | Up_shtest_col($B_l, B_r, itsc[0], T_1, P$);
if $itsc[\eta] \notin T\_m$ **then**
  | Down_shtest_col($B_l, B_r, itsc[\eta], T_m, P$);

3.2 Constructing Local Optimal Column Using A-Star Algorithm

Now we describe the proposed algorithm called LOC_A (local optimal column by A-star algorithm), which is used to find a local optimal column in $A[B_l, B_r]$. As mentioned in previous section, we start constructing local optimal column from the common PEs of B_l and B_r. Generally, there is not just one common PE of B_l and B_r. Assume $itsc(B_l, B_r)$ means the set of common PEs between B_l and B_r, and the elements of $itsc(B_l, B_r)$ are in an ascending order of row index, thus we have the following definition:

Definition 3. *Let* $u, v \in itsc(B_l, B_r)$, *if* $|row(u) - row(v)| = 1$, *then* u *and* v *are continuous. Otherwise,* u *and* v *are discontinuous.*

Suppose the set $itsc(B_l, B_r)$ has η elements, and let $s_\gamma \in itsc(B_l, B_r)$ be the γth element for $1 \leq \gamma \leq \eta$. Thus, the algorithm is divided into three parts to construct a local optimal column:

(1) Reroute the path from the first common PE s_1 to T_1.
(2) Reroute the path from the last common PE s_η to T_m.
(3) Reroute the path between two adjacent common PEs $s_\gamma, s_{\gamma+1} \in itsc(B_l, B_r)$, which are not continuous.

Firstly, we scan $itsc(B_l, B_r)$ to reroute two adjacent discontinuous PEs. If s_γ and $s_{\gamma+1}$ is discontinuous, then we reroute the path between s_γ and $s_{\gamma+1}$ (denoted as Shtest_link); Secondly, if s_1 does not belong to T_1, then we reroute the path between s_1 and T_1 (denoted as Up_shtest_col); Lastly, if s_η does not belong to T_m, then we reroute the path between s_η and T_m (denoted as Down_shtest_col). The formal description of this procedure is as Algorithm 1.

Algorithm 2. Shtest_link(B_l, B_r, S, T, P)

Input: two PEs S and T.
Output: P-the shortest path between S and T.
$open\_list :=$ includes S;
$close\_list :=$ the empty list;
$g[S] := 0, h[S] := h\_calc[S, T], f[S] := h[S]$;
while $open\_list \neq \phi$ **do**
 $curr := Extract\_Min\_F(open\_list)$;
 if $curr = T$ **then**
  return *the shortest path between S and T stored in P*;
 Add $curr$ to $close\_list$;
 while $adj^+(curr) \neq \phi$ **do**
  /*Expand the upper adjacent PEs.*/
  $u :=$ one PE in $adj^+(curr)$;
  if $u \in close\_list$ **then**
   do nothing;
  if $u \in open\_list$ **then**
   recalculate g, h, f of u;
  else
   calculate g, h, f of u;
   u's parent $:= curr$;
   Add u to $open\_list$;
  $adj^+(curr) := adj^+(curr)/u$;

In this algorithm, these three subprocedures are three independent A^* procedures, although they have different destinations. Shtest_link has a specific destination, while the destinations of Up_shtest_col and Down_shtest_col are a set of destinations. When we calculate the estimated cost of the optimal path from the current PE to the destination PE, Shtest_link has a specific value, while the values of Up_shtest_col and Down_shtest_col are the minimum value of the current PE to the each element of the destination set.

We treat the host array as a grid map and use diagonal distance as the heuristic function $h(n)$. Assume d is the length of a short interconnect and D is the length of a long interconnect, and let $D = \lfloor \sqrt{2} \times d \rfloor$. Thus, $h(n)$ can be defined as follows, where t is the destination PE:

$$\Delta x = |row(n) - row(t)|,$$
$$\Delta y = |col(n) - col(t)|, \tag{3}$$
$$h(n) = \Delta x \times d + (D - d) \times \Delta y.$$

Algorithm 2 shows the detailed descriptions of Shtest_link. It maintains two lists, an OPEN list and a CLOSE list. The OPEN list is a priority queue and keeps track of the nodes in it to find out the next node with that least evaluating function to pick. The CLOSE list keeps track of nodes that have already been examined.

h_calc indicates the calculation of the estimated cost from the current PE u to the target PE T, and $Extract\_Min\_F$ denotes finding and extracting the element with the lowest value of $f(n)$ in the OPEN list. Starting with the initial PE S, at each step of the algorithm, the PE u with the lowest $f(n)$ value is removed from the queue, then the $f(n)$, $g(n)$ and $h(n)$ values of the PEs in $adj^+(u)$ are updated accordingly, and these PEs are added to the queue. The procedure continues until the target PE T is reached (or until the queue is empty).

It is worth noting that when we expand the adjacent PEs of the current PE u, we just need to expand the PE of $adj^+(u)$ or $adj^-(u)$. For Up_shtest_col and Down_shtest_col, we expand the PE of $adj^-(u)$ and $adj^+(u)$, respectively. However, since the Shtest_link procedure can expand adjacent PEs in two directions from the two different common PEs, one is from $itsc[i]$ to $itsc[i+1]$, while the other is from $itsc[i+1]$ to $itsc[i]$. We choose the former to be the direction to expand adjacent PE in this paper. Furthermore, when the $f(n)$ values of all elements in $adj^+(u)$ or $adj^-(u)$ are identical, we select the middle adjacent PE to have the higher priority in the OPEN list. For the detailed descriptions of the Up_shtest_col and Down_shtest_col procedures, see appendixes.

Figure 3 shows an instance for constructing the local optimal column by dynamic programming and the proposed algorithm LOC_A. It is obvious from Fig. 3(a) that the dynamic programming method visits all 19 fault-free PEs in $A[X_2, Z_2]$, while LOC_A (Fig. 3(b)) only needs to visit 13 PEs. In Fig. 3(b), the local optimal column is divided into two parts, because there is only one intersection P. Suppose d is 10, then D is 14. Thus, the $f(n)$ value of each adjacent PE of P can be computed. For example, the $f(w)$ value of the lower

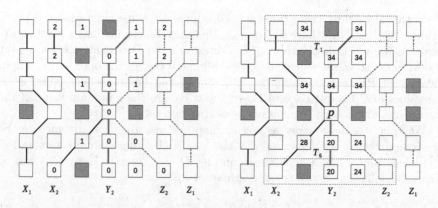

(a) Logical optimal logical column by dynamic programming. The digits indicate the cost of the shortest path from the current PE to the bottom row.

(b) Logical optimal logical column by A*. The digits indicate the value of the evaluation function $f(n)$.

Fig. 3. Construct local optimal logical column by dynamic programming and by A*.

middle adjacent PE w of P is 20, where the value of $g(w)$ is 10, and the value of $h(w)$ is 10.

4   Experimental Results

We have implemented ALG14 and the version of ALG14 with the proposed algorithm (denoted as NEW_CMTA) in C for performance comparison. As mentioned in previous section, the proposed algorithms depend on a priority queue. In our experiment, we have implemented a binary heap as the priority queue. These algorithms all ran on a Intel(R) Xeon(R) E5607 2.27 GHz computer with 4 GB RAM. In order to make a fair comparison, we maintain the same assumptions as in [9,10]. The fault density in the host arrays in varied from 0.1 % to 20 % for a comprehensive comparison. All of them were tested and compared with each other on the same random input instances.

In this section, the following notations are utilized for the performance evaluation of the algorithms.

- $tot.vis$. The total number of visited PE.
- $imp.vis$. Improvement over ALG14 in terms of $tot.vis$, which is calculated by

$$(1 - \frac{tot.vis\_of\_\text{NEW\_CMTA}}{tot.vis\_of\_\text{ALG14}}) \times 100\,\%.$$

- $imp.time$. Improvement over ALG14 in terms of runtime, which is calculated by

$$(1 - \frac{Runtime\_of\_\text{NEW\_CMTA}}{Runtime\_of\_\text{ALG14}}) \times 100\,\%.$$

Table 1 shows the $tot.vis$ and runtime performance comparison between ALG14 and NEW_CMTA. The data are collected for the host array with different sizes from 64×64 to 256×256. The fault distribution in the whole array is in a uniform way, which corresponds to a random fault model. Clearly, the propose algorithm greatly reduces the number of visits to the fault-free PEs in the local area for constructing a local optimal column. From Table 1, $imp.vis$ is increased with the increase of fault density, since the size of the local area is also increased with the increase of fault density. Moreover, the total number of visited PE ($tot.vis$) can be decreased by 50 % with 20 % fault density for all host arrays with different size. On a 128×128 host array with 20 % faults, for example, the value of $tot.vis$ is 21,479 for ALG14, while it is 10,177 for NEW_CMTA, such that the improvement on $tot.vis$ is over 52 %. Moreover, the runtime of the reconfiguration is significantly reduced. As can be seen from Table 1, the algorithm NEW_CMTA is very fast. The runtime improvement of NEW_CMTA over ALG14 is up to 38.64 % on a 128×128 host arrays with fault density 20 %.

However, the improvement in runtime is not increased in the same proportion as $imp.vis$. For a host array of size 128×128, the $imp.vis$ and $imp.time$ is 8.69 % and 31.40 % for a random fault with density of 0.1 %, respectively, while they are 52.62 % and 38.64 % with 20 % faults. This is because, the proposed

Table 1. The runtime performance comparison of the algorithms ALG14 and NEW_CMTA for random faults of uniform distribution, averaged over 20 random instances.

| Host array | | Target array | Performance | | | | | |
| Size | Fault | Size | tot.vis | | imp.vis(%) | Runtime(ms) | | imp.time(%) |
$m \times n$	(%)	$m \times k$	ALG14	NEW_CMTA		ALG14	NEW_CMTA	
64 × 64	0.1	64 ×63	4392	4055	7.67	1.190	0.969	18.57
	1	64 ×61	4852	4024	17.07	1.344	1.084	19.35
	5	64 ×55	5467	3822	30.09	1.415	1.089	23.04
	10	64 ×48	5574	3457	37.98	1.382	1.096	20.69
	15	64 ×41	5493	3009	45.22	1.371	1.124	18.02
	20	64 ×33	5284	2513	52.44	1.243	0.979	21.24
128 × 128	0.1	128 × 126	17766	16222	8.69	5.300	3.636	31.40
	1	128 × 123	19245	16238	15.62	6.031	4.028	33.21
	5	128 × 110	21817	15452	29.17	5.933	4.067	31.45
	10	128 × 96	23028	13888	39.69	5.656	3.660	35.29
	15	128 × 81	22881	12026	47.44	5.843	3.915	33.00
	20	128 × 68	21479	10177	52.62	5.652	3.468	38.64
256 × 256	0.1	256 × 254	69389	65297	5.90	20.078	15.933	20.64
	1	256 × 247	75404	65396	13.27	20.880	17.127	17.97
	5	256 × 221	88043	62087	29.48	22.097	17.684	19.97
	10	256 × 194	90605	56252	37.92	20.611	17.073	17.17
	15	256 × 164	90801	48763	46.30	19.297	16.085	16.65
	20	256 × 136	86090	41207	52.13	17.747	14.848	16.34

algorithm depends on a priority queue, the frequent operations of the insertion and extraction of the elements of the priority queue are time consuming. In addition, the calculation of heuristic function h is also a time-consuming operation, especially when there are more elements in the set T_1 or T_m, since it takes more time to obtain the minimum $h(n)$. Furthermore, it is noteworthy that the quality of heuristic function $h(n)$ directly affects the runtime of reconfiguration. If the value of the heuristic function $h(n)$ is exactly equal to the cost of moving from the current node to the goal, then A* will only follow the best path and never expand anything else, which makes it very fast.

5   Conclusions

We have proposed an efficient algorithm, called LOC_A, based on the A-star algorithm for accelerating reconfiguration of the power efficient degradable VLSI arrays. The problem of constructing a local optimal logical column is regarded as a shortest path problem. Thus, the local optimal logical column can be constructed by utilizing the A-star algorithm. Compared with the-state-of-art, the proposed algorithm greatly reduced the number of visits to the fault-free PEs for constructing a logical array and significantly reduced the reconfiguration time without loss of harvest.

Acknowledgments. This work is supported by the National Natural Science Foundation of China under grant No. 61262008, 61363030, 61562015, U1501252 and 61572146, the High Level Innovation Team of Guangxi Colleges and Universities and Outstanding Scholars Fund, Guangxi Natural Science Foundation of China under grant No. 2014GXNSFAA118365, 2015GXNSFDA139038, Innovation Project of GUET Graduate Education No. YJCXS201537, Guangxi Key Laboratory of Trusted Software Focus Fund, Program for Innovative Research Team of Guilin University of Electronic Technology.

Appendix

Algorithm. Down_Shtest_col(B_l, B_r, S, T_m, P)

/*Find the shortest path between S and T_m in area $A[B_l, B_r]$ using A-star algorithm.*/

Input: two PEs S and T_m.

Output: P-the shortest path between S and T_1.

$open\_list :=$ includes S;

$close\_list :=$ the empty list;

$g[S] := 0, h[S] := h\_calc[S, T_m], f[S] := h[S]$;

while $open\_list \neq \phi$ **do**

    $curr := Extract\_Min\_F(open\_list)$;

    if $curr \in T_m$ **then**

        return *the shortest path between S and T_m stored in P*;

    Add $curr$ to $close\_list$;

    while $adj^+(curr) \neq \phi$ **do**

        $u :=$ one PE in $adj^+(curr)$;

        if $u \in close\_list$ **then**

            do nothing;

        if $u \in open\_list$ **then**

            recalculate g, h, f of u;

        else

            calculate g, h, f of u;

            u's parent $:= curr$;

            Add u to $open\_list$;

        $adj^+(curr) := adj^+(curr)/u$;

Algorithm. Up_Shtest_col(B_l, B_r, S, T_1, P)

/*Find the shortest path between S and T_1 in area $A[B_l, B_r]$ using A-star algorithm.*/
Input: two PEs S and T_1.
Output: P-the shortest path between S and T_m.
open_list := includes S;
close_list := the empty list;
$g[S] := 0, h[S] := h\_calc[S, T_m], f[S] := h[S]$;
while *open_list* $\neq \phi$ **do**
 curr := $Extract\_Min\_F(open\_list)$;
 if *curr* $\in T_1$ **then**
  ⌊ **return** *the shortest path between S and T_1 stored in P*;
 Add *curr* to *close_list*;
 while $adj^-(curr) \neq \phi$ **do**
  u := one PE in $adj^+(curr)$;
  if $u \in close\_list$ **then**
   ⌊ do nothing;
  if $u \in open\_list$ **then**
   | recalculate g, h, f of u;
  else
   calculate g, h, f of u;
   $u's$ parent := $curr$;
   ⌊ Add u to *open_list*;
 ⌊ $adj^-(curr) := adj^-(curr)/u$;

References

1. Negrini, R.M., Sami, M., Stefanelli, R.: Fault-tolerance through Reconfiguration of VLSI and WSI Awards. The MIT Press, Cambridge (1989)
2. Kuo, S.Y., Chen, I.Y.: Efficient reconfiguration algorithms for degradable VLSI/WSI arrays. IEEE Trans. Comput. Aided Des. Integr. Circ. Syst. **10**, 1289–1300 (1992)
3. Low, C.P., Leong, H.W.: On the reconfiguration of degradable VLSI/WSI arrays. IEEE Trans. Comput. Aided Des. Integr. Circ. Syst. **10**, 1213–1221 (1997)
4. Low, C.P.: An efficient reconfiguration algorithm for degradable VLSI/WSI arrays. IEEE Trans. Comput. **6**, 553–559 (2000)
5. Jigang, W., Srikanthan, T.: An improved reconfiguration algorithm for degradable VLSI/WSI arrays. J. Syst. Archit. **1**(2), 23–31 (2003)
6. Jigang, W., Srikanthan, T.: Accelerating reconfiguration of degradable VLSI arrays. Proc. IEEE Circ. Dev. Syst. **4**, 383–389 (2006)
7. Jigang, W., Srikanthan, T., Wang, X.: Integrated row and column re-routing for reconfiguration of VLSI arrays with 4-port switches. IEEE Trans. Comput. **10**, 1397–1400 (2007)
8. Jigang, W., Srikanthan, T., Han, X.: Preprocessing and partial rerouting techniques for accelerating reconfiguration of degradable VLSI arrays. IEEE Trans. Very Large Scale Integr. Syst. **2**, 315–319 (2010)

9. Jigang, W., Srikanthan, T.: Reconfiguration algorithms for power efficient VLSI subarrays with 4-port switches. IEEE Trans. Comput. **3**, 243–253 (2006)
10. Jigang, W., Srikanthan, T., Jiang, G.Y., Wang, K.: Constructing sub-arrays with short interconnects from degradable VLSI arrays. IEEE Trans. Parallel Distrib. Syst. **4**, 929–938 (2014)
11. Qian, J.Y., Zhou, Z.D., Gu, T.L., Zhao, L.Z., Chang, L.: Optimal reconfiguration of high-performance VLSI subarrays with network flow. IEEE Trans. Parallel Distrib. Syst. (Online) (2016). doi:10.1109/TPDS.2016.2539958
12. Dechter, R., Pearl, J.: Generalized best-first search strategies and the optimality of A*. J. Assoc. Comput. Mach. **3**, 505–536 (1985)

Performance Evaluation of MAC Protocols in Energy Harvesting Wireless Sensor Networks

Vladimir Shakhov[✉]

Institute of Computational Mathematics and Mathematical Geophysics,
Prospect Akademika Lavrentjeva, 6, 630090 Novosibirsk, Russia
shakhov@rav.sscc.ru
http://www.sscc.ru

Abstract. Wireless Sensor Networks is considered as one of the most important elements in the upcoming Internet of Things. As sensors based applications are widely deployed, limited battery power of the sensor nodes becomes a serious problem. Intrusions or malfunction of legitimate sensor protocols can lead to the quick depletion of sensors batteries and a network failure. Energy harvesting facilities provide an attractive solution to this problem. The potential of the energy harvesting wireless sensor networks can be properly applied if the corresponding efficient network operations protocols will be implemented. Development of this protocols requires the corresponding mathematical tools for system performance evaluation. Most papers in this area focus on some concrete technical problem, but there is lack of papers analyzing common principles of energy harvesting wireless sensor networks operating. In this paper we partially fill this gap.

Keywords: Wireless sensor network · Energy harvesting · MAC protocols

1 Introduction

The Internet of Things (IoT) is evolved from a stage of research and innovation to a stage of market deployment. In the near future, networked wireless devices appreciably outnumber traditional electronic appliances. IoT will enable a plethora of new applications in environmental monitoring, agricultural industry, medical diagnostics, security and transportation business, equipment management, troubleshooting and monitoring. Cisco CEO John Chambers predicted the Internet of Things would become about USD 20 trillion market in the near future.

Wireless Sensor Networks (WSNs) is considered as one of the most important elements in the upcoming IoT. Small and cheap wireless sensors, which consist of sensing modules, data processing, and communicating components, are combined into context aware, self-governing, flexible and reliable networks [1]. However, sensors resources are quite limited. Thus, a battery power is usually a scare

© Springer International Publishing Switzerland 2016
O. Gervasi et al. (Eds.): ICCSA 2016, Part II, LNCS 9787, pp. 344–352, 2016.
DOI: 10.1007/978-3-319-42108-7_26

component in wireless devices. Intrusions or malfunction of legitimate sensor protocols can lead to the quick depletion of sensors batteries and a network failure [2]. For these reasons, the problem of WSNs modeling is still an important research direction since it allows to optimize system performance under limited recourses.

In response to growing market demand for practical wireless sensor networks applications, the concept of Energy Harvesting Wireless Sensor Networks (EH-WSNs) has been produced. An essential progress in wireless communications, electro-mechanical and digital electronics technologies allows to deploy a large-scale EH-WSNs. In EH-WSNs each sensor is supplied with an energy harvesting device. There are many potential sources of renewable energy in the environment at which EH-WSNs are deployed, such as solar energy, vibration, thermal energy, sound energy, wind power and so on. The energy harvesting mechanism allows to reduce the effect of DDoS attacks. And it can increase an efficiency of MAC protocols. Indeed, the sleeping mode can be reduced or eliminated. There are many sources for the material on MAC protocol for WSNs (e.g. [3,4]). Generally, the researchers try to increase the packet delivery ratio, reduce the end-to-end delay, optimize the sensor lifetime and energy consumptions. At the same time, the energy harvesting facilities have not been considered properly. Hence, new models and methods are required for MAC protocol optimization taking into account the properties of EH-WSNs.

The rest of the paper is organized as follows. In the next Section we provides a conceptual framework and provide a general description of MAC protocols. Next, we introduce some designations, model assumptions and then, we propose mathematical models for performance analysis of EH-WSNs MAC protocols. Finally, we conclude the paper.

2 Preliminaries

The duty of a sensor node is to monitor the sensor's environment and to partic-ipate in data transmission. Generally, the set of WSNs nodes is divided on the sender-receiver couples by an initialization protocol and MAC protocols define rules for sender-receiver cooperative work. The data transmission process con-sumes more energy than any other activities. For example, the energy consumed for the calculation operations is very low as compared with the packets transmis-sion energy. The energy needed to transmit 1 KB over a 100 m distance approxi-mately equals the energy necessary to carry out 3 million instructions at a speed of 100 million instructions per second [5]. By this reason WSNs MAC protocols consider the activities schedule efficiency as one of the most important factors. In WSNs most of the time a sensor has to be in sleep mode. MAC protocol defines a time, when each sensor turns off and wakes up to listen the medium. A sender can prepare data for transmission, but a receiver can be in sleep mode. So, the sender has to wait and vice-versa. It generates some latency in the network. An efficient MAC protocol gets a trade-off between the packets latency and energy consumption.

Thus, asynchronous sender initiated protocols reduce the idle time of receivers sensors. A sender transmits a preamble. If receiver wakes up and recognizes a preamble then one waits data packets, sends ACK packets etc. Otherwise, the receiver comes to the sleep mode. In B-MAC [6], a carrier sense media access protocol for WSNs that provides a flexible interface to obtain low power operations, effective collisions avoidance, and high channels utilization, the preamble is continuous and lengthy. B-MAC implementation is similar to preamble sampling in ALOHA [7]. Each time a sensor wakes up, it turns on the wireless communication component and checks for radio activity. If activity is detected, the node powers up and stays awake for the time required to receive the incoming data. After packets receiving, the node returns to sleep. If no packet is received, a timeout forces the node back to sleep. In X-MAC [8] it is a set of shorts preambles. This preambles contain some additional information, e.g. receiver ID, so other adjacent nodes do not wait the data. Also, X-MAC inserts pauses into the series of preamble bits, creating a space, which enables the targeted receiver to stop the preambles sending process via an early acknowledgement. It allows to get additional energy savings at both the sender and receive and to reduce per-hop latency. WiseMAC [9] mechanism predicts a wake up time of receiver. The preamble is very short, however the mismatching probability can be high. Both service bits transmissions and packets retransmissions increase the power consumption and reduce sensors lifetime. Energy consumption of communication sessions between a sender and a receiver is defined by the offered load as well. Therefore, it is necessary to take into account the intensities of outgoing and incoming traffic.

Energy harvesting facilities can essentially increase MAC protocols performance. However, the energy source behavior is generally irregular (solar, wind). The sleep mode can be reduced. But it can be unreasonable to reject it. Anyway, MAC protocols have to be revised taking into account the EH-WSNs concept. Efficiency of modified MAC protocols requires a novel mathematical tools for theirs performance analysis and improvement.

3   System Models

3.1   Assumptions and Designations

Assume, the traffic flows and energy harvesting procedure are random processes. These process satisfy the properties of the simple stream (Poisson). It is not guaranteed that the a sensor complete the recharging session before the following packets transmission. So, if a sensor has got enough energy for at least one packet transmission, then the sensor transmits a packet. We offer a model of the senders behaviour based on Continuous Time Discrete States Markov process. The model parameters can be used for MAC protocol analysis and modification. Let us make the following designations.

- λ : the intensity of data transmission by sender. It is defined by the results of monitoring, retransmission flow, the receiver readiness and MAC details.

- μ: the intensity of packets receiving by sender. It is defined by receiver activities and MAC protocols.
- δ: the intensity of energy harvesting. It is defined by the properties of concrete energy source.
- E_T: energy consumption for packets transmission.
- E_R: energy consumption for other activities, e.g. packets receiving.
- C: the sensor node battery capacity. The provided model is based on the sender battery consideration.
- n: the maximal number of packets transmissions without energy consumption facilities, i.e. $C = nE_T$.
- k: the ration of E_T and E_R, $E_T = kE_R$. Hence, $C = nkE_T$.
- N: the maximal number of receiving packets without energy consumption, i.e. $N = nk$, and hence $C = NE_R$.

In some investigations it is assumed that the parameter E_T fully defines energy consumption without loss of generality. Energy consumption of other activity is relatively small. Other investigators take into account the energy of packets receiving. Both cases are considered in this paper.

3.2   Transmission Availability

First, let us consider the case $E_T >> E_R$. Here, the number of model states equals $n + 1$. We focus on residual energy of sender node battery.

The sensor behaviour is described by the random process of energy consumption for packets transmissions and by the random process of energy harvesting. If no any activity then the sensor keeps the state. Otherwise, sensor gets to adjacent states.

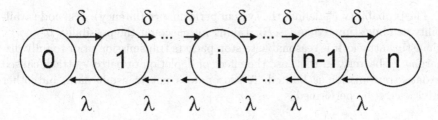

Fig. 1. The case $E_T >> E_R$.

From our consideration we get the following states.

- The state n, i.e. full battery.
- The state 0, here, the battery is fully discharged.
- Intermediate states $i, 0 < i < n$. The current battery state is enough for i packets transmissions.

The state diagram is given in Fig. 1. It is looking like the birth-death process.

The Kolmogorov equations for this system and the normalization condition are as follows

$$\frac{dP_0(t)}{dt} = -\delta P_0 + \lambda P_1,$$

$$\cdots$$

$$\frac{dP_i(t)}{dt} = -(\delta + \lambda)P_i + \lambda P_{i+1} + \delta P_{i-1}, 1 < i < n,$$

$$\cdots$$

$$\frac{dP_n(t)}{dt} = \delta P_{n-1} - \lambda P_n,$$

$$\sum_{i=0}^{n} P_i = 1.$$

Here and below P_i is the state i probability. For steady-state occupancy probabilities we get

$$\delta P_0 = \lambda P_1,$$

$$\cdots$$

$$(\delta + \lambda)P_i = \lambda P_{i+1} + \delta P_{i-1}, 1 < i < n,$$

$$\cdots$$

$$\delta P_{n-1} = \lambda P_n.$$

Solving the system we receive

$$P_0 = \frac{\rho - 1}{\rho^n - 1}, \quad \rho = \frac{\delta}{\lambda}. \tag{1}$$

The probability P_0 defines the system performance (latency). The node availability can be estimated as $1 - P_0$. Taking into account the possibility of energy exhausting attacks it is reasonable to stop packets transmission prior to fully discharging of battery. In this case the effect of depletion of battery attack caused by data transmission is limited. A sensor does not transmit data, but other activities can be performed.

3.3   Advanced Model

Next, let us take into account E_R. Here, the number of model states equals $N+1$. For description of sensors activity the additional random process of energy consumption for the packets transmission has to be considered. In this case, the intermediate states i means energy for i packets receiving. It is possible to transmit from the stage i to the stage $i - 1$ by receiving a packet. Also, it is possible to transmit from the stage i to the stage $i - k$ by sending a packet.

A fragment of the state diagram is given in Fig. 2. All possible transmission facilities are shown for the state i only.

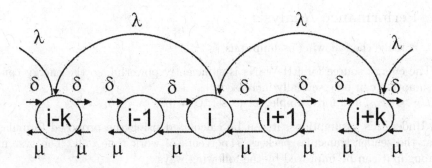

Fig. 2. The advanced model

Here, the Kolmogorov equations are as follows

$$\frac{dP_0(t)}{dt} = -\delta P_0 + \mu P_1 + \lambda P_k,$$

$$\dots$$

$$\frac{dP_i(t)}{dt} = -(\delta + \mu)P_i + \mu P_{i+1} + \delta P_{i-1}, 1 < i < k,$$

$$\dots$$

$$\frac{dP_i(t)}{dt} = -(\delta + \mu + \lambda)P_i + \mu P_{i+1} + \delta P_{i-1} + \lambda P_{i+k}, k \leq i \leq N - k,$$

$$\dots$$

$$\frac{dP_i(t)}{dt} = -(\delta + \mu + \lambda)P_i + \mu P_{i+1} + \delta P_{i-1}, N - k < i < N,$$

$$\dots$$

$$\frac{dP_N(t)}{dt} = \delta P_{N-1} - (\lambda + \mu)P_N.$$

All states and all subsets of states are transitive. Therefore, the set of all states is ergodic. Hence, steady-state regime for this system is reached. Using the corresponding steady-state conditions and normalization equation, we get the system of liner algebraic equations and receive formulas for the states probabilities. It is not place here due to the paper format.

Now, P_0 defines the system unreliability. It is the probability of incoming packet rejection. Hence, the loss rate equals μP_0.

A range of network management tasks can be solved by methods based on random graphs [10,11]. To use this approach we need a method for the node availability estimation. The probability P_0 gives the required node reliability.

A sender cannot transmit a packet if it is in the states $0, 1, 2, \dots, k-1$. Thus, the formula

$$\sum_{j<k} P_j.$$

provides the transmission delay probability.

The proposed models can be applied for analysis of MAC protocols efficiency and performance comparison of strategies for MAC protocol improvement.

4   Performance Analysis

Let us make the following assumptions

- The energy source for EH-WSNs is sufficiently powerful, so the energy consumption can be essentially increased.
- $E_T >> E_R$, and for example, let us set: $C = 3E_T, \lambda = 10, \delta = 7$.

Under this assumptions, it can be calculated using the proposed formulas that the sender transmits packets 54 percents of work time only. The system throughput can be improved by the following ways.

First, the intensity of energy harvesting δ can be increased. For this purposes, a sensor node can be equipped by additional modules permitting to use alternative energy sources. As an alternative, if magnetic resonant coupling are used for wireless power transfer [12] then an external energy source can increase the power transmission rate and hence, it increases δ. In this case, complication of sensor node structure and additional energy sources are not required. Next, the battery capacity can be increased. Taking into account the proposed model, it means that n is increased.

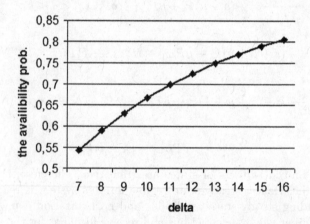

Fig. 3. Increasing of energy harvesting intensity.

In the long term, the first way is better than the next one. The results of calculations are shown on the Figs. 3 and 4. Here, to calculate P_0 for the case $\delta = \lambda$ we use

$$\lim_{\rho \to 1} P_0 = \lim_{x \to 0} \frac{x}{(x+1)^n - 1} = \lim_{x \to 0} \frac{x}{x^n + \ldots + nx + 1 - 1} = \frac{1}{n}.$$

Improvement of energy harvesting facilities can potentially increase availability almost 100 percents.

$$\lim_{\delta \to \infty} P_0 = \lim_{\rho \to \infty} \frac{\rho - 1}{\rho^n - 1} = 0.$$

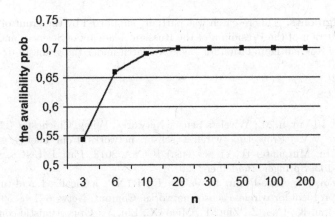

Fig. 4. Battery capacity increasing.

And here, any large investment in battery capacity increasing is unreasonable. The effect of this action is limited

$$\frac{\delta}{\lambda} < 1, \lim_{n \to \infty} P_0 = 1 - \frac{\delta}{\lambda}.$$

Therefore, in this example the transmission probability is limited by the value 0.7.

If $\rho > 1$ then both approaches lead to high availability. For large n system improvement is not required. In some cases the second way can provide a better performance. Sleep mode parameters, rules for wake up, conflict resolution policy, and other details of MAC protocols can be aggregated in the parameters of presented models.

Let us remark that the first approach is usually expensive. For performance comparison it is reasonable to introduce a cost function and consider the problem under limited budget. It is the direction of future work.

5 Conclusion

It is clear that the potential of the EH-WSNs technologies can be properly applied if the corresponding efficient MAC protocols will be designed. For those purposes, it is necessary to develop the mathematical tools for MAC performance analysis. Some of these issues have been proposed on this paper. We focus on the senders nodes. The models of sender behaviour are based on Continuous Time Discrete States Markov process. First, we offer a simple model, which is very close to general birth-death process. In this case, transmission energy consumption is only considered. Next, we offer the advanced model. Here, energy consumption of packet receiving is considered as well. The proposed models can be applied for performance comparison of EH-WSNs MAC protocols.

Acknowledgments. This research was partially supported by the grant of the Basic Research Program of the Presidium of the Russian Academy of Sciences and partially supported by Russian Foundation for Basic Research under the grant 14-07-00769.

References

1. Akyildiz, I., Vuran, M.: Wireless Sensor Networks. Wiley, Chichester (2010)
2. Shakhov, V.V.: Protecting wireless sensor networks from energy exhausting attacks. In: Murgante, B., et al. (eds.) ICCSA 2013, Part I. LNCS, vol. 7971, pp. 184–193. Springer, Heidelberg (2013)
3. Mouradian, A., Auge-Blum, I., Valois, F.: RTXP: a localized real-time MAC-routing protocol for wireless sensor networks. Comput. Netw. **67**, 43–59 (2014)
4. Ma, Q., Liu, K., Cao, Z., Zhu, T., Miao, X., Liu, Y.: Opportunistic concurrency: a MAC protocol for wireless sensor networks. IEEE Trans. Parallel Distrib. Syst. **26**(7), 1999–2008 (2015)
5. Zheng, J., Jamalipour, A.: Wireless Sensor Networks: A Networking Perspective, pp. 41–42. Wiley, Chichester (2008)
6. Polastre, J., Hill, J., Culler, D.: Versatile low power media access for wireless sensor networks. In: The 2nd ACM Conference on Embedded Networked Sensor Systems (SenSys), pp. 95–107 (2004)
7. El-Hoiydi, A.: Aloha with preamble sampling for sporadic traffic in ad hoc wireless sensor networks. In: Proceedings of IEEE International Conference on Communications (2002)
8. Buettner, M., Yee, G., Anderson, E., Han, R.: X-MAC: a short preamble MAC protocol for duty-cycled wireless sensor networks. In: SenSys 2006, pp. 307–320 (2006)
9. El-Hoiydi, A., Decotignie, J.: Low power downlink MAC protocols for infrastructure wireless sensor networks. ACM Mob. Netw. Appl. **10**(5), 675–690 (2005)
10. Migov, D.A., Shakhov, V.: Reliability of ad hoc networks with imperfect nodes. In: Jonsson, M., Vinel, A., Bellalta, B., Belyaev, E. (eds.) MACOM 2014. LNCS, vol. 8715, pp. 49–58. Springer, Heidelberg (2014)
11. Shakhov, V., Sokolova, O., Yurgenson, A.: A fast method for network topology generating. In: Jonsson, M., Vinel, A., Bellalta, B., Belyaev, E. (eds.) MACOM 2014. LNCS, vol. 8715, pp. 96–101. Springer, Heidelberg (2014)
12. Kurs, A., Karalis, A., Moffatt, R., Joannopoulos, J.D., Fisher, P., Soljacic, M.: Wireless power transfer via strongly coupled magnetic resonances. Science **317**(5834), 83–86 (2007)

Fog Networking: An Enabler for Next Generation Internet of Things

Saad Qaisar[✉] and Nida Riaz[✉]

SEECS, National University of Sciences and Technology, Islamabad, Pakistan
{saad.qaisar,nriaz.msee15seecs}@seecs.edu.pk

Abstract. Fog networking, an emerging concept in the context of cloud computing, is an idea to bring computation, communication and storage near to edge devices. Fog computing can offer low latency, geographically distributed mobile applications, and distributed control systems. On the other side, Software defined networking is a concept to make networking flexible and programmable. These two technologies together can create flexible and scalable networks to handle heterogeneous and massively increasing applications of IoT. In this paper we discuss how these two technologies can interplay with each other to be enabler of next generation IoT. We discuss that why these technologies are important for IoT and what current architectures are available to support IoT.

Keywords: Fog networking · IoT/IoE · Industrial automation · Autonomous vehicles · SDN · Open source SDN controllers

1 Introduction

IoT, as name suggests, is a network of physical objects. In its true manifestation, it resides on internet connected sensors and nodes, at scale. Every device is embedded with sensing capabilities, monitoring the targeting phenomena and providing unique information. Internet of things envisions nodes and devices to be interconnected and their information be shared to create intelligent and smarter systems.

Emerging technologies like 5G and IoT/IoE, needs the modification of existing networking techniques to support them. Today's cloud models are not feasible for the variety, volume and velocity of data that IoT generates. According to [1] in about 5 years the approximate number of devices connected to the internet will be around 50-70 billion, and sending all the data to the cloud may not be feasible. A new data handling paradigm is required which not only meets all the features of cloud computing but also supports mobility and its proximity to the end-users. Fog networking is one such paradigm. Fog networking result in better data management, reduced service latency, high computational power, and better QoS (Quality of service).

Fog networking envision bringing unique features of cloud close to the users by using the resources already available at the edge. This allows lowering the costs and improving the system efficiency by saving the bandwidth and power, otherwise utilized in sending data to the cloud.

© Springer International Publishing Switzerland 2016
O. Gervasi et al. (Eds.): ICCSA 2016, Part II, LNCS 9787, pp. 353–365, 2016.
DOI: 10.1007/978-3-319-42108-7_27

SDN is the concept of separating control and data plane. Concept of SDN can also be integrated in the discussed system as SDN provides programmability and flexibility in a network. We can integrate fog and SDN together to provide better connected systems with required properties. Hence making them an enabler of next generation Internet of things.

1.1 Organization of Paper

Section 2 of this paper provides an overview of next generation networks, its use cases, and specific demands. In Sect. 3 we discuss about fog networking paradigm, features it offers and motivations that how fog computing can be used to fulfill the demands of next generation networks. Section 4 will be about architecture of fog networking already available. Section 5 will discuss about the integration of SDN in this system. We conclude this paper by providing a brief literature survey of existing work in this domain and highlighted research issues in this domain.

2   Next Generation Networks: A Connected World Ahead

The advent of 5G technologies and IoT/IoE mark the start of a connected era. Much research has been done already and many of the applications of IoT are in use. IoT enables the vision of connected smart cities, energy efficient grid systems, automated industries and better health care facilities. Next generation networks conceive a super connected and super intelligent world, providing heterogeneous applications each with a different requirements. Some applications require low latency and high reliability, some requires high throughput, while other require efficient processing of huge volumes of data. These networks drive the evolution of M2 M communication. The next wave of digital world is characterized by massive networks of machines and devices envisioning a world where machines will interact with themselves and with the humans. But it still faces many design challenges. Three key fundamental requirements for building 5G networks as highlighted by Huawei [2] are as follows:

1. Support for massive scalability and massive connectivity.
2. Flexible and efficient use of available resources (Bandwidth and Power).
3. Supporting diverse set of applications having different requirements using a single architecture.

Coming text will provide insight about some IoT application in 5G networks their specific needs and then focus on the reasons that how fog networking can help realizations of these applications.

2.1   Industrial and Vehicular Automation (Latency-Sensitive and Reliable)

5G and IoT brings with themselves concept of industrial as well as vehicular automation. The road ahead is becoming clear as we are moving towards world of automation. Smart transportation and connected vehicles will allow safer traffic circulation and agility.

Vehicles can communicate with each other as well with road side units for safety messages exchange, allowing them to move smoothly over long distances without any limitations as well as other value added services. We can introduce smart traffic light system as discussed by F. Bonomi [3, 4] and by N.B Truong et al. [8] or transport management like lane change service.

On the other side portable robot arm in future automated industries will perform flexible product line functions based on varying process. Various sensing nodes will collaborate with each other to monitor pressure, temperature and other critical measurements to make industrial environment safe and reduce man power. Both of these scenarios require security, precision, accuracy and massive collaboration.

Key requirements of such IoT systems are:

1. *Minimum Latency.* Autonomous vehicles require low latency communication between themselves and with roadside units. Similarly industrial automation requires low latency communication between various sensing nodes and between nodes and actuators. For next generation networks to work, latency should be less than 1 ms to provide highly mobile communication links.
2. *High reliability.* Industrial automation and smart traffic systems need highly available and highly reliable networks to provide 24/7 monitoring service.
3. *Power constrained.* This feature becomes more significant in case of industrial automation where we have battery powered sensing nodes installed to monitor the various characteristics. These small node or motes, as we call them, are highly power constrained.
4. *Highly distributed nodes.* In scenarios of traffic management system, we not only have large number of nodes but highly distributed nodes. We need architecture to centrally control all these nodes.

2.2 HD Videos and Live Streaming (High Throughput)

5G envisions providing high speed streaming of data for HD quality videos and augmented reality. It envisions connecting the world using holograms and introducing fiber like user experience in wireless networks. It visualizes providing immersive experience of at least 1 Gbps or more data rates to support ultra-high definition video and virtual reality applications [2]. Key features that these systems require:

1. *High Throughput.* The only feature that such systems require is that of high throughput in wireless scenarios. Current network architectures are not capable of supporting such high speeds.

2.3 Better Agriculture System (Data Analysis and Large Number of Nodes)

Increasing crop cultivation in a limited land is a huge challenge, but this challenge can be resolved using next generation networks which envision supporting massive connections. Peoples can monitor the topography of land, weather updates, humidity in the environment, crop status using sensing nodes and data analytics. These features can help increasing the productivity and agriculture advancement. Key features of such a system are:

1. **Huge bulks of data.** As large amount of data will be collected from different nodes at various times, a data handling mechanism is required to perform data analysis and provide useful insights.
2. **Distributed nodes.** All of these nodes will be geographically distributed
3. **Power constrained.** Most of these nodes will be wireless sensor nodes hence these are power constrained.

2.4   Why We Need Fog Networking for IoT?

Though the requirements of various use cases of IoT are already been discussed, given below is the summary of characteristic features of IoT in general that necessitate the use of fog.

1. **Huge volumes of data:** IoT is expected to generate huge bulks of data from its connected network of trillion of devices. Most of this data is consumed locally A scalable networking technology is required to incorporate this data. All of this data cannot be sent to the cloud for storage and computation as this would require massive bandwidth consumption.
2. **Geo-distribution and need of massive cooperation:** Today IoT environment not only include large number of devices but devices are massively distributed [3]. In visions like smart cites, vehicles and industrial automation distributed sensing nodes each with a capability of processing and storage are required. We need a middleware platform to perform the task of cooperation and to maintain service consistency between these nodes.
3. **Latency minimization:** *Most of the* applications envisioned by next generation networks are latency sensitive applications. This feature requires the presence of network intelligence close to the network edge.
4. **High mobility applications:** In an IoT application to enable a driverless car scenario for example, we need a networking architecture that can work and interact with highly mobile devices, allowing them to operate smoothly and safely over long distances. This feature cannot be provided by a Cloud.
5. **Security issues:** In todays connected world, security has always remained a paramount issue. In many scenarios IoT are used to deliver critical and confidential information, which needs to be protected both in transit and then at rest. Fog nodes provide solution for this security concern as these nodes are deployed in close vicinity of users and we can monitor and control the devices that collect and analyze our data.
6. **Scalability:** We need scalable architecture to support the dense network of massively increasing devices.

3   An Overview of Fog Networking: A Potential Platform for IoT

Terms similar to fog computing like Cloudlets, cyber foraging and edge computing were introduced way before fog; but the word fog computing was first introduced by Cisco in 2012 [1]. Fog computing was introduced as practical and efficient solution to cater the needs of emerging IoT (Internet of things) technology and for the processing of huge bulks of data. Fog is as diverse as IoT itself. Fog computing provides a decentralized system in which we can share computing and storage resources at the edge, allowing real time data processing within the limitation of given bandwidth and power. Presently we have more powerful edge devices having immense storage and computation resources. Moreover, mobile devices are the major source of information whereas cloud is becoming the place to centralize and manage all the information. Fog computing was introduced to provide integration between cloud and mobile services, which seemed unscalable in scenarios of IoT. Fog does not replace cloud but complements by allocating tasks between edge devices. New applications require certain properties which cloud cease to provide them hence necessitating the need to introduce Fog networking.

3.1   Characteristics of Fog Networking Paradigm

Fog nodes not only provide benefits of cloud computing but opens up a complete path for new breed of applications. Given below are some of the distinguished characteristics of Fog computing as highlighted in various papers [4, 5, 6]

1. *Low Latency Networks and real time processing of data.* Latency-sensitivity is an established feature of fog computing, as these nodes are deployed closer to the user end hence can provide network services with low latency and high data quality as compared to cloud networking. This feature of fog computing makes it suitable to be uses in different IoT applications as well as to enable virtual reality interface as envisioned by 5G.
2. *Pooling of local resources.* Fog networks consist of independent distributed nodes which can work together in a cooperative manner to increase the value of a network. These nodes can work in a collaborative manner to increase throughput of a system [7].
3. *Location Awareness.* Fog nodes are deployed locally so these nodes are aware of user's location and requirements on the clients. This unique information of locality can be used in various applications.
4. *Heterogeneity and Agility.* Support for heterogeneous technologies is another distinguished feature of fog networking. Different applications and different communication technologies can be used together in fog nodes according to users' requirement, making these nodes heterogeneous in nature. Fog nodes support multiple applications and multiple protocols suitable for multiple scenarios. Moreover we can enable users to select network of their own choice out of different heterogeneous networks. Fog networking also introduces agility by providing rapid innovation. It is usually much cheaper and easier to experiment with edge devices than to wait for vendors to adopt an innovation [13].

5. *Support for mobility.* Fog nodes support mobility techniques. In many scenarios Fog nodes have to communicate directly with mobile devices.
6. *Interplay with cloud.* Some applications require globalization of cloud for business intelligence analytics as well as analysis of critical information close to the edge. Fog networks provide such a system by a 2 tier hierarchy. Time sensitive calculations are done at 1st tier whereas filtered data from this tier is sent to 2nd tier of centralized cloud where this data is stored for a longer duration of time which helps in big data analysis.

4   Architecture of Fog Networking

This section discuss about the envisioned fog networking architecture. Fog architecture provides both the data plane and control plane protocols. The basic architecture as explained by F. Bonomi [3] is shown in Fig. 1. This architecture consists of an abstraction layer and an orchestration layer.

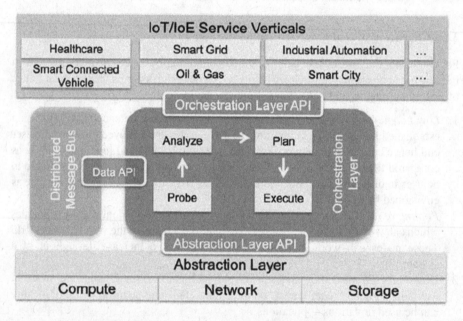

Fig. 1. Fog Architecture envisioned by F. Bonomi

As discussed earlier, fog nodes comprises of different devices each with its own resource of computation and storage. Abstraction layer deals with the data plane issues of the network. It hides the heterogeneity of fog nodes by hiding the complexities of these devices and providing APIs to manage the resources across the devices. This layer employs the virtualization techniques to allow multi-tenancy as well as provide APIs to manage various hypervisors and VMs across the network.

Orchestration layer provides control plane functionalities in the fog architecture by managing services across the whole network using distributed foglet agents. It consists of distributed storage to store control information as well as message service to synchronize all components together.

5 Integrating SDN into the System

SDN opens up a whole new path for the development of networking. It manifests the idea of separating control plane from the data forwarding plane. Communication networks are built from many devices (routers, switches, other middle boxes), each with its own complex protocols. SDN is the concept to free these forwarding devices from the task of route management and to perform it separately in a logically central control plane. This simplifies the process of network control and configuration and has added many capabilities in the network by adding a component of virtualization.

Here in this tutorial we propose that SDN when integrated with fog networking can make massive networks, flexible and programmable. Hence Fog and SDN together can be enabler for future communication networks.

In above sections we have discussed that how Fog networks can resolve many issues related to IoT. Here we suggest two advantages that make SDN suitable to be integrated with Fog for IoT. These are:

1. *Programmable networks are flexible.* Fog networking is network of various heterogeneous devices. Making it programmable by use of SDN simplifies the network management, both in deployment and also for development and testing of new protocols and applications.
2. *Global view of network.* IoT are diverse networks, with massive devices and nodes interconnected. Integrating SDN into the system will help in having a global centralized control of the whole network, and can help in controlling the network infrastructure in vendor independent manner.

5.1 Different SDN Controllers Available

SDN controllers are often referred to as the brain of the network, as they centrally manage the whole network. Concept of SDN controller was born with the SDN itself. Orchestration of network can be simplified and automated by using a central controller providing various API's for controlling purposes and hiding the complexities of heterogeneous devices. This enables increase agility, intelligence and scalability of a network. A controller provides south bound APIs to control different switches and routers and northbound APIs to manage business applications.

Presently different open source SDN controllers as well as large number of commercial SDN controllers are present. In this tutorial we will focus on four open source SDN controllers, will compare the features and architectures and try to figure out that which controller will be best for scenarios of IoT. Table 1 summarizes these key features.

Table 1. Summarizing key features of Open source SDN controllers which can be exploited in IoT

SDN Controllers	Programming Language	Key features supporting IoT
OpenDaylight	Java	• Model driven-service abstraction layer providing flexibility.
		• Strong Northbound abstractions: Intent based network Composition, Network virtualization
		• South bound abstractions: support both OpenFlow and non-OpenFlow switches, OVSDB,BGP, NETCONF
		• Secure. AAA mechanism
Floodlight	Java	• High performance due to multithreading
		• Compatible with OpenStack
		• Strong northbound APIs: Contains representational state transfer APIs that make it easier to program interface with the product
		• South Bound Abstractions: OpenFlow
		• Apache licensed: open to be used for any purpose
Ryu	Python	• Component Based Architecture
		• Northbound Abstractions: VLAN support, topology discovery
		• Multiple southbound protocols for managing devices, such as OpenFlow, NETCONF, OF-CONFIG, and others.
		• Apache licensed: Open to be used for any purpose
ONOS	Java	• Distributed Core which provides scalability and High availability
		• Rich Northbound APIs, like Intent framework and Network Graph
		• Southbound abstractions: OpenFlow, BGP, TL1

5.1.1 OpenDaylight ODL [11]

OpenDaylight is a java based modular, highly available and scalable SDN controller infrastructure built for deployment in networks of multivendor heterogeneous devices. ODL provides a platform based on open source and open standard APIs to make our networks more flexible, programmable and intelligent. Some key features of ODL that make it suitable for deployment in IoT are as follows.

Model Driven-service Abstraction layer. In the ODL Model Driven Service Abstraction Layer (MD-SAL), any application or function can be bundled into a service that is then loaded into the controller. This allows the users to write apps that can work easily across a wide variety of southbound protocols and devices. This also provides network managers an ability to combine multiple services and protocols to solve more complex problems as need arise. The architecture also enables the user to select the required feature and to create controllers according to their requirements. This feature of ODL makes it suitable to be used with IoT, which requires different complex functionalities.

ODL provides security. Another distinguished feature of ODL, useful in scenarios of IoT, is its mechanism for security. ODL provides a framework for AAA (Authentication, authorization and accounting) as well as secure network elements.

Strong north and southbound Abstractions. Besides AAA service ODL also provides various other services like intent based network composition and network virtualization. These APIs can be useful in deployment scenarios of IoT, as one can easily view the whole network and can manage network accordingly. It also provides various south-bound APIs to support standard network management interfaces. It can support both Open-flow and non-open flow devices.

5.1.2   Floodlight [12]

Floodlight is an enterprise class, java based open source SDN controller, developed by Big Switch Networks and is supported by community of developers. Floodlight is designed to work with large number of switches and routers that support OpenFlow protocol. OpenFlow is a standard protocol to exchange messages between controller and edge devices. Floodlight controller is easy to deploy as it provides a web based GUI and a java based UI. Some of the features that make it useful in scenarios of IoT are:

High Performance using multi-threading. Floodlight is designed to be at high perform-ance using multithreading, which allows its performance to increase linearly as we increase number of cores.

Network virtualization and Cloud support. It can support open stack cloud platform as well as network virtualization functions. Virtualization is an important feature required for IoT as we need devices to support multi-tenancy.

Northbound APIs. Floodlight SDN controller can be used to deploy required networking applications using its north bound APIs creating northbound abstractions, hiding complexities from the users and allowing them to develop useful applications. Floodlight provides many APIs Like it can be used in application of circuit pusher using its REST APIs.

5.1.3   Ryu [10]

RYU is another component based open source SDN controller. This is Python based controller, which provides software components with well-defined APIs. Though RYU is mainly used for cloud data center offerings there are some key features of it which make it useful in scenarios of IoT.

Component Based Architecture. This feature of Ryu makes it scalable and easy to deploy. Software components provided by Ryu with well-defined APIs make it easier to develop new network management applications.

Integration with Zookeeper for high availability. Ryu provides integration with zookeeper to keep two copies of management data so that when master server is down, slave can provide the functionalities.

Rich libraries. Ryu SDN controller has a rich collection of libraries containing many southbound protocols and network management applications. This makes it easy to deploy. Ryu packet library also helps to build various protocols like MPLS and VLAN, hence making it easy to configure networks. It supports many versions of OpenFlow like 1.1, 1.3, 1.5 and Nicira extensions.

5.1.4   Open Network Operating System (ONOS) [9]

ONOS, as said, is the first OS for the control of networks developed in concert with many leading service providers, vendors and ONF (Open Networking Foundation) [9]. ONOS is targeted specifically for the service providers and is gaining success in WAN use cases. This is the first open source controller focusing on WAN and service providers' needs. Scalability, high availability and high performance are the main aim behind ONOS. The key features of ONOS that make it suitable to be deployed in IoT are as follows.

Scalability using distributed core. ONOS is deployed as a service on cluster of servers. It has a distributed core. Same software runs on each server and they remain synchronized with each other providing state management and leader election services to and between instances. All of these servers collectively form a logical single entity controlling the whole network. This distributed core feature of ONOS makes it scalable. Network operator can add server any time according to requirement without any disruption to the remaining network.

High Availability. The distributed core architecture of ONOS provides highly available networks. For IoT application this is a necessary feature as many applications require continuous monitoring. ONOS software running on cluster of servers ensures that applications do not experience network related downtime. This also adds the capability that network can easily be updated by taking one instance down, updating it and then bringing it back online.

Northbound Abstractions. ONOS provides two strong northbound APIs which can be very useful in scenarios of IoT. These are intent framework and the global network view. Using Global network view API network manager can easily view the whole network graph and can easily manage network of different devices. On the other hand using intent framework API different business applications can be created in which user can easily ask for its requirements by simply specifying their intent.

South-bound abstractions. The southbound abstraction of ONOS represents each network element as an object in a generic form. Hence allowing distributed core to maintain the state of network elements without knowing the complexities of device. This feature will be useful in case of IoT, which will be consisting of heterogeneous devices.

6 Conclusion

6.1 Literature Survey

Research is been going on in integrating SDNs and fog networks with IoT networks. Though SDN has been a popular technique in context of wired networks like in data-centers where massive interconnectivity is supported by the flexibility of SDNs, it's still in its early stage in context of wireless networks. Researchers are envisioning integrating SDN into the IoT system to make efficient and flexible networks. They are working to develop wireless supportive architecture of SDN to be integrated in internet of things scenario. Surveying the available related literature in chronological order: [14] proposed integration of SDN in smart grid systems. According to them flexibility provided by OpenFlow can outperform MPLS networks in providing the required performance.

[15] Provided a complete layered architecture of SDN to be used in context of IoT. Layered architecture proposed by them assumes that layering creates abstractions hence hiding the complexity which arise due to heterogeneity of devices at scale, present in IoT environment. This layered architecture can provide required efficiency, and flexibility. They proposed a four layer architecture in which we can map task to specific resources using semantic modeling approach and provide optimization of flow scheduling using GA-based algorithm. But up to our knowledge this model is only in research phase and has not been implemented yet. [16] Presented a complete SDN framework using free scale VortiQa Open Network Director and Switch solution to be incorporated in IoT networks. From commercial point of view Huawei has launched first software defined base IoT platform to create more agile network for services [19]

Works has also been done to integrate fog with SDN in different application scenarios. [8] Proposed FSDN (Fog based SDN) to be used in VANET scenarios. In this paper they provide an architecture based on both fog networking and SDN controller to effectively control the communications in VANETs. They proposed that SDN can be deployed to provide a central overview of the network whereas Fog networks can help sharing the load of network management with SDN controller by operating as Road side unit controller. These RSUCs can perform important network analysis at the edge, which can help in simplifying the task of SDN controller. Similar work has also been presented in [17], in which they proposed the concept of RSUs cloud which consists of RSUs and some specialized SDN based micro datacenters RSUs. These RSUs micro datacenters are capable of performing virtualization and communication using SDN. Programmability provided by SDN can help dynamic instantiation, replication and migration of services across various RSUs, whereas micro datacenters strengthen these RSU to provide non-safety communication to vehicles. In this paper they also proposed a novel approach to reduce the cost incurred due to reconfigurations in such system so that system will be commercially viable. SDN based fog computing system to be used in context of IoT is also proposed recently in [18], in which they proposed an architecture for fog computation based on SDN, which can enable programmability to the network switches as well as centralized control plane to properly utilize the underlying network resources. They also analyzed the delivery throughput of such SDN based fog nodes.

So the work is in research phase in integrating SDN and fog together to create complete architectural support for emerging IoT.

6.2 Final Word

In this tutorial we discussed how fog networking and SDN together can provide an architectural basis for next generation internet of things. Both technologies provide certain features to support the volume, variety and diversity of envisioned next generation networks. We highlighted core advantages of Fog networks as well as summarized different features provided by different SDN controllers which can be used in IoT scenarios. Research is already been going on in the mention areas and many research paths are still open for studies. By surveying the literature it has been found that, though some architectures has been proposed which incorporate both SDN and fog networks but they cover a single application area. We still have to formulate a complete general architecture, with well-defined protocols for interaction between different components to support next generation networks. We need standard protocols to perform resource pooling (managing different edge devices), protocols for fog-cloud interactions and protocols for fog to fog communications. Similarly available SDN controllers are not specifically designed for wireless scenarios. We need to work that how existing controllers can be modified to become adaptable in wireless context of IoT.

References

1. Cisco White paper: Fog Computing and the Internet of Things: Extend the Cloud to Where the Things Are. http://www.cisco.com/web/solutions/trends/iot/docs/computing-overview.pdf
2. Huawei white paper: 5G A Technology vision http://www.huawei.com/5gwhitepaper/
3. Bonomi, F., Zhu, J., et al.: Fog computing: a platform for internet of things and analytics. In: Bessis, N., Dobre, C. (eds.) Big Data and Internet of Things: A Roadmap for Smart Environments. Studies in Computational Intelligence, vol. 546, pp. 169–186. Springer International Publishing, Switzerland (2014)
4. Bonomi, F., Milito, R., Zhu, J., Addepalli, S.: Fog computing and its role in the internet of things. In: Proceedings of the First Edition of the MCC Workshop on Mobile Cloud Computing, MCC 2012, pp. 13–16. ACM (2012)
5. Kumar, V.A., Prasad, E.: Fog computing: characteristics, advantages and security – privacy. Int. J. Comput. Sci. Manag. Res. **3**(11), 4211–4215 (2014)
6. Stojmenovic, I., When, S.: The fog computing paradigm: scenarios and security issues. In: Proceedings of the 2014 Federated Conference on Computer Science and Information Systems, ACSIS, vol. 2, pp. 1–8 (2014)
7. Zhu, J., Chan, D.S., Prabhu, M.S., Bonomi, F, et al.: Improving web sites performance using edge servers in fog computing architecture. In: Presented at 2013 IEEE Seventh International Symposium on Service-Oriented System Engineering (2013)

8. Truong, N.B., et al.: Software defined networking-based vehicular adhoc network with fog computing. In: Published in 2015 IFIP/IEEE International Symposium on Integrated Network Management (IM), pp. 1202–1207, May 2015
9. ONOS Whitepaper: Introducing ONOS - a SDN network operating system for Service Providers. http://onosproject.org/wp-content/uploads/2014/11/Whitepaper-ONOS-final.pdf
10. https://osrg.github.io/ryu/
11. https://www.opendaylight.org/
12. http://www.projectfloodlight.org/floodlight/
13. Chiang, M.: Fog networking: an overview on research opportunities. https://arxiv.org/ftp/arxiv/papers/1601/1601.00835.pdf
14. Sydney, A.: The evaluation of software defined networking for communication and control of cyber physical systems. Ph.D. dissertation, Department of Electrical and Computer Engineering College of Engineering, Kansas State University, Manhattan, Kansas (2013)
15. Qin, Z., et al.: A software defined networking architecture for the internet-of-things. In: 2014 IEEE Network Operations and Management Symposium (NOMS). IEEE (2014)
16. Tadinada, V.R.: Software defined networking: redefining the future of internet in IoT and Cloud Era. In: 2014 International Conference on Future Internet of Things and Cloud (FiCloud). IEEE (2014)
17. Salahuddin, M.A., Al-Fuqaha, A., Guizani, M.: Software-defined networking for RSU clouds in support of the internet of vehicles. IEEE Internet Things J. 2(2), 133–144 (2015)
18. Xu, Y., Mahendran, V., Radhakrishnan, S.: Towards SDN-based fog computing: MQTT broker virtualization for effective and reliable delivery. In: 2016 8th International Conference on Communication Systems and Networks (COMSNETS). IEEE (2016)
19. http://telecoms.com/422331/huawei-launches-sdn-based-iot-platform/

Application of Optimization of Parallel Algorithms to Queries in Relational Databases

Yulia Shichkina[1]([⊠]), Alexander Degtyarev[2]([⊠]), Dmitry Gushchanskiy[2], and Oleg Iakushkin[2]

[1] Department of Computer Science and Engineering,
Saint Petersburg Electrotechnical University "LETI", St. Petersburg, Russia
strange.y@mail.ru
[2] Saint Petersburg State University, St. Petersburg, Russia
deg@csa.ru, nyapko@gmail.com, oleg.jakushkin@gmail.com

Abstract. All known approaches to parallel data processing in relational client-server database management systems are based only on inter-query parallelism. Nevertheless, it's possible to achieve intra-query parallelism by consideration of a request structure and implementation of mathematical methods of parallel calculations for its equivalent transformation. This article presents an example of complex query parallelization and describes applicability of the graph theory and methods of parallel computing both for query parallelization and optimization.

Keywords: Parallel computing · Optimization methods · Relational database · Query · Information graph

1 Introduction

Nowadays the use of databases and information management systems is an integral part of business processes of enterprises and organizations. The growth of database volumes imposes new and quite strong requirements for the developing hardware infrastructure of data centers. Mark Hurd, co-CEO of Oracle Corporation, believes that the volume of accumulated data will increase fiftyfold by 2020. In 8 years more than 100,000 companies and corporations will be using databases, and the volume of each such database will surpass 1 petabyte [3].

Even today organizations already need powerful tools for transferring, processing and analyzing accumulated information.

The survey performed by Gartner in 2010 showed that the problem of data growth makes the top three most serious problems for 47 % commercial companies. System performance and scalability worry 37 % of organizations, while 36 % of organizations believe that the potential issues are related to network congestion and connectivity. Six in ten questioned organizations announced plans to invest in data retirement efforts. According to April Adams, research director at Gartner, while all the top data center

© Springer International Publishing Switzerland 2016
O. Gervasi et al. (Eds.): ICCSA 2016, Part II, LNCS 9787, pp. 366–378, 2016.
DOI: 10.1007/978-3-319-42108-7_28

hardware infrastructure challenges impact cost to some degree, data growth is particularly associated with increased costs relative to hardware, software, associated maintenance, administration and services [2]. There is one way out: the methods of data processing should be updated in according to available equipment.

During the past twenty years development of databases has been forging ahead. The new branch of database technology called NoSQL emerged, which provides new simple ways of data scaling. And, for processing large amounts of data, for example, those received for IoT with GLONNAS, many companies tend to choose NoSQL solutions. And if yesterday some mechanisms applicable for relational databases were not available for NoSQL, today, for example, fuzzy slices are easily implemented in databases such as MongoDB. However, not every NoSQL technology, especially high scalable ones, has solved the issues related to correspondence of operations to requirements of ACID (atomicity, consistency, isolation, durability) – the standard which guarantees accuracy of operational transactions performance by resources of database management systems in case of a system failure. Today this problem is being handled by a new database systems technology class NewSQL, members of which have combined new approaches of distributed systems from NoSQL with relational data presentation model and data query language SQL. The developers of the NewSQL system VoltDB affirm that it is approximately fiftyfold faster than the traditional OLTP RDBMS.

However, considering NoSQL and NewSQL, it must be noted that:

1. For the absolute majority of big projects related to implementation of database management systems relational DBMS are still preferred, and the market continues maintaining traditional approaches to solving such issues;
2. In spite of the long history of relational databases development, due to the development of new supercomputing approaches it turns out that not all mechanisms of relational databases are actively used and new opportunities open up.
3. Data in relational databases is created, updated, deleted, and queried through the API method calls. API is open to users and allows them to perform maintenance and scaling of its database instances. The most common clients API are JDBC, ODBC, OLE-DB, etc. API-functions "know" how to send a request to the database and how to process the returned cursor. They provide conversion of user queries transmitted over the network as packets, which are processed by a dedicated server. They do not contain embedded parallelization mechanisms though.
4. All modern relational databases use CBO (Cost Based Optimization). It suggests that each operation is determined by its "cost", and the total cost of the query is reduced through the use of the most "cheap" chain operations. Although this mechanism significantly speeds up query processing, it does not take into the possibility of parallel execution of parts of a query.

Considering the above and the fact that data growth in the existing databases continues, as well as the need of fast data processing increases, the task of creation of an interactive mechanism for parallel query processing becomes more and more relevant. If one manages to create a device which will provide effective parallelizing of queries in relational databases, it will not just extend their lifecycle and expand their sphere of application, but also will reduce expenses of organizations with informational systems, which are successfully created long ago and based exactly on relational databases.

It should be noted, that, in spite of intensive development of computer systems, new methods, and programming languages, parallel data processing is still the area, where the problem of dividing computation into processes falls on end users (programmers, database managers, etc.). Quality and speed of derivable solutions on data processing most often depend on qualification of the user of a database management system. It is possible to reduce the impact of human factor on the parallel data processing by means of transfer the function of computing the number of CPU needed for handling applied problems and modeling the future parallel query for ECM.

The software for DBMS can provide the following types of parallelism (Fig. 1) [9]:

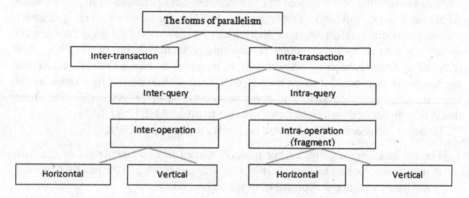

Fig. 1. Types of parallelism

Inter-transaction parallelism is parallel execution of many separate transactions for the same database.

Intra-transaction parallelism is parallel execution of a separate transaction [1].

Inter-query parallelism is a parallel processing of separate SQL queries in the same transaction [6].

Intra-query parallelism is parallel execution of a separate SQL query. This type of parallelism is common in relational databases. This is due to the fact that relational operations on series of tuples are well-adapted for efficient parallelization [4].

Inter-operation parallelism is parallel processing of relational operations on the same query. Inter-operation parallelism can be implemented either as horizontal parallelism or vertical parallelism [5].

Horizontal parallelism is parallel processing of separate operations in a query [7].

Vertical parallelism is a parallel processing of different operations in a query based on pipeline mechanism.

Fragment parallelism is splitting of the relation, which is an argument of a relational operation, into disjoint parts. A single relational operation is performed as several parallel processes (agents), either of which handles an independent fragment of the relation. The obtained resulting fragments merge into combined resulting relation [8].

Fragmentation in relational database systems can be vertical or horizontal. Vertical fragmentation provides splitting of the relation into fragments of columns (attributes). Horizontal fragmentation provides splitting of the relation into fragments of rows (tuples). Almost all parallel database management systems, which sustain fragment parallelism, use only horizontal fragmentation.

This article deals with inter-transaction parallelism based on the mixture of inter-query, intra-query inter-operation horizontal parallelism, and inter-operation parallelism with horizontal fragmentation.

The degree of inter-query parallelism is restricted both by the number of SQL queries, which compose this transaction, and by precedence constraints between separate SQL queries. Therefore the inter-query approach to query parallelization can be successfully combined with the intra-query one for increase of the degree of complex query parallelism in general.

Theoretically intra-query fragment parallelism can provide the arbitrarily high degree of relational operations parallelization. In practice, however, the degree of fragment parallelism can be significantly restricted by two factors. In the first place, fragmentation of the relation always depends on operation semantics. In the second place, a failed fragmentation can lead to significant imbalance in CPU load.

Both inter-query and intra-query types of parallelism require good parallelization skills, understanding of inter-query dependences, and significant time and labor costs for their implementation from a programmer. Particularly for this reason inter-query parallelism is not sustained by majority of modern database management systems. However, research shows that the information graph of a query, through simple modifications in accordance with the principles of inter-query, horizontal, and fragment horizontal parallelism, by its structure is identical to the information graph of an algorithm, and therefore for its analysis and optimization one can implement mathematical methods of parallel computing based on graph theory and adjacency lists.

2 Query Parallelization in Client-Server Databases

For all mass computer models and different operating systems the PC software market offers many commercial database management systems, which vary in their functionality and capabilities. The most popular client-server database management systems are: Microsoft SQL Server, Oracle, Firebird, PostgreSQL, and MySQL. Despite of some of these DBMS have more powerful functional set and others are built with less various data processing functions, the operation principle of all database management systems listed above is the same.

One way to achieve higher efficiency is to use task parallelization algorithms. There are three application areas for such algorithms in DBMS:

- Parallel input/output,
- Parallel administration tools and utilities,
- Parallel processing of database queries.

Parallelization of input/output in conjunction with optimal task planning allows accomplishing quite efficient simultaneous access to fragmented tables and indexes located on several physical disks, thus boosting the operations with comparatively slow external devices manyfold.

In contrast with parallel input/output and administration, parallelism implementation in request processing is considered more difficult. A theoretical foundation of the possibility of query parallelization in relational database management systems is the property of relational closeness. The result of each relational operator: SELECT is selection of subsets of relation (table) rows; PROJECT is selection of subsets of fields (columns); JOIN is combination of two tables – is a new relation, and, as far as any query can be divided into hierarchy of elementary operators, it is rational to try to execute them in parallel. Undoubtedly, parallelism is inherent in SQL internally. Query processing consists of a set of atomic operations, and their structure and sequence are determined by the performance enhancer after the examination of several options.

In client-server DBMS data processing is performed on a server where data is stored. Client applications send requests for processing and receiving data from database management system and receive the answers. Client applications do not have immediate access to data files. Database server is a multi-user version of DBMS with parallel processing of queries coming from all workstations. Its task is to implement the transaction manipulation logic using necessary synchronization methods – maintaining locking protocols for the resources and providing prevention and/or elimination of deadlocks. In response to a user query, a workstation will receive not "raw material" for future processing, but complete results. By such architecture the workstation software represents only the front-end of central database management system. This allows reduction of network traffic, shortening the time of waiting locked data resources in multi-user mode, unloading workstations, and, provided that the central machine is powerful enough, utilization of cheaper equipment for them. However, this does not allow distributing the parts of a query between hardware cores for their parallel execution.

For example, in MySQL queries are made in a parallel way only if they are from different clients. It means that MySQL cannot parallelize execution of a query on several processing nodes. Therefore, for increase of query efficiency it is necessary to optimize complex queries by means of their decomposing into smaller ones and executing from different clients followed by merge of the results.

For instance, a query for output of user id, email, post id, post message, topic name, topic id, and full name of the user takes 32,77 ms:

```
SELECT   u.id_user,   u.email,   p.id_post,   p.message,
t.topic_name, t.id_topic, u.name FROM users u, topics t,
posts  p  WHERE  u.id_user  BETWEEN  100  and  900  and
t.id_topic = p.id_topic GROUP BY u.name.
```

After dividing this query into 2 separate ones, the time shortens to 20.6 ms – 16.17 ms faster than the first query:

```
SELECT  u1.id_user,  u1.email,  p.id_post,  p.message,
t.topic_name, t.id_topic, u1.name FROM users u1, topics t,
```

```
posts  p  WHERE  u1.id_user  BETWEEN  100  and  900  and
t.id_topic = p.id_topic GROUP BY u1.name LIMIT 0,10000;
SELECT  u2.id_user,  u2.email,  p.id_post,  p.message,
t.topic_name, t.id_topic, u2.name FROM users u2, topics t,
posts  p  WHERE  u2.id_user  BETWEEN  100  and  900  and
t.id_topic = p.id_topic GROUP BY u2.name LIMIT 10001, 20000;
```

Firstly, there are different ways of implementation for the majority of queries with embedded SELECT constructions, and among them there could be ones with faster as well as slower execution speed. Secondly, the example above shows that decomposition of a complex query into several interconnected simple ones allows running the part of selected mutually independent subsequent queries in a parallel way as if they are from different clients. This leads to data processing speed boost.

In PostgreSQL it is easy to parallelize queries with intra-query with horizontal fragmentation parallelism by using a simple and evident trick: creating an index with the function "the remainder on dividing the id by a number of processing nodes".

For example, the table with the data from some transducers has the following format:

```
CREATE  TABLE  device  (id  BIGINT,  time  TIMESTAMP,
device_id INTEGER, indication INTERVAL);
```

The data from devices comes into this table at a high rate, and must be processed in reasonable time by the function:

```
FUNCTION calculate(IN device_id INTEGER, IN indication
INTERVAL, OUT status_code text) RETURNS void;
```

The processing is executed by the query:

```
SELECT calculate(device_id, indication) FROM device;
```

With continuous data reading from the devices, server processing node will soon fail to perform its work properly while the amount of data will be growing.

PostgreSQL, as well as MySQL, cannot parallelize queries on its own. This situation can be handled by creating the following index:

```
CREATE  INDEX  idx  ON  device  USING  btree  ((device_id %
4));
```

and executing a query in four threads:

```
SELECT  calculate(device_id,  indication)  FROM  device
WHERE device_id % 4 = @rank;
```

where @rank is equal to 0, 1, 2 or 3 (its own for each thread). As a result this action solves problems with locks, which can appear if two different threads get a signal from a single device. Moreover these processes will work faster in the context of parallel execution than in case of parallelizing the database itself, for instance, as in Oracle. This method is applicable for any database with support of function-based index (Oracle, PostgreSQL). In MS SQL one can create calculated column and base index on it. There is no support of functional indexes in MySQL, but as an alternative it is possible to create a new column with its index and renew it with a trigger.

Therefore, implementation of intra-query parallelism with horizontal fragmentation can be very effective, but it requires high qualification of programmers in two spheres: applied software (good knowledge of functions and capabilities of the used SQL server) and good understanding of the query structure including detecting the ways of query decomposition into a number of subqueries or table decomposition followed by merging the results into the final output buffer. It is possible to simplify work of programmers in improving both parallel performance and quality of work, as well as to reduce the cost of the work, if the function of query optimization is put in a simple and light program wrapper above the SQL server. The wrapper implements equivalent query transformation by mathematical methods in the form that allows parallel execution on a set of processing nodes.

3   Query Decomposition and Visualization with Graphs

Query decomposition can be made in vertically or horizontally, as well as combining these two approaches.

Horizontal query decomposition is a decomposition of relations, to which the query is applied (horizontal fragmentation). The query itself stays unchanged (Fig. 2b). The Fig. 2a shows the original query.

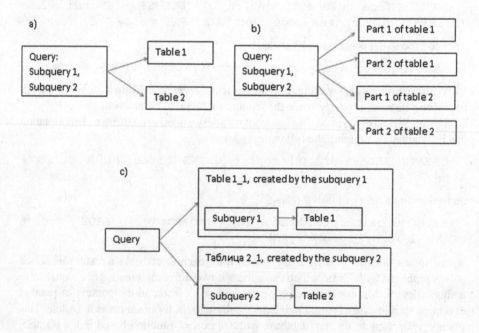

Fig. 2. Queries modification

Vertical request decomposition is a decomposition of the query itself into separate subqueries, through relations, to which the original query was applied, stay unchanged (Fig. 2c). Depending on the structure of the primary query, it is possible to execute either horizontal or vertical decomposition, or, in the best case, their combined variant of simultaneous implementation both vertical and horizontal decomposition (Fig. 3).

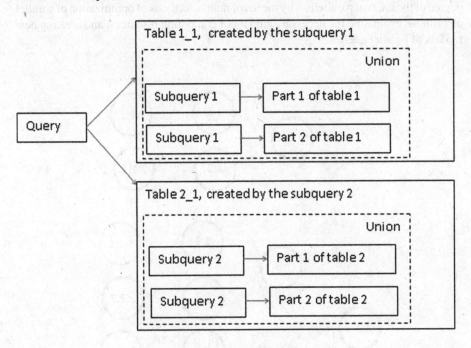

Fig. 3. Combined query modification

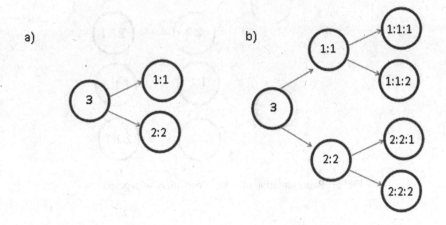

Fig. 4. Visualization of queries with graphs

Graph theory gives more vivid visualization options for presenting interconnections between parts of a query and corresponding relations. On the Fig. 4a one can see a graph that corresponds to the vertical query transformation from the Fig. 2c. The Fig. 4b corresponds to the combined decomposition (Fig. 3). It is possible to decompose the relations even deeper (Fig. 5). Graphs allow, besides query visualization, defining the degree of the internal parallelism by means of matrix methods of optimization of parallel algorithms, evaluating the necessary amount of computing resources, and creating new models of parallel queries.

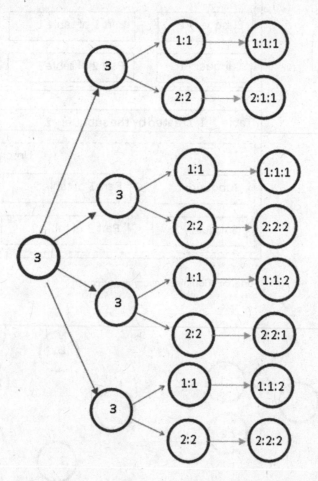

Fig. 5. Representation of a combined query with graphs

The Fig. 6 shows a query, which consists of several subqueries. The same query can be executed in some other ways, for example, by the code shown in the Fig. 7.

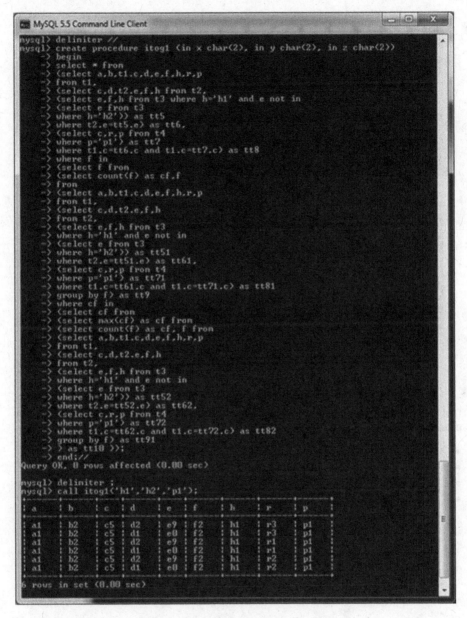

Fig. 6. An example of a "heavy" query

The degree of query parallelism depends on how successfully its structure will be set up. However, it is quite hard to analyze such queries manually. Moreover, even if some faster and more efficient solution will be found, this won't guarantee that the obtained solution will represent the global extremum of all solutions.

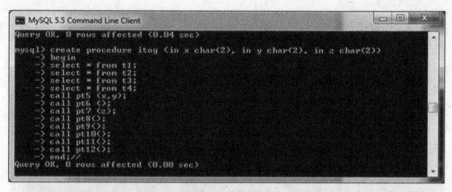

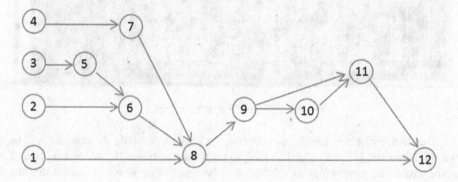

Fig. 7. The modified query

Figure 8 shows query information graph. The searching methods of parallel branches, elaborated for classic algorithms, also could be applied to such graph, for example:

- Methods of finding early and late terms;
- Methods based on scheduling theory;
- Matrix methods based on information graph;
- List-based methods.

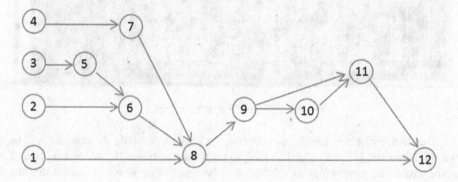

Fig. 8. Query information graph

The optimization methods, elaborated for classic algorithms, could be applied to such graph too:

- By the amount of computing resources;
- By the runtime length;
- Taking the communication into account;
- Methods of multiparametric optimization.

Classic evaluation methods of acceleration, parallelization efficiency, compute density, and others, which are also elaborated for classical algorithms, fit queries represented by information graphs as well.

In conclusion it is important to note that horizontal fragmentation appends additional vertices and edges to an information graph. This also corresponds to one of the stages of classical parallel algorithms design called algorithm decomposition. After obtaining the estimates of acceleration and compute density of the query, it is possible to inverse the process through one of the known methods – fragment upsizing – for those parts of the query, which will allow doing this without any loss of computational efficiency. However only the information graph and mathematical methods of its analysis and equivalent transformation can answer the question, which fragments will it be.

4   Conclusion

The research shows that, if the parts of a complex query are represented as independent client queries, parallelization of complex queries can be achieved by capabilities of standard SQL regardless of a SQL server. Optimization methods and methods of parallel algorithms modeling based on matrix algebra and graph theory can be applied to such queries with success. Moreover, by implementing a special program wrapper it is possible to analyze the structure and to transform any complex user query written in standard SQL.

It should be noted that, for instance, CouchBase DBMS already has capabilities for automated parallel computing, including on multicore processors. But, firstly, Couch-Base is a document-oriented DBMS, and it works in NoSQL approach. Conversion of existing large relational databases to CouchBase in order to accelerate query perform-ance is ineffective. Secondly, it has not been fully investigated, how effective is its mechanism for automatic parallelization compared to efficient manual parallelization. The proposed approach is implemented in an interactive form that allows users to improve their work by creating queries based on the options offered by the parallelization program.

Another advantage of the proposed approach is the ability to use it effectively for sparse databases. In this case, the approach can be effectively implemented on the basis of an ideology used in dataspaces. Thus during the fragmentation of a query empty slots in database will fall into separate fragments. The following application of known opti-mization algorithms for runtime length or data size to the resulting information graph will completely remove the empty fragments from the processing. As a result, the acceleration will be achieved not only by a query execution on the computing system with parallel processing, but also due to consolidation of data in the structure and elimination of query processing operations on empty fragments.

References

1. Feng, L., Li, Q., Wong, A.: Mining inter-transactional association rules: generalization and empirical evaluation. In: Kambayashi, Y., Winiwarter, W., Arikawa, M. (eds.) DaWaK 2001. LNCS, vol. 2114, pp. 31–40. Springer, Heidelberg (2001)
2. http://www.computerworld.com/article/2513954/data-center/data-growth-remains-it-s-biggest-challenge–gartner-says.html
3. http://www.eweek.com/enterprise-apps/scaleout-introduces-analytics-server-to-crunch-big-data
4. Akal, F., Böhm, K., Schek, H.-J.: OLAP query evaluation in a database cluster: a performance study on intra-query parallelism. In: Manolopoulos, Y., Návrat, P. (eds.) ADBIS 2002. LNCS, vol. 2435, pp. 218–231. Springer, Heidelberg (2002)
5. Sanjay, A., Narasayya, V.R., Yang, B.: Integrating vertical and horizontal partitioning into automated physical database design. In: Proceedings of the ACM SIGMOD International Conference on Management of Data, pp. 359–370, June 2004
6. https://eden.dei.uc.pt/~pnf/publications/Furtado_survey.pdf
7. Chen, Y., Dehne, F., Eavis, T., Rau-Chaplin, A.: Parallel ROLAP data cube construction on shared-nothing multiprocessors. Distrib. Parallel Databases 15(3), 219–236 (2014)
8. DeWitt, D.J., Gray, J.: Parallel database systems: the future of high performance database systems. Commun. ACM 35(6), 85–98 (1992)
9. Valduriez, P.: Parallel database systems: open problems and new issues. Distrib. Parallel Databases 1(2), 137–165 (1993). doi:10.1007/BF01264049

Factory: Master Node High-Availability for Big Data Applications and Beyond

Ivan Gankevich[✉], Yuri Tipikin, Vladimir Korkhov, Vladimir Gaiduchok,
Alexander Degtyarev, and Alexander Bogdanov

Department of Computer Modelling and Multiprocessor Systems,
Saint Petersburg State University, Universitetskaia emb. 7-9,
199034 Saint Petersburg, Russia
{i.gankevich,y.tipikin,v.korkhov}@spbu.ru,
gvladimiru@gmail.com, {deg,bogdanov}@csa.ru
http://spbu.ru/

Abstract. Master node fault-tolerance is the topic that is often dimmed in the discussion of big data processing technologies. Although failure of a master node can take down the whole data processing pipeline, this is considered either improbable or too difficult to encounter. The aim of the studies reported here is to propose rather simple technique to deal with master-node failures. This technique is based on temporary delegation of master role to one of the slave nodes and transferring updated state back to the master when one step of computation is complete. That way the state is duplicated and computation can proceed to the next step regardless of a failure of a delegate or the master (but not both). We run benchmarks to show that a failure of a master is almost "invisible" to other nodes, and failure of a delegate results in recomputation of only one step of data processing pipeline. We believe that the technique can be used not only in Big Data processing but in other types of applications.

Keywords: Parallel computing · Big data processing · Distributed computing · Backup node · State transfer · Delegation · Cluster computing · Fault-tolerance

1 Introduction

Fault tolerance of data processing pipelines is one of the top concerns in development of job schedulers for big data processing, however, most schedulers provide fault tolerance for subordinate nodes only. These types of failures are routinely mitigated by restarting the failed job or its part on healthy nodes, and failure of a master node is often considered either improbable, or too complicated to handle and configure on the target platform. System administrators often find alternatives to application level fault tolerance: they isolate master node from the rest of the cluster by placing it on a dedicated machine, or use virtualisation technologies instead. All these alternatives complexify configuration and maintenance, and by decreasing probability of a machine failure resulting in a whole system failure, they increase probability of a human error.

© Springer International Publishing Switzerland 2016
O. Gervasi et al. (Eds.): ICCSA 2016, Part II, LNCS 9787, pp. 379–389, 2016.
DOI: 10.1007/978-3-319-42108-7_29

From such point of view it seems more practical to implement master node fault tolerance at application level, however, there is no generic implementation. Most implementations are too tied to a particular application to become universally acceptable. We believe that this happens due to people's habit to think of a cluster as a collection of individual machines each of which can be either master or slave, rather than to think of a cluster as a whole with master and slave roles being dynamically assigned to a particular physical machine.

This evolution in thinking allows to implement middleware that manages master and slave roles automatically and handles node failures in a generic way. This software provides an API to distribute parallel tasks on the pool of available nodes and among them. Using this API one can write an application that runs on a cluster without knowing the exact number of online nodes. The middleware works as a cluster operating system overlay allowing to write distributed applications.

2   Related Work

Dynamic role assignment is an emerging trend in design of distributed systems [3,5,8,21,26], however, it is still not used in big data job schedulers. For example, in popular YARN job scheduler [29], which is used by Hadoop and Spark big data analysis frameworks, master and slave roles are static. Failure of a slave node is tolerated by restarting a part of a job on a healthy node, and failure of a master node is tolerated by setting up standby reserved server [22]. Both master servers are coordinated by Zookeeper service which itself uses dynamic role assignment to ensure its fault-tolerance [25]. So, the whole setup is complicated due to Hadoop scheduler lacking dynamic roles: if dynamic roles were available, Zookeeper would be redundant in this setup. Moreover, this setup does not guarantee continuous operation of master node because standby server needs time to recover current state after a failure.

The same problem occurs in high-performance computing where master node of a job scheduler is the single point of failure. In [10,27] the authors use replication to make the master node highly-available, but backup server role is assigned statically and cannot be delegated to a healthy worker node. This solution is closer to fully dynamic role assignment than high-availability solution for big data schedulers, because it does not involve using external service to store configuration which should also be highly-available, however, it is far from ideal solution where roles are completely decoupled from physical servers.

Finally, the simplest master node high-availability is implemented in Virtual Router Redundancy Protocol (VRRP) [18,20,23]. Although VRRP protocol does provide master and backup node roles, which are dynamically assigned to available routers, this protocol works on top of the IPv4 and IPv6 protocols and is designed to be used by routers and reverse proxy servers. Such servers lack the state that needs to be restored upon a failure (i.e. there is no job queue in web servers), so it is easier for them to provide high-availability. In Linux it is implemented in Keepalived routing daemon [6].

In contrast to web servers and HPC and Big Data job schedulers, some distributed key-value stores and parallel file systems have symmetric architecture, where master and slave roles are assigned dynamically, so that any node can act as a master when the current master node fails [3,5,8,21,26]. This design decision simplifies management and interaction with a distributed system. From system administrator point of view it is much simpler to install the same software stack on each node than to manually configure master and slave nodes. Additionally, it is much easier to bootstrap new nodes into the cluster and decommission old ones. From user point of view, it is much simpler to provide web service high-availability and load-balancing when you have multiple backup nodes to connect to.

Dynamic role assignment would be beneficial for Big Data job schedulers because it allows to decouple distributed services from physical nodes, which is the first step to build highly-available distributed service. The reason that there is no general solution to this problem is that there is no generic programming environment to write and execute distributed programmes. The aim of this work is to propose such an environment and to describe its internal structure.

The programming model used in this work is partly based on well-known actor model of concurrent computation [2,17]. Our model borrows the concept of actor—an object that stores data and methods to process it; this object can react to external events by either changing its state or producing more actors. We call this objects *computational kernels*. Their distinct feature is hierarchical dependence on parent kernel that created each of them, which allows to implement fault-tolerance based on simple restart of a failed subordinate kernel.

However, using hierarchical dependence alone is not enough to develop high-availability of a master kernel—the first kernel in a parallel programme. To solve the problem the other part of our programming model is based on bulk-synchronous parallel model [28]. It borrows the concept of superstep—a sequential step of a parallel programme; at any time a programme executes only one superstep, which allows to implement high-availability of the first kernel (under assumption that it has only one subordinate at a time) by sending it along its subordinate to a different cluster node thus making a distributed copy of it. Since the first kernel has only one subordinate at a time, its copy is always consistent with the original kernel. This eliminates the need for complex distributed transactions and distributed consensus algorithms and guarantees protection from at most one master node failure per superstep.

To summarise, the framework developed in this paper protects a parallel programme from failure of any number of subordinate nodes and from one failure of a master node per superstep. The paper does not answer the question of how to determine if a node failed, it assumes a failure when the network connection to a node is prematurely closed. In general, the presented research goes in line with further development of the virtual supercomputer concept coined and evaluated in [4,12,13].

3   Methods

3.1   Model of Computation

To infer fault tolerance model which is suitable for big data applications we use bulk-synchronous parallel model [28] as the basis. This model assumes that a parallel programme is composed of several sequential steps that are internally parallel, and global synchronisation of all parallel processes occurs after each step. In our model all sequential steps are pipelined where it is possible. The evolution of the computational model is described as follows.

Given a programme that is sequential and large enough to be decomposed into several sequential steps, the simplest way to make it run faster is to exploit data parallelism. Usually it means finding multi-dimensional arrays and loops that access their elements and trying to make them parallel. After transforming several loops the programme will still have the same number of sequential steps, but every step will (ideally) be internally parallel.

After that the only possibility to speedup the programme is to overlap execution of code blocks that work with different hardware devices. The most common pattern is to overlap computation with network I/O or disk I/O. This approach makes sense because all devices operate with little synchronisation, and issuing commands in parallel makes the whole programme perform better. This behaviour can be achieved by allocating a separate task queue for each device and submitting tasks to these queues asynchronously with execution of the main thread. So, after this optimisation, the programme will be composed of several steps chained into the pipeline, each step is implemented as a task queue for a particular device.

Pipelining of otherwise sequential steps is beneficial not only for code accessing different devices, but for code different branches of which are suitable for execution by multiple hardware threads of the same core, i.e. branches accessing different regions of memory or performing mixed arithmetic (floating point and integer). In other words, code branches which use different modules of processor are good candidates to run in parallel on a processor core with multiple hardware threads.

Even though pipelining may not add parallelism for a programme that uses only one input file (or a set of input parameters), it adds parallelism when the programme can process multiple input files: each input generates tasks which travel through the whole pipeline in parallel with tasks generated by other inputs. With a pipeline an array of files is processed in parallel by the same set of resources allocated for a batch job, and possibly with greater efficiency for busy HPC clusters compared to executing a separate job for each input file, because the time that each subsequent job after the first spends in a queue is eliminated.

Computational model with a pipeline can be seen as *bulk-asynchronous model*, because of the parallel nature of otherwise sequential execution steps. This model is the basis of the fault-tolerance model developed here.

3.2   Fail over Model

Although, fault-tolerance and high-availability are different terms, in essence they describe the same property—an ability of a system to switch processing from a failed component to its live spare or backup component. In case of fault-tolerance it is the ability to switch from a failed slave node to a spare one, i.e. to repeat computation step on a healthy slave node. In case of high-availability it is the ability to switch from a failed master node to a backup node with full restoration of execution state. These are the core abilities that constitute distributed system's ability to *fail over*.

The key feature that is missing in the current parallel programming and big data processing technologies is a possibility to specify hierarchical dependencies between parallel tasks. When one has such dependency, it is trivial to determine which task should be responsible for re-executing a failed task on a healthy node. To re-execute the root of the hierarchy, a backup root task is created and executed on a different node. There exists a number of engines that are capable of executing directed acyclic graphs of tasks in parallel [1,19], but graphs are not good to infer master-slave relationship between tasks, because a node in the graph may have multiple parent nodes.

3.3   Programming Model

This work is based on the results of previous research: In [15,16] we developed an algorithm that allows to build a tree hierarchy from strictly ordered set of cluster nodes. The sole purpose of this hierarchy is to make a cluster more fault-tolerant by introducing multiple master nodes. If a master node fails, then its subordinates try to connect to another node from the same or higher level of the hierarchy. If there is no such node, one of the subordinates becomes the master. In [14] we developed a framework for big data processing without fault tolerance, and here this framework is combined with fault-tolerance techniques described in this paper.

Each programme that runs on top of the tree hierarchy is composed of computational kernels—objects that contain data and code to process it. To exploit parallelism a kernel may create arbitrary number of subordinate kernels which are automatically spread first across available processor cores, second across subordinate nodes in the tree hierarchy. The programme is itself a kernel (without a parent as it is executed by a user), which either solves the problem sequentially on its own or creates subordinate kernels to solve it in parallel.

In contrast to HPC applications, in big data applications it is inefficient to run computational kernels on arbitrary chosen nodes. More practical approach is to bind every kernel to a file location in a parallel file system and transfer the kernel to that location before processing the file. That way expensive data transfer is eliminated, and the file is always read from a local drive. This approach is more deterministic compared to existing ones, e.g. MapReduce framework runs jobs on nodes that are "close" to the file location, but not necessarily the exact node where the file is located [7]. However, this approach does not come without disadvantages: scalability of a big data application is limited by the strategy that was employed to distribute its input files across cluster nodes. The more nodes used to store input files, the more read performance is achieved. The advantage of our approach is that the I/O performance is more predictable, than one of hybrid approach with streaming files over the network.

3.4 Handling Master Node Failures

A possible way of handling a failure of a node where the first kernel is located (a master node) is to replicate this kernel to a backup node, and make all updates to its state propagate to the backup node by means of a distributed transaction. This approach requires synchronisation between all nodes that execute subordinates of the first kernel and the node with the first kernel itself. When a node with the first kernel goes offline, the nodes with subordinate kernels must know what node is the backup one. However, if the backup node also goes offline in the middle of execution of some subordinate kernel, then it is impossible for this kernel to discover the next backup node to return to, because this kernel has not discovered the unavailability of the master node yet. One can think of a consensus-based algorithm to ensure that subordinate kernels always know where the backup node is, but distributed consensus algorithms do not scale well to the large number of nodes and they are not reliable [11]. So, consensus-based approach does not play well with asynchronous nature of computational kernels as it may inhibit scalability of a parallel programme.

Fortunately, the first kernel usually does not perform operations in parallel, it is rather sequentially launches execution steps one by one, so it has only one subordinate at a time. Such behaviour is described by bulk-synchronous parallel programming model, in the framework of which a programme consists of sequential supersteps which are internally parallel [28]. Keeping this in mind, we can simplify synchronisation of its state: we can send the first kernel along with its subordinate to the subordinate node. When the node with the first kernel fails, its copy receives its subordinate, and no execution time is lost. When the node with its copy fails, its subordinate is rescheduled on some other node, and in the worst case a whole step of computation is lost.

Described approach works only for kernels that do not have a parent and have only one subordinate at a time, and act similar to manually triggered checkpoints. The advantage is that they

- save results after each sequential step when memory footprint of a programme is low,
- they save only relevant data,
- and they use memory of a subordinate node instead of stable storage.

4 Results

Master node fail over technique is evaluated on the example of wave energy spectra processing application. This programme uses NDBC dataset [24] to reconstruct frequency-directional spectra from wave rider buoy measurements and compute variance. Each spectrum is reconstructed from five variables using the following formula [9].

$$S(\omega, \theta) = \frac{1}{\pi} \left[\frac{1}{2} + r_1 \cos(\theta - \alpha_1) + r_2 \sin(2(\theta - \alpha_2)) \right] S_0(\omega).$$

Here ω denotes frequency, θ is wave direction, $r_{1,2}$ and $\alpha_{1,2}$ are parameters of spectrum decomposition and S_0 is non-directional spectrum; $r_{1,2}$, $\alpha_{1,2}$ and S_0 are acquired through measurements. Properties of the dataset which is used in evaluation are listed in Table 1.

Table 1. NDBC dataset properties.

Dataset size	144 MB
Dataset size (uncompressed)	770 MB
No. of wave stations	24
Time span	3 years (2010–2012)
Total no. of spectra	445422

The algorithm of processing spectra is as follows. First, current directory is recursively scanned for input files. Data for all buoys is distributed across cluster nodes and each buoy's data processing is distributed across processor cores of a node. Processing begins with joining corresponding measurements for each spectrum variables into a tuple, then for each tuple frequency-directional spectrum is reconstructed and its variance is computed. Results are gradually copied back to the machine where application was executed and when the processing is complete the programme terminates (Table 2).

Table 2. Test platform configuration.

CPU	Intel Xeon E5440, 2.83 GHz
RAM	4 Gb
HDD	ST3250310NS, 7200 rpm
No. of nodes	12
No. of CPU cores per node	8

In a series of test runs we benchmarked performance of the application in the presence of different types of failures:

– failure of a master node (a node where the first kernel is run),
– failure of a slave node (a node where spectra from a particular station are reconstructed) and
– failure of a backup node (a node where the first kernel is copied).

Table 3. Benchmark parameters.

Experiment no.	Master node	Victim node	Time to offline, s
1	root		
2	root	leaf	30
3	leaf	leaf	30
4	leaf	root	30

A tree hierarchy with sufficiently large fan-out value was chosen to make all cluster nodes connect directly to the first one so that only one master node exists in the cluster. In each run the first kernel was launched on a different node to make mapping of kernel hierarchy to the tree hierarchy optimal. A victim node was made offline after a fixed amount of time early after the programme start. To make up for the node failure all data files have replicas stored on different cluster nodes. All relevant parameters are summarised in Table 3 (here "root" and "leaf" refer to a node in the tree hierarchy). The results of these runs were compared to the run without node failures (Fig. 1).

The benchmark showed that only a backup node failure results in significant performance penalty, in all other cases the performance is roughly equals to the one without failures but with the number of nodes minus one. It happens because a backup node not only stores the copy of the state of the current computation step but executes this step in parallel with other subordinate nodes. So, when a backup node fails, the master node executes the whole step once again on arbitrarily chosen healthy subordinate node.

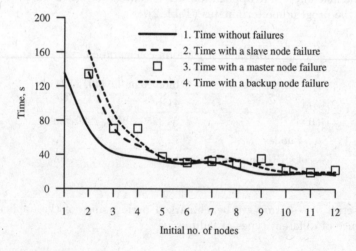

Fig. 1. Performance of spectrum processing application in the presence of different types of node failures.

5   Discussion

Described algorithm guarantees to handle one failure per computational step, more failures can be tolerated if they do not affect the master node. The system handles simultaneous failure of all subordinate nodes, however, if both master and backup nodes fail, there is no chance for an application to survive. In this case the state of the current computation step is lost, and the only way to restore it is to restart the application.

Computational kernels are means of abstraction that decouple distributed application from physical hardware: it does not matter how many nodes are online for an application to run successfully. Computational kernels eliminate the need to allocate a physical backup node to make master node highly-available, with computational kernels approach any node can act as a backup one. Finally, computational kernels can handle subordinate node failures in a way that is transparent to a programmer.

The disadvantage of this approach is evident: there is no way of making existing middleware highly-available without rewriting their source code. Although, our programming framework is lightweight, it is not easy to map architecture of existing middleware systems to it: most systems are developed keeping in mind static assignment of server/client roles, which is not easy to make dynamic. Hopefully, our approach will simplify design of future middleware systems.

6   Conclusion

Dynamic roles assignment is beneficial for Big Data applications and distributed systems in general. It decouples architecture of a distributed system from underlying hardware as much as possible, providing highly-available service on top of varying number of physical machines. As much as virtualisation simplifies management and administration of a computer cluster, our approach may simplify development of reliable distributed applications which run on top of the cluster.

Acknowledgements. The research was carried out using computational resources of Resource Centre "Computational Centre of Saint Petersburg State University" (T-EDGE96 HPC-0011828-001) within frameworks of grants of Russian Foundation for Basic Research (projects no. 16-07-01111, 16-07-00886, 16-07-01113) and Saint Petersburg State University (project no. 0.37.155.2014).

References

1. Acun, B., Gupta, A., Jain, N., Langer, A., Menon, H., Mikida, E., Ni, X., Robson, M., Sun, Y., Totoni, E., et al.: Parallel programming with migratable objects: Charm++ in practice. In: SC14: International Conference for High Performance Computing, Networking, Storage and Analysis, pp. 647–658. IEEE (2014)
2. Agha, G.A.: Actors: a model of concurrent computation in distributed systems. Technical report, DTIC Document (1985)

3. Anderson, J.C., Lehnardt, J., Slater, N.: CouchDB: The Definitive Guide. O'Reilly Media, Inc., Sebastopol (2010)

4. Bogdanov, A., Degtyarev, A., Korkhov, V., Gaiduchok, V., Gankevich, I.: Virtual Supercomputer as Basis of Scientific Computing. Horizons in Computer Science Research, vol. 11, pp. 159–198 (2015)

5. Boyer, E.B., Broomfield, M.C., Perrotti, T.A.: Glusterfs one storage server to rule them all. Technical report, Los Alamos National Laboratory (LANL) (2012)

6. Cassen, A.: Keepalived: Health checking for lvs & high availability (2002). http://www.linuxvirtualserver.org

7. Dean, J., Ghemawat, S.: MapReduce: Simplified data processing on large clusters. Commun. ACM **51**(1), 107–113 (2008)

8. Divya, M.S., Goyal, S.K.: Elasticsearch: an advanced and quick search technique to handle voluminous data. Compusoft **2**(6), 171 (2013)

9. Earle, M.D.: Nondirectional and directional wave data analysis procedures. Technical report, NDBC (1996)

10. Engelmann, C., Scott, S.L., Leangsuksun, C.B., He, X.B., et al.: Symmetric active/active high availability for high-performance computing system services. J. Comput. **1**(8), 43–54 (2006)

11. Fischer, M.J., Lynch, N.A., Paterson, M.S.: Impossibility of distributed consensus with one faulty process. J. ACM (JACM) **32**(2), 374–382 (1985)

12. Gankevich, I., Gaiduchok, V., Gushchanskiy, D., Tipikin, Y., Korkhov, V., Degtyarev, A., Bogdanov, A., Zolotarev, V.: Virtual private supercomputer: design and evaluation. In: CSIT 2013–9th International Conference on Computer Science and Information Technologies, Revised Selected Papers, pp. 1–6 (2013)

13. Gankevich, I., Korkhov, V., Balyan, S., Gaiduchok, V., Gushchanskiy, D., Tipikin, Y., Degtyarev, A., Bogdanov, A.: Constructing virtual private supercomputer using virtualization and cloud technologies. In: Murgante, B., et al. (eds.) ICCSA 2014, Part VI. LNCS, vol. 8584, pp. 341–354. Springer, Heidelberg (2014)

14. Gankevich, I., Degtyarev, A.: Efficient processing and classification of wave energy spectrum data with a distributed pipeline. Comput. Res. Model. **7**(3), 517–520 (2015). http://crm-en.ics.org.ru/journal/article/2301/

15. Gankevich, I., Tipikin, Y., Degtyarev, A., Korkhov, V.: Novel approaches for distributing workload on commodity computer systems. In: Gervasi, O., Murgante, B., Misra, S., Gavrilova, M.L., Rocha, A.M.A.C., Torre, C., Taniar, D., Apduhan, B.O. (eds.) ICCSA 2015. LNCS, vol. 9158, pp. 259–271. Springer, Heidelberg (2015)

16. Gankevich, I., Tipikin, Y., Gaiduchok, V.: Subordination: cluster management without distributed consensus. In: International Conference on High Performance Computing & Simulation (HPCS), pp. 639–642. IEEE (2015)

17. Hewitt, C., Bishop, P., Steiger, R.: A universal modular actor formalism for artificial intelligence. In: Proceedings of the 3rd International Joint Conference on Artificial Intelligence, pp. 235–245. Morgan Kaufmann Publishers Inc. (1973)

18. Hinden, R., et al.: Virtual router redundancy protocol (vrrp); rfc3768. txt. IETF Standard, Internet Engineering Task Force, IETF, CH, pp. 0000–0003 (2004)

19. Islam, M., Huang, A.K., Battisha, M., Chiang, M., Srinivasan, S., Peters, C., Neumann, A., Abdelnur, A.: Oozie: towards a scalable workflow management system for Hadoop. In: Proceedings of the 1st ACM SIGMOD Workshop on Scalable Workflow Execution Engines and Technologies, p. 4. ACM (2012)

20. Knight, S., Weaver, D., Whipple, D., Hinden, R., Mitzel, D., Hunt, P., Higginson, P., Shand, M., Lindem, A.: Rfc2338. Virtual Router Redundancy Protocol (1998)

21. Lakshman, A., Malik, P.: Cassandra: a decentralized structured storage system. ACM SIGOPS Oper. Syst. Rev. **44**(2), 35–40 (2010)

22. Murthy, A.C., Douglas, C., Konar, M., OMalley, O., Radia, S., Agarwal, S., Vinod, K.V.: Architecture of next generation apache hadoop mapreduce framework. Apache Jira (2011)
23. Nadas, S.: Rfc 5798: Virtual router redundancy protocol (vrrp) version 3 for ipv4 and ipv6. Internet Engineering Task Force (IETF) (2010)
24. NDBC directional wave stations. http://www.ndbc.noaa.gov/dwa.shtml
25. Okorafor, E., Patrick, M.K.: Availability of jobtracker machine in hadoop/mapreduce zookeeper coordinated clusters. Adv. Comput. Int. J. (ACIJ) **3**(3), 19–30 (2012)
26. Ostrovsky, D., Rodenski, Y., Haji, M.: Pro Couchbase Server. Apress, Berkeley (2015)
27. Uhlemann, K., Engelmann, C., Scott, S.L.: Joshua: symmetric active/active replication for highly available hpc job and resource management. In: 2006 IEEE International Conference on Cluster Computing, pp. 1–10. IEEE (2006)
28. Valiant, L.G.: A bridging model for parallel computation. Commun. ACM **33**(8), 103–111 (1990)
29. Vavilapalli, V.K., Murthy, A.C., Douglas, C., Agarwal, S., Konar, M., Evans, R., Graves, T., Lowe, J., Shah, H., Seth, S., et al.: Apache hadoop yarn: yet another resource negotiator. In: Proceedings of the 4th annual Symposium on Cloud Computing, p. 5. ACM (2013)

Petri Nets for Modelling of Message Passing Middleware in Cloud Computing Environments

Oleg Iakushkin[1]([✉]), Yulia Shichkina[2], and Olga Sedova[1]

[1] Saint-Petersburg University, 7/9 Universitetskaya nab.,
St. Petersburg 199034, Russia
o.yakushkin@spbu.ru
[2] Department of Computer Science and Engineering,
Saint Petersburg Electrotechnical University "LETI", St. Petersburg, Russia

Abstract. Cloud systems allow to run parallel applications using solutions with distributed heterogeneous architecture. Software development for heterogeneous distributed environment requires a module-based design. The components in such module system are connected by means of telecommunications network enabling message passing. This article describes an interaction model for components in distributed applications. The model was designed based on the paradigm of Variable Speed Hybrid Petri Nets and allows to analyse system performance at various tiers: selection of the optimum approach to load balancing between components; making scaling decisions to enhance performance of certain modules; fine-tuning the interaction between system components. The model is not contingent on particular tools a user might employ to implement a solution; it also provides a monitoring data integration functionality.

The model contains descriptions of standard messaging patterns linking components of distributed applications. These patterns include request-reply and publish-subscribe. Load balancing algorithms for various schemes of these patterns usage have been developed for a cloud environment.

Keywords: Cloud services · Messaging patterns · Systems architecture · Petri nets · Messaging middleware · Distributed applications

1 Introduction

Cloud platforms have become a flexible computing environment that can run a wide range of applications. They allow dynamic allocation and control of heterogeneous computing resources, which means computing can be provided as a service. This enables scaling of active applications within a common network of data centres located all around the world.

O. Iakushkin—This research was partially supported by SPbU (Saint Petersburg State University) grants 9.37.157.2014, 0.37.155.2014 and Russian Foundation for Basic Research grant (project no. 16-07-01113 and no. 16-07-01111).

O. Gervasi et al. (Eds.): ICCSA 2016, Part II, LNCS 9787, pp. 390–402, 2016.
DOI: 10.1007/978-3-319-42108-7_30

There are tasks that can be solved on an individual cloud platform. However, creation of a hybrid cloud computing environment is also relevant [4,5,9,10]. Such environment pools together the resources of private data processing centres and commercial suppliers of cloud services.

Cloud platforms can solve a host of tasks which are fundamentally different. At first approximation, we can single out the following three types: 1. handling many small tasks (e.g., mass calculation); 2. processing large data arrays; 3. calculation of a single big task, represented be a coherent system of control commands that changes its state after each new command is introduced. These types of tasks set out different requirements for load balancing and scaling. Remarkably, the first type is the only one that allows easy scaling even within a grid. On the whole, however, cloud environment is impossible to master without tools enabling interaction between components [12,15–19].

This paper is focused on components of distributed applications that run on different computing resources united in a single computing network. These components must function in a continuous and uninterrupted way; moreover, they require constant interaction and information exchange. As a rule, modular solutions allow usage of computing resources with adjustable capacity which can be altered while the task is being solved. Coordination and load balancing are a major challenge for communication systems that connect modules with each other. This paper offers a mathematical model allowing formal description of such systems. The validity of the model is not contingent on particular technological solutions.

2 Related Work

Petri nets are employed to describe and analyze the features of distributed communication systems.

The use of Generalised Stochastic Petri Nets (GSPN) for messaging middleware in broker-based architecture was discussed by Fernandes et al. based on the case of IBM Web Server solution [8]; the major attention was given to publish-subscribe and message queuing patterns. The Coloured Petri Net (CPN) was later used by Fahland and Gierds [7] to develop models allowing analysis of Enterprise Integration Patterns (EIP) described in [14]. Some of these models can classify as middleware for message passing: Pipes and Filter, Message Router, Message Translator, Message Endpoint, Recipient List, Aggregator, Request-Reply, Channel Adapter. However, load balancing and coordination of multiple sending/receiving nodes were not addressed; neither were the temporal aspects of message passing. We should note, that most of modern message-passing systems by default employ Round-Robin to perform load balancing. The technique utilizes the algorithm that partly correlates with the one presented in [21], where it is used to describe Ethernet packet switching. In this paper, we took into account the load balancing description model contained in [21]. The approach to formalization of middleware systems description was mentioned in [22]; the work offered a transition from the concept 'API plus informal prose' to the concept

'API plus formal description'. In [22], there was also provided an example of formal description for Common Object Request Broker Architecture (CORBA). However, this method described only the behaviour of the system, without regard to its mathematical dimension. Therefore, it was not supported by the expert community.

Wester-Ebbinghaus [2] used a web-service to give an example of a full-fledged programme whose code is based on a Petri nets model. The stability problem in two particular classes of queuing systems was analysed by Konigsberg [20]; the analysis involved timed Petri nets, Lyapunov methods, and max-plus algebra.

So, we demonstrated there is a wide range of research works using Petri nets to design a formalized approach to the architecture of message-passing software. The task is handled in terms of both the internal structure and an outside observer which analyzes the system's components. However, most of the models of message passing and load balancing limit themselves to one of the following options: they either examine broker-based architecture (i.e., the tasks are solved externally), or analyze separate client/server pairs. As a result, the failure of interacting components remains largely neglected. Moreover, there is one more issue that is overlooked: the functioning of systems which include many modules communicating with each other and concurrently running various message-passing patterns, scaling methods, and balancing strategies.

3   The Interaction Model for Components of Distributed Application Architecture in a Cloud Environment

This paper describes a model specifying components interaction in a distributed application architecture in a cloud environment. The model was built using message-passing patterns allowing to directly link computing nodes. The patterns used in the model are available in ZeroMQ[1] [13,23] and NamoMsg[2] systems: request-response, publish-subscribe and pipeline. These solutions are widely used by large companies to solve a broad variety of calculation tasks: from high-energy physics (e.g., by European Organization for Nuclear Research, CERN) to independent business applications [1,6]. Our model is mainly distinguished by the opportunity to involve an outside controller. The controller can allocate additional computing load, invoke or stop service system modules, and monitor interaction in the service network. Introduction of the controller allowed a major optimization of the model's logic.

The model accommodates patterns enabling forwarding of large messages split into sets of fragments. In essence, we are speaking of data streaming supported by load balancing functionality. We also examine load balancing for the cases requiring inversion of control (i.e., when a client needs to maintain connection to a particular server for an unspecified number of request/reply cycles).

The formal description of the systems in question is generally performed using Petri nets, whose apparatus provides a rich arsenal of means. To select the

---

[1] http://zeromq.org/.

[2] http://nanomsg.org/.

ones suited for our task, we analyzed typical features of distributed systems: the same channels are able to pass various types of messages; the time needed for message passing correlates with the capacity of communication channels between the nodes; possible data loss while delivering messages; temporal unavailability óf certain components; connection loss between modules according to Brewer's conjecture [11]. Petri nets allow to produce a valid formal model reflecting this particularities. A detailed mechanism of such description is available in Colored Extended Variable Speed Hybrid Petri Nets (VHPN).

We designed a model based on the principles of formalism and graphical representation outlined in [3].

Noteworthily, there is a crucial issue that should be resolved prior to the development of the model: the mark symbol for VSPN must be bound to any message forwarded within the system, unless explicitly stated otherwise. This binding is of paramount importance for the case under review.

3.1 The Model of the Architecture Component 'The Choice of a Message Receiver'

Let us we consider message sending from a client to one of the equivalent servers. The examined libraries employ the Round-Robin algorithm to balance the load between servers at the side of the client. The algorithm for automatic balancing using Petri nets is given in Fig. 1.

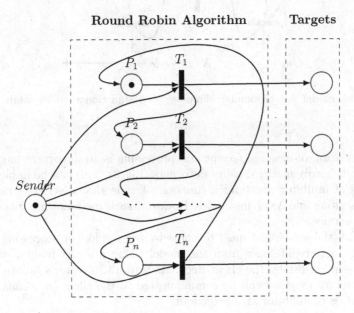

Fig. 1. The algorithm for automatic selection of a receiving node from an array of servers.

3.2   Models of the Architecture Component 'Controlling Message Passing When Nodes are Tuned Off or Overloaded'

Clouds offering IaaS allow to scale computing resources on running systems. However, the budgets of end users are not unlimited. Therefore, it is sound to make an estimate of a distributed system's capabilities that is, to calculate the maximum number of server computing nodes available to the end users; and to analyze the system operation in case of potential shutdown of nodes.

The design of a module incorporated into the client's load balancing system for message queuing is shown in Fig. 2. The design provides for a possible shut-down or failure of a remote server. Here the mark is responsible for the state of the remote node. If the mark is 'turned off', the Round-Robin algorithm will skip this node when sending a message.

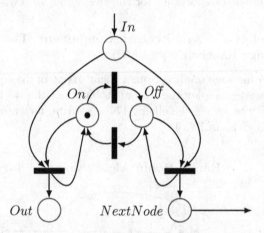

Fig. 2. The model for automatic skipping of dysfunctional nodes while sending messages.

Confirmation of message receipt and processing is an important mechanism which can be easily modeled using Petri nets (Fig. 3). This can be implemented by means of inhibitor arcs and a function of time that describes transitions between sending and receiving nodes. However, such mechanism is not easy to put into practice.

Here it is the server that must response to message loss or processing failure. This is the major difference from the model where a server node is excluded from the sending list by the client itself. As a result, a server's failure or slow response due to overload can be communicated to the client by means of only two strategies of inhibitor arcs behaviour:

– Temporal waiting for confirmation of message receipt by a server node;
– Outside control mechanism that monitors the client's and the server's nodes; it should be capable of initiating the message resend.

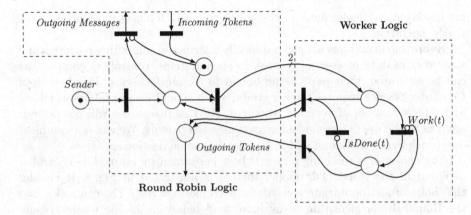

Fig. 3. The model for confirmation of message receipt and successful processing.

The former strategy can result in multiple resending of one and the same message to all server nodes available. This might be the case, if the server node or the network are overloaded and cannot promptly notify the client about receipt and processing. This situation can be critical for the whole system.

The strategy described is implemented in broker-based messaging systems. Let us examine how it works in RabbitMQ[3] broker-based solution [24] which uses the wide-spread Advanced Message Queuing Protocol, or AMPQ[4] [25]. The client and the server are connected through a broker which forwards the client's messages to the server. If we use the receipt confirmation strategy, the broker subsequently resends the client's query to each of subscriber servers. The broker utilizes the Round-Robin balancing algorithm to go over the servers until the confirmation is received within the timespan set by the client. In other words, if no confirmation is received in a due time, the message is sent to be processed by another potential node. The system developer can set the waiting time before the messages are sent, but this approach reduces the system's throughput by more than two orders of magnitude[5].

The latter strategy requires an external controller and cannot be used as an out-of-the-box solution within the message-passing systems under review. The core features of such controller are described below as a general model, flexible enough to suit a variety of concrete solutions.

3.3 The Model of External Controller in Message Passing

The peer-to-peer networks provide little practical opportunities for using an out-side controller. The reasons include their unlimited size and unlimited remoteness of computing nodes. Cloud platforms are more well-suited for the purpose: they

---

[3] http://www.rabbitmq.com/.

[4] http://www.amqp.org/.

[5] http://www.rabbitmq.com/confirms.html.

are hosted by particular data processing centres, which allows monitoring of the nodes' operation.

Hybrid cloud systems use geographically distributed computing resources and a network of data processing centres. In such systems, controlling components can be separated, their parts being bound to certain locations. This branch of Computer Science requires further study, which should involve laborious efforts and a large number of versatile resources. For this reason, we will restrict our analysis to the use of our model for mass message passing. We will not comment upon such issues as scaling and bidirectional communications analysis.

The choice of a receiving server is best performed by an outside controller deployed in the cloud. The model thereof is illustrated in Fig. 4. It enables the choice of an appropriate receiver for each message sent. The controller uses the Round-Robin algorithm to evaluate load balancing in the whole system. In the model of the architecture component 'choice of a message receiver' (see Picture 1), each client independently performs the choice of the receiving server. If the number of servers is limited, such solution may result in an uneven load and a system failure. The problem is solved, if we use a controller that monitors message passing from clients to server nodes.

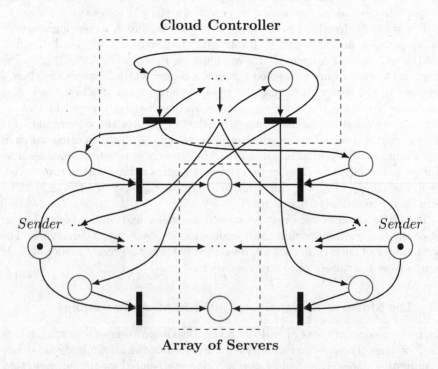

Fig. 4. The controller-based model of message passing from clients to a bunch of server nodes.

This solution has a drawback: the system relying on an outside controller and the Round-Robin load balancing algorithm will send each message at least three times as slow as compared to client-side balancing. The external controller node will need to collect data from each sending node in every case of message passing.

However, the outside controller opens up new load balancing opportunities. There are strategies allowing to optimize the number of messages required by the controller for proper operation, and its impact on the system in general:

- heartbeat tool used to communicate the statistics of the nodes' functioning;
- external (system-based) and internal utility programmes monitoring the nodes' work;
- loading one node to a full capacity before the load is distributed to the next node.

These external tools can be introduced to the load balancing system as an extension to the model in Fig. 4.

3.4   The Load Balancing Model for Data Streaming

The model in Fig. 5 contains a mark that is forwarded from the client to the server. The mark is a positive number, not necessarily an integer, that reflects the state of the data stream. The load balancing in this model relies on a counter of the streams transmitted to each server. The data are streamed to the least loaded server.

The streams balancing system can be also included into the cloud controller model displayed in Fig. 4. Moreover, the final model can accommodate all the extensions mentioned above.

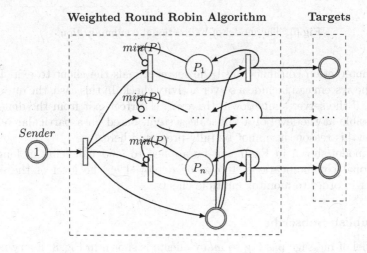

Fig. 5. The load balancing model for data streaming.

3.5   Bidirectional Communication Model

A message exchange dialogue is an important tool of nodes interaction. For example, the control over a client's movements is given to a navigator that is, a particular instance of an external server. Until the client arrives to the specified point, it notifies the server about every command executed. Such long-term bilateral communication does not comply with the Round-Robin algorithm going from one server to another. In this case, the initial task is known to the first server only. Therefore, it is this server that should track the changes in the object it interacts with.

The dialogue model between the client and the server is provided in Fig. 6. Two colors are used:

- *EndT* the color indicating a message ending the dialogue;
- *T* the color indicating any other message.

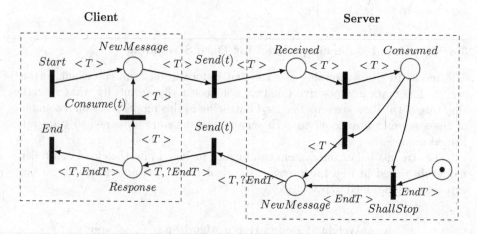

Fig. 6. The model of bidirectional communication.

The interaction continues until the server signals the client to exit. Importantly, the system can include a server activity check. In this case, the interaction will end, if the server shuts down. However, a correct exit from the dialogue is only possible at the level of a full-scale system, based on a particular business logic. For this reason, it cannot be fully presented here.

The model shown in Fig. 7 can be applied to load balancing in long-term bidirectional communications. It can be deployed at the level of the outside controller in order to monitor multiple clients

3.6   Publish-Subscribe

The model of message passing to many clients is shown in Fig. 8. Every node in the system has two options: it is either a subscriber, or does not receive messages.

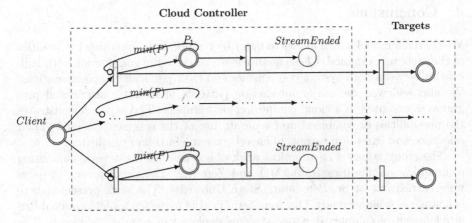

Fig. 7. The load balancing model for long-term dialogues.

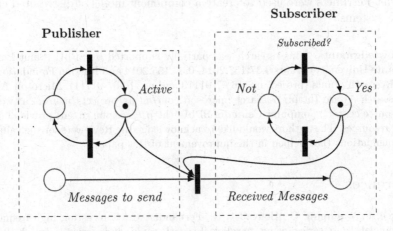

Fig. 8. The model of Publish-Subscribe pattern.

For hybrid cloud systems, it is important to allow message passing to network segments which are very remote from the publisher or require pre-processing of messages.

The model including a broker which forwards messages beyond the cloud system can be easily presented using a publisher with a subscriber which is retransmitting received messages to peers local to its cloud.

3.7 Remarks

The system's work can be modeled using the function of time and the function of message length, provided we know data transmission speed. The system can be described using several layers corresponding to data transmission channels distinguished by name or by message type.

4   Conclusions

We offered a model which allows to describe temporal and quantitative qualities of the systems examined. These qualities are described in conjunction with logical ties between both system components and their particular implementations. We also reviewed the major interaction patterns of modular distributed programmes executed in a cloud computing environment. The paper demonstrates the possibilities of combined and separate use of the schemes analysed. Their relevance and extension prospects have been explained and justified.

The study is based on practical work of service systems development using communication libraries RabbitMQ and ZeroMQ. The development projects were carried out at St. Petersburg State University. The work presented here is a stage in a larger study. Our long-term goal is to design a platform enabling development of complex dynamic systems deployed in a variety of clouds. The platform will employ both standard and user- defined messaging and interaction patterns. Petri nets were used to create a component model of distributed cloud service systems.

Acknowledgments. This research was partially supported by SPbU (Saint Petersburg State University) grants 9.37.157.2014, 0.37.155.2014 and Russian Foundation for Basic Research grants (project no. 16-07-01113 and no. 16-07-01111). Microsoft Azure for Research Award (http://research.microsoft.com/en-us/projects/azure/) as well as the resource center "Computer Center of SPbU" (http://cc.spbu.ru/en) provided computing resources. The authors would like to acknowledge the Reviewers for the valuable recommendations that helped in the improvement of this paper.

References

1. Arce, P., Maureira, C., Bonvallet, R., Fernandez, C.: Forecasting high frequency financial time series using parallel ffn with cuda and zeromq. In: 2012 9th Asia-Pacific Symposium on Information and Telecommunication Technologies (APSITT), pp. 1–5, November 2012
2. Betz, T., Cabac, L., Duvigneau, M., Wagner, T., Wester-Ebbinghaus, M.: Software engineering with petri nets: a web service and agent perspective. In: Haddad, S., Yakovlev, A. (eds.) ToPNoC IX. LNCS, vol. 8910, pp. 41–61. Springer, Heidelberg (2014)
3. David, R., Alla, H.: Discrete, Continuous, and Hybrid Petri Nets. Springer, Heidelberg (2010)
4. Degtyarev, A., Gankevich, I.: Balancing load on a multiprocessor system with event-driven approach. In: Gavrilova, M.L., Tan, C.J.K. (eds.) Trans. on Comput. Sci. XXVII. LNCS, vol. 9570, pp. 35–52. springer, Heidelberg (2016). doi:10.1007/978-3-662-50412-3_3
5. Degtyarev, A.B., Logvinenko, Y.V.: Agent system service for supporting river boats navigation. Procedia Comput. Sci. **1**(1), 2717–2722 (2010). ICCS 2010
6. Dworak, A., Charrue, P., Ehm, F., Sliwinski, W., Sobczak, M.: Middleware trends and market leaders 2011. In: Conference Proceedings C111010 (CERN-ATS-2011-196), FRBHMULT05, 4 p., October 2011

7. Fahland, D., Gierds, C.: Analyzing and completing middleware designs for enterprise integration using coloured petri nets. In: Salinesi, C., Norrie, M.C., Pastor, Ó. (eds.) CAiSE 2013. LNCS, vol. 7908, pp. 400–416. Springer, Heidelberg (2013)
8. Fernandes, S., Silva, W., Silva, M., Rosa, N., Maciel, P., Sadok, D.: On the generalised stochastic petri net modeling of message-oriented middleware systems. In: 2004 IEEE International Conference on Performance, Computing, and Communications, pp. 783–788 (2004)
9. Gankevich, I., Gaiduchok, V., Gushchanskiy, D., Tipikin, Y., Korkhov, V., Degtyarev, A., Bogdanov, A., Zolotarev, V.: Virtual private supercomputer: design and evaluation. In: Computer Science and Information Technologies (CSIT), 2013, pp. 1–6 (2013)
10. Gankevich, I., Korkhov, V., Balyan, S., Gaiduchok, V., Gushchanskiy, D., Tipikin, Y., Degtyarev, A., Bogdanov, A.: Constructing virtual private supercomputer using virtualization and cloud technologies. In: Murgante, B., et al. (eds.) ICCSA 2014, Part VI. LNCS, vol. 8584, pp. 341–354. Springer, Heidelberg (2014)
11. Gilbert, S., Lynch, N.: Brewer's conjecture and the feasibility of consistent, available, partition-tolerant web services. SIGACT News 33(2), 51–59 (2002)
12. Grishkin, V., Iakushkin, O.: Middleware transport architecture monitoring: topology service. In: 2014 20th International Workshop on Beam Dynamics and Optimization (BDO), pp. 1–2, June 2014
13. Hintjens, P.: ZeroMQ: Messaging for Many Applications. O'Reilly, Sebastopol (2013)
14. Hohpe, G., Woolf, B.: Enterprise integration patterns. In: 9th Conference on Pattern Language of Programs, pp. 1–9 (2002)
15. Iakushkin, O.: Intellectual scaling in a distributed cloud application architecture: a message classification algorithm, pp. 634–637 (2015) (cited By 0)
16. Iakushkin, O.: Cloud middleware combining the functionalities of message passing and scaling control, vol. 108 (2016) (cited By 0)
17. Iakushkin, O., Grishkin, V.: Messaging middleware for cloud applications: Extending brokerless approach. In: 2014 2nd International Conference on Emission Electronics (ICEE), pp. 1–4, June 2014
18. Iakushkin, O., Grishkin, V.: Unification of control in P2P communication middleware: towards complex messaging patterns. In: Simos, T.E., Tsitouras, C. (eds.) Proceedings of the International Conference of Numerical Analysis and Applied Mathematics 2014 (ICNAAM-2014), AIP Conference Proceedings, vol. 1648. Amer Inst Physics, Melville (2015)
19. Iakushkin, O., Sedova, O., Valery, G.: Application control and horizontal scaling in modern cloud middleware. In: Gavrilova, M.L., Tan, C.J.K. (eds.) Transactions on Computational Science XXVII. LNCS, vol. 9570, pp. 81–96. Springer, Heidelberg (2016). doi:10.1007/978-3-662-50412-3_6
20. Konigsberg, Z.: Timed petri nets modeling and lyapunov/max-plus algebra stability analysis for a type of queuing systems. Int. J. Pure Appl. Math. 86(2), 301–323 (2013)
21. Pedroso, C.M., Fonseca, K.: Modeling weight round robin packet scheduler with petri nets. In: International Conference on Communication Systems, vol. 1, pp. 342–345. IEEE (2002)
22. Rosa, N.S., Cunha, P.R.F.: Behavioural specification of middleware systems. J. Braz. Comput. Soc. 12, 63–74 (2006)

23. Sstrik, M.: The Architecture of Open Source Applications, vol. 2. CreativeCommons, Mountain View (2012)
24. Videla, A., Williams, J.J.: RabbitMQ in Action. Manning, Shelter Island (2012)
25. Vinoski, S.: Advanced message queuing protocol. IEEE Internet Comput. **10**(6), 87–89 (2006)

Geometric Modeling, Graphics and Visualization

A Comparative Study of LOWESS and RBF Approximations for Visualization

Michal Smolik[✉], Vaclav Skala, and Ondrej Nedved

Faculty of Applied Sciences, University of West Bohemia,
Univerzitni 8, 30614 Pilsen, Czech Republic
smolik@kiv.zcu.cz

Abstract. Approximation methods are widely used in many fields and many techniques have been published already. This comparative study presents a comparison of LOWESS (Locally weighted scatterplot smoothing) and RBF (Radial Basis Functions) approximation methods on noisy data as they use different approaches. The RBF approach is generally convenient for high dimensional scattered data sets. The LOWESS method needs finding a subset of nearest points if data are scattered. The experiments proved that LOWESS approximation gives slightly better results than RBF in the case of lower dimension, while in the higher dimensional case with scattered data the RBF method has lower computational complexity.

Keywords: Radial basis functions · LOWESS · Approximation

Notation used

D: dimension
K: k-nearest points
M: number of radial basis functions for approximation
N: number of all input points
R: number of points at which the approximation is calculated
ξ: point where to calculate the approximation
d: degree of polynomial
r: $r = d + 2$
q: $q = d + 1$

1 Introduction

Interpolation and approximation techniques are often used in data processing. Approximation methods of values y_i in the given $\{\langle x_i, y_i \rangle\}_1^N$ data set lead to a smooth function which minimizes the difference between given data and the determined function [13]. It can be used for visualization of noisy data [1, 2], visualization of the basic shape of measured/calculated data [9], for prediction, and other purposes. Many methods have been described together with their properties. This paper describes LOWESS (Locally weighted scatterplot smoothing) and RBF (Radial basis functions) methods and their experimental comparison.

© Springer International Publishing Switzerland 2016
O. Gervasi et al. (Eds.): ICCSA 2016, Part II, LNCS 9787, pp. 405–419, 2016.
DOI: 10.1007/978-3-319-42108-7_31

2 Lowess

The locally weighted scatterplot smoothing method (LOWESS) [3] is often used, especially in statistical applications. The value of an approximated function at a point x_0 is calculated from the formula of a curve which minimizes a sum S in the k-nearest neighborhood (KNN) points of the given point ξ.

$$S = \sum_{i=1}^{K} \omega_i \cdot \left(y_i - P_{(d)}(x_i)\right)^2, \tag{1}$$

where $P_{(d)}(x) = a_0 + a_1 x + a_2 x^2 + \ldots + a_d x^d$ is a d degree of a polynomial function with unknown coefficients $a = [a_0, a_1, a_2, \ldots, a_d]^T$. We can rewrite the sum S in a matrix form as:

$$S = (b - Aa)^T \cdot W \cdot (b - Aa), \tag{2}$$

where $b = [y_1, y_2, \ldots, y_K]^T$ is a vector of function values, matrix A is equal to:

$$A = \begin{bmatrix} 1 & x_1 & \cdots & x_1^d \\ 1 & x_2 & \cdots & x_2^d \\ \vdots & & \ddots & \vdots \\ 1 & x_K & \cdots & x_K^d \end{bmatrix} \tag{3}$$

and matrix W is a diagonal matrix:

$$W = \begin{bmatrix} \omega(\|x_1 - \xi\|) & & & 0 \\ & \omega(\|x_2 - \xi\|) & & \\ & & \ddots & \\ 0 & & & \omega(\|x_K - \xi\|) \end{bmatrix} = \begin{bmatrix} \omega_1 & & & 0 \\ & \omega_2 & & \\ & & \ddots & \\ 0 & & & \omega_K \end{bmatrix}, \tag{4}$$

where $\omega(r)$ are weighting functions, which have to satisfy the following conditions defined as:

$$\forall a, b \in [0; 1], a < b : \omega(a) \geq \omega(b) \wedge \omega(0) = 1 \wedge \forall c \geq 1 : \omega(c) = 0. \tag{5}$$

One such example of a weighting function ω can be the tricube function:

$$\omega(r = \|x_i - \xi\|) = \omega_i = \begin{cases} (1 - r^3)^3 & r \in \langle 0; 1\rangle \\ 0 & r > 1 \end{cases}. \tag{6}$$

Equation (2) can be modified as:

$$\begin{aligned} S &= b^T W b - b^T W A a - (Aa)^T W b + (Aa)^T W A a \\ &= b^T W b - b^T W A a - a^T A^T W b + a^T A^T W A a. \end{aligned} \tag{7}$$

The sum S is minimal if the partial derivative of S with respect to a is equal to zero:

$$\frac{\partial S}{\partial a} = -\left(b^T W A\right)^T - A^T W b + 2A^T W A a = 0 \tag{8}$$

as $W = W^T$ and therefore:

$$A^T W A a = A^T W b$$
$$a = \left(A^T W A\right)^{-1} A^T W b. \tag{9}$$

The numerical stability of calculations is influenced by the position of the interval of the k-nearest neighborhood points of the point ξ. The LOWESS approximation is "locally" based, as only k-nearest points are used and thus r is actually computed as $r = \|x_i - \xi\|$. To solve problems with the numerical stability of calculations and independence of absolute position, we have to use relative position of all the k-nearest neighborhood points of the point ξ such that the matrix A from (3) is defined as:

$$A = \begin{bmatrix} 1 & (x_1 - \xi) & \cdots & (x_1 - \xi)^d \\ 1 & (x_2 - \xi) & & (x_2 - \xi)^d \\ \vdots & & \ddots & \vdots \\ 1 & (x_K - \xi) & \cdots & (x_K - \xi)^d \end{bmatrix} \tag{10}$$

2.1 LOWESS with Linear Regression

Linear regression, i.e. choosing $d = 1$, appears to strike a good balance between computational simplicity and the flexibility needed to reproduce patterns in the data. In such a case, we can rewrite (9) as:

$$a = \begin{bmatrix} \sum\limits_{i=1}^{K} \omega_i & \sum\limits_{i=1}^{K} \omega_i x_i \\ \sum\limits_{i=1}^{K} \omega_i x_i & \sum\limits_{i=1}^{K} \omega_i x_i^2 \end{bmatrix}^{-1} \cdot \begin{bmatrix} \sum\limits_{i=1}^{K} \omega_i y_i \\ \sum\limits_{i=1}^{K} \omega_i x_i y_i \end{bmatrix} \tag{11}$$

and after some adjustments we can get a final formula for unknown coefficients a:

$$\begin{bmatrix} a_0 \\ a_1 \end{bmatrix} = \frac{1}{\left(\sum_{i=1}^{K} \omega_i\right) \cdot \left(\sum_{i=1}^{K} \omega_i x_i^2\right) - \left(\sum_{i=1}^{K} \omega_i x_i\right)^2}$$
$$\begin{bmatrix} \left(\sum\limits_{i=1}^{K} \omega_i y_i\right)\left(\sum\limits_{i=1}^{K} \omega_i x_i^2\right) - \left(\sum\limits_{i=1}^{K} \omega_i x_i\right)\left(\sum\limits_{i=1}^{K} \omega_i x_i y_i\right) \\ -\left(\sum\limits_{i=1}^{K} \omega_i y_i\right)\left(\sum\limits_{i=1}^{K} \omega_i x_i\right) + \left(\sum\limits_{i=1}^{K} \omega_i\right)\left(\sum\limits_{i=1}^{K} \omega_i x_i y_i\right) \end{bmatrix} \tag{12}$$

2.2 LOWESS with Constant Regression

Constant regression, i.e. choosing $d = 0$, is the most computationally simple, but from a practical point of view, an assumption of local linearity seems to serve far better than an assumption of local constancy because the tendency is to plot variables that are related to one another. Thus, the linear LOWESS regression produces better results than the constant LOWESS regression, which is very simple. In this case, we can rewrite it from (9) as:

$$a_0 = \frac{\sum_{i=1}^{K} \omega_i y_i}{\sum_{i=1}^{K} \omega_i}. \tag{13}$$

Comparing formulas from (13) and (12), it can be seen that LOWESS with constant regression is computationally much easier than LOWESS with linear regression.

3 Radial Basis Functions

Radial basis functions (RBF) [4, 10–12] is based on distances, generally in D-dimensional space. The value of an approximated function at a point x is calculated from the formula:

$$f(x) = \sum_{i=1}^{M} \lambda_i \Phi(\|x - \xi_i\|) + P_d(x), \tag{14}$$

where $P_{(d)}(x) = a_0 + a_1 x + a_2 x^2 + \ldots + a_d x^d$ is a d degree polynomial function with unknown coefficients $a = [a_0, a_1, a_2, \ldots, a_d]^T$, M is the number of radial basis functions, and $\lambda = [\lambda_1, \ldots, \lambda_M]$ are weights of radial basis functions $\Phi(\|x - \xi_i\|)$. The function Φ is a real-valued function whose value depends only on the distance from some other point ξ_i, called a center, so that:

$$\Phi_i(x) = \Phi(\|x - \xi_i\|). \tag{15}$$

As the values $f(x_i)$ at a point x_i are known, Eq. (14) represents a system of linear equations that has to be solved in order to determine coefficients λ and a, i.e.

$$f(x_j) = \sum_{i=1}^{M} \lambda_i \Phi(\|x_j - \xi_i\|) + P_d(x_j) \text{ for } \forall j \in \{1, \ldots, N\}. \tag{16}$$

Using matrix notation we can rewrite (16) as:

$$\begin{bmatrix} \Phi(\|x_1 - \xi_1\|) & \cdots & \Phi(\|x_1 - \xi_M\|) & 1 & x_1 & \cdots & x_1^d \\ \vdots & & \vdots & \vdots & \vdots & & \vdots \\ \Phi(\|x_N - \xi_1\|) & \cdots & \Phi(\|x_N - \xi_M\|) & 1 & x_N & \cdots & x_N^d \end{bmatrix} \cdot \begin{bmatrix} \lambda_1 \\ \vdots \\ \lambda_M \\ a_0 \\ \vdots \\ a_d \end{bmatrix} = \begin{bmatrix} f(x_1) \\ \vdots \\ f(x_N) \end{bmatrix}. \tag{17}$$

We can create a "simple" RBF formula, see (18), using (17) with only one radial basis function, i.e. $M = 1$. This formula can be used in the same manner as the LOWESS method for calculating approximated value at the point ξ, using only the k-nearest neighborhood points of the point ξ, which is the center of radial basis function $\phi(\|x - \xi\|)$, too.

$$\begin{bmatrix} \Phi(\|x_1 - \xi\|) & 1 & x_1 & \cdots & x_1^d \\ \vdots & & \vdots & \vdots & \vdots \\ \Phi(\|x_K - \xi\|) & 1 & x_K & \cdots & x_K^d \end{bmatrix} \cdot \begin{bmatrix} \lambda_1 \\ a_0 \\ \vdots \\ a_d \end{bmatrix} = \begin{bmatrix} f(x_1) \\ \vdots \\ f(x_K) \end{bmatrix} \rightarrow A \cdot \lambda = f \cdot \qquad (18)$$

The coefficients $\eta = [\lambda_1, a^T]^T$ in overdetermined system of linear Eq. (18) are computed by the least squares error method:

$$\eta = (A^T A)^{-1} \cdot (A^T f). \qquad (19)$$

As the numerical stability of calculations is influenced by the position of the interval of the k-nearest neighborhood points of the point ξ and the RBF approximation is "locally" based, only k-nearest points are used. To solve problems with the numerical stability of calculations, we have to move all the k-nearest neighborhood points of the point ξ such that the matrix A from (18) is defined as:

$$A = \begin{bmatrix} \Phi(\|x_1 - \xi\|) & 1 & (x_1 - \xi) & \cdots & (x_1 - \xi)^d \\ \vdots & \vdots & \vdots & & \vdots \\ \Phi(\|x_K - \xi\|) & 1 & (x_K - \xi) & \cdots & (x_K - \xi)^d \end{bmatrix} \qquad (20)$$

and $f(x)$ is defined as:

$$f(x) = \lambda_1 \Phi(\|x - \xi\|) + P_d(x - \xi). \qquad (21)$$

For locally-based approximation, any compactly supported radial basis function (CSRBF) [8, 12] can be used. CSRBF is a function defined on $r \in \langle 0; 1 \rangle$, is equal to 0 for all $r > 1$, and has to satisfy the conditions in (5). In the tests presented here, the $\Phi(r)$ function was selected as:

$$\Phi(r) = \begin{cases} (1 - r^3)^3 & r \in \langle 0; 1 \rangle \\ 0 & r > 1 \end{cases}, \qquad (22)$$

which is exactly the same function as weighting function (6) for LOWESS approximation.

3.1 Simplified RBF with a Constant Polynomial

Choosing $d = 0$, we will get a polynomial of zero degree which is only a constant, i.e.;
$P_d = a_0$.

$$
\begin{bmatrix}
\Phi(\|x_1 - \xi\|) & 1 \\
\vdots & \vdots \\
\Phi(\|x_K - \xi\|) & 1
\end{bmatrix}
\cdot
\begin{bmatrix}
\lambda_1 \\
a_0
\end{bmatrix}
=
\begin{bmatrix}
f(x_1) \\
\vdots \\
f(x_K)
\end{bmatrix}
\rightarrow A \cdot \eta = f \cdot
\tag{23}
$$

It leads to overdetermined system of linear equations. Using the method of least squares, we can calculate η:

$$
\eta = \left(A^T A\right)^{-1} \cdot \left(A^T f\right),
\tag{24}
$$

where $\eta = [\lambda_1, a_0]^T$.

$$
\begin{bmatrix}
\lambda_1 \\
a_0
\end{bmatrix}
=
\begin{bmatrix}
\sum_{i=1}^{K}(\Phi(\|x_i - \xi\|))^2 & \sum_{i=1}^{K}\Phi(\|x_i - \xi\|) \\
\sum_{i=1}^{K}\Phi(\|x_i - \xi\|) & \sum_{i=1}^{K} 1
\end{bmatrix}^{-1}
\cdot
\begin{bmatrix}
\sum_{i=1}^{K}\Phi(\|x_i - \xi\|) \cdot f(x_i) \\
\sum_{i=1}^{K} f(x_i)
\end{bmatrix},
\tag{25}
$$

where $\sum_{i=1}^{K} 1 = K$ and after adjustments:

$$
\begin{bmatrix}
\lambda_1 \\
a_0
\end{bmatrix}
=
\frac{1}{\left(\sum_{i=1}^{K}(\Phi(\|x_i - \xi\|))^2\right) \cdot K - \left(\sum_{i=1}^{K}\Phi(\|x_i - \xi\|)\right)^2}
$$
$$
\cdot
\begin{bmatrix}
K & -\sum_{i=1}^{K}\Phi(\|x_i - \xi\|) \\
-\sum_{i=1}^{K}\Phi(\|x_i - \xi\|) & \sum_{i=1}^{K}(\Phi(\|x_i - \xi\|))^2
\end{bmatrix}
\begin{bmatrix}
\sum_{i=1}^{K}\Phi(\|x_i - \xi\|) \cdot f(x_i) \\
\sum_{i=1}^{K} f(x_i)
\end{bmatrix}.
\tag{26}
$$

The value $f(\xi)$ is calculated as:

$$
f(\xi) = \lambda_1 \Phi(\|\xi - \xi\|) + a_1 = \lambda_1 \Phi(0) + a_0.
\tag{27}
$$

3.2 Simplified RBF Without a Polynomial

In the case of using simplified RBF without polynomial P_d, we get the following equation:

$$
\begin{bmatrix} \Phi(\|x_1 - \xi\|) \\ \vdots \\ \Phi(\|x_K - \xi\|) \end{bmatrix} \cdot [\lambda_1] = \begin{bmatrix} f(x_1) \\ \vdots \\ f(x_K) \end{bmatrix} \rightarrow A \cdot \lambda_1 = f, \tag{28}
$$

where A and f are column vectors. Using the method of least squares, we can calculate λ_1:

$$
\lambda_1 = \frac{A^T \cdot f}{A^T A}. \tag{29}
$$

Equation (29) can be rewritten as:

$$
\lambda_1 = \frac{\sum_{i=1}^{K} \Phi(\|x_i - \xi\|) \cdot f(x_i)}{\sum_{i=1}^{K} (\Phi(\|x_i - \xi\|))^2}. \tag{30}
$$

The value $f(\xi)$ is calculated as:

$$
f(\xi) = \lambda_1 \Phi(\|\xi - \xi\|) = \lambda_1 \Phi(0). \tag{31}
$$

4 Comparison of Time Complexity

In the following, a comparison of LOWESS and RBF will be made. The main criteria for comparison are:

- The computational complexity, which is critical if many points have to be approximated.
- The quality of the final approximation (see Sect. 5).

4.1 LOWESS

The size of matrix A is $k \times q$, where the number of used nearest points is k and q is equal to the degree of the polynomial plus 1. The size of diagonal matrix W is $k \times k$, the size of vector b is $k \times 1$ and the size of vector $\mathbf{x}$ is $k \times 1$. The time complexity of LOWESS using Eq. (9) can be calculated in the following way:

$$
\begin{aligned}
A^T W A &\rightarrow O(q^2 k + qk) \\
(A^T W A)^{-1} &\rightarrow O(q^2 k + qk + q^3) \\
A^T W b &\rightarrow O(2qk) \\
(A^T W A)^{-1} A^T W b &\rightarrow O(k(q^2 + 3q) + q^3 + q^2)
\end{aligned} \tag{32}
$$

As the size k of matrix A is much larger than the size q of matrix A, the time complexity from (32) will become:

$$\begin{array}{ll} O(3qk) & for \quad q = \{1,2\} \\ O(q^2 k) & for \quad q \geq 3 \end{array} \tag{33}$$

The time complexity of LOWESS when calculating the approximation value in R points will become:

$$\begin{array}{ll} O(N \log N + R \cdot 3qk) & for \quad q = \{1,2\} \\ O(N \log N + R \cdot q^2 k) & for \quad q \geq 3 \end{array} \tag{34}$$

where N is the number of input points and $O(N \log N)$ is the time complexity of the sorting algorithm for 1&1/2 dimensional data. In the case of higher dimensions D&1/2, i.e. $D > 1$, the total time complexity of selecting K-nearest points from N points increases (see Sect. 7 for more details).

4.2   Simplified RBF

The size of matrix A is $k \times r$, where the number of used nearest points is k and r is equal to the degree of the polynomial plus 2. The size of vector f is $k \times 1$ and the size of vector $\eta = \left[\lambda^T, a^T\right]^T$ is $k \times 1$. The time complexity of RBF using Eq. (24) can be calculated in the following way:

$$\begin{array}{ccc} A^T A & \rightarrow & O(r^2 k) \\ (A^T A)^{-1} & \rightarrow & O(r^2 k + r^3) \\ A^T f & \rightarrow & O(rk) \\ (A^T A)^{-1}(A^T f) & \rightarrow & O(k(r^2 + r) + r^3 + r^2) \end{array} \tag{35}$$

As the size k of matrix A is much larger than the size r of matrix A, the time complexity from (35) will become:

$$O(r^2 k) \tag{36}$$

The time complexity of simplified RBF when calculating the approximation value in R points can be estimated:

$$O(N \log N + R \cdot r^2 k) \tag{37}$$

where N is the number of input points and $O(N \log N)$ is the time complexity of the sorting algorithm for 1&1/2 dimensional data.

5   Comparison of Measured Errors

For a demonstration of LOWESS and RBF approximation properties, the standard testing function, which is considered by Hickernell and Hon [5], has been selected:

$$\tau(x) = e^{\left[-15\left(\left(x-\frac{1}{2}\right)^2\right)\right]} + \frac{1}{2}e^{\left[-20\left(\left(x-\frac{1}{2}\right)^2\right)\right]} - \frac{3}{4}e^{\left[-8\left(\left(x+\frac{1}{2}\right)^2\right)\right]} \tag{38}$$

This function was sampled at $\langle -1, 1 \rangle$. We added random noise with uniform distribution from interval $\langle -0.1, 0.1 \rangle$ and used it as input for both methods. The following graphs present the behavior of the LOWESS and RBF approximations.

Function values are on the vertical axis; values on the horizontal axis are sample indices. Since the function is sampled at $\langle -1, 1 \rangle$ and the number of samples is 2000, the sampling rate is 1000 samples per 1 unit.

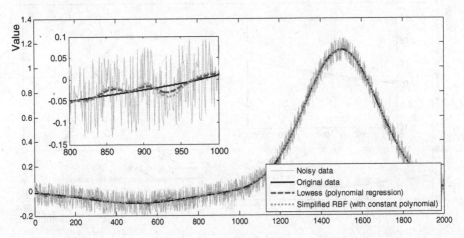

Fig. 1. Comparison of LOWESS with Simplified RBF. 100 nearest samples out of 2000 total were used as values for approximation; sampled interval: $\langle -1, 1 \rangle$ (Color figure online)

The error of approximation can be measured in different ways. The first one is to measure the change of the first derivative, which is the curvature of the resulting curve. If the first derivative changes too much, then the curve is jagged; on the contrary, if the first derivative does not change too much, then the curve is smooth. The absolute error can be calculated using the formula:

$$E_c = \sum_{i=1}^{N} \|f''(x_i)\|, \tag{39}$$

where $f''(x_i)$ is calculated using the formula:

$$f''(x_i) = \frac{f(x_{i+1} - 2x_i + x_{i-1})}{(x_{i+1} - x_i)(x_i - x_{i-1})}, \tag{40}$$

Let $\boldsymbol{p} = [x, f(x)]$ be the approximated point in current space (2D for 1&1/2 dimensions) and $\boldsymbol{\kappa} = [x, \tau(x)]$ be a point of the sampled function (38), which is a set

$K = \{\kappa_1, \ldots, \kappa_N\} = \{[x_1, \tau(x_1)], \ldots, [x_N, \tau(x_N)]\}$, then the distance error from the original curve without noise can be calculated as:

$$E_d = \sum_{i=1}^{N} \|\boldsymbol{p}_i - \boldsymbol{\xi}_j\|, \text{ where } \|\boldsymbol{p}_i - \boldsymbol{\kappa}_j\| \text{ is minimal } \forall j \in \{1, \ldots, N\} \text{ for given } i. \quad (41)$$

Let us note that the distance is not measured vertically to the curve but "orthogonally" to the curve. Using Formulas (39) and (41), we can show the following table of calculated errors.

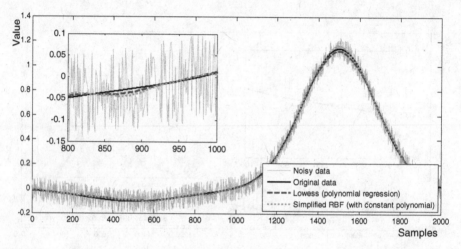

Fig. 2. Comparison of LOWESS with Simplified RBF. 200 nearest samples out of 2000 total were used as values for approximation; sampled interval: $\langle -1, 1 \rangle$ (Color figure online)

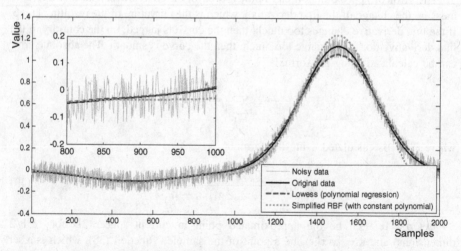

Fig. 3. Comparison of LOWESS with Simplified RBF. 500 nearest samples out of 2000 total were used as values for approximation; sampled interval: $\langle -1, 1 \rangle$. (Color figure online)

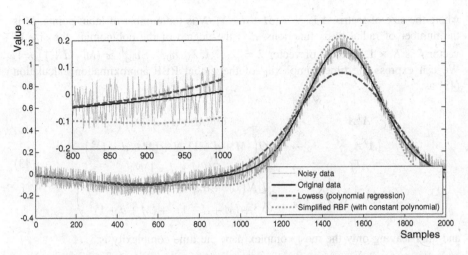

Fig. 4. Comparison of LOWESS with Simplified RBF. 1000 nearest samples out of 2000 total were used as values for approximation; sampled interval: $\langle -1, 1 \rangle$. (Color figure online)

Table 1. Measured errors for Figs. 1, 2, 3 and 4 (*for N = 2000*)

k-nearest samples	E_c		E_d	
	LOWESS	Simplified RBF	LOWESS	Simplified RBF
100	0.0721	1.4585	7.2997	12.5647
200	0.0212	0.7689	10.5378	14.7898
500	0.0132	0.3103	15.6759	40.5985
1000	0.0091	0.1618	45.0717	70.8979

Some comparison results can be seen using (Table 1). The LOWESS approximation is always smoother (according to measured error E_c) and closer to the original data without noise (according to measured error E_d) when using the same k-nearest samples.

6 Global RBF Approximation

Global RBF approximation can be calculated using (14). In this case, the whole data set has to be processed at once. Compared to the simplified version of RBF approximation, we only get one λ vector for all input samples and thus we solve a linear system only once. Moreover, we do not need to sort the input points in any way, unlike LOWESS and simplified RBF approximations, which were presented in previous sections. The global RBF approximation is calculated using the following formula (from (17)):

$$A\lambda = f \rightarrow \lambda = \left(A^T A\right)^{-1} \cdot \left(A^T f\right) \tag{42}$$

where the size of matrix A is $N \times (M + d + 1)$, N is the number of input points, M is the number of radial basis functions, d is the degree of the polynomial, the size of vector f is $N \times 1$, the size of vector $\lambda = [\lambda_1, \ldots, \lambda_M, a_0, \ldots, a_d]^T$ is $(M + d + 1) \times 1$. We can express the time complexity of the global RBF approximation calculation (42) as:

$$
\begin{aligned}
A^T A &\rightarrow & O\left((M + d + 1)^2 N\right) \\
(A^T A)^{-1} &\rightarrow & O\left((M + d + 1)^2 N + (M + d + 1)^3\right) \\
A^T f &\rightarrow & O((M + d + 1)N) \\
(A^T A)^{-1} \cdot (A^T f) &\rightarrow & O\left(\begin{array}{c} N\left((M + d + 1)^2 + (M + d + 1)\right) \\ + (M + d + 1)^3 + (M + d + 1)^2 \end{array}\right)
\end{aligned}
\tag{43}
$$

and after leaving only the most complex part, the time complexity is:

$$
O(M^2 N)
\tag{44}
$$

We sampled function (38), added random noise with uniform distribution from interval $\langle -0.1, 0.1 \rangle$, and used that data as input for both methods mentioned in previous sections (LOWESS and simplified RBF) and for global RBF approximation as well. The following graph presents the behavior of the LOWESS, simplified RBF and global RBF approximations. Function values are on the vertical axis, values on the horizontal axis are sample indices. Since the function is sampled at $\langle -1, 1 \rangle$ and the number of samples is 2000, the sampling rate is 1000 samples per 1 unit.

The following table presents calculated errors using Formulas (39) and (41) (for Fig. 5) for all approximation methods described in this paper.

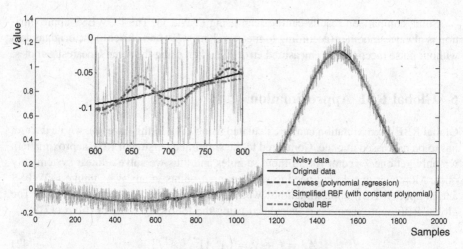

Fig. 5. Comparison of LOWESS and Simplified RBF with global RBF. 100 nearest samples out of 2000 total were used as values for local approximation, which gives 20 pivots (lambdas) for global RBF; sampled interval: $\langle -1, 1 \rangle$. (Color figure online)

Table 2. Measured errors for Fig. 5.

E_c			E_d		
LOWESS	Simplified RBF	Global RBF	LOWESS	Simplified RBF	Global RBF
0.0718	1.5266	0.0168	10.6785	16.0734	6.0123

It can be seen, that global RBF approximation is closer to the original data and smoother than simple RBF or even LOWESS approximation. For the situation in Fig. 5 and Table 2, the time complexity of global RBF is exactly the same as the time complexity of both other methods when calculating the approximation at all input points.

7 Approximation in Higher Dimensions

Let as assume that a scattered data approximation [6, 7] in 2&1/2 or 3&1/2 dimensions, i.e. D&1/2 dimensions have to be made. In the following, we describe the expansion of LOWESS and RBF approximation algorithms into higher dimensions.

In higher than 1&1/2 dimensions, we have to deal with the fact that there is no ordering defined in general. Thus, we cannot sort all input points at once in the beginning and then choose k-nearest points with $O(1)$ time complexity. The time complexity of selecting k-nearest points from N points is $O(N \log N)$, and thus the time complexity of LOWESS or Simple RBF approximation can be estimated as:

$$O\left(R \cdot \left(N \log N + \left\{ \begin{array}{c} O_{LOWESS} \\ \text{or} \\ O_{RBF} \end{array} \right\} \right) \right), \tag{45}$$

where O_{LOWESS} is the same time complexity as the time complexity of LOWESS approximation in 1&1/2 dimensions and O_{RBF} is the same time complexity as the time complexity of Simple RBF approximation in 1&1/2 dimensions.

7.1 LOWESS

In the case of D&1/2 dimensional approximation, we have to change the notation in (1) as x is a D-dimensional position vector:

$$S = \sum_{i=1}^{N} \omega_i \cdot \left(h_i - P^{(D)}(x_i) \right)^2, \tag{46}$$

where $h = P^{(D)}(x)$ is a D dimensional hypersurface function with unknown coefficients $a = [a_0, a_1, a_2, \ldots, a_k]^T$. For $D = 2$, we can write $P^{(D)}(x)$, for example, like:

$$P^{(2)}(x) = a_0 + a_1 x + a_2 y + a_3 x^2 + a_4 y^2 + a_5 xy, \tag{47}$$

where $x = [x, y]^T$. The matrix A is then equal to:

$$A = \begin{bmatrix} 1 & x_1 & y_1 & x_1^2 & y_1^2 & x_1y_1 \\ 1 & x_2 & y_2 & x_2^2 & y_2^2 & x_2y_2 \\ \vdots & \vdots & \vdots & \vdots & \vdots & \vdots \\ 1 & x_N & y_N & x_N^2 & y_N^2 & x_Ny_N \end{bmatrix} \tag{48}$$

We can omit some coefficients a_i and corresponding columns in matrix A, where $i \in \{0, 1, \ldots, 5\}$. All other computations remain the same.

The computation complexity will increase as the size of matrix A increases. However, if we use a constant hypersurface function with only one coefficient a_0, then the time complexity does not change with different dimensions D.

7.2 Simplified RBF

The RBF approximation is formally independent from the dimension D. Therefore, all the computations remain the same as described above. The computation complexity increases slightly as the complexity of polynomial/hypersurface $P^{(D)}(x)$ increases. However, if we use a constant hypersurface function with only one coefficient a_0, then the time complexity will not change with different dimensions D. The polynomial $P^{(D)}(x)$ is actually a data approximation using a basic function and $\sum_{i=1}^{M} \lambda_i \Phi_i(r)$ controls the perturbation from $P^{(D)}(x)$.

8 Conclusion

We have introduced the LOWESS method of approximation and modified RBF approximation, which is comparable with LOWESS. Both methods use the same number of nearest samples for approximation and the time complexity of both these methods is the same. We calculated the distance of approximated noisy data to the original data. In all cases, for the same number of nearest samples for approximation, LOWESS gives better results. Another comparison of both methods is calculation of the smoothness. The LOWESS approximation gives us smoother results than the Simple RBF approximation. However, both these methods use a different approach for approximation than global RBF approximation; we compared them with global RBF approximation as well. Using global RBF approximation we can achieve better results (closer distance to original data and smoother approximation) when having the same time complexity of calculation. Moreover, we get one simple continuous formula and not only function values at discrete points. On the other hand, both methods can be used in higher dimensions, but the time complexity will increase for both of them compared to the situation in 1&1/2 dimensions. Due to this fact, in higher dimensions, global RBF approximation has lower time complexity than either LOWESS or Simple RBF approximation due to necessity of finding k-nearest neighbor points.

Therefore, the global RBF approximation is recommendable for approximation of scattered data in higher dimensions, i.e. $2\&1/2$ dimensions and higher.

All methods for approximation compared in this paper were implemented and tested in MATLAB.

Acknowledgements. The authors would like to thank their colleagues at the University of West Bohemia, Plzen, for their comments and suggestions, and anonymous reviewers for their valuable critical comments and advice. The research was supported by MSMT CR projects LH12181 and SGS 2016-013.

References

1. Bellochino, F., Borghese, N.A., Ferrari, S., Piuri, V.: 3D surface Reconstruction. Springer, New York (2013). ISBN:978-1-4614-5632-2
2. Chen, L.M.: Digital Functions and Data Reconstruction. Springer, New York (2013). ISBN:978-1-4614-5637-7
3. Cleveland, W.S.: Robust locally weighted regression and smoothing scatterplots. J. Am. Stat. Assoc. **74**(368), 829–836 (1979). doi:10.2307/2286407
4. Fasshauer, G.E.: Meshfree Approximation Methods with MATLAB. World Scientific Publ., Singapore (2007). ISBN:978-981-270-633-1
5. Hickernell, F.J., Hon, Y.C.: Radial basis function approximation as smoothing splines. Appl. Math. Comput. **102**, 1–24 (1999)
6. Lazzaro, D., Montefusco, L.B.: Radial basis functions for the multivariate interpolation of large scattered data sets. J. Comput. Appl. Math. **1040**, 521–536 (2002). Elsevier
7. Narcowich, F.J.: Recent developments in error estimates for scattered data interpolation via radial basis functions. Numer. Algorithms **39**, 307–315 (2005). Springer
8. Pan, R., Skala, V.: A two level approach to implicit modeling with compactly supported radial basis functions. Eng. Comput. **27**(3), 299–307 (2011). doi:10.1007/s00366-010-0199-1. ISSN:0177-0667. Springer
9. Pan, R., Skala, V.: Surface reconstruction with higher-order smoothness. Vis. Comput. **28**(2), 155–162 (2012). ISSN:0178-2789. Springer
10. Skala, V: Progressive RBF interpolation. In: 7th Conference on Computer Graphics, Virtual Reality, Visualisation and Interaction, Afrigraph 2010, pp. 17–20. ACM (2010). ISBN:978-1-4503-0118-3
11. Skala, V., Pan, R., Nedved, O.: Making 3D replicas using a flatbed scanner and a 3D printer. In: Murgante, B., et al. (eds.) ICCSA 2014, Part VI. LNCS, vol. 8584, pp. 76–86. Springer, Heidelberg (2014). ISBN:978-3-319-09152-5
12. Wendland, H.: Piecewise polynomial, positive definite and compactly supported radial functions of minimal degree. Adv. Comput. Math. **4**(1), 389–396 (1995). doi:10.1007/BF02123482
13. Yao, X., Fu, B., Lü, Y., et al.: Comparison of four spatial interpolation methods for estimating soil moisture in a complex terrain catchment. PLoS ONE **8**(1), e54660 (2013). doi:10.1371/journal.pone.0054660

Improving the ANN Classification Accuracy of Landsat Data Through Spectral Indices and Linear Transformations (PCA and TCT) Aimed at LU/LC Monitoring of a River Basin

Antonio Novelli, Eufemia Tarantino[✉], Grazia Caradonna,
Ciro Apollonio, Gabriella Balacco, and Ferruccio Piccinni

Politecnico di Bari, via Orabona 4, Bari 70125, Italy
eufemia.tarantino@poliba.it

Abstract. In this paper an efficient Artificial Neural Networks (ANN) classification method based on LANDSAT satellite data is proposed, studying the Cervaro river basin area (Foggia, Italy). LANDSAT imagery acquisition dates of 1984, 2003, 2009 and 2011 were selected to produce Land Use/Land Cover (LULC) maps to cover a time trend of 28 years. Land cover categories were chosen with the aim of characterizing land use according to the level of surface imperviousness. Nine synthetic bands from the PC, Tasseled Cap (TC), Brightness Temperature (BT) and vegetation indices (Leaf area Index LAI and the Modified Soil Adjusted Vegetation Index MSAVI) were identified as the most effective for the classification procedure. The advantages in using the ANN approach were confirmed without requiring a priori knowledge on the distribution model of input data. The results quantify land cover change patterns in the river basin area under study and demonstrate the potential of multitemporal LANDSAT data to provide an accurate and cost effective means to map and analyze land cover changes over time that can be used as input for subsequent hydrological and planning analysis.

Keywords: Artificial Neural Networks · Spectral indices · Tasseled cap transformation · Principal component analysis · Land Use/Land Cover · River basin

1 Introduction

Many studies all over the world have demonstrated that conversion of rural landscapes to urban and suburban land uses is directly related to increasing amounts of impervious areas [1–4]. Such surfaces need to be monitored both for planning activities and hydrological hazard prevention [5–8].

LCLU mapping using satellite images has become widely popular in the last decades [9–11] and many methods have been studied for mapping large areas with a high level of accuracy. One of the most widespread methods is the artificial neural networks (ANNs) approach that can simulate the decision making processes of the human brain. Such method is independent from the statistical distribution of data,

© Springer International Publishing Switzerland 2016
O. Gervasi et al. (Eds.): ICCSA 2016, Part II, LNCS 9787, pp. 420–432, 2016.
DOI: 10.1007/978-3-319-42108-7_32

allowing classes to be searched mainly in correspondence of their distribution within the domain of each data source [12].

The implementation of ANNs on satellite imagery have been progressively increasing with main applications using the multilayer perceptron neural network trained with a backpropagation algorithm [13, 14]. ANNs have been used for land cover mainly based on raw band data, obtaining a higher classification accuracy than standard classification techniques [15]. However, since vegetation classes may vary with seasons, the results of classifications also vary consequently.

Spectral indices are suitable in identifying specific features on the ground such as water, vegetation or impervious and, therefore, offering more information than raw band data results in higher classification accuracy [16]. Patel et al. [17] demonstrated that ANNs, when used with the spectral indices, can reach higher accuracy. Erbek et al. [18] proved that multisource data may easily be added into the classification process with ANN to improve performance. Sehgal [15] chose Backpropagation Neural Network (BPNN) as the most accurate among the ANN techniques under study to classify LANDSAT multispectral data using textural, spatial, and spectral information of images. Li et al. [19] simulated grassland above-ground biomass (AGB) using multi-spectral reflectance derived from ANN to investigate grazing intensity in China.

This paper focuses on the improvement of ANN classification accuracy of LANDSAT- TM data by implementing synthetic bands based on Spectral Indices (LAI and MSAVI), Tasseled Cap transformation and Principal Component Analysis in the pre-processing phase. This combination demonstrated its suitability with the highest accuracy in extracting impervious pixel classes considering as case study the basin area of Cervaro, Italy.

2 Data and Method

Cervaro river basin is one of largest in the Puglia region, with an extension of about 775 km^2. It crosses 4 towns located within the province of Avellino and 15 towns within the province of Foggia. The basin is also important from a naturalist viewpoint, as it includes a ZPS area (Special Protection Zone), a SIC area (Site of Community Interest), an Important Bird Area as well as a Regional Natural Park.

For the purposes under this study, four cloud-free LANDSAT-TM scenes were selected (acquisition dates: 06/27/1984; 07/18/2003; 06/16/2009; 06/22/2011), by using the USGS EROS web site (http://glovis.usgs.gov/). LANDSAT-TM scenes are provided to users as Level 1 terrain corrected (L1T) data. L1T processing includes radiometric and geometric correction, precision correction using ground control chips and the use of a digital elevation model to correct parallax error due to local topographic relief [20]. Moreover, data are distributed under the GEOTIFF format with the Universal Transverse Mercator (UTM) projection and the World Geodetic System 84 (WGS84) datum.

2.1 Pre-processing of LANDSAT-TM Data

Pre-processing of LANDSAT-TM data was aimed at enhancing the classification performance. For this purpose, the input of the Artificial Neural Network Classifier (ANNC) was only derived from synthetic bands (i.e. vegetation indices, Principal Components (PC), etc.) due to their capability to highlight specific attributes and to make use of redundant information. Nine synthetic bands were extracted from the PC, Tasseled Cap (TC), Brightness Temperature (BT) and vegetation indices.

LANDSAT-TM digital numbers were firstly converted to Top of Atmosphere (TOA) reflectance and corrected for sun elevation angle by using the standard procedures suggested by the USGS [21]. The parameters required for the conversions were found within each scene metadata file. The Dark Object Subtraction (DOS) procedure was implemented to reduce atmospheric artifacts. Vegetation indices, PC and TC bands were computed from the DOS output of each scene.

Vegetation indices were used to enhance the detection of vegetated areas. The Leaf Area Index (LAI) and the Modified Soil Adjusted Vegetation Index (MSAVI) were chosen as two of the nine input bands for the ANNC. Both the implemented LAI and MSAVI expressions contain the Normalized Difference Vegetation Index (NDVI) as input, thus it was computed for each scene (Table 1).

Table 1. $NDVI_{max}$, $NDVI_{back}$ and k (De Jong 1994) implemented in LAI computation.

$NDVI_{max}$	$NDVI_{back}$	k
0.224	0.859	0.213

LAI is commonly used in agronomic and environmental modeling. Unlike normalized difference indices LAI is not normalized and, although zero is the defined value for bare soils, an upper LAI bound is not defined [22–25]. In this study LAI was evaluated through a semi-empirical model derived from the Lambert-Beer Law (1)

$$LAI = -\frac{1}{k}\ln\left(\frac{NDVI_{max} - NDVI}{NDVI_{max} - NDVI_{back}}\right) \tag{1}$$

where $NDVI_{max}$ and $NDVI_{back}$ are respectively an upper and a lower bound of the NDVI and k is an extinction coefficient. Since LAI was used as ancillary data, and not to estimate real LAI values, $NDVI_{max}$, $NDVI_{back}$ and k were found in literature [26] for Mediterranean areas and LANDSAT-TM data (Table 2).

Table 2. Values of a_{ki} and b_k coefficients.

TM	Band1	Band2	Band3	Band4	Band5	Band7	B
Brightness	0.2909	0.2493	0.4806	0.5568	0.4438	0.1706	10.3695
Greenness	−0.2728	−0.2174	−0.5508	0.7221	0.0733	−0.1648	−0.731
Wetness	0.1446	0.1761	0.3322	0.3396	−0.621	−0.4186	−3.3828

Equation 1 shows that the condition $NDVI < NDVI_{back}$ results in a negative LAI. The LAI distribution of each scene features a bimodal shape with one of the two local maximum within the region of negative LAI values. Pixels belonging to the neighbourhood of the negative local maximum are distinctive of lakes, sea and salt planes. This shows that the evaluated LAI is useful also in water surfaces detection.

MSAVI is one of the vegetation indices that makes use of the soil line [27, 28]. The soil line is the regression line of the distribution of soil pixels within the RED-NIR plane of the feature space with slope "a". MSAVI mathematical expression is showed through Eqs. 2–4.

$$MSAVI = \frac{\rho_{NIR} - \rho_{RED}}{\rho_{NIR} + \rho_{RED} + M}(1 + M) \qquad (2)$$

$$M = 1 - 2a * (NDVI * WDVI) \qquad (3)$$

$$WDVI(WeightedDifferenceVegetationIndex) = \rho_{NIR} - a\rho_{RED} \qquad (4)$$

The soil slope a = 1.47 was evaluated from a group of pixels manually selected and chosen as representative of bare soils. For each subset the WDVI, the M parameter and the MSAVI were respectively evaluated.

The enhancement of the classification process was also achieved by substituting the original channels with their linear transformations. Tasseled Cap (TC) transform and Principal Component (PC) were implemented for this purpose. The TC transform Cap (TC) transform [29] is one of the techniques designed to enhance the informative content of multiband imagery. TC synthetic channels highlight the evolution of the vegetation spectral response in a multitemporal study. Analytically TC bands are computed through Eq. 5:

$$Z_k = \sum_{i=1,N} a_{ki}X_i + b_k \qquad (5)$$

Where Z_k and X_i are the k-th synthetic band and the i-th original band respectively, while a_{ki} and b_k are the transformation coefficients (Table 2).

PC analysis is aimed at the reduction of the dimensionality of a multiband dataset through the computation of uncorrelated synthetic bands [30]. Only the first three PC were used for each scene since within they enclose more than the 99 % of the total variance of each scene (6).

$$\% \, of \, variance \, within \, the \, first \, three \, PC = \frac{100}{\sum_{i=1}^{k} \lambda_i} * \sum_{i=1}^{3} \lambda_i \qquad (6)$$

Where λ_i is the i-th eigenvalue of the variance-covariance matrix for the considered scene and $\lambda_i \geq \lambda_{i+1}$.

The last synthetic band was derived from the thermal band. Commonly thermal information has been used for Land Surface Temperature (LST) retrieval [31–34]. In this study, the thermal band was used as ancillary data [35, 36] to create a further input for the ANNC. Digital numbers were converted to at-satellite BT for each scene using the logarithmic formula and the thermal constants respectively provided by the USGS EROS web site and the metadata files.

Lastly, the value range of LAI and BT were rescaled through Eq. 7 in order to avoid great numerical difference among the values of the input layer of the ANNC.

$$y = \frac{x - x_{min}}{x_{max} - x_{min}} \tag{7}$$

Where y is the rescaled input, x_{max} and x_{min} are respectively the maximum and the minimum value of the considered x input.

2.2 Output Classes

The identification of output classes was an iterative procedure. Ground reference for the most recent scenes were extracted from Apulian Land Use Cartography and from Google Earth. Apulian Cartography is distributed under the UTM WGS84 zone 33 N reference system, created from an aerial orthophoto dated 2006-2007 with 50 cm of geometric resolution, in compliancy with Corine Land Cover project standards. The classes considered as the most representative of the land cover in the study area were extracted from LANDSAT-TM scenes. Training and ground reference pixels were first collected through random and stratified sampling from Land Use cartography and validated with Google Earth. The total of collected pixels was then divided in two groups to verify classification accuracy without using ANNC training pixels. The tests carried out showed that a simple stratified sampling scheme was not adequate to achieve an elevate spectral separability. Particularly, the problem encountered with the use of Apulian Cartography was that areas belonging to the same land cover were characterized by different spectral signature and by a consequent different distribution within the feature space. This problem was solved with a double approach involving both the decomposition of the selected training class and the use of a synthetic channel.

The decomposition was carried out by using binary masks. The aim was to enhance the classification performance removing pixels of the same land use but with different radiometric content (i.e. arable crops senescent and luxuriant) from mixed training and reference areas. Binary masks were used to select problematic pixels from the original channels and rearrange them. The general scheme followed was the creation of a small Region Of Interest (ROI) related to a specific category of pixels, the extraction of band statistics for ROI pixels, and the creation of the i-th mask (8)

$$(b_i < b_{i,max}) \; AND \; (b_i > b_{i,min}) \; AND \; (b_{i+1} < b_{i+1,max}) \; AND$$
$$(b_{i+1} > b_{i+1,min}) \ldots AND \; (b_n < b_{n,max}) \; AND \; (b_n > b_{n,min}) \, with \; i = 1, 2, ..n \tag{8}$$

Where b_i is the TOA reflectance value for the i-th channel, $b_{i,max}$ and $b_{i,min}$ are respectively the maximum and the minimum for the i-th channel of the selected ROI and n number of channel considered.

This process ended with thirteen output classes (Table 3).

Table 3. Legend

0. Clouds/ unclassified zones.

1. Bare soil with high reflectance in RGB channels.

2. Arable Crop (e.g. cereals; forage; grains; tomatoes; cabbage); including vineyards; with senescent or low density planting.

3. Shrub andlow density orchard.

4. River channel and wetland.

5. Orchard or vegetation.

6. Greenhouse or plastic covered vineyard.

7. Olive grove or Orchard.

8. Built-up Area.

9. Arable Crop (e.g. cereals; forage; grains; tomatoes; cabbage); includingvineyards; with high leaf area.

10. Forest (broadleaved and coniferous).

11. Marsh.

12. Salt plane.

13. Lake.

For each scene the same training pixel locations were used. When ground reference was not available (i.e., lacking ancillary data) selected training pixels were considered as belonging to the same class of scenes with availability of ancillary data.

The positive effect of synthetic channels was proved by evaluating the Jeffries Matusita Distance (JMD) [37] for each i-th and j-th class. JMD_{ij} was computed through Eqs. 9-11:

$$JMD_{ij} = \left[2\left(1 - e^{-B_{ij}}\right)\right]^{0.5} \tag{9}$$

$$B_{ij} = \frac{1}{8}MH_{ij} + \frac{1}{2}\ln\left(\frac{\left|\frac{C_i + C_j}{2}\right|}{\sqrt{|C_i| * |C_j|}}\right) \tag{10}$$

$$MH_{ij} = \left[\left(\mu_i - \mu_j\right)^T * \left(\frac{C_i + C_j}{2}\right)^{-1} * \left(\mu_i - \mu_j\right)\right]^{0.5} \tag{11}$$

Where B_{ij} and MH_{ij} are the Bhattacharrya and the Mahalanobis distances; μ_i, μ_j, C_i, C_j are the vectors of the means and variance-covariance matrices for pixels associated to the i-th and the j-th training classes. JMD takes into account the overlapping of training class pixels distributions (by means of C_i and C_j). Figures 1 and 2 show examples of the great enhancement of spectral separability achieved by using the described combination of synthetic bands.

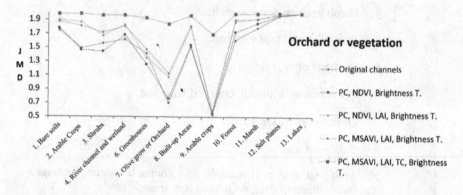

Fig. 1. JMD for different combinations of bands related to the 5th class for the 2011 scene (Color figure online)

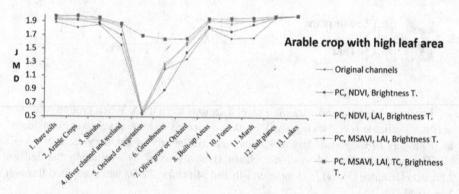

Fig. 2. JMD for different combinations of bands related to the 9th class for the 2011 scene (Color figure online)

2.3   Neural Net Classification

An ANNC allows for the analysis of complex relationships among variables without making a priori assumptions about data (i.e. normality or linearity) [38, 39].

Important features of an ANNC are the architecture, the neurons and the learning rule [40]. Neurons are computational units characterized by weights, biases and a transfer function. The architecture is the distribution of neurons and the organization of connections within different layers: common architectures for ANNC are constituted by one input layer, one output layer and one or two hidden layers. Learning capability is achieved by training data constituted by input vectors and associated output vectors. By comparing the current output layer results to the user imposed output response, the difference can be obtained and used to adjust weights connecting neurons in the network. The goal is to achieve a set of weights that will produce the output that mostly resembles the imposed output vectors. Once trained, the ANNC can recall stored knowledge to perform a classification [41]. In this study the ANNC was implemented

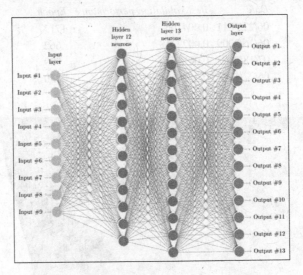

Fig. 3. ANNC architecture

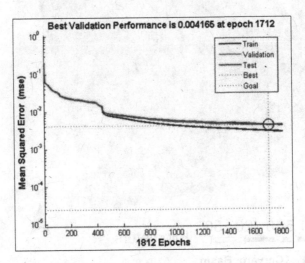

Fig. 4. Training iterations for the 2011 training data: Train subset (blue), Validation subset (green), Test subset (red). (Color figure online)

through the Matlab environment by Mathworks [42]. Each neuron implements a logsig transfer function (12) and a bias.

$$logsig(n) = \frac{1}{1+e^{-n}} \qquad (12)$$

Table 4. Training result for each scene.

Date	Best validation performance	Epoch
06/22/2011	0.004165	1712
06/16/2009	0.0022218	662
07/18/2003	0.0014576	829
06/27/1984	0.0035561	2430

0 5 10 20   30 Kilometers

☐ Cervaro Basin

N

Fig. 5. ANNC output for the 2011 scene on the Cervaro river basin (Foggia, Italy).

With $n \in (-\infty; +\infty)$ and $logsig(n) \in (0; 1)$. The log-sigmoid transfer function is commonly used in multilayer networks that are trained using the family of back-propagation algorithms because differentiable [43]. The implemented architecture is the result of several attempts. Lastly the best configuration found is a four layer ANNC with one input layer with nine neurons (one for each synthetic band), two hidden layers with respectively twelve and thirteen neurons and one output layer with thirteen neurons (one for each output class). Figure 3 shows the described architecture.

The learning algorithm used was a scaled conjugate gradient algorithm [44] since with this architecture it performs faster than the classic backpropagation algorithm. The learning process was tested and validated dividing training data into three subsets [42]: 70 % of training data (train subset) were used to train the ANNC, 15 % of training data (validation subset) were used to validate the generalization of the ANNC and 15 % of training data (test subset) were used to test the ANNC against unknown data. Four ANNC, one for each scene, were trained achieving elevate convergence. Figure 4 shows the mean squared error evolution for the learning process of the 2011 scene and (Table 4) summarizes training results for each scene. Figure 5 shows the ANNC output for the 2011 scene on the Cervaro basin.

In order to evaluate ANNC output accuracy four confusion matrices were computed using 150 ground truth pixels for each class. Table 5 shows the overall accuracy from each confusion matrix.

Table 5. Overall accuracies achieved for each classification output.

Scene (date)	Overall accuracy [%]
06/22/2011	89
06/16/2009	88
07/18/2003	86
06/27/1984	79

3 Conclusion

In this work a classification procedure using the ANN method on LANDSAT data over a river basin area was tested, choosing land use categories according to surface imperviousness level. The analysis demonstrated that using synthetic bands improved the accuracy of the classification, if compared to past studies tested through raw bands over similar study areas. Some drawbacks were found on 1984 data, probably due to inefficient comparison with ground reference data, lacking sufficient true information on surface covers.

References

1. Naik, P.K., Tambe, J.A., Dehury, B.N., Tiwari, A.N.: Impact of urbanization on the groundwater regime in a fast growing city in central India. Environ. Monit. Assess. **146**, 339–373 (2008)

2. Sharma, R., Joshi, P.: Monitoring urban landscape dynamics over Delhi (India) using remote sensing (1998–2011) inputs. J. Indian Soc. Remote Sens. **41**, 641–650 (2013)
3. Park, S., Hepcan, Ç.C., Hepcan, Ş., Cook, E.A.: Influence of urban form on landscape pattern and connectivity in metropolitan regions: a comparative case study of Phoenix, AZ, USA, and Izmir, Turkey. Environ. Monit. Assess. **186**, 6301–6318 (2014)
4. Sallustio, L., Munafò, M., Riitano, N., Lasserre, B., Fattorini, L., Marchetti, M.: Integration of land use and land cover inventories for landscape management and planning in Italy. Environ. Monit. Assess. **188**, 1–20 (2016)
5. Gioia, A., Manfreda, S., Iacobellis, V., Fiorentino, M.: Performance of a theoretical model for the description of water balance and runoff dynamics in Southern Italy. J. Hydrol. Eng. **19**(6), 1113–1123 (2013)
6. Manfreda, S., Samela, C., Gioia, A., Consoli, G.G., Iacobellis, V., Giuzio, L., Cantisani, A., Sole, A.: Flood-prone areas assessment using linear binary classifiers based on flood maps obtained from 1D and 2D hydraulic models. Nat. Hazards **79**(2), 735–754 (2015)
7. Iacobellis, V., Castorani, A., Di Santo, A.R., Gioia, A.: Rationale for flood prediction in karst endorheic areas. J. Arid Environ. **112**, 98–108 (2015)
8. Iacobellis, V., Claps, P., Fiorentino, M.: Climatic control on the variability of flood distribution. Hydrol. Earth Syst. Sci. Discuss. **6**(2), 229–238 (2002)
9. Yousefi, S., Khatami, R., Mountrakis, G., Mirzaee, S., Pourghasemi, H.R., Tazeh, M.: Accuracy assessment of land cover/land use classifiers in dry and humid areas of Iran. Environ. Monit. Assess. **187**, 1–10 (2015)
10. Lasaponara, R., Lanorte, A.: Satellite time-series analysis. Int. J. Remote Sens. **33**(15), 4649–4652 (2012)
11. Lasaponara, R.: Geospatial analysis from space: advanced approaches for data processing, information extraction and interpretation. Int. J. Appl. Earth Obs. Geoinf. **20**, 1–3 (2013)
12. Zhou, W.: Verification of the nonparametric characteristics of backpropagation neural networks for image classification. IEEE Trans. Geosci. Remote Sens. **37**, 771–779 (1999)
13. Aitkenhead, M., Aalders, I.: Classification of landsat thematic mapper imagery for land cover using neural networks. Int. J. Remote Sens. **29**, 2075–2084 (2008)
14. Tarantino, E., Novelli, A., Aquilino, M., Figorito, B., Fratino, U.: Comparing the MLC and JavaNNS approaches in classifying multi-temporal LANDSAT Satellite Imagery over an ephemeral river area. Int. J. Agric. Environ. Inf. Syst. (IJAEIS) **6**(4), 83–102 (2015)
15. Sehgal, S.: Remotely sensed LANDSAT image classification using neural network approaches. Int. J. Eng. Res. Appl. **2**, 43–46 (2012)
16. Xu, H.: Extraction of urban built-up land features from Landsat imagery using a thematic oriented index combination technique. Photogram. Eng. Remote Sens. **73**, 1381–1391 (2007)
17. Patel, N., Mukherjee, R.: Extraction of impervious features from spectral indices using artificial neural network. Arab. J. Geosci. **8**, 3729–3741 (2015)
18. Erbek, F.S., Özkan, C., Taberner, M.: Comparison of maximum likelihood classification method with supervised artificial neural network algorithms for land use activities. Int. J. Remote Sens. **25**, 1733–1748 (2004)
19. Li, F., Zheng, J., Wang, H., Luo, J., Zhao, Y., Zhao, R.: Mapping grazing intensity using remote sensing in the Xilingol steppe region, Inner Mongolia. China Remote Sens. Lett. **7**, 328–337 (2016)
20. Roy, D.P., Ju, J., Kline, K., Scaramuzza, P.L., Kovalskyy, V., Hansen, M., Loveland, T.R., Vermote, E., Zhang, C.: Web-Enabled Landsat Data (WELD): landsat ETM + composited mosaics of the conterminous United States. Remote Sens. Environ. **114**, 35–49 (2010)
21. Chander, G., Markham, B.: Revised Landsat-5 TM radiometric calibration procedures and postcalibration dynamic ranges. IEEE Trans. Geosci. Remote Sens. **41**, 2674–2677 (2003)

22. Gao, F., Anderson, M.C., Kustas, W.P., Houborg, R.: Retrieving leaf area index from landsat using MODIS LAI products and field measurements. IEEE Geosci. Remote Sens. Lett. **11**, 773–777 (2014)
23. Aquilino, M., Novelli, A., Tarantino, E., Iacobellis, V., Gentile, F.: Evaluating the potential of GeoEye data in retrieving LAI at watershed scale. In: SPIE Remote Sensing, pp. 92392B-92392B-92311. International Society for Optics and Photonics (2014)
24. Balacco, G., Figorito, B., Tarantino, E., Gioia, A., Iacobellis, V.: Space–time LAI variability in Northern Puglia (Italy) from SPOT VGT data. Environ. Monit. Assess. **187**, 1–15 (2015)
25. Tarantino, E., Novelli, A., Laterza, M., Gioia, A.: Testing high spatial resolution WorldView-2 imagery for retrieving the leaf area index. In: Third International Conference on Remote Sensing and Geoinformation of the Environment, p. 95351N-95351N-95358. International Society for Optics and Photonics (2015)
26. De Jong, S.M.: Derivation of vegetative variables from a Landsat TM image for modelling soil erosion. Earth Surf. Proc. Land. **19**, 165–178 (1994)
27. Qi, J., Chehbouni, A., Huete, A., Kerr, Y., Sorooshian, S.: A modified soil adjusted vegetation index. Remote Sens. Environ. **48**, 119–126 (1994)
28. Zhang, C., Pan, Z., Dong, H., He, F., Hu, X.: Remote estimation of leaf water content using spectral index derived from hyperspectral data. In: First International Conference on Information Science and Electronic Technology (ISET 2015). Atlantis Press (2015)
29. Crist, E.P., Laurin, R., Cicone, R.C.: Vegetation and soils information contained in transformed Thematic Mapper data. In: Proceedings of IGARSS 1986 Symposium, pp. 1465–1470. European Space Agency Publications Division Paris (1986)
30. Canty, M.J.: Image Analysis, Classification and Change Detection in Remote Sensing: With Algorithms for ENVI/IDL and Python. CRC Press, Boca Raton (2014)
31. Muthulakshmi, A., Natesan, U., Ferrer, V.A., Deepthi, K., Venugopalan, V., Narasimhan, S.: A novel technique to monitor thermal discharges using thermal infrared imaging. Environ. Sci. Process. Impacts **15**, 1729–1734 (2013)
32. Ozelkan, E., Bagis, S., Ozelkan, E.C., Ustundag, B.B., Ormeci, C.: Land surface temperature retrieval for climate analysis and association with climate data. Eur. J. Remote Sens. **47**, 655–669 (2014)
33. Tarantino, E.: Monitoring spatial and temporal distribution of sea surface temperature with TIR sensor data. Ital. J. Remote Sens. **44**, 97–107 (2012)
34. Labbi, A., Mokhnache, A.: Derivation of split-window algorithm to retrieve land surface temperature from MSG-1 thermal infrared data. Eur. J. Remote Sens. **48**, 719–742 (2015)
35. Novelli, A., Tarantino, E.: The contribution of Landsat 8 TIRS sensor data to the identification of plastic covered vineyards, p. 95351E-95351E-95359 (2015)
36. Novelli, A., Tarantino, E.: Combining ad hoc spectral indices based on LANDSAT-8 OLI/TIRS sensor data for the detection of plastic cover vineyard. Remote Sens. Lett. **6**, 933–941 (2015)
37. Bruzzone, L., Roli, F., Serpico, S.B.: An extension of the Jeffreys-Matusita distance to multiclass cases for feature selection. IEEE Trans. Geosci. Remote Sens. **33**, 1318–1321 (1995)
38. Ingram, J.C., Dawson, T.P., Whittaker, R.J.: Mapping tropical forest structure in southeastern Madagascar using remote sensing and artificial neural networks. Remote Sens. Environ. **94**, 491–507 (2005)
39. Jensen, J., Qiu, F., Ji, M.: Predictive modelling of coniferous forest age using statistical and artificial neural network approaches applied to remote sensor data. Int. J. Remote Sens. **20**, 2805–2822 (1999)
40. Haykin, S.: Neural Network-a Comprehensive Foundation; a Computational Approach to Learning and Machine Intelligence. Macmillan, New York (1994)

41. Lloyd, R.: Spatial Cognition: Geographic Environments. Springer, Netherlands (1997)
42. Demuth, H., Beale, M., Hagan, M.: Neural network toolbox™ 6 user's guide (2008)
43. Dorofki, M., Elshafie, A.H., Jaafar, O., Karim, O.A., Mastura, S.: Comparison of artificial neural network transfer functions abilities to simulate extreme runoff data. Int. Proc. Chem. Biol. Environ. Eng. **33**, 39–44 (2012)
44. Møller, M.F.: A scaled conjugate gradient algorithm for fast supervised learning. Neural Netw. **6**, 525–533 (1993)

Automatic Temporal Segmentation of Articulated Hand Motion

Katharina Stollenwerk[1]([✉]), Anna Vögele[2], Björn Krüger[3], André Hinkenjann[1], and Reinhard Klein[2]

[1] Institute of Visual Computing, Bonn-Rhein-Sieg University of Applied Sciences, Sankt Augustin, Germany
`katharina.stollenwerk@h-brs.de`
[2] Insitute for Computer Science II, Bonn University, Bonn, Germany
[3] Gokhale Method Institute, Stanford, CA, USA

Abstract. This paper introduces a novel and efficient segmentation method designed for articulated hand motion. The method is based on a graph representation of temporal structures in human hand-object interaction. Along with the method for temporal segmentation we provide an extensive new database of hand motions. The experiments performed on this dataset show that our method is capable of a fully automatic hand motion segmentation which largely coincides with human user annotations.

1 Introduction

Motion Capture has become a standard technique for motion data recording in the past decades. Easy access to improved and relatively low-cost systems makes motion capture possible to a wider community and for numerous applications.

There is an increased focus on recording facial movement and hand gestures, since both are significant parts of communication and daily life. Recent work [1,2] has brought to attention the importance of correctly-timed hand motions and details in face and hand movement.

There is a need for high quality data in order to enable both motion analysis and synthesis. Recent technologies allow for fast capturing of highly resolved motion data [3–5]. However, in order to make use of the recordings, appropriate tools for data processing are needed. The segmentation of motions into simple data units is a crucial step in processing [6]. Recent developments of segmentation methods produce good results for full-body motion [6,7]. Our goal is presenting techniques which work for gestures and grasping and are similarly efficient as the above-mentioned. We introduce a method for temporal segmentation of hand motions which enables the isolation of primitive data units. We show that these units coincide with perceptive motor primitives by comparing the results to those which have been manually segmented by different users. Moreover, we discuss a method for clustering the motion segments achieved by the segmentation, thus resulting in a compact representation of hand motion.

The main contributions of this paper are:

© Springer International Publishing Switzerland 2016
O. Gervasi et al. (Eds.): ICCSA 2016, Part II, LNCS 9787, pp. 433–449, 2016.
DOI: 10.1007/978-3-319-42108-7_33

- A database of hand motions
- A technique for fully automatic and accurate segmentation
- A method to cluster segments

The remainder of this paper is organised as follows. Section 2 gives an overview of the work related to temporal segmentation and processing of hand motion. Our recording setup and the database that our experiments are based on are described in Sect. 3. Section 4 introduces the segmentation technique. Section 5 discusses our clustering approach. Finally, an evaluation as well as a comparison to other segmentation approaches are presented in Sect. 6.

2   Related Work

Hand Motion Capturing. Possibilities for capturing finger motion data include marker-based optical and image-based video tracking with or without depth information as well as glove-based systems with or without tactile sensors. An overview of the main approaches also covering advantages and drawbacks of the respective techniques was surveyed by Wheatland et al. [8].

Current approaches combine multiple capturing methods to overcome limitations of individual techniques. Zhao et al. [3] describe how to record high-fidelity 3D hand articulation data with a combination of an optical marker-based motion capture system and a kinect camera. Arkenbout et al. [9] integrate a 5DT Data Glove into the kinect-based Nimble VR system using Kalman filter. This resolved visual self-occlusion of the hand and fingers and improved precision and accuracy of the joint angles' estimates. Ju and Liu [10] capture joint angle data, finger and hand (contact) force data and sEMG data of forearm muscles in order to study correlations of different sensory information.

This variety of sensor fusion for capturing hand motions indicates that the acquisition of high-quality hand motion data has not yet been satisfactorily resolved. All of the above mentioned approaches point out that they are increasing the quality of the recorded data. Nevertheless, we have decided to only use a CyberGlove data glove as, on the one hand, it was not important that the user's hand is unencumbered and on the other hand we believe that the main challenge in recording hand motion and manipulation data is occlusion from an object or the hand itself, both of which are handled effortlessly by a data glove.

Databases. While there are a number of high-quality, full-body motion capture databases that can be used for academic purposes (e.g. the CMU and HDM05 motion capture databases [11,12]), only a few such data collections exist for articulated hand motions.

In the field of robotics, Goldfeder et al. [13] presented algorithms for automatic generation of a database of precomputed stable grasps, i.e. a single pose, for robotic grasping. This resulted in *The Columbia grasp database* which contains computed grasp configurations of different (robotic) hands along with a set of graspable objects. Thus, the database only contains single-hand poses and no finger movements. Feix et al. [14] provide a small dataset of human grasping used as a basis for evaluation of the motion capabilities of artificial hands. The dataset contains 31 motions, each performed by five subjects twice. The motion data contains only the 3D position and orientation of each fingertip with no information on the specific underlying hand model. Notable also due to its size, the *NinaPro database* [15] contains 52 full-hand and wrist motions collected from 27 subjects. The data recorded consists of surface electromyography (sEMG) data together with the 8-bit valued raw output of a 22-sensor CyberGlove (kinematic data). The 8-bit valued raw kinematic recordings of each sensor only roughly represent joint angles and fail to account for cross coupled sensors in the glove.

Segmentation and Re-use of Motion Capture Data. Manual segmentation and annotation of motion data into meaningful phases is a tedious and daunting task. But it is segmentation and annotation that makes the data re-usable.

For full-body motion data, unsupervised, temporal segmentation techniques have been developed. Among them is the work of Beaudoin et al.'s [16] who focus on visualising the structure of motion datasets. They propose to partition motion data streams into motion-motifs organised in a graph structure useful for motion blending and motion data compression. Zhou et al. [7,17] segment human motion data based on (hierarchical) aligned cluster analysis (H(ACA)). They frame the task of motion segmentation as temporal clustering of (motion capture) data into classes of semantically-similar motion primitives. Min and Chai's Motion Graph++ [18] is not only capable of segmenting motions into basic motion units, but also of motion recognition and synthesis. Their segmentation automatically extracts keyframes of contact transitions and semi-automatically extracts keyframes exhibiting salient visual content changes. Vögele et al. [6] employ (backward and forward) region growing in order to identify start and end frames of activities (groups of motion primitives). These activities are split into motion primitives by taking advantage of how repetitive activity patterns manifest in self-similarity matrices (SSSM). Recently, Krüger et al. [19] further improved the outcomes of [6] by aligning and grouping segregated feature trajectories and exploiting the symmetric nature of motion data.

For temporal segmentation of captured hand motion data, we have mainly found motion streams of conversational hand gestures to have been automatically segmented into different phases and synthesised into new motions, often based on an accompanying audio stream. Examples thereof include Levine et al. [20], Jörg et al. [2] and Mousas et al. [21] who solely use features derived from the wrist joint's position over time for segmentation into gesture phases. While the first two works synthesise new gestures based on the whole hand at once, the authors of the last paper estimate finger motion separately for each finger. The estimation is limited to adjusting frame times for creating optimally timed transitions.

There is little published on temporal segmentation of hand motion data; no research has been found that analysed automatic or unsupervised segmentation of such data. For robot programming and teaching from example in, e.g. pick and place scenarios, researchers have looked into temporal segmentation of recorded human hand and finger motions. This is often needed to characterise grasp phases (e.g. pre-grasp, grasp, manipulation). However, usually this information is included only implicitly in models used for grasp classification: Ekvall and Kragic [22], for instance, classify grasp movements using 5-state HMMs for grouping fingertip positions and transitions. A noteworthy exception is the work of Kang and Ikeuchi [23]. They segment grasp motions of a human demonstrator using motion profiles and volume sweep rates. Their results, however, are limited to two segmented exemplary motion sequences without specification of reference data or ground truth data. In the area of computer animation, Zhao et al. [24] combined recorded 3D hand motion capture data of ten different grip modes and physics-based simulation with the aim of achieving physically-plausible interaction between a hand and a grasped object. Each motion was then manually segmented into three phases: reaching, closing and manipulation, the last of which is assumed to be a static pose. From their paper, it is unclear whether the motion data was recorded from interacting with a real or a virtual object.

3   Database

To the best of our knowledge there still is no database of articulated human hand motions covering a wide variety of actions and actors usable for motion analysis and data-driven synthesis. We therefore decided to create one.

We chose to include two main setups in the database: *Uncontrolled transport* and *controlled transport*, both of which will be described below. To ensure inter-person consistency in the recording setups and for later reproducibility, a protocol was written detailing each step of the recording. The setups can be summarised as follows:

In *uncontrolled transport*, each person is presented with a variety of objects in random order with no object being presented twice in a row. The task is to pick up the object, move it to a different location, and put it down. The setup was designed in order to obtain a high diversity of possible grasps per object. Each object had to be moved five times. In *controlled transport* each person is again presented with a variety of objects. The task is to pick up the object, move it to a different location, and put it down. Only here, a picture illustrates how to hold the object during transport and the task has to be executed five consecutive times on each object. Contrary to the first setup, the focus here lies on reproducing consistent, predefined grasp motions.

In all setups the hand was placed flat on a table before and after task execution. The data were recorded at an acquisition rate of 60[Hz] using an 18-sensor right-hand ImmersionSquare CyberGlove as depicted in Fig. 1. Each hand pose (a set of 18 data points per sample) is represented by a joint angle configuration of the joints in each finger, wrist and palm.

Choice of Grasp Types in Controlled Transport. Several fields of research (e.g. biomechanics, robotics, medicine) have introduced grasp taxonomies for grouping different grasps by common criteria seeking simplification of the hand's complex prehensile capabilities.Due to the wide range of application and a resulting

 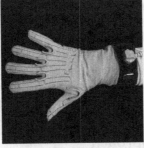 

(a) Objects in the database   (b) CyberGlove   (c) Grasping a tennis ball

Fig. 1. (a) Objects used for grasping: classic notebook, oval jar, tennis ball, mug, cube, bottle crate, small cylinder, pen, bottle, glass, business card, bowl, cylinder large, carton ($7.6 \times 26 \times 37.5[\text{cm}^3]$). Objects in red were used in *controlled transport*. (Color figure online)

lack of consensus in naming and classifying grasp types Feix et al. [25, 26] collected and consolidated the vast amount of grasp examples found in literature. Their taxonomy comprises 33 grasp types grouped into 17 basic types by merging equivalent grasps. The chosen taxonomy defines a grasp as *a static configuration of one hand able to securely hold an object* (in that hand). This explicitly rules out intrinsic movements in the hand during grasp, bi-manual interaction and gravity-dependent grasps such as an object lying in equilibrium on a flat hand.

Out of the list of 17 basic grasp types we have chosen to cover 13, omitting speciality configurations such as holding chopsticks (tripod variation), a pair of scissors (distal type), a cigarette between index finger and middle finger (abduction grip), and holding a lid of a bottle after unscrewing it (lateral tripod). Some

Table 1. List of grasps in controlled transport grouped by basic grasp type. It includes the grasp type's number and name based on Feix et al. [25] and the objects grasped.

№ Name	Object grasped	№ Name	Object grasped
1 large diameter	bottle (8cm diameter)	7 prismatic 3 finger	cylinder 10mm diameter
2 small diameter	cylinder 25mm diameter	27 quadpod	tennis ball (64mm diameter)
3 medium wrap	bottle crate	6 prismatic 4 finger	cylinder 10mm diameter
10 power disk	lid of jar (7cm diameter)	13 precision sphere	tennis ball (64mm diameter)
11 power sphere	tennis ball (64mm diameter)	20 writing tripod	pen (2cm diameter)
31 ring	glass (6cm diameter)	17 index finger extension	cylinder 25mm diameter
28 sphere 3 finger	tennis ball (64mm diameter)	4 abducted thumb	cylinder 25mm diameter
18 extension type	classic notebook (15mm thick)	15 fixed hook	bottle crate
26 sphere 4 finger	tennis ball (64mm diameter)	30 palmar	classic notebook (15mm thick)
9 palmar pinch	business card	16 lateral	business card
33 inferior pincer	tennis ball (64mm diameter)	32 ventral	cylinder 25mm diameter
8 prismatic 2 finger	cylinder 10mm diameter	22 parallel extension	classic notebook (15mm thick)
14 tripod	bottle at cap (3cm diameter)		

of the basic grasp types were covered more often, hence extending the number of grasp types to 25 also using different sized objects. For a complete list of grasps and objects used for controlled transport see Table 1 and Fig. 1.

The final database contains approximately 2000 grasp motions of ten different persons interacting with 15 different objects represented by 25 grasp types (13 basic grasp types). Each motion is annotated with the object that was grasped and – for controlled transport – the grasp type that was used.

4   Segmentation Approach

We present a novel method for temporal segmentation of articulated hand motion. Since our method is related to techniques in image processing on self-similarity adjacency matrices, it belongs to the same category of methods as the technique introduced in [6] on segmentation of full-body motion. However, there is a specific focus on the demands of hand motion segmentation.

In the following subsections, the process is outlined by discussing pre-processing and feature computation, segmentation, and merging.

4.1   Pre-processing

A motion consists of a sequence of n frames each containing a hand pose p_i, $i = 1, \dots, n$ which itself is encoded in a feature vector $f_i = (f_{i_j})_{j=1..N}$, of dimension N. The feature vector used for our segmentation method consists of the frame-based positions in $\mathbb{R}^3$ inferred from the outermost recorded joint in each finger and thumb (i.e. the thumb's tip and other fingers' distal interphalangeal joint) with respect to the position of the wrist joint. These positions are derived from the recorded angle data by defining a schematic hand model as depicted in Fig. 4. For each frame the recorded joint angles are mapped onto the model hand yielding 3D positions for the joints.

In order to convey temporal information in a single feature vector and to emphasise motion consistency over a certain period of time, features are stacked in the time domain. This leads to a vector $[f_{i-t_1}, f_i, f_{i+t_2}]$ of features where f_{i-t_1} is temporally located t_1 frames before f_i and f_{i+t_2} is t_2 frames after f_i.

4.2   Segmentation

The segmentation process can be broken down into two main stages: *construction of the local neighbourhood* and *identification of primitive data units*.

The *local neighbourhood* of a pose p_i is the set S_i of nearest neighbours of p_i. We construct the local neighbourhood based on the Euclidean distance d_{ij} between pairs of feature vectors (f_i, f_j) of hand poses p_i and p_j and a predefined *search radius* r. The *search radius* is a constant R depending solely on the dimensionality N of the feature vector, $r = R \cdot \sqrt{N}$. A pose p_j is added to the set of neighbours of p_i if d_{ij} is below r. This can be efficiently achieved by building a kd-tree from all feature vectors and reporting the subset of vectors located

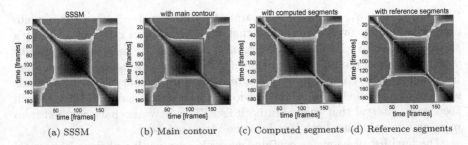

(a) SSSM   (b) Main contour   (c) Computed segments (d) Reference segments

Fig. 2. Overview of results of the steps in our segmentation method. The main contour in (b) is divided into its upper (red) and lower (green) part. (Color figure online)

within the *search radius* r for each feature vector f_i (representing pose p_i). As a result we obtain a set S_i of nearest neighbours for each pose p_i.

These sets are subsequently converted into a sparse self-similarity matrix (SSSM). In our case, this matrix holds the pairwise distances d_{ij} for all pairs of poses p_i, p_j of all sets S_i (see Fig. 2). The SSSM can thus be divided into populated regions representing pairs of poses, that are within the *search radius* (greyscale regions in Fig. 2), and empty regions in which the pairwise pose distance is outside the *search radius* (blue regions in Fig. 2).

Hands, during the grasp phase, do not exhibit major intrinsic movements. This is reflected by a large set of local neighbours and (were it ideal data) expressed in the SSSM by a square region along its main diagonal. We exploit this fact for *identification of primitive data units* and search for square-like shapes along the main diagonal of the matrix. To this end, we first extract start and end indices, t_{s_i} and t_{e_i}, $i = 1, \ldots, n$, of populated regions along the diagonal of the SSSM in a row-wise fashion. For each new index t_j we ensure that the sequence found up to t_j is monotonically increasing, i.e. $t_j \geq t_{j-1} \; \forall \, j \leq i$. That way the two index sequences each form a path contouring the upper and lower (list of end and start indices) populated region along the diagonal of the SSSM (Fig. 2 (b)). This contour does not contain any neighbour outside of the defined *search radius*.

In the following step, the sequence of end indices is inspected for significant increases, noting the end index t_{e_i} and the row i in which this increase occurs as interval boundaries $[i, t_{e_i}]$. An increase is considered significant if it is larger than a fixed parameter B, the *ignoreband*. The list of start indices is processed similarly, traversing it from back to front and seeking significant decreases. Eventually, this will result in two preliminary lists of intervals marking candidate primitive data units (rests and grasps) in the motion.

4.3 Merge Step

As we have posed only few constraints on finding square-like shaped segments in the SSSM there is sometimes a significant overlap of the preliminary segments which were identified. While minor overlaps merely illustrate that the fixed *search radius* is too high to separate two distinct structures blending into each other, a large overlap may show that the *search radius* is too low, causing a single structure to split into two. To remedy the latter, we merge segment intervals in each list of preliminary segments if they overlap one another by more than half their widths. Lastly, the two lists of merged preliminary segment intervals are merged into one list under the previously mentioned condition. Here, we will keep every interval that existed in both lists. This may also have resulted from merging intervals from both lists into one bigger segment.

The final list of segments contains intervals representing phases of rests and grasps. The phases inbetween two consecutive segment intervals contain the motion transitioning from the one segment into the next.

In our extended version of the merge step, we additionally reproject candidate intervals from a merge back into the SSSM and check if the region covered by the corresponding square is populated by at least 95 %. A merge is only performed if it does. For exemplary results of the segmentation and merge step see Fig. 2 (c).

5   Clustering Approach

Once a set of motion trials has been segmented into primitive data units by extracting square-like regions from the main diagonal of a SSSM, we can group similar segments and moreover group similar motions within the set of trials.

5.1   Clustering of Primitives

Primitive data units are represented by squares along the main diagonal of a SSSM. Looking more closely at the populated areas in the SSSM, we find an interesting structure mainly consisting of off-diagonal blobs. These blobs indicate similarity between motion segments (see Fig. 2). We will avail ourselves of this structure for clustering motion segments based on their similarity with respect to *segment area coverage* and *path coverage*, respectively, as described below.

Consider a set of motion trials segmented into primitive data units $\mathcal{I} = I_1, \ldots, I_K$. In order to keep track of pairwise similarities within $\mathcal{I}$, we build a similarity graph $G_{\mathcal{I}}$. Each primitive $I_k, k = 1, \ldots, K$, is associated with one node in a similarity graph $G_{\mathcal{I}}$. Two nodes I_v, I_w will be connected by an edge if they are considered similar (based on the similarity measures described below). The final graph will consist of several strongly connected components representing clusters of similar types of primitive data units.

As the segmentation is performed per motion trial, but clustering is aimed at inter-trial comparison of primitive data units, we have to construct the local neighbourhood for each pair of primitives. The comparison of the primitives is based on the resulting sets of local neighbours which, for clustering, are not further converted into a SSSM.

Segment Clustering by Segment Area Coverage. A simple straightforward approach for comparing primitive data units uses the area covered by each off-diagonal blob restricted to the range of the compared primitives in their

SSSM. If this area is sufficiently covered/populated, we add an edge between the nodes representing the compared primitives. This approach, however, is incapable of representing a temporal alignment of the segments needed for re-use of the data in, e.g. motion synthesis. Also note that while the SSSM here is immensely useful for explaining the underlying concept of this approach, we in fact count the number of nearest neighbours in each of the relevant sets.

Segment Clustering by Path Coverage. A classical way of searching for the best temporal alignment of two time series $Q = \{q_1, \ldots, q_N\}$ and $V = \{v_1, \ldots, v_M\}$ is dynamic time warping (DTW). The alignment is given as an optimum cost warping path $P_{Q,V}$ between Q and V. A path $P_{Q,V}$ of length L is constituted by a sequence $\pi = \{\pi_1, \ldots, \pi_L\}$ of index pairs $\pi_l = (n_l, m_l) \in [1, N] \times [1, M] \subset \mathbb{N} \times \mathbb{N}$ for $l \in [1, L] \subset \mathbb{N}$ into Q and V subject to constraints such as $q_{n_l} \leq C \cdot q_{n_{l+1}}$ and $v_{m_l} \leq C \cdot v_{m_{l+1}}$ thus limiting the path's *slope*.

This kind of motion matching has been elegantly solved by Krüger et al. [27] using the sets of local neighbours in a neighbourhood graph. They represent subsequence DTW as kd-tree-based fixed-radius nearest neighbour search in the motions' feature set. For details see [27].

We compute the warping path for each pair of motion segments. Such a path is considered to be valid if it sufficiently covers the segments and does not fall below a certain length. If we find a valid warping path between two segments we record this information in the graph $G_\mathcal{I}$ by adding an edge between the two nodes representing the compared primitives.

This algorithm was adapted from Krüger et al. [19] and Vögele et al. [6].

5.2 Clustering of Motion Sequences

After clustering of the segments, we are able to represent each trial by a sequence of IDs. Each ID represents a specific cluster of primitive data units. These sequences of IDs are subsequently used to group similar motion trials: First, each motion trial's sequence of IDs is processed such that it is free from successive identical IDs, e.g. 11123331 becomes 1231. For all pairs of sequences we compute their weighted longest common subsequence (LCS), i.e. we divide the pairwise LCS by the minimum length of the compared sequences. Based on this similarity measure, motion trials are grouped into final motion trial clusters.

Finally, each cluster is assigned to the class of the motion trial occurring most frequently in the cluster. Clusters containing exactly one motion trial will be disregarded and later counted as being unidentified.

6   Results

This section summarises achieved results for segmentation and clustering. We will start with an evaluation of different values for the two important segmentation parameters. Subsequently, we will compare our segmentation approach with the method proposed by Vögele et al. [6]. Concluding this section we will present results from clustering found motion segments.

6.1   Parameter Evaluation

The parameter space of the discussed method is determined by the *search radius constant R* and the *ignoreband B*. In our evaluation, we have iterated over both values with respect to the two different quality criteria, *overlap union ration* and *cut localisation*, which our segmentation evaluation is based on. The results are given in Fig. 3. Each parameter has a separate contribution in the overall segmentation results as the *search radius* is responsible for the structure of the sets of local neighbours, and hence determines the structure of populated and empty regions in the SSSM. The *ignoreband* has a major influence on the minimum detectable segment interval width.

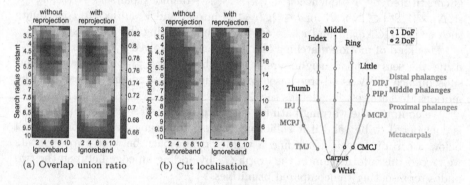

(a) Overlap union ratio   (b) Cut localisation

Fig. 3. Results of the parameter evaluation with respect to the presented segmentation quality measures. *Search radius constant* was plotted against *ignoreband* without (each left) and with (each right) reprojecting candidate intervals from merges. The optimum value for *overlap union ration* is 1.0 and for *cut localisation* 0.0.

Fig. 4. Schematic kinematic chain of the hand model used in this work. Abbreviated joints spell out (proximal, distal) interphalangeal joints (PIPJ, DIPJ, IPJ), metacarpophalangeal joint (MCPJ), carpometacarpal joint (CMCJ) and trapeziometacarpal joint (TMCJ).

The results demonstrate that the method performs well for a range of parameters, i.e. for $R \in [3.5, 5.5]$ and $B \in \{2, 3, 4, 5, 6\}$. Based on optimisation in this parameter space we have chosen $R = 4.25$ and $B = 4$ for all the experiments presented below.

6.2   Segmentation

To evaluate our segmentation algorithm, 50 prehensile motions from our database of controlled grasps were randomly chosen and annotated as belonging to one of five phases: rest, reaching, grasp, retraction, and rest. Reaching and retraction are transitional phases. These manually segmented motions will be referred to as *reference (segmentation)* as opposed to *automatic* or *computed segmentation*. In terms of features, we have chosen to stack five frames

$[f_{i-6}, f_{i-1}, f_i, f_{i+1}, f_{i+6}]$ leading to a 75-dimensional feature vector and a *search radius* of $r = 4.25 \cdot \sqrt{75}$ [cm] ≈ 37 [cm]. For better comparability with [6] we have converted the intervals found by our segmentation into a sequence of cuts and regard each two consecutive cuts as an interval. In order to assess the quality of a computed segment interval I_c with respect to a reference segment interval I_r we use the following measures:

Quality Measures. The *overlap union ratio* relates the overlap of two segment intervals I_r, I_c to the width of their union, $\frac{|I_r \cap I_c|}{|I_r \cup I_c|}$. In case of multiple computed segment intervals overlapping a reference interval we use the largest computed segment and disregard the others. Overlap union ratio is invariant to the position within a complete overlap. It ranges from 0 (no overlap) to 1 (exact overlap).

The second measure (*cut localisation*) considers detected segments as a linear list of segment cuts. It computes the distance of start/end frames in a reference segment to the closest computed cut. The conversion from segment intervals to cuts is straightforward and is computed by simply concatenating the interval boundaries and dropping the motion's first and last frame. Additionally, and to avoid favouring over-segmentation, the *total number of cuts* in the complete motion trial for both the computed and reference segmentation is reported.

Discussion of Results. For a visual comparison of segmentation results refer to Fig. 5. Segments are colour-coded according to the best matched reference segment. Segments found by our methods highly coincide with the reference segments in both, number of found segments and position of these segments within each motion trial. Our approach outperforms that of Vögele et al. [6] which often misses segments with respect to the reference segmentation.

This tendency is confirmed in the evaluation of the segmentation quality based on the aforementioned measures. Figure 6 shows histograms of absolute

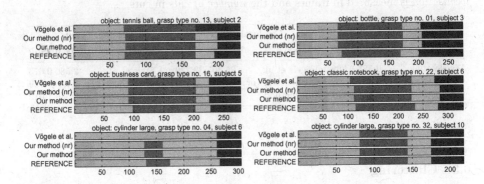

Fig. 5. Segmentation results for a number of annotated motion trials. For each trial, the bottom row displays the reference annotation. The two middle rows show our results and the top row depicts results of Vögele et al. [6]. We use '(nr)' throughout figures and tables to abbreviate 'no reprojection' (of candidate segment intervals).

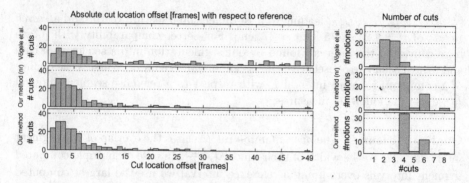

Fig. 6. Cut localisation (absolute number of frame offsets) and number of cuts with respect to to the reference segmentation. Reference segmentations consist of four cuts. Top row depicts results from the approach of Vögele et al., the second row depicts our approach without reprojection and the last row is our approach with reprojection.

cut location offsets and number of cuts identified in the motions from the three methods. Our methods not only mainly find the correct number of (four) cuts for each motion but also with little frame offset with respect to the reference segmentation. Mean values and standard deviations of the quality measures used are listed in Table 2. Here, our methods reach an *overlap union ratio* of almost 0.84 with a very low standard deviation of 0.15. The combination of mean *cut localisation* of our method with and without reprojection (5.09 frames and 4.85 frames with a standard deviation of 6.85 and 6.43 frames) and mean *number of found cuts* (4.58 and 4.74 cuts with a standard deviation of 1.01 and 1.14 cuts) further confirm that our segmentation closely matches the reference.

Table 2. Mean and standard deviations for different methods and different evaluation measures. *Overlap union ratio* ranges from 0 (no overlap) to 1 (exact overlap), *cut localisation* is measured in frames and the *number of cuts* in cuts.

Overlap union ratio		Cut localisation			Number of cuts		
Method	Mean Std.	Method	Mean	Std.	Method	Mean	Std.
Vögele et al.	0.525 0.369	Vögele et al.	25.305	32.229	Vögele et al.	2.580	0.673
Our method (nr)	0.837 0.153	Our method (nr)	4.845	6.429	Our method (nr)	4.740	1.139
Our method	0.838 0.151	Our method	5.090	6.851	Our method	4.580	1.012

6.3   Clustering

For clustering motion trials we used an excerpt of the controlled transport setup in our database. This excerpt covers all objects featured in the experiment and was divided into ten sets based on the test person. Hence, each class in this section represents an object. Unlike Krüger et al. [27], we chose to allow

$\{(1,1),(1,2),(2,1),(1,4),(4,1)\}$ as warping steps to account for the mainly static nature of our primitive data units. A segment I_a is considered to sufficiently cover a second segment I_b if the off-diagonal area of I_a and I_b in their SSSM is populated by at least 0.66 %. For path coverage this refers to coverage in the horizontal and vertical extent of a computed warping path.

As a side effect of how clusters are affiliated with motion trial classes, each cluster represents one class, but classes may spread multiple clusters. In this section, we will present measures for assessing the quality of our clustering process and discuss the results.

Quality Measures. In order to evaluate the accuracy of this assignment we use *cluster purity* as well as *precision, recall,* and F_1*-score* of the clustering. *Purity* measures the quality of a cluster by putting the number of correctly assigned motion trials in relation to the total number of motion trials. This does not take into account the number of clusters with respect to the number of actual classes, so we also give the total number of clusters and the number of classes (objects) to be represented by them.

Two trials should only be in the same cluster if they are similar and should be in different clusters if they are dissimilar. Based on this, we can derive the *precision* of the clustering as the number of (pairs of) trials correctly-grouped into the same cluster with regard to the total number of (pairs of) trials in these clusters. This quantifies the correctness of the separation of dissimilar trials into different clusters, or, put differently, the amount of correct predictions. Conversely, *recall* puts the number of trials correctly grouped into the same cluster in proportion to the number of trials that should have been grouped into the same cluster. This measures the success of avoiding separation of similar trials into different clusters or the ability to group trials by similarity. Finally, the F_1*-score* combines precision and recall through their harmonic mean, thus conveying the balance between the both.

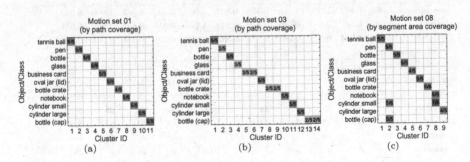

Fig. 7. Results of clustering motion sequences. Colouring is based on the number of motions in a class grouped together and ranges from green (similar trials grouped together) over yellow to red (trial grouped with dissimilar trials). Non-zero entries contain the number of motions of a class in a cluster (left) and the number of motions in that class (right). Multiple entries in a column represent a cluster covering multiple classes, multiple entries in a row indicate a class split into multiple clusters.

Discussion of Results. Figure 7 illustrates results obtained from clustering motion trials while Table 3 provides quantitative results obtained from evaluating the motion trial clustering. As can be seen from the table all test sets reached high cluster purity while the number of clusters nearly match the number of classes/objects. It should be noted that because we are basing computation of precision (and recall) on pairs of motion trials, incorrectly clustered trials strongly influence precision. This effect is less pronounced for recall.

The bottom half of Table 3 illustrates that we can reach high precision and recall values in many cases for clustering by path coverage. The minimum value for precision (recall) amounts to 0.71 (0.832). The top half of the table summarises results for clustering by segment area coverage. Overall, values are slightly lower than for clustering by path coverage. Minimum precision drops to 0.56, which is

Table 3. Results of the clustering for all sets. The last column contains the mean (or, where appropriate, the total) of the quality measures. We use # to abbreviate 'number of'. In the listing *found* counts the number of motion trials grouped into clusters with other trials, *correct* is the number of correctly grouped similar trials, *incorrect* denotes the number of incorrectly grouped dissimilar trials, and *unidentified* lists the number of trials that could not be grouped with other trials. Because we do not measure precision and recall based on single class division but based on pairs of motion trials, these values cannot be directly derived from the number of found, correct, etc. trials.

Criterion	Set 01	Set 02	Set 03	Set 04	Set 05	Set 06	Set 07	Set 08	Set 09	Set 10	Mean/Total
Segment area coverage											
# clusters	11	10	14	12	9	11	10	9	12	11	10.9
# classes	11	11	11	11	9	11	11	11	11	11	10.8
purity	1.000	0.811	1.000	1.000	0.860	0.926	1.000	0.796	0.944	1.000	0.934
# trials	55	55	55	55	44	56	55	57	55	55	542
found	53	53	48	52	43	54	42	54	54	50	503
correct	53	43	48	52	37	50	42	43	51	50	469
incorrect	0	10	0	0	6	4	0	11	3	0	34
unidentified	2	2	7	3	1	2	13	3	1	5	39
precision	1.000	0.662	1.000	1.000	0.709	0.835	1.000	0.560	0.862	1.000	0.863
recall	1.000	0.961	0.837	0.939	0.890	0.944	1.000	0.953	0.887	1.000	0.941
F_1-score	1.000	0.784	0.911	0.969	0.789	0.886	1.000	0.706	0.874	1.000	0.892
Path coverage											
# clusters	11	11	14	11	10	13	12	10	11	11	11.4
# classes	11	11	11	11	9	11	11	11	11	11	10.8
purity	1.000	0.906	1.000	1.000	0.976	0.963	1.000	0.889	0.855	0.980	0.957
# trials	55	55	55	55	44	56	55	57	55	55	542
found	54	53	48	50	42	54	48	54	55	51	509
correct	54	48	48	50	41	52	48	48	47	50	486
incorrect	0	5	0	0	1	2	0	6	8	1	23
unidentified	1	2	7	5	2	2	7	3	0	4	33
precision	1.000	0.797	1.000	1.000	0.934	0.918	1.000	0.766	0.710	0.949	0.907
recall	1.000	0.961	0.837	1.000	0.899	0.832	0.929	0.916	0.891	0.959	0.922
F_1-score	1.000	0.871	0.911	1.000	0.916	0.873	0.963	0.834	0.790	0.954	0.911

due to the fact that three different classes are identified as equivalent, and hence share the same cluster and heavily influence precision (compare Fig. 7 (c)).

7 Limitations

Discussion of Segmentation Limitations. While the approach of Vögele et al. [6] tends to miss short segments in particular, both our proposed methods are more likely to over-segment the motions (see Fig. 8). This is due to the fact that our algorithm has posed a relatively strict condition on how to find the segments during extraction of the diagonal contour. This leads to every gap in the SSSM within a square-like region along the diagonal causing our algorithm to start a new segment (segments stretch from the interior of the diagonal outwards). By contrast, the approach by Vögele et al. [6] introduce cuts whenever the main diagonal band is interrupted (segments wrap the around the diagonal structure from the exterior).

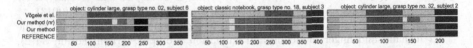

Fig. 8. Segmentation results illustrating over-segmentations by our algorithm. Reprojecting candidate intervals can help alleviating this issue (left and right).

Discussion of Clustering Limitations. By basing our clustering essentially on feature similarity we implicitly assume that these features are able to discriminate well between classes. For grasping this is not entirely true as the configuration of the hand strongly depends on the size and shape of the object as well as on the grasp applied to hold the object. This, on the one hand, can lead to our clustering separating single classes into multiple clusters (Fig. 7 (b)) and, on the other hand, to grouping multiple classes of similar objects into the same cluster (Fig. 7 (c)).

8 Conclusion and Future Work

In this paper, we presented a database of prehensile movements and a novel method for temporal segmentation of articulated hand motion. One of our goals was to present an effective method for segmentation and clustering of hand data. Our experiments confirm a high coincidence of our results with manual segmentation (cf. Sect. 6.2). Also, comparison to the clustered results of Vögele et al. [6] shows that both our evaluation methods (path coverage and segment coverage) yield higher accuracy scores (refer to Table 3). Particularly, the recall values are convincing compared to the relatively poor results by Vögele et al.

Acknowledgement. We would like to thank Fraunhofer IAO for providing us with the CyberGlove used to record the motion data. We also thank the authors of [6] for providing source code of their method for comparison.

References

1. Jörg, S., Hodgins, J., O'Sullivan, C.: The perception of finger motions. In: Proceedings of the APGV, pp. 129–133 (2010)
2. Jörg, S., Hodgins, J.K., Safonova, A.: Data-driven finger motion synthesis for gesturing characters. ACM Trans. Graph. **31**(6), 189:1–189:7 (2012)
3. Zhao, W., Chai, J., Xu, Y.Q.: Combining marker-based mocap and RGB-D camera for acquiring high-fidelity hand motion data. In: Proceedings of the ACM SCA, pp. 33–42 (2012)
4. Tognetti, A., Carbonaro, N., Zupone, G., Rossi, D.D.: Characterization of a novel data glove based on textile integrated sensors. In: IEEE EMBS, pp. 2510–2513 (2006)
5. Dipietro, L., Sabatini, A.M., Dario, P.: A survey of glove-based systems and their applications. IEEE Trans. SMC-C **38**(4), 461–482 (2008)
6. Vögele, A., Krüger, B., Klein, R.: Efficient unsupervised temporal segmentation of human motion. In: Proceedings of the ACM SCA (2014)
7. Zhou, F., De la Torre, F., Hodgins, J.K.: Hierarchical aligned cluster analysis for temporal clustering of human motion. IEEE Trans. PAMI **35**, 582–596 (2013)
8. Wheatland, N., Wang, Y., Song, H., Neff, M., Zordan, V., Jörg, S.: State of the art in hand and finger modeling and animation. Comput. Graph. Forum **34**(2), 735–760 (2015)
9. Arkenbout, E.A., de Winter, J.C.F., Breedveld, P.: Robust hand motion tracking through data fusion of 5DT data glove and Nimble VR kinect camera measurements. Sensors **15**(12), 31644–31671 (2015)
10. Ju, Z., Liu, H.: Human hand motion analysis with multisensory information. IEEE/ASME Trans. Mechatron. **19**(2), 456–466 (2014)
11. CMU: Carnegie Mellon University Graphics Lab: Motion Capture Database (2013)
12. Müller, M., Röder, T., Clausen, M., Eberhardt, B., Krüger, B., Weber, A.: Documentation Mocap database HDM05. Technical report CG-2007-2, Universität Bonn (2007)
13. Goldfeder, C., Ciocarlie, M.T., Dang, H., Allen, P.K.: The columbia grasp database. In: IEEE ICRA, pp. 1710–1716 (2009)
14. Feix, T., Romero, J., Ek, C.H., Schmiedmayer, H.B., Kragic, D.: A metric for comparing the anthropomorphic motion capability of artificial hands. IEEE Trans. Robot. **29**(1), 82–93 (2013)
15. Atzori, M., Gijsberts, A., Heynen, S., Hager, A.G.M., Deriaz, O., van der Smagt, P., Castellini, C., Caputo, B., Müller, H.: Building the ninapro database: a resource for the biorobotics community. In: Proceedings of the IEEE/RAS-EMBS BioRob, pp. 1258–1265 (2012)
16. Beaun, P., Coros, S., van de Panne, M., Poulin, P.: Motion-motif graphs. In: Proceedings of the ACM SCA, pp. 117–126 (2008)
17. Zhou, F., la Torre, F.D., Hodgins, J.K.: Aligned cluster analysis for temporal segmentation of human motion. In: Proceedings of the IEEE CAFGR (2008)
18. Min, J., Chai, J.: Motion graphs++: a compact generative model for semantic motion analysis and synthesis. ACM Trans. Graph. **31**(6), 153:1–153:12 (2012)
19. Krüger, B., Vögele, A., Willig, T., Yao, A., Klein, R., Weber, A.: Efficient unsupervised temporal segmentation of motion data. CoRR abs/1510.06595 (2015)
20. Levine, S., Theobalt, C., Koltun, V.: Real-time prosody-driven synthesis of body language. ACM Trans. Graph. **28**(5), 172:1–172:10 (2009)

21. Mousas, C., Anagnostopoulos, C.N., Newbury, P.: Finger motion estimation and synthesis for gesturing characters. In: Proceedings of the SCCG, pp. 97–104 (2015)
22. Ekvall, S., Kragic, D.: Grasp recognition for programming by demonstration. In: Proceedings of the IEEE ICRA, pp. 748–753 (2005)
23. Kang, S.B., Ikeuchi, K.: Determination of motion breakpoints in a task sequence from human hand motion. In: Proceedings of the IEEE ICRA, vol. 1, pp. 551–556 (1994)
24. Zhao, W., Zhang, J., Min, J., Chai, J.: Robust realtime physics-based motion control for human grasping. ACM Trans. Graph. 32(6), 207:1–207:12 (2013)
25. Feix, T., Pawlik, R., Schmiedmayer, H.B., Romero, J., Kragic, D.: A comprehensive grasp taxonomy. In: Robotics, Science and Systems: Workshop on Understanding the Human Hand for Advancing Robotic Manipulation (2009)
26. Feix, T., Romero, J., Schmiedmayer, H.B., Dollar, A.M., Kragic, D.: The grasp taxonomy of human grasp types. IEEE Trans. HMS 46(1), 66–77 (2016)
27. Krüger, B., Tautges, J., Weber, A., Zinke, A.: Fast local and global similarity searches in large motion capture databases. In: Proceedings of the ACM SCA, pp. 1–10 (2010)

A Method for Predicting Words by Interpreting Labial Movements

Osvaldo Gervasi[1(✉)], Riccardo Magni[2], and Matteo Ferri[3]

[1] Department of Mathematics and Computer Science,
University of Perugia, Perugia, Italy
osvaldo.gervasi@unipg.it
[2] Pragma Engineering SrL, Perugia, Italy
[3] Prometeia SpA, Bologna, Italy

Abstract. The study of lips movements is relevant for a series of interesting applications in real world to enhance the communication means and in medical applications. In the present paper we illustrate a method we implemented with the purpose of helping Amyotrophic Lateral Schlerosys (ALS) patients to communicate, once the progress of the disease requires to intubate the patient and the voice is lost.

The Method uses several subsystems to carry out a so complex task and the results are really promising. However the method need to be improved in order to make the system more easy to use and more reliable in the prediction of pronounced words.

1 Introduction

The progress made in recent years in the medical field and the enormous technological developments that are having a big impact in our daily lives pushing research towards increasingly ambitious and challenging frontiers.

At the same time we must consider how the aging of modern society imposes choices to streamline healthcare costs, trying to keep unchanged the services provided to end-users. This is an exciting challenge in which we have been involved for years trying to provide answers increasingly suitable to modern needs, realizing the digitization, automation and optimization of resources and services that will surely be the expected response considering the modern requirements. [6–8]

In the present paper we afforded a very challenging task: the identification of the words pronounced based on the lips movements, so that we can understand the words of patients who, because of their disease, have lost their voice. The problem is very complex and involves several disciplines. However we can assert that the system implemented, although far from being used in the every day life, has interesting features and stimulates further research to consolidate and optimize the results achieved so far.

The present paper is organized as follows: in Sect. 2 the most promising lip tracking algorithms are presented and discussed. After the lips identification and tracking, the main features are extracted considering four Points of Interests (POI), as discussed in Sect. 3. In Sect. 4 is presented the algorithm implemented

O. Gervasi et al. (Eds.): ICCSA 2016, Part II, LNCS 9787, pp. 450–464, 2016.
DOI: 10.1007/978-3-319-42108-7_34

for the classification of visemes, based on neural networks approach. The algorithm used for the prediction of the words associated to the detected visemes is illustrated in Sect. 5. Preliminary results of the words prediction made by the system are illustrated in Sect. 6 and finally in Sect. 7 some conclusions and the future works are presented.

2 Lip Tracking Techniques

The identification of lips in an image is a complex task, which involves the segmentation of the image and the object detection techniques. The problem is well known in Computer Vision, since the complexity of such a process influences the general performances and the quality of results of all components in which this task is involved. The target of our study is the implementation of a method performing a stable and reliable lips tracking that allows the recognition of words. To this purpose we analyzed the most promising algorithms of object tracing and object detection available in literature and we report our evaluation.

2.1 Active Shape Model

The Active Shape Model (ASM) is a statistical localization technique introduced by Tim Cootes [4], which in recent years attracted with success a lot of interest. The method has been formulated with the aim of providing a very accurate localization of an object of interest; it provides a representation for the shapes to be searched, using the model defined by a set of points which characterize the objects. It has been applied successfully in various applications. In our case, we do not simply need the definition of the bounding box containing the lips: we need to define the exact position of the external and internal contours. In other words, we need the *lips localization*, instead of the lips detection.

The main drawbacks we encounter developing methods based on ASM are the following: (a) the laying of faces that may appear frontally, slightly in profile, etc., and may result in partial or complete occlusion of some components; (b) some properties that may be present or not, such as mustache and beard, which may have a large variability, in terms of shape, size and color; (c) different color or shape of the lips; (d) poor image quality, due to special lighting conditions or other characteristics, make that the acquisition devices can introduce noise.

The ASM methods are based on a construction of a shape that can be modified and deformed to map the object that we are looking in the picture. A priori knowledge of the form type constrains the deformation of the model and makes that the method is robust with respect to noise or to partial occlusions. The model is characterized by a set of points that identify the relevant characteristics of the shape and, at each iteration, one must determine how to move them so that the shape is adapted to the object in the image. For each point, one must identify what is the most suitable position in which to move it. The iterative method used for this research is the weak point of the algorithm, as it requires a relatively long time.

To decide how to move the points, we have to select an area where looking for the most promising points. To identify the best candidates, some color models are used, estimated during the construction of the shape model, describing the appearance of the image nearby the target points.

For each point in the search area is extracted the same representation used to estimate the color model. This representation is compared with the estimated model, by means of the Mahalanobis [15] distance: the point for which is obtained the minimum distance is used for moving. The experimental results showed that the algorithm is very sensitive to the position within the image from which is started the search: if the starting position is selected far from the correct zone, the use of the segments will not lead to the desired positions, even after many iterations.

Even if very promising, the method has not being used because of the high computational time and the low rate of frame processing (7 frames/sec). The method is useful for single images, but when applied to a streaming video it is inadequate.

2.2   Active Contour - Snake

The deformable models are modeled using geometrical representations with some degrees of freedom, like the splines. The deformations are then placed according to physical principles with which it is prevented that the model evolves in an uncontrolled way; from a physical point of view the deformable models are nothing more than elastic bodies that deform in response to forces and constraints applied on them. The deformable models are associated to energy functions defined according to the degrees of freedom that the bodies may exhibit; the energy function is constructed to include constraints súch as stiffness or the symmetry of form, and grows as the model deforms away from the desired shape. The optimization theory is used to minimize the energy, iteratively forcing the deformation of the model to obtain the optimum matching with real data.

Ones of the approaches based on deformable models that had greater success are the Snakes (also known as Active Contour) proposed in 1987 by Kass, Witkin and Terzopoulos [11]. The snakes are based on modeling shape through deformable splines, which are associated to energy functional whose minimization binds the spline to move toward the desired features. The spline deforms consequently to internal forces and external forces. The internal forces are independent from the image in which it searches, act only on the spline and impose constraints on the rigidity or smoothness of form. External forces instead are used to direct the snake to certain features of the analyzed image; varying the functional which defines the external energy one snake can be used to identify lines or edge.

The use of snakes is useful for the detection of contours in cases where the shapes have a high variability from one another and you do not have a priori information sought to impose restrictions on the form, since the spline subject to various forces. It is able to adapt independently to the image. In practice, snakes do not provide for the construction of any shape model to be deformed, but simply they fit to the image in which they are initialized. This model has been successfully adopted in Lip Tracking [14].

However, the use of this method in our context has proved to be not very stable and accurate; in fact, the contour points, tend to move slightly even in absence of true movements, making it difficult the construction of the necessary features forces the classifier. The cause of this instability should not however be attributed to the method, but to the poor quality of the Webcam, which makes unclear the contours. In fact the contour points tend to move even in absence of movement.

2.3 Space Color Based

One of the most popular techniques in Computer Vision is the segmentation of the image based on the color. The main issue related to this method is the high variability of the color in the RGB space.[1] This method is heavily affected by small variations of luminosity and color.

To assess the goodness of this method applied to our needs, we considered several implementations of the method in the case of lips recognition (see for example [10,16]) and since each method has advantages and disadvantages we combined the best features of each method. Our need is to classify all pixels of the image based on the characteristics "lip" and "non-lip", so that we may be able to get only "lip" classified pixels. In addition we need a method that is independent from external conditions like lighting, the lips color which varies from an individual to another, presence of beard, and noise created by the acquisition system.

The adopted method consists in evaluating the QSI channels, which definition is provided in the next paragraphs, that are extracted from RGB in order to emphasized the lip detection.

The channel Q is the one of the YIQ color classification introduced by Walter Buchsbaum in [1] and defined by Eq. 1. The YIQ system is intended to take advantage of human color-response characteristics.[2] We apply the YIQ transformation to increase the differences between the pixels of the lips and those of the skin. In Q channel the lips are represented with pixel lighter than the skin.

$$\begin{bmatrix} Y \\ I \\ Q \end{bmatrix} = \begin{bmatrix} 0.299 & 0.587 & 0.114 \\ 0.595 & -0.274 & -0.321 \\ 0.211 & -0.522 & 0.311 \end{bmatrix} \begin{bmatrix} R \\ G \\ B \end{bmatrix} \tag{1}$$

The channel S uses a non-visible to the human eye property, which is significant in the composition of the color of the lips; in fact, observing the RGB space of an image, skin pixels appear always more yellow than the pixels of the lips. The S channel is evaluated as shown in Eq. 2.

$$S = \frac{R - G}{R} \tag{2}$$

---

[1] RGB is a classification for the colors expressed in terms of the triple expressing the amount of the Red, Green and Blue colors, each ranging from 0 to 255.

[2] YIQ is the color space used by the NTSC color TV system used mainly in North America, Central America and Japan.

The I channel, introduced by Canzlerm and Dziurzyk [3], considers that the Blue channel (B) in the RGB representation plays a marginal role in the discrimination of lips from the skin. I is defined in Eq. 3.

$$I = \frac{2G - R - 0.5B}{4} \tag{3}$$

Three empirical functions have been introduced to take into account the hardware characteristics of the camera used to capture the images, described in Eqs. 4, 5 and 6, where α, β and γ are empirical constants that incorporate the characteristics of the camera.

$$f(Q) = \begin{cases} -2 * (\alpha - Q) & \text{if } Q \leq \alpha \\ 4 * Q & \text{otherwise} \end{cases} \tag{4}$$

$$f(S) = \begin{cases} -3 * (\beta - S) & \text{if } S \leq \beta \\ 3 * S & \text{otherwise} \end{cases} \tag{5}$$

$$f(I) = \begin{cases} -(\gamma - I) & \text{if } I \leq \gamma \\ I & \text{otherwise} \end{cases} \tag{6}$$

These values are then combined according to Eq. 7, and the obtained result is shown in Fig. 1. The lips pixel are easily identified using a threshold function that selects the lighter pixels.

$$QSI = \frac{f(Q) + f(S) + f(I)}{5} \tag{7}$$

This type of image has been used as the input for the Snake method described in Sect. 2.2, obtaining very good results: the identification method is robust, working well in presence of beard and mustache and over a wide range of lighting conditions.

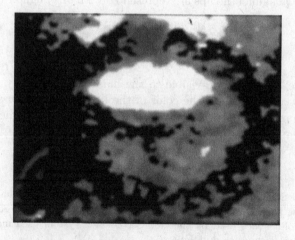

Fig. 1. Outcome of combining channels Q,S and I.

However, this method has not been used since the contour tends to slightly change its position, also by standing firm in front of the camera. This is due to the noise introduced by the input device, and the compulsory choice of introducing thresholds for filtering, giving slightly different results between a frame and the following one. These small movements make the classification of the detection algorithm unstable and inaccurate. For this reason we preferred to use the technique described in Sect. 2.4, that is more stable and flexible.

2.4   Template Matching

Template matching is a method for object recognition based on a target image (template) associated to the object or area to be detected that is compared to the current image. We may use a distance metric which assesses them similarity between the template and an area of the current image.

If t is the target image and i is the search window centered in (u, v), we may use as a metric the euclidean distance, shown in Eq. 8.

$$d_{i,t}(u, v) = \sum_{x,y} [i(x, y) - t(x - u, y - v)]^2 =$$

$$= \sum_{x,y} i(x, y)^2 + \sum_{x,y} t(x - u, y - v)^2 - 2 \sum_{x,y} i(x, y) t(x - u, y - v) \quad (8)$$

The first two terms are constant, while the third one expresses the cross-correlation between the two images: the template and the search windows.

$$c_{i,t}(u, v) = \frac{\sum_x \sum_y i_{u+x, v+y} t_{x,y}}{\sqrt{\sum_x \sum_y i^2_{(u+x, v+y)} \sum_x \sum_y t^2_{(x,y)}}} \quad (9)$$

At each new frame captured by webcam, the template is moved in all possible points within the image, looking one that minimizes the above function. In addition to the template position it is also possible to have a measure of the likelihood of the template and the current frame.

This technique, although the theory on which it is based may seem simpler than other techniques, in practice behaves very well offering a good compromise between computational speed, accuracy and stability.

2.5   Template Acquisition

To solve the labial comprehension problem, we applied the Template matching method at the same time to four points, called *Points of Interest* (POI), that represent the vertex of the lips, illustrated in Fig. 2.

It should be noted that the template size has a significant effect on both the stability and the performance. Larger is the template, more precise will be the method, since a larger surface is available to calculate the cross-correlation, and more computational time will be required. The optimal size of the template has been evaluated empirically: 25×25 pixels. The templates vary on the individual characteristics, so that they have to be selected before using the system.

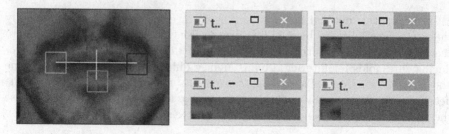

Fig. 2. Acquisition of the labial templates of the *Point of Interest*: Top, Bottom, Left and Right

2.6   Tracing Lips Movements

Once defined the templates of the Points of Interest, we have to define a method for tracing the lips movements. The analysis has been limited to a portion of the frame, to improve the stability and the speed of the process. For each template of dimensions 25×25 pixels, a *search area* of 35×35 pixels is considered.

By limiting the area to be investigated, it is vital to select appropriately the search area of each POI for each new frame. We have created a system of constraints between the POI by which for each frame is chosen in an appropriate manner in which zone of the image the search area has to be moved. To this purpose we used the information known a priori on the shape of the lips, which remains the same for every individual, and the location of each POI respect to the others. For example the *top* POI is always centered respect to the *right* and *left* POIs, while the *bottom* POI is always positioned below the axis defined by the *right* and *left* POIs.

To coordinate the repositioning of the search area the cross-correlation coefficient between the template and the image is considered. Furthermore, a series of constraints between POIs have been defined, to take into account the POIs relative positions.

3   Lips Movements Reading

Once having defined a reliable method able to follow the lips movements, we have to extract the features that will enable the neural network engine to classify the visemes[3] during the pronunciation of words. According to the literature [2] we decide to consider 4 parameters: **HD** (horizontal distance: distance in pixels between *left* and *right*), **VD** (vertical distance: distance between *top* and *bottom*), **DA** (dark area internal to the lips), **PD** (proportional distance: distance between *top* and the axis *left-right*).

---

[3] A viseme is a generic facial image that can be used to describe a particular sound. A viseme is the visual equivalent of a phoneme or unit of sound in spoken language. Using visemes, the hearing-impaired can view sounds visually - effectively, "lip-reading" the entire human face.

To evaluate **DA** the image is transformed from RGB to QSI and the internal zone of the rhomboid delimited by the four POI as vertex is analyzed. The internal area is evaluated considering a dynamical threshold function that identify the darkest area. The dynamical threshold is evaluated according to Eq. 10.

$$S_{dark} = \alpha * \frac{\sum_{i=1}^{n} I(x_i, y_i)}{n} \tag{10}$$

where n is the number of pixels of the image, $I(x, y)$ is the intensity of the pixel in position x, y and is a constant set to 0.5, in agreement with the experimental results. Discrimination between dark pixels and non-dark pixels occurs according to Eq. 11.

$$\begin{cases} if I(x,y) \leq S_{dark} & \text{dark pixel} \\ otherwise & \text{non-dark pixel} \end{cases} \tag{11}$$

The information gathered by the video acquisition system are then passed to the neural network that will provide the classification of visemes detected in the frames.

4   Classification of Visemes

The subsystem implemented for the classification of visemes involves the use of a neural network to classify in one of the possible visemes the movement of the lips, processing the parameters obtained from the video acquisition subsystem.

A viseme is the minimum unit of representation for a particular posture of the lips during the articulation of the words. To each viseme is associated one or more phoneme and moreover the visemes depend on the language.

In the present work we implemented the various classes of Italian visemes and then we trained the neural network in the visemes classification. In their work [2], Magno et al. identified the visemes of the Italian language, only for the consonants, observing the spatio-temporal characteristics and the movements of the lips relative to the displacement of the upper and the lower lip. They identified five visemes associated to different phonemes, which are shown in Table 1.

Table 1. The phonemes associated to each of the five visemes of the Italian language considering only consonants.

Viseme	Associated phonemes
1	— p, m, b —
2	— f, v —
3	— t, d, z, s, ts, dz —
4	— N, L, S, tS, dZ —
5	— k, g, n, r, l —

Table 2. The phonemes associated to the ten visemes of the Italian language useful for ALS patients.

Viseme	Associated phonemes
0	a, l, n
1	e, ca, GA, tee
2	i, chi
3	o, go, co
4	u, k, cu, GU
5	f, v
6	t, d, z, s, r
7	p, m, b
8	s, GI, Ce, GE, SC
9	neut

This outcome partially solves our problem, since the vowels are not considered. We extended the set of visemes, *limiting the definition of visemes to a subset of words that will be important in the case of ALS patients.* The obtained result cannot be considered the exhaustive set of visemes of the Italian language. In Table 2 the complete set of the identified visemes is presented; the last class *neut* is related to the lips position in case of silence.

In Fig. 3 the mean values of HD, VD, DA and PD related to each of the ten visemes are shown.

Neural networks work in two phases: a first training phase in which one provides to the system many observations and the related belonging class, and a second phase in which one offers to the classifier a single observation and claims

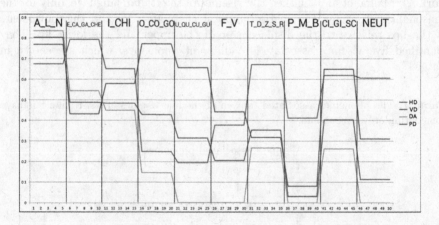

Fig. 3. Mean values of HD, VD, DA, PD for each of the ten visemes. (Color figure online)

the output class. Before using the neural network, therefore, we must create a dataset with the values provided by the lips tracking system, repeating the same viseme several times and collecting the output data of HD, VD, DA, and PD. To increase the accuracy of the classification we must create a data set for each user who will have to use the system. In agreement with the theory underlying neural networks the values obtained to provide as input to the network are normalized in the range $[0, 1]$.

In the present version of the system, the identification of the visemes from the values of the features HD, VD, DA, and PD is heavily dependent from the distance face-camera. If such distance varies, the classifiers will make errors in the classifications.

In Fig. 4 the classification of visemes for the Italian word *ciao* is shown. The classification is satisfactory, even if the frames from 24 to 26 are wrong, in fact the error is due to a variation in the distance face-camera.

FRAME	VISEMA		VISEMA		VISEMA		VISEMA
1	NEUT	15	CI_GI_CE_GE_SC	29	A_L_N	43	O_CO_GO
2	NEUT	16	CI_GI_CE_GE_SC	30	I_CHI	44	O_CO_GO
3	NEUT	17	CI_GI_CE_GE_SC	31	CI_GI_CE_GE_SC	45	O_CO_GO
4	NEUT	18	CI_GI_CE_GE_SC	32	CI_GI_CE_GE_SC	46	O_CO_GO
5	CI_GI_CE_GE_SC	19	CI_GI_CE_GE_SC	33	O_CO_GO	47	CI_GI_CE_GE_SC
6	CI_GI_CE_GE_SC	20	CI_GI_CE_GE_SC	34	O_CO_GO	48	CI_GI_CE_GE_SC
7	CI_GI_CE_GE_SC	21	I_CHI	35	O_CO_GO	49	O_CO_GO
8	CI_GI_CE_GE_SC	22	I_CHI	36	O_CO_GO	50	O_CO_GO
9	CI_GI_CE_GE_SC	23	A_L_N	37	O_CO_GO	51	CI_GI_CE_GE_SC
10	CI_GI_CE_GE_SC	24	A_L_N	38	O_CO_GO	52	CI_GI_CE_GE_SC
11	CI_GI_CE_GE_SC	25	A_L_N	39	O_CO_GO	53	CI_GI_CE_GE_SC
12	CI_GI_CE_GE_SC	26	A_L_N	40	O_CO_GO	54	NEUT
13	CI_GI_CE_GE_SC	27	A_L_N	41	O_CO_GO	55	NEUT
14	CI_GI_CE_GE_SC	28	A_L_N	42	O_CO_GO	56	NEUT

Fig. 4. Classification of visemes for the Italian word *ciao*.

5   Algorithm for Words Prediction

The last implemented subsystem has the role to predict the pronounced word on the base of the visemes identified by the neural network. In fact for each viseme there are several phonemes associated, and consequently the algorithm has to identify the most appropriated word. We adopted the popular XT9 algorithm.[4] The main difference with XT9 is due to the fact that the neural network sometimes makes errors, introducing some noise that has to be removed.

Another problem is due to the fact that when we are analyzing a word, there are two or more visemes that are equal. It is hard to detect the correct number of visemes that compose the word.

---

[4] XT9 is a text predicting and correcting system for mobile devices with full keyboards. It is a successor to T9, a popular predictive text algorithm for mobile phones with only numeric pads.

Sequence	9	9	9	4	7	7	7	4	4	4	6	6	1	0	0	0	0	0	0	3	3	3	3	3	3	3	3	3	8	6	9	9	9	9
Value			9	4			7			4		6	1						0										3	8	6			9
Observation			3	1			3			3		2	1						6										8	1	1			4

Fig. 5. Grouping equal visemes.

The proposed algorithm takes as input a sequence of visemes and provides in output possible modular words basing its decision on a dictionary of the Italian language. To optimize the necessary computational time and to limit failures we used an abridged version of about 2000 words of the Italian dictionary, which includes the most used Italian words.

In the first phase, the algorithm removes the noise introduced by casual modifications of a face-camera distance. The list of visemes is analyzed to group all visemes equal that occur consecutively associating to each grouping the number of occurrences. The output of such process is a table similar to that one represented in Fig. 5.

The first row reports the type of viseme detected, numbered from 0 to 9 (see Fig. 2). This step provides a double benefit, by shortening the vector that contains the data and eliminating the noise due to the sequence of visemes lower than a predetermined threshold. In the example shown in Fig. 5 we can assume that the visemes that occur only once does not really belong to the word. We empirically set this threshold to 3, and even if in some cases this threshold seemed to be too low, we preferred to leave some infrequently wrong viseme, rather than risk eliminating from the sequence a correct viseme.

In the second step the algorithm detects the visemes with $neut$, identifying the beginning and the end of a word. Then the blocks of visemes are analyzed, each corresponding to a word. Since to each viseme correspond several letters, in order to identify the correct word, it is necessary to analyze all possible combinations of letters. The algorithm visit a tree whose level l corresponds to the l-th viseme of the sequence and each node n corresponds to the n-th character of the viseme. For example the node 2 of the level 3 corresponds to the second letter of the third viseme. Thus each path going form a root to a leaf specifies a word. In Fig. 6 a graphical representation of an example of this approach is illustrated.

After having evaluated all possible words, the algorithm verifies their presence in the dictionary. A simple comparison between the words found by the neural network with the ones present in the vocabulary, is rarely successful. In fact, it is sufficient that the neural network incorrectly classify only a viseme, to fail the identification of the right word. If, for example in the case shown in Fig. 4 related to the Italian word $ciao$, the neural network interprets as a u the last visemes, the resulting word will be $ciaou$, which is wrong. Thus, it is evident we need a spell checker algorithm. A possible solution is to use a metric which measures the distance between two words, like the Levenshtein distance [5, 12, 13], which is obtained by finding the cheapest way to transform one string into another.

Applying the Levenshtein algorithm to all modular words implies the measure of their distance with the words of the dictionary, and the selection of those that

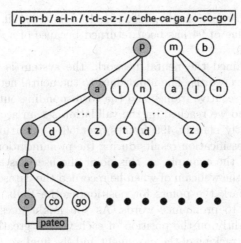

Fig. 6. Algorithm for searching words

have distance less than or equal to 1. If there are more words that respect this constraint, those with the shortest distance are shown before the other; in case of equal distance, they are shown in alphabetical order. One of the possible improvements is to create a custom dictionary that remembers the user's most frequently used words to suggest first the most used words. In this way it is possible to overcome most of the issues arising from an erroneous classification of visemes, as for example no classification of a viseme and especially when are present in the word two letters belonging to the same viseme. Rarely it occurs that the letters belonging to the same viseme are more than two.

6   Results

We are presenting a complete example that starts from the acquisition of the template up to the prediction of the word in order to provide a summary and an overview of the whole system, commenting on the results of the various phases and related issues. The input device used is a webcam with a 640×480 resolution and an update rate of 30 frames per second.

The first step is to create the dataset needed for training the neural network. Once detected a suitable user-webcam distance, which has to be kept as much constant as possible even during routinary use, the user has to pronounce individually each of the ten visemes. The lip tracking module will provide in output a series of values for the parameters HD, VD, DA and PD for each captured frame.

To create a complete dataset each viseme should be repeated for about 15 times. The error rate of the algorithm is 1.8 % and has been calculated during the recording of 2000 visemes, analyzing consecutive frames, and considering as an error the incorrect placement of a single POI, identified by observing the sequence of values returned in the output. So, if we receive the following sequence related to the HD parameter: 100, 100, 101, 100, 94, 100, 99, 100, ..., 100 it is evident that the value of 94 has been returned because of a wrong positioning of one or more POIs.

Once having trained the neural network, the system is ready to classify visemes and recognize words. The input data to the neural network are built by concatenating 3 consecutive frames and the corresponding output vector. With the proposed method we reached on the validation set an accuracy of 99.02%, managing to correctly classify 499 frames over 507. Unfortunately we do not achieve the same classification results during the pronunciation of words.

Continuing with the example, at this point the user has to position at the same distance from the webcam of when he recorded the dataset for the training phase, he has to mark the points for positioning the POIs for the matching template and start to pronounce words. As you might guess, mark well the word and pause slightly on the position of each viseme greatly reduces errors, improving both the tracking of the movement and the final word prediction. This example presents the case in which the user pronounce the Italian word *ciao*. The values obtained in output from the lip tracking module are passed to the neural network for classification that returns the values shown in Fig. 7.

frame	viseme		viseme		viseme		viseme
1	9=NEUT	19	3=O_CO_GO	37	0=A_L_N	55	3=O_CO_GO
2	9=NEUT	20	3=O_CO_GO	38	0=A_L_N	56	3=O_CO_GO
3	9=NEUT	21	3=O_CO_GO	39	0=A_L_N	57	3=O_CO_GO
4	9=NEUT	22	3=O_CO_GO	40	0=A_L_N	58	3=O_CO_GO
5	9=NEUT	23	3=O_CO_GO	41	0=A_L_N	59	3=O_CO_GO
6	5=F_V	24	3=O_CO_GO	42	0=A_L_N	60	3=O_CO_GO
7	3=O_CO_GO	25	3=O_CO_GO	43	0=A_L_N	61	3=O_CO_GO
8	3=O_CO_GO	26	3=O_CO_GO	44	0=A_L_N	62	3=O_CO_GO
9	3=O_CO_GO	27	3=O_CO_GO	45	2=I_CHI	63	3=O_CO_GO
10	3=O_CO_GO	28	3=O_CO_GO	46	3=O_CO_GO	64	3=O_CO_GO
11	3=O_CO_GO	29	8=CI_GI_CE_	47	3=O_CO_GO	65	4=U_QU_CU_GU
12	3=O_CO_GO	30	8=CI_GI_CE_	48	3=O_CO_GO	66	5=F_V
13	3=O_CO_GO	31	8=CI_GI_CE_	49	3=O_CO_GO	67	4=U_QU_CU_GU
14	3=O_CO_GO	32	8=CI_GI_CE_	50	3=O_CO_GO	68	9=NEUT
15	3=O_CO_GO	33	2=I_CHI	51	3=O_CO_GO	69	9=NEUT
16	3=O_CO_GO	34	2=I_CHI	52	3=O_CO_GO	70	9=NEUT
17	3=O_CO_GO	35	0=A_L_N	53	3=O_CO_GO	71	9=NEUT
18	3=O_CO_GO	36	0=A_L_N	54	3=O_CO_GO	72	9=NEUT

Fig. 7. Classification of visemes acquired from a camera and related to the pronunciation of the Italian word *ciao*.

As shown in Fig. 7, 45 frames are correctly classified over 72 frames, that is 62.5%. This result has not a relevant influence on the deduction of the spoken word. In the next step the algorithm, which has the task of predicting the word, takes as input the string containing the sequence of visemes. The first step counts how many consecutive times appears the same viseme and eliminates all

Table 3. Visemes forming the Italian word *ciao*.

Viseme	Associated phonemes
viseme_0	o, co, go
viseme_1	GI, ci, ce, ge, sc
viseme_2	a, l, n
viseme_3	o, co, go
viseme_4	neut

the blocks of consecutive visemes less or equal to 3, reducing the long string in Fig. 7 to the sequence 3, 8, 0, 3, 9. Therefore we have the visemes shown in Table 3.

Now we have to identify all words that can be formed by combining the visemes, excluding the last one which indicates the end of the word. Every word so formed is searched in a short dictionary of the Italian language and those who have a Levenshtein distance less than or equal to 1 are selected as possible. In this case the word *o-c-i-a-o* has a distance 1 compared to *c-i-a-o* and is also the only which fulfills this requirement. So the final output of the system is *ciao*.

7   Conclusions

The implemented system is still at the prototypical stage, however it promise very interesting results that may increase the quality of life of some categories of patients who have lost the voice for various types of diseases.

We will work in order to increase the quality of the various subsystems, making the system more reliable and easy to use. We have to find a solution to make independent the classification from the user-webcam distance. For facilitating the use by the user it is important to introduce an algorithm which selects automatically and precisely the templates of the four points of interest used in the template matching algorithm. We may improve the classification algorithm in several ways: from introducing new parameters to better represent the viseme pronounced, to extending the observation area from the lips to the rest of the face, especially the area of the cheeks and of the chin.

The use of visemes seems to be a good choice, however it has to be confirmed with extended tests and research.

The word deduction algorithm can be enhanced considering a broader context that considers the entire sentence in which the word has to be inserted.

A possible improvement of the performance may be also obtained implementing the neural network calculations on General Purpose GPU Computing, which has been demonstrated a changeling and efficient approach to improve the computing performances of many classes of algorithms [9].

References

1. Buchsbaum, W.H.: Color TV Servicing, 3rd edn. Prentice Hall, Englewood Cliffs (1975)
2. Magno Caldognetto, E., Zmarich, C., Cosi, P., Ferrero, F.: Italian consonantal visemes: Relationships between spatial/temporal articulatory characteristics and coproduced acoustic signal. In: Proceedings of AVSP-97, Tutorial and Research Workshop on Audio-Visual Speech Processing: Computational and Cognitive Science Approaches, Rhodes (Greece), pp. 5–8 (1997)
3. Canzler, U., Dziurzyk, T.: Extraction of non manual features for videobased sign language recognition. In: lAPK Workshop on Machine Vision Applications, MVA2002, Nara, Japan, pp. 318–321 (2002)
4. Cootes, T., Taylor, C., Cooper, D., Graham, J.: Active shape models-their training and application. Comput. Vis. Image Underst. **61**, 61 (1995)
5. Gale, W.A., Church, K.W.: A program for aligning sentences in bilingual corpora. In: Proceedings of the 29th Annual Meeting on Association for Computational Linguistics, ACL 1991, Stroudsburg, PA, USA, pp. 177–184. Association for Computational Linguistics (1991)
6. Gervasi, O., Magni, R., Macellari, S.: A brain computer interface for enhancing the communication of people with severe impairment. In: Murgante, B., et al. (eds.) ICCSA 2014, Part VI. LNCS, vol. 8584, pp. 709–721. Springer, Heidelberg (2014)
7. Gervasi, O., Magni, R., Riganelli, M.: Mixed reality for improving tele-rehabilitation practices. In: Gervasi, O., Murgante, B., Misra, S., Gavrilova, M.L., Rocha, A.M.A.C., Torre, C., Taniar, D., Apduhan, B.O. (eds.) ICCSA 2015. LNCS, vol. 9155, pp. 569–580. Springer, Heidelberg (2015)
8. Gervasi, O., Magni, R., Zampolini, M.: Nu!rehavr: virtual reality in neuro tele-rehabilitation of patients with traumatic brain injury and stroke. Virtual Real. **14**(2), 131–141 (2010)
9. Gervasi, O., Russo, D., Vella, F.: The aes implantation based on opencl for multi/many core architecture. In: Proceedings of the 2010 International Conference on Computational Science and Its Applications, ICCSA 2010, Washington, DC, USA, pp. 129–134. IEEE Computer Society (2010)
10. Pan, S.W.J., Guan, Y.: A new color transformation based fast outer lip contour extraction. J. Inform. Comput. Sci. **9**(9), 2505–2514 (2012)
11. Kass, M., Witkin, A., Terzopoulos, D.: Snakes: active contour models. Int. J. Comput. Vis. **1**(4), 321–331 (1988)
12. Kruskal, J.B.: An overview of sequence comparison. In: Sankoff, D., Kruskal, J.B. (eds.) Time Warps, String Edits, and Macromolecules: The Theory and Practice of Sequence Comparison, pp. 1–44. Addison-Wesley, Reading (1983)
13. Levenshtein, V.I.: Binary codes capable of correcting deletions, insertions and reversals. Sov. Phy. Dokl. **10**, 707 (1966)
14. Lievin, M., Delmas, P., Coulon, P.Y., Luthon, F., Fristol, V.: Automatic lip tracking: Bayesian segmentation and active contours in a cooperative scheme. In: IEEE International Conference on Multimedia Computing and Systems, 1999, vol. 1, pp. 691–696, Jul 1999
15. Mahalanobis, P.C.: On the generalised distance in statistics. Proc. Natl. Inst. Sci. India **2**(1), 49–55 (1936)
16. Saeed, U., Dugelay, J.-L.: Combining edge detection and region segmentation for lip contour extraction. In: Perales, F.J., Fisher, R.B. (eds.) AMDO 2010. LNCS, vol. 6169, pp. 11–20. Springer, Heidelberg (2010)

A New 3D Augmented Reality Application for Educational Games to Help Children in Communication Interactively

Chutisant Kerdvibulvech[1(✉)] and Chih-Chien Wang[2]

[1] Graduate School of Communication Arts and Management Innovation,
National Institute of Development Administration, 118 SeriThai Road, Klong-Chan, Bangkapi,
Bangkok 10240, Thailand
chutisant.ker@nida.ac.th
[2] Graduate Institute of Information Management, National Taipei University,
151, Daxue Road, Sanxia District, New Taipei City 23741, Taiwan
wangson@mail.ntpu.edu.tw

Abstract. In recent years, the use of technology to help children for augmented and alternative communication (AAC) is extremely a vital task. In this paper, a novel three-dimensional human-computer interaction application is presented based on augmented reality (AR) technology for assisting children with special problems in communication for social innovation. To begin with, three-dimensional human hand model is constructed to estimate and track the hand's position of users. An extended particle filter is applied for calculating the pose of background and the positions of children. The likelihoods based on the edge map of the image and pixel color values are utilized to estimate the joint likelihood in three-dimensional model. A flexible real-time hand tracking framework using the 'golden energy' scoring function is integrated for capturing region of interests. An inertial tracking technique is used for calculating the quaternion. Three-dimensional models from Google SketchUp are employed. We then use a built QR-code for scanning to access the system, and then utilize for selecting a character three-dimensional designed cartoon by applying the Vidinoti image application. After that, representative three-dimensional cartoons and augmented environments are overlaid, so that it is able to entertain children. A printed coloring photo, called Augmented Flexible Tracking is designed and provided in the system for visualization. The process of the system is done in real-time. Our experiments have revealed that the system is beneficial both quantitatively and qualitatively for assisting children with special needs in communication interactively.

Keywords: Augmented reality · Augmented and mixed reality · Educational games · Communication · Children with special problems · Human-computer interaction · Three-dimensional interaction

1 Introduction and Related Works

In the 21st century, computer with innovative technology has emergently evolved from conventional teletypewriters to smart technological tools in an interdisciplinary aspect for social innovation and connecting people to people, even people with special needs

© Springer International Publishing Switzerland 2016
O. Gervasi et al. (Eds.): ICCSA 2016, Part II, LNCS 9787, pp. 465–473, 2016.
DOI: 10.1007/978-3-319-42108-7_35

in communication to limitless opportunity. The emerging innovative technology has transformed the way we communicate markedly. Augmented reality (AR) is a vitally new field of computational science, computer science and computer engineering. In recent years, it has being explored to enhance, help and assist children with special problems in spoken and written communication with social interaction. Several interesting augmented reality applications have been recently presented to help children in communication.

For example, Tanner et al. [1] built an augmented reality classroom, presented in 2014, using educational tablet technology. Their goal is to attempt for helping and educating students with their comprehension of a certain procedural task, such as creating Lego robotic devices. They used the Aurasma application to interact with an animated Lego robotic device of the same static manual. Next, Bhatt et al. [2], proposed similarly in 2014, created an augmented reality game for assisting children by enhancing hand-eye coordination and social interaction. In their case, they specifically focused only on children with Asperger syndrome, a neurobiological autism spectrum disorder on the higher-functioning end. They used Adobe Flash CS6 and Actionscript 3.0 to create this emotion game to interact with only the faces of children. Therefore, although this augmented reality game has a limitation for an automatic facial recognition, it is able to allow children for dragging and dropping features interactively onto their faces to build freely an expression. Furthermore, Bai et al. from the University of Cambridge [3] built an augmented reality system, presented lately in 2015, for conceptualizing the representation of pretense visually for eliciting pretend play for small children with special needs. Their main aim is to help small children with autism spectrum condition only aged lower than 7 to be able to interact with the system. Moreover, Magnenat et al. [4], proposed recently in 2015, from Disney Research Zurich developed a printed coloring book method for children. It is done by texturing and displaying the color characters in three dimensions using a smartphone in a children's coloring book. In their work, a texturing algorithm was also presented in real-time for transforming the input texture from a two-dimensional colored drawing to a three-dimensional character. Furthermore in 2015, Persefoni and Tsinakos [5] described a good recent overview for the use of augmented reality in modern education in many different context aware technologies. In their case, they use augmented reality technology in tablets and smartphones for providing interactively and uniquely educational experiences. Several open course projects using augmented reality in the Eastern Macedonia and Thrace Institute of Technology are included and discussed for a modern educational way. Also, one of the pioneering research works in the study of augmented reality technology for rehabilitation of cognitive children with special problems is the work of [6] proposed preliminarily by Richard et al. They defined the term of Augmented Reality applied to Vegetal. The term is shortly called ARVe. They implemented the application for allowing people, focusing on disabled children and pupils with special problems, to handle plant entities in both two dimensions and three dimensions generated graphics. Different senses of olfactory, vision or auditory are used for helping disabled pupils in an elementary school in France for decision making. A similar work using augmented reality technology for exploring the feasibility of utilizing augmented reality technology in early educational school was also presented in the very recent work of [7] by Huang et al. in 2016. But this design-based work was used and tested in different location (i.e., in Hong Kong kindergarten) Besides in 2015, Zünd et al. from ETH Zurich used the term "augmented creativity" [8] to describe their augmented reality-based work for enhancing creative play for children. Their research aim is to musically allow people to make their own cartoon using augmented reality on smartphones. Cunha et al. [9] also presented recently on February 2016 an augmented reality-based application called GameBook for helping young people with autism spectrum disorders (ASD). It is aimed for recognizing and acquiring emotions of children with special needs by involving interactively their motivation and attention. In addition, augmented reality is not only technology to possibly help children in communication, but it also includes some related technologies such as virtual reality (VR) and multimedia. Multimedia is basically content combined from a number of different types of forms, including sounds, texts, images, animated video, and interactive contents. Due to the variety of forms of multimedia interactively and virtually, it is able to be applied to support people recently in different fields, including disabled people with special rehabilitation. Significantly, it is also able to help developing assistive technology (AT) for contributing to the improvement of the life of people with disabilities. There are some interesting multimedia-based assistive technology applications for supporting people with special needs and children with disabilities to perform tasks that they were previously unable to achieve, or had difficulty achieving extremely.

However, to the best of our knowledge, in every previous system, they have different goals for our presented work in this paper. Our main purpose of the research described here is to assist people, especially children, who have problems in communication at school for educational games, focusing in primary school aged between 8 and 12 for augmented and alternative communication (AAC). Some representative three-dimensional cartoons and augmented environments for visualization are shown during running the system. The rest of our presented paper is organized as follows. The next section (Sect. 2) will explain about our proposed system configuration and how does it work. We then show our representative experimental results in Sect. 3. Finally in the last section (Sect. 4), we conclude the paper and points out the directions of our future work.

2   System Configuration and Proposed System

This section describes the system configuration and the presented system. We, in this paper, propose a novel three-dimensional human-computer interaction application based on augmented reality technology for helping children with special problems in non-verbal communication in a unique way for social innovation. We first track a camera to estimate the pose of background and the positions of people by using a robust particle filter extended as described in [10] in real-time. In this system, we use and adapt three-dimensional models from Google SketchUp [11] to create several new three-dimensional cartoon models for educational games. A built QR-code is also utilized to scan to access the system. This QR-code is then used to choose and select a character three-dimensional designed cartoon by applying the Vidinoti image application [12]. After that, we apply the hand tracking using extended distance transform and our geometric hand model we created using a set of quadrics, roughly representing the position and anatomy of a real hand of human.

It is used for locating the pose of 27 DOF human hand model for overlaying as explained in [13]. We also extend a flexible real-time hand tracking method proposed and presented by Sharp et al. [14] from Microsoft Research in 2015 to capture region of interests. We estimate the quaternion for obtaining the better results.

As shown in Fig. 1, the hand model is built from 39 truncated quadrics with 27 DOF: 4 for the pose of four fingers, 5 for the position of the thumb, and 6 for the global hand pose using OpenGL. In this system, we use the likelihoods based on the edge map of the image and pixel color values in each input image. Therefore, the joint likelihood of is calculated using

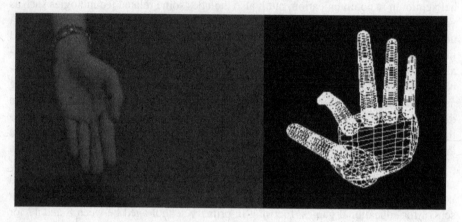

Fig. 1. The hand model we constructed from the input hand images using OpenGL and the 'golden energy' scoring function

$$p(z \mid x) = p(z^{edge}, z^{sil} \mid x)$$
$$p(z^{edge}, z^{sil} \mid x) \approx p(z^{edge} \mid x)p(z^{col} \mid x),$$

(1)

while the p(z|x) is denoted as observations z in both images to the unrecognized state x, z^{edge} is the edge map in each image, and z^{sil} is the pixel color values of silhouette in an image. By analyzing the edge likelihood, we use the chamfer distance function to estimate and extract feature in the image. This distance function is calculated and estimated for various model templates by employing and utilizing a distance transform of the edge image. The silhouette likelihood is then calculated by utilizing a Bayesian classifier as presented and described in [15]. The likelihood function is adaptively computed using

$$p(z^{edge} \mid x) = \frac{1}{Z} \exp(-\lambda d(A(x), B(z^{edge}))).$$

(2)

In this likelihood equation, A(x) is the set of template points in the shape template P, while B is denoted as the set of edge points received from the edge image. After that, a flexible real-time hand tracking algorithm [14] according to the 'golden energy' scoring function is integrated and utilized to capture region of interests where a model pixel we used is tracked and determined using

$$E^{au}(Z \mid R) = \sum_{ij} \rho(\overline{z} - r) + C, \tag{3}$$

where Z is a tight region of interest in an image and C in the equation is a constant value. We attempt to calculate the quaternion for simplifying the results. By applying the inertial tracking method as presented in [16] by Baldi et al., we then calculate and compute the quaternion, called r(t), using

$$r(t) = \alpha * g(t) + (1 - \alpha)q(t), \tag{4}$$

where the value of α we used in this process is between 0 and 1. Representative three-dimensional cartoons and augmented environments are then immediately overlaid onto the background we created to help and entertain children for educational games in a limited amount of time. In this presented system, we also design and provide a printed coloring photo, called Augmented Flexible Tracking, for providing and giving it to entertain children. User participation is also done in our system development process. Finally, experiment from this human-computer interaction system is done to evaluate and assess the performance of the presented work with real children with special needs in the real situation using a mobile phone. The experimental results have revealed that our application is beneficial for children with special needs and specific learning difficulties and assessment in non-verbal communication interactively and effectively. Figure 2 depicts the system configuration of the proposed human-computer interaction work using augmented reality for educational games. In this configuration, it is composed of a smartphone which can be alternatively any device from the internet of things. After tracking the position of camera, a QR-code is used for scanning to access and choose a character three-dimensional cartoon using a mobile phone. After children

Fig. 2. System configuration of the proposed human-computer interaction work using augmented reality technology

enter in the scene, three-dimensional cartoons and augmented environments are then immediately overlaid onto the background for entertaining children in real-time for visualization. Augmented Flexible Tracking is ultimately used for printing a coloring photo for people focusing on children for educational games.

3   Representative Results

In this section, we show our representative experimental results. We conducted a user study to test and evaluate the robustness of the presented system both quantitatively and qualitatively. The processing speed (i.e., the computation time) for this presented educational application is about 14 frames per second, which is quite real-time. For qualitative results, we randomly selected ten children with special needs in primary school aged between 8 and 12. Each child was asked to test our presented system. It is important to note that, in the experiment, we did not know each child personally beforehand. This is to avoid some possibly unbiased answers, so that most children were chosen randomly to make sure that they are unbiased. Each child took about 10 min to run the study for educational games. Almost children were able to use the application after a brief explanation in 2-3 min., even though there was one child who took slightly long time to understand how to use the system (i.e., 5 min.). After the individual tests, every child was asked to give qualitative feedback. Table 1 shows and illustrates representative qualitative feedback from these ten participants. This included interest in the proposed application, user satisfaction, smoothness of the presented application, ease of use of the designed interface, and overall system impression. General comments on the test were also collected from every participant.

Table 1. Qualitative results from users who were randomly chosen to test the system

	Minimum score (10 points)	Maximum score (10 points)	Average score (10 points)
Interest in the application	7	10	8.24
User satisfaction	8	10	9.23
Smoothness of the system	8	10	9.01
Ease of use of the interface	6	10	7.72
Overall impression	8	10	8.41

According to the qualitative user study we conducted, we received mainly positive comments. Many participants agreed that they are satisfied with the smoothness of the proposed system. The range of the application scores for smoothness of the system is 8 through 10, with the average score at 9.01 from 10. Regarding about interest in this application from each participant, the range of the application scores is 7 through 10, with the average score at 8.24 from 10. They reasoned that this was since the application is able to perform in real time. Moreover, many participants indicated that they were impressed by the system, especially by the idea of developing this application. The range of the application scores for user satisfaction is 8 through 10, with the average score at 9.23 from 10. This average score for user satisfaction is the highest. Generally, the range of the application scores for overall impression is 8 through 10. The average score for overall impression is at 8.41 from 10. Nevertheless, the most common non-positive feedback was about ease of use of the application. They commented that the current system may not extremely be easy to use, if no one teaches them some giving instructions of how to use the system. There is one participant who suggested that the system should provide some short instruction manual of how to use the application for every new user. This was the reason why they gave lower scores to ease of use of the interface than other individual criterion scores. The range of the application scores for ease of use of the interface is 6 through 10 with the average score at 7.72 from 10. Although this score for ease of use of the interface was not too low, it could be improved in the future with some new technical methods. However, we believe that these numbers from qualitative results of participants are quite suitable enough to make the presented application practical for helping children with special needs in communication for educational games. Also, several representative experimental results are shown. Figure 3 illustrates some examples of our augmented reality application in several representative parts showing and overlaying some three-dimensional cartoons for educational games and augmented environments onto the background we built. After testing and conducting the experiment both qualitatively and quantitatively, it has revealed that children with specific learning difficulties are satisfied with this new augmented reality based-system interactively.

Fig. 3. Examples of our augmented reality application in some representative parts for helping children with special needs in communication interactively for educational games

4   Conclusions and Future Works

In computational science, augmented and mixed reality has been recently an extremely popular topic. In recent years, many emerging applications usually bring augmented and mixed reality to make them novel, interesting and robust in an advanced manner differently. In this paper, we have developed a novel vision-based application that assists children with special problems in communication based on augmented and mixed reality approaches. We construct three-dimensional human hand model for detecting and tracking the hand's position using a particle filter extended. The likelihoods based on the edge map of the image and pixel color values are applied for calculating and estimating the joint likelihood in three-dimensional model. A flexible real-time hand tracking technique based on the 'golden energy' scoring function is then used to determine some region of interests. We also use three-dimensional models from Google SketchUp for creating educational games. A QR-code for scanning to access the system is achieved to allow users to choose a character three-dimensional designed cartoon freely using Vidinoti image application for visualization. Therefore, several cartoons and augmented environments in three-dimensions can be overlaid for entertaining children. In the proposed system, a printed coloring photo is provided also for three-dimensional interaction. The process of the system is done quite quickly and performed in real-time.

After conducting a user study, we believe that we can achieve a robust current system output to assist children in a novel way. However, the system has still some limitations for improvements. Therefore, our future work is aimed to deal with the problem of the occlusion of tracking. We also plan to cope with the problem of ease of use of the interface while using the system automatically. In the future, we intend to make technical improvements to apply and potentially extend some continuously adaptive mean shift techniques such as [17] and several adaptive randomized ensemble tracking algorithms such as [18] to further refine these problems in a robust manner.

Acknowledgments. This research presented herein was partially supported by a research grant from the Research Center, NIDA (National Institute of Development Administration).

References

1. Tanner, P., Karas, C., Schofield, D.: Augmenting a child's reality: using educational tablet technology. J. Inf. Technol. Educ. Innovations Pract. (JITE) **13**, 45–54 (2014)
2. Bhatt, S.K., De Leon, N.I., Al-Jumaily, A.: Augmented reality game therapy for children with autism spectrum disorder. Int. J. Smart Sens. Intell. Syst. **7**, 519–536 (2014). Massey University
3. Bai, Z., Blackwell, A.F., Coulouris, G.: Using augmented reality to elicit pretend play for children with autism issue. IEEE Trans. Vis. Comput. Graph. **21**(05), 598–610 (2015)
4. Magnenat, S., Ngo, D.T., Zund, F., Ryffel, M., Noris, G., Röthlin, G., Marra, A., Nitti, M., Fua, P., Gross, M., Sumner, B.: Live texturing of augmented reality characters from colored drawings. In: IEEE International Symposium on Mixed and Augmented Reality (ISMAR 2015), Fukuoka, Japan, 10 p., 29 September 2015
5. Giannakos, M.N., Divitini, M., Iversen, O.S., Koulouris, P.: Making as a pathway to foster joyful engagement and creativity in learning. In: Chorianopoulos, K., Divitini, M., Hauge, J.B., Jaccheri, L., Malaka, R. (eds.) ICEC 2015. LNCS, vol. 9353, pp. 566–570. Springer, Heidelberg (2015). doi:10.1007/978-3-319-24589-8_58

6. Richard, E., Billaudeau, V., Richard, P., Gaudin, G.: Augmented reality for rehabilitation of cognitive disabled children: a preliminary study. IEEE Virtual Rehab. IEEE Eng. Med. Biol. Soc. pp. 102–108, Venice, Italy, 27–29 September 2007

7. Huanga, Y., Lia, H., Fonga, R.: Using augmented reality in early art education: a case study in Hong Kong kindergarten. Early Child Dev. Care **186**(6), 879–894 (2016). doi: 10.1080/03004430.2015.1067888

8. Zünd, F., Ryffel, M., Magnenat, S., Marra, A., Nitti, M., Kapadia, M., Noris, G., Mitchell, K., Gross, M., Sumner, R.W.: Augmented creativity: bridging the real and virtual worlds to enhance creative play. In: Proceeding of ACM SIGGRAPH Asia 2015, Mobile Graphics and Interactive Applications, Article No. 21, Kobe, Japan, 2–6 November 2015

9. Cunha, P., Brando, J., Vasconcelos, J., Soares, F., Carvalho, V.: Augmented reality for cognitive and social skills improvement in children with ASD. In: 13th International Conference on Remote Engineering and Virtual Instrumentation (REV), pp. 334–335. IEEE Publisher, Madrid, Spain, 24–26 February 2016

10. Kerdvibulvech, C.: Human hand motion recognition using an extended particle filter. In: Perales, F.J., Santos-Victor, J. (eds.) AMDO 2014. LNCS, vol. 8563, pp. 71–80. Springer, Heidelberg (2014)

11. Grover, C.: Google SketchUp: The Missing Manual, 1 edn., 602 p. O'Reilly Media, Sebastopol, 1 June 2009

12. Perdikakis, A., Araya, A., Kiritsis, D.: Introducing augmented reality in next generation industrial learning tools: a case study on electric and hybrid vehicles. In: Procedia Engineering, The Manufacturing Engineering Society International Conference (MESIC), vol. 132, pp. 251–258. Elsevier Publishing, ScienceDirect (2015)

13. Kerdvibulvech, C.: Hand tracking by extending distance transform and hand model in real-time. Pattern Recogn. Image Anal. **25**(3), 437–441 (2015). Springer Publisher

14. Sharp, T., Keskin, C., Robertson, D., Taylor, J., Shotton, J., Kim, D., Rhemann, C., Leichter, I., Vinnikov, A., Wei, Y., Freedman, D., Kohli, P., Krupka, E., Fitzgibbon, A., Izadi, S.: Accurate, robust, and flexible real-time hand tracking. In: ACM Conference on Human Factors in Computing Systems (CHI), Seoul, Republic of Korea, 18–23 April 2015

15. Papoutsakisa, K.E., Argyros, A.A.: Integrating tracking with fine object segmentation. Image Vis. Comput. **31**(10), 771–785 (2013)

16. Baldi, T.L., Mohammadi, M., Scheggi, S., Prattichizzo, D.: Using inertial and magnetic sensors for hand tracking and rendering in wearable haptics. In: 2015 IEEE World Haptics Conference (WHC), Evanston, IL, pp. 381–387, 22–26 June 2015

17. Im, J., Jung, J., Paik, J.: Single camera-based depth estimation and improved continuously adaptive mean shift algorithm for tracking occluded objects. In: Ho, Y.S., Sang, J., Ro, Y.M., Kim, J., Wu, F. (eds.) PCM 2015. LNCS, vol. 9315, pp. 246–252. Springer International Publishing, New York (2015)

18. Li, W., Lin, Y.: Adaptive randomized ensemble tracking using appearance variation and occlusion estimation. Math. Prob. Eng. **2016**, 1–11 (2016). Article ID 1879489

An Improved, Feature-Centric LoG Approach for Edge Detection

Jianping Hu$^{(\boxtimes)}$, Xin Tong, Qi Xie, and Ling Li

College of Science Northeast Dianli University, Jilin, China
neduhjp307@163.com

Abstract. Gaussian filter is used to smooth an input image to prevent false edge detection caused by image noises in the classic LoG edge detector, but it weakens the image features at the same time which results in some edges cannot be detected efficiently. To ameliorate, this paper presents an improved, feature-centric LoG approach for edge detection. It firstly uses non-local means filter based on structural similarity measure to replace Gaussian filter to smooth an input image which enables the image features to be preserved better, and then image edges can be extracted efficiently by the zero-crossing method for the smoothed image operated by Laplacian operator. Experimental results show that the proposed method can improve the edge detection precision of the classic LoG edge detector, and the non-local means filter used in the presented method achieves better results than the other two typical filters with edge-preserving ability.

Keywords: Non-local means filter · LoG edge detector · Zero-crossing method · Feature-centric

1 Introduction

An image edge is the area where the image gray values become discontinuous. It either marks the end of a region or the beginning of another region [1]. Edge detection aims at identifying image edges by calculating the changing rate of the image gray values of each pixel's neighborhood. It is the precondition of image segmentation, feature extraction, image compression and image recognition. The investigation of edge methods has been an active research area in the digital image processing field [2–4] and it still is.

The classical local differential edge detectors extract image edges by observing the changes of first or second order directional derivatives around edges in a certain neighborhood of each pixel, such as Roberts operator [5], Sobel operator [6], Prewitt operator [7], Laplacian operator [1], etc. Although local differential edge detectors can describe the local characteristics of each pixel, but it is

J. Hu—This work is supported by National Natural Science Foundation of China (61202261), Scientific and Technological Development Plan of Jilin Province (20130522113JH) and the 13th Five-Year Scientific and Technological Research Foundation of Education Department of Jilin Province(No. 97 in 2016).

O. Gervasi et al. (Eds.): ICCSA 2016, Part II, LNCS 9787, pp. 474–483, 2016.
DOI: 10.1007/978-3-319-42108-7_36

affected by noises easily and tends to extract many false image edges. In order to reduce the effects of obvious noise on the edge detector, LoG edge detector uses Gaussian filter to filter out the noise to prevent false edge detection firstly, and then extracts image edges by Laplacian operator. However, Gaussian filter weakens the image features when smoothing out image noises, which results in the loss of image edge points and reduces the accuracy of edge detection.

In view of the disadvantage of Gaussian filter in LoG detector, a natural idea is to use another filter with edge-preserving ability to smooth images which can get rid of image noises and preserve image features efficiently at the same time. In recent years, the edge-preserving image smoothing techniques has been widely studied in digital image processing, such as bilateral filter [8,9], weighted least square filter [10], L_0 gradient minimization filter [11], non-local means filter [12–14], etc.

In this paper, the non-local means filter based on structural similarity measure is used to replace Gaussian filter to smooth an input image in the LoG edge detector, and then image edges can be extracted efficiently by the zero-crossing method for the smoothed image operated by Laplacian operator. Experimental results show that the proposed method can improve the edge detection accuracy of the classic LoG edge detector. Furthermore, the non-local means filter used in our improved, feature-centric LoG edge detection method achieves better results than the other two typical edge-preserving filters (bilateral filter [8] and L_0 gradient minimization filter [11]).

2   LoG Edge Detector

LoG edge detector uses Gaussian filter to smooth an input image first, and then uses the zero-crossing method to extract edges [1] for the smoothed image operated by Laplacian operator.

2.1   Gaussian Filter

A Gaussian filter modifies the input image $f(x, y)$ by convolution with a Gaussian function $G(x, y)$ in order to get a smoothed image $I(x, y)$, i.e.,

$$I(x, y) = G(x, y) * f(x, y), \tag{1}$$

where $*$ denotes a convolution operator, and $G(x, y)$ is a two-dimensional Gaussian function with the following form

$$G(x, y) = \frac{1}{2\pi\sigma^2} exp(-\frac{x^2 + y^2}{2\sigma^2}), \tag{2}$$

where σ is a parameter to control to the smoothing level and a larger parameter means a higher smoothing level.

2.2　Laplacian Operator

Laplacian operator is a second order differential operator given by the divergence of the gradient of a function defined in Euclidean space, and can be denoted as ∇^2. According to operating the smoothed image $I(x,y)$ by Laplacian operator, we can get the second order directional derivative image $M(x,y)$, i.e.,

$$M(x,y) = \nabla^2[I(x,y)] = \nabla^2[G(x,y) * f(x,y)] = \nabla^2[G(x,y)] * f(x,y). \qquad (3)$$

Laplacian operator of $G(x,y)$ can also be denoted as

$$\nabla^2 G(x,y) = \frac{\partial^2 G(x,y)}{\partial^2 x} + \frac{\partial^2 G(x,y)}{\partial^2 y}, \qquad (4)$$

where

$$\frac{\partial^2 G(x,y)}{\partial^2 x} = \frac{1}{2\pi\sigma^4}(\frac{x^2}{\sigma^2} - 1)exp(-\frac{x^2+y^2}{2\sigma^2}),$$
$$\frac{\partial^2 G(x,y)}{\partial^2 y} = \frac{1}{2\pi\sigma^4}(\frac{y^2}{\sigma^2} - 1)exp(-\frac{x^2+y^2}{2\sigma^2}). \qquad (5)$$

Now combining Eqs. (4) and (5), we can get

$$\nabla^2 G(x,y) = (\frac{1}{2\pi\sigma^4})(\frac{x^2+y^2}{\sigma^2} - 2)exp(-\frac{x^2+y^2}{2\sigma^2}). \qquad (6)$$

Equation (6) can be regarded as the combination of Gaussian filter and Laplacian operator, which is called LoG operator. It looks like a hat and is also called Mexican hat as shown in Fig. 1.

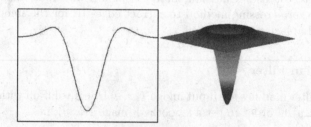

Fig. 1. 1D and 2D representations of LoG operator.

2.3　Zero-Crossing Processing

Zero-crossing processing is commonly used in digital image processing and analysis. Its purpose is to extract the cross section where the value of a function changes from positive to negative (or from negative to positive), namely, zero-crossing points [1]. If an image is operated by Laplacian operator, the zero crossing points in the directional derivative image correspond to the image edges.

Consequently, the image edges of the input image $f(x, y)$ can be obtained by extracting the zero-crossing points of the image $M(x, y)$.

Thanks to the use of Gaussian filter in the LoG edge detector, false edge detection caused by image noises can be reduced greatly. However, the true edges of image are alsox weakened at the same time, which results in the loss of image edge points and reduces the accuracy of edge detection.

3   Improved, Feature-Centric LoG Edge Detection

This paper proposes an improved, feature-centric LoG edge detection method, which uses the non-local means filter based on the structure similarity measure (SSIM) to smooth images instead of the Gaussian filter in the classical LoG edge detector. This non-local means filter can preserve image features efficiently which are weakened by Gaussian filter in the smoothing processing.

3.1   Non-local Means Filter Based on Structural Similarity Measure (SSIM)

The Non-local means filter has been considered as a renovation method in the image denoising field. It takes a mean of all pixels in the image, weighted by how similar these pixels are to the target pixel. The classic non-local means filters [12] use the Euclidean distance of the corresponding pixels of two image blocks to measure the pixel similarity, which may result in the blurring of the image areas with rich texture features. In this paper, the structure similarity measure (SSIM) [15] of the image blocks is introduced to calculate the similarity weights to improve the edge-preserving ability of the classical non-local means filter.

Given an image $f(x, y)$ with N pixels, the new gray value $f'(x_i, y_i)$ of the i-th pixel (x_i, y_i) can be obtained in the non-local means filter by

$$f'(x_i, y_i) = \sum_j \omega(i, j) f(x_i, y_i) i = 1, ..., N, \tag{7}$$

where $\omega(i, j)$ denotes the similarity weight between the i-th pixel and the j-th pixel calculated by combing structure similarity measure and Gaussian weighted Euclidean distance in the classic non-local means filter, i.e.,

$$w(i, j) = \frac{1}{C(i)} exp(-S(i, j)|d(N_i - N_j)|^2 / h^2), \tag{8}$$

where $C(i) = \sum_j exp(-S(i, j)|d(N_i - N_j)|^2 / h^2)$ is a normalizing constant, h controls the smoothing intensity, N_i and N_j represent the neighborhood blocks with the window size $w \times w$, whose centers are the i-th pixel and the j-th pixel respectively, $d(N_i - N_j)$ is the Euclidean distance of the corresponding pixel gray values between the two neighborhood blocks, and $S(i, j)$ is a measure computed by SSIM.

In the following, we discuss how to compute the measure $S(i,j)$ of two image neighborhood blocks N_i and N_j by their structural similarity measure (SSIM) [15] which is defined as

$$SSIM(i,j) = l(i,j) * c(i,j) * s(i,j), \tag{9}$$

where $l(i,j)$, $c(i,j)$ and $s(i,j)$ represent the brightness, contrast and structure information, respectively, and they can be computed by

$$l(i,j) = \frac{2\mu_i\mu_j + C_1}{\mu_i^2 + \mu_j^2 + C_1},$$

$$c(i,j) = \frac{2\sigma_i\sigma_j + C_2}{\sigma_i^2 + \sigma_j^2 + C_2},$$

$$s(i,j) = \frac{\sigma_{ij} + C_3}{\sigma_i\sigma_j + C_3},$$

where μ_k and σ_k are the mean intensity and variance of the image blocks $N_k, k = i, j$, respectively, σ_{ij} is the covariance of two image neighborhood blocks N_i and N_j, $C_k(k = 1, 2, 3)$ is set into a very small constant to prevent unstable results when the denominator is close to zero. By setting $C_3 = \frac{1}{2}C_2$, the structure similarity measure can be simplified as

$$SSIM(i,j) = \frac{(2\mu_i\mu_j + C_1)(2\sigma_{ij} + C_2)}{(\mu_i^2 + \mu_j^2 + C_1)(\sigma_i^2 + \sigma_j^2 + C_2)}. \tag{10}$$

Then we can get the measure $S(i,j)$ in Eq. (8) to make the structure similarity measure and the Euclidean distance have the same monotonicity by changing Eq. (10) into

$$S(i,j) = \frac{1}{T(i)} * \frac{1 - SSIM(i,j)}{2}, \tag{11}$$

where $T(i) = \sum_j \frac{1-SSIM(i,j)}{2}$ is a normalization factor. Obviously, $S(i,j)$ varies from 0 and 1 and a smaller value means two image neighborhood blocks has a higher similar structure.

In practice, the computation in Eq. (7) is performed over a fixed size searching window instead of the whole image are due to the spatial concentration of the structure similarity of image blocks.

3.2 Implementation Steps of Improved LoG Edge Detection

The implementation steps of the proposed improved, feature-centric LoG edge detection method can be described as follows:

Step 1: Use the non-local means filter based on SSIM to smooth the image $f(x,y)$ to get the smoothed image $I(x,y)$.

Step 2: Generate the second order directional derivative image $M(x,y)$ of the image $I(x,y)$ by Laplacian operator.

Step 3: Extract the image edges according to the zero-crossing method for the image $M(x,y)$.

4   Experimental Results and Comparisons

We have implemented the above improved, feature centric LoG edge detection approach in Matlab 2012 and tested it on some typical 256×256 images, including Lena image with many rich texture features, Rice image only with high-intensity edges and Circuit image with many low intensity edges. In order to test the robustness of our algorithm for noises, a certain degree of Gaussian noise is added into each tested image (Fig. 2). In our experiments, we set the neighborhood window's size to be 7×7, and set the searching window's size to be 21×21. Besides, we selected the smoothing parameter h according to the noise level of each image carefully in order to obtain satisfied edge detection results.

In order to illustrate the efficiency of the proposed method, we compared it with the classical LoG edge detection algorithm. Obviously, the image edges are weakened when the noises are suppressing by the Gaussian filter, which causes the loss of true edge points and reduces the accuracy of edge detection (Figs. 3(a), 4(a), 5(a) and 6(a)). The proposed method uses the non-local means filter based on SSIM to smooth images instead of Gaussian filter, which can preserve image edges more efficiently and can improve the edge detection accuracy of the classic LoG edge detector (Figs. 3(d), 4(d), 5(d) and 6(d)).

(a)   (b)   (c)   (d)

Fig. 2. The tested images with different degree of Gaussian noises (σ is the standard deviation) in our experiments. (a)Lena($\sigma=10$). (b)Lena($\sigma=20$). (c)Rice($\sigma=20$). (d)Circuit($\sigma=20$).

In recent years, many other edge-preserving filtering methods have also proposed in the image denoising field, such as bilateral filtering [8,9], weighted least square filtering [10], L_0 gradient minimization filtering [11], etc. Therefore, we also compared our method with the methods using other two typical edge-preserving filters (bilateral filter [8] and L_0 gradient minimization filter [11]) to replace the Gaussian filter in our improved, feature-centric LoG edge detection frame.

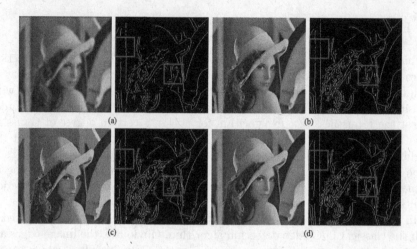

Fig. 3. Comparisons of the filtering results (Left) and the edge detection results (Right) by the improved LoG method based on different filtering techniques for the Lena image in Fig. 2(a). (a)Gaussian filtering(σ=2). (b)Bilateral filtering(σ_L=3,σ_R=0.1). (c)L_0 filtering(λ=0.005). (d)The non-local means filtering based on SSIM($h = 12$).

4.1   Improved LoG Method Based on Bilateral Filter

Given an image $f(x,y)$ with N pixels, the new gray value $f'(x_i, y_i)$ of the i-th pixel (x_i, y_i) can be obtained by bilateral filtering as follows,

$$f'(x_i, y_i) = \frac{\sum\limits_{j \in N(i)} L(i,j)R(i,j)f(x_j, y_j)}{\sum\limits_{j \in N(i)} L(i,j)R(i,j)}, i = 1, ..., N, \quad (12)$$

where $N_{(i)}$ represents the $w \times w$ image neighbor block, whose center is the pixel (x_i, y_i), $L(i,j)$ and $R(i,j)$ represent the spatial position and the gray weight coefficients respectively which can be computed by

$$L(i,j) = exp(-(|x_j - x_i|^2 + |y_j - y_i|^2)/2\sigma_L^2), \quad (13)$$

$$R(i,j) = exp(-|f(x_j, y_j) - f(x_i, y_i)|^2/2\sigma_R^2), \quad (14)$$

where σ_L and σ_R represent the smoothing intensity of the spatial position and image gray, respectively. σ_L and σ_R are selected carefully according to the noise level of each image in order to obtain satisfied edge detection results in our experiments.

The improved LoG method based on bilateral filter achieves much better results in extracting image edges than the classical LoG operator (Figs. 3, 4, 5 and 6), especially for the Lena image with small noises (Fig. 3(b)) and the Rice image only with high intensity edges (Fig. 5(b)). However, the bilateral filter only uses the local gray information of each pixel to smooth in the Eq. (12), which results in the failure of edge detection in some areas for the images with

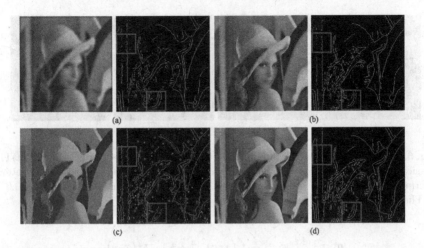

Fig. 4. Comparisons of the filtering results (Left) and the edge detection results (Right) by the improved LoG method based on different filtering techniques for the Lena image in Fig. 2(b). (a)Gaussian filter(σ=2.5). (b)Bilateral filtering(σ_L=3,σ_R=0.2). (c)L_0 filtering (λ=0.03). (d)The non-local means filtering based on SSIM($h = 24$).

large noises or rich texture features (Figs. 4(b), 5(b) and 6(b)). Obviously, the improved LoG method based on the non-local means filter incorporates the structure information of each pixel into the weighted averaging and obtains better results (Figs. 3(d), 4(d), 5(d) and 6(d)).

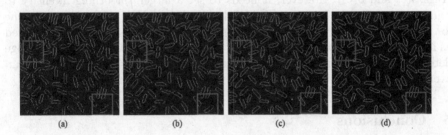

Fig. 5. Edge detection comparison for the Rice image in Fig. 2(c). (a)The classical LoG operator(σ=2.5). (b)Improved LoG method based on bilateral filtering(σ_L=3,σ_R=0.2). (c)Improved LoG method based on L_0 filtering(λ=0.02). (d)The proposed method($h = 24$).

4.2   Improved LoG Method Based on L_0 Gradient Minimization Filter

The L_0 gradient minimization filter is devised according to the sparse feature of image edges by Xu et al. [11]. Given an input image $f(x, y)$, it generates the filtered image $f'(x, y)$

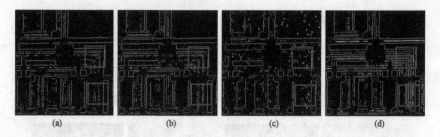

Fig. 6. Edge detection comparison for the Circuit image in Fig. 2(d). (a)The classical LoG operator(σ=2.5). (b)Improved LoG method based on bilateral filtering(σ_L=3,σ_R=0.2). (c)Improved LoG method based on L_0 filtering(λ=0.03). (d)The proposed method($h = 24$).

$$\min_{f'(x,y)} (\| f(x,y) - f'(x,y) \|_2^2 + \lambda \| \triangledown f(x,y) \|_0),$$

where $\triangledown$ is the gradient operator, $\|.\|_0$ denotes the L_0 norm λ controls the smoothing intensity and a larger value means a higher smoothing level. λ is selected carefully according to the noise level of each image in order to obtain satisfied edge detection results in our experiments.

The L_0 gradient minimization filter can smooth out noises and detect image edges because it can focus the image energy into the image edges. As for the Lena image with small noises and the Rice image only with high intensity edges, the improved LoG method based on this filtering technique can obtain good results (Figs. 3(c) and 5(c)). However, it needs to set a larger smoothing intensity to remove noises for the images with high level noises, which cannot extract some low intensity edges (Figs. 4(c) and 6(c)). Besides, the noises are also regarded as image features in this smoothing processing, which results in some false edge detection (Figs. 4(c) and 6(c)). According to Figs. 4(d) and 6(d), our method can overcome these limitations and generates better edge detection results.

5   Conclusions

To overcome the limitation that the true edges of image are weakened in the smoothing processing based on Gaussian filter in the LoG edge detector. This paper proposes an improved, feature-centric LoG edge detection method. It uses the non-local means filter based on the structure similarity measure (SSIM) to smooth images. Experimental results show that the proposed method improves the edge detection accuracy of the classic LoG operator. Moreover, the non-local means filter used in our improved, feature centric LoG edge detection method achieves better results than the other two typical edge-preserving filters.

References

1. Gonzalez, R., Woods, R.: Digital Image Processing. Publishing house of electronics industry, Beijing (2011)
2. Teng, J., Zhou, Y., Yan, J., Yang, Y.: The research of serial morphological filters based on multiple structuring elements. J. Northeast Dianli Univ. **28**, 38–41 (2008)
3. Song, Q., Zhou, L.: An improved approach to denoising of image based on wavelet threshold and its application on the gas-liquid two-phase flow. J. Northeast Dianli Univ. **32**, 41–44 (2012)
4. Trier, Q.D., Jain, A.K., Taxt, T.: Feature extraction methods for character recognition-a survey. Pattern Recogn. **29**, 641–662 (2010)
5. Moshe, P., Alexander, T.: Exact optimality of the shiryaev-roberts procedure for detecting changes in distributions. In: International Symposium on Information Theory and its Applications, pp. 1–6 (2008)
6. Yasri, I., Hamid, N.H.: Performance analysis of FPGA based Sobel edge detection operator. In: International Conference on Electronic Design, vol. 10, pp. 1–4 (2008)
7. Maini, R., Sohal, J.S.: Performance evaluation of prewitt edge detector for noisy image. Int. J. Comput. Appl. **9**, 39–46 (2006)
8. Tomsi, C., Manduchi, R.: Bilateral filtering for gray and color images. In: Proceedings of Sixth International Conference on Computer Vision (ICCV 1998), pp. 839–846. IEEE Computer Society, Washington, D.C. (1998)
9. Prudhvi, V., Venkateswarlu, V.: Ultrasound medical image denoising using hybrid bilateral filtering. Int. J. Comput. Appl. **56**, 44–51 (2012)
10. Farbman, Z., Fattal, R., Lischinsk, D.: Edge-preserving decompositions for multi-Scale tone and detail manipulation. ACM Trans. Graph. (TOG) **27**, 67–71 (2008). ACM
11. Xu, L., Lu, C., Xu, Y.: Image smoothing via L_0 gradient minimization. ACM Trans. Graph **30**, 174–177 (2011)
12. Buades, A., Coll, B., Morel, J.M.: A non-local algorithm for image denoising. In: Proceedings of IEEE Computer Society Conference on Computer Vision and Pattern Recognition, vol. 2, pp. 60–65 (2005)
13. Peter, J.D., Govindan, V.K., Mathew, A.T.: Robust estimation approach for NL-means filter. In: Bebis, G., Boyle, R., Parvin, B., Koracin, D., Remagnino, P., Porikli, F., Peters, J., Klosowski, J., Arns, L., Chun, Y.K., Rhyne, T.-M., Monroe, L. (eds.) ISVC 2008, Part II. LNCS, vol. 5359, pp. 571–580. Springer, Heidelberg (2008)
14. Salin, M.E., Zhang, X., Ding, M.: Two modifications of weight calculation of the non-local means denoising method. Engineering **5**, 522–526 (2013)
15. Wang, Z., Bovik, A.H., Sheikh, E.: Image quality assessment: from error visibility to structural similarity. IEEE Trans. Image Process. **13**, 600–612 (2004)

The Use of Geoinformation Technology, Augmented Reality and Gamification in the Urban Modeling Process

Miłosz Gnat[(✉)], Katarzyna Leszek[(✉)], and Robert Olszewski[(✉)]

Department of Cartography, Warsaw University of Technology, Warsaw, Poland
milosz.gnat@gmail.com, kasia.leszek@yahoo.com,
r.olszewski@gik.pw.edu.pl

Abstract. The aim of the paper is to present the concept of a geoinformation technology, extended by so-called augmented reality (AR) module to support the social geoparticipation process with respect to spatial planning. The authors propose to increase the level of residents' activity by using gamification tools adapted to the cultural, historical, economic and social reality of the given city as well as precise 3D modelling techniques and VR/AR tools. The authors strongly believe that the implementation of this idea will enable not only the improvement of spatial planning process but also to develop an open geoinformation society that will create "smart cities" of the future.

Keywords: Gamification · Geoinformation technology · Geoparticipation · Augmented reality · 3D city model · Smart cities · Geoinformation society

1 Introduction

As Manuel Castels pointed out in "The Information Age", 'the paradox of the great civilization change consists in the fact that we have practically unlimited access to information and data and yet we are nearly unable to use it in any way'. Knowledge acquisition based on available information is essential in the era of network society when economic value is not generated in factories any more but is produced by media and IT and telecommunication networks instead. Industrialism has given way to informationism, and the industrial society has been replaced by network society.

The process of social digitalisation is characterized not only by the creation of a physical broadband network infrastructure but also on unlimited access to data (also spatial data), and first of all, on education related to the ability to convert source information into useful knowledge. The so-called DIKW Pyramid (data-to-information-to-knowledge-to-wisdom transformation) proposed in the 1970s may constitute a useful analogy for the development of geoinformation infrastructure (Fig. 1).

The application of modern geoinformation technologies may be used not only for the effective gathering of source data about existing objects (such as the Open Street Map created spontaneously by millions of users) but also to obtain spatial knowledge about phenomena and processes – both physical and social ones. Transforming "raw" data into useful information is particularly important for the development of modern urban agglomerations that aspire to be "smart cities". The key factor for this process is

© Springer International Publishing Switzerland 2016
O. Gervasi et al. (Eds.): ICCSA 2016, Part II, LNCS 9787, pp. 484–496, 2016.
DOI: 10.1007/978-3-319-42108-7_37

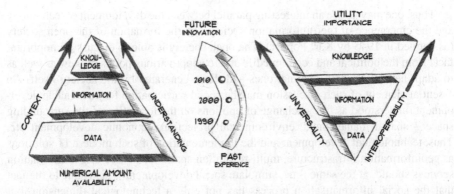

Fig. 1. Data-to information-to knowledge pyramid. (Source: based on the DIKW Pyramid – data-to-information-to-knowledge-to-wisdom transformation)

the development of appropriate geoinformational technologies enabling social participation. However, this requires not only to use sophisticated high-tech tools but also to persuade local communities to use them widely. In the opinion of the authors of this study, the impulse that triggers social energy and releases the creative potential of inhabitants is gamification. Gamification is an emerging form of engaging people into active participation in cities life through playing game. The use of this approach will enable not only to collect enormous sets of data, but also to process them and to develop models of optimal use and development of space. The process of social (geo) participation initiated in this way may be used both for the purposes of creating a vision of smart city and (in a longer perspective) for the development of (geo)information society.

2 Gamification, Playable City and Their Role in Spatial Planning and Social Participation

2.1 Geoinformation Society

The globalization process redefines the way in which we perceive both social-cultural and economic relations. It also contributes to the evolution of information society based on technological knowledge and common access to information. The objective set by the information society is to provide every member of the society with the possibility to create, obtain, use and share information and knowledge. In modern post-industrial states this objective is achieved by means of applying information and communication technologies (ICT). According to the EU Directorate General Information Society, more than 50 % of the economic value of public information in the EU is generated by geoinformation. On the other hand, according to the estimations of the US Federal Geographical Data Committee, approx. 80 % of public data contain a spatial component. This means that we are witnessing the process of emergence of geoinformation society, which "widely uses information obtained by means of generally available services of the geoinformation infrastructure" [24].

Thus, one may notice an interesting parallel between the development of technology and the emergence of (geo)information society and the formation of the open society (as defined in 1945 by Karl Popper). The open society is able to discuss all important facts from the political and economic life and to adopt various points of view as well as to adapt new ideas, both external ones and those generated by the society itself. An essential element of such discussion may be social participation for sustainable development, understood as free exchange of opinions on the formation of the surrounding space – spatial management, environmental protection, economic development etc. Thus, technological development and the implementation of such modern IT solutions as geoinformation infrastructure, multi-resolution spatial databases or geoinformation services should, at the same time, stimulate social development. This is due to the fact that the social informatisation process has not only a technological dimension, but mainly a civilizational one. The development and popularization of the Internet, mobile devices and geoinformation technologies, including spatial data, make social informatisation process easier than ever before. Moreover, this refers to all citizens, not just selected social groups. The inevitable situation where we, as users of the Internet and mobile devices are becoming (whether consciously or not) suppliers of spatially localized information, forces us to ask about the possibilities to use the growing popularity of the crowdsourcing and VGI[1] ideas [12]. In the opinion of the authors, the activity of inhabitants of developing smart cities may be used in the social participation process for the purposes of creation of spatial development plans by the local community.

2.2   Spatial Planning and Social Participation

The involvement of local residents and social consultations in the spatial planning process allow for the elimination, already on early stages of the works, of potential planning mistakes and for solving conflicts. They also enable the expression of individual needs and opinions about the inhabited space which is the common good, and thus to influence the way in which it is shaped. Social acceptance for the adopted planning solutions and subsequent investment activities fosters, among others, the creation of spatial order, growing satisfaction of inhabitants with the changes that occur in the surrounding space and the shortening of the duration of the procedures related to the local spatial development plan. Many countries, including Poland, are still searching for new, more effective methods and platforms for dialogue between local inhabitants and decision-makers, planners or investors [8]. The list of legal means with respect to the use of social participation in the spatial planning process is quite wide, and, although the legal regulations concerning social participation in the spatial planning process still require specification, this is not the main reason of the lack of involvement of local communities in the creation of their own surroundings [22]. This problem results mainly from low social awareness of the spatial planning issues, poor information policies related to participation or disclosing planning documents to the public, the form and course of social consultations or the form of planning studies, which are often unclear for the inhabitants. Thus, the aim of the authors is to propose an innovative method of

---

[1]   Volunteered Geographic Information.

social participation based on the application of the gamification concept and modern geoinformation technologies.

2.3 Gamification and Playable City

Gamification means the "application of game mechanics, aesthetics and thinking in game categories in order to increase human involvement, motivate people to act, promote learning and practice problem-solving". Gamification is a relatively new tendency, which emerged as a result of the reflections of scientists who study the influence of games on human life and behavior and of the practical experiences of companies that use game mechanisms for shaping the behavior of their employees. Gamification can be used in numerous areas of social activity. The authors of this study believe that gamification may also be an extremely effective and powerful tool supporting the spatial planning process.

When proposing a gamification solution based on geoinformation technology in the context of social participation, one should refer to a more general idea, so-called "Playable City". "Playable city" is a new way of thinking about the city, which should result in initiating social dialogue by means of the application of games and motivating users to share various experiences related to the city. The aim of such approach is to create and strengthen relationships not only between human players but also between people and places. By introducing the element of fun and steering the game plot in a competent way it is possible to initiate discussion on specific city solutions, both negative and positive ones. Thus, the city becomes not only scenery where we live our daily life, but also an important element or even a "participant" of this life.

The city and its users interact with each other. Users give the city their own senses and meanings. The city is flexible by nature, so that it can be adapted to one's own needs. However, this works both ways: people influence the city, but the city forms them too, even by the resistance that they encounter when trying to impose a personal form. A well-planned gamification process, that takes into account this mutual relationship and the "spiritus urbis" characteristic for the given city, may contribute to the improvement of civic activity, both with respect to the immediate surroundings of individual users and to other spaces that are important for social functioning. Currently existing educational games can be analyzed in two aspects:

(1) the topic of the game, related to spatial planning, management of city space, creating efficient and functional urban development areas and increasing social participation, which is the main objective of the authors' project "Creative Urbanizer",
(2) technologies used in the game, including:
   - Game platform (PC or such mobile devices as smartphones or tablets);
   - The way of presentation of the world in the game (2D graphics, 3D graphics, elements of augmented reality – AR);
   - The way in which the game interacts with the physical space (a fully computer game, or a game using elements existing in the real world, e.g. with use of GPS technology).

In the following part we will discuss the possible technologies used in our concept of a playable city achieved by geoparticipation of networking society through gamification.

3   Technologies of Gamification: Geoinformation, 3D Modeling, VR and AR

Reflections on the technologies that will be crucial while developing applications for games in the city are mainly considering matters connected to the acquisition of data describing urban tissue, its visualization and, if needed, the way of positioning the player.

City as a scenery for computer games is a common practice, known for a long time. Since the 90s, SimCity, among others, presented a city not only as scenery, but also as a subject of a game. Those simulations, dedicated for entertainment were mainly created on fictional data. Compliance of data with reality had no meaning in this approach – it did not affect the satisfaction of a player. The nature of serious games is different. In this case, to achieve required assumptions we need to use real data. And this aspect is directing the reflections on technologies in the field of geoinformatics.

3.1   Geoinformation Technologies

Geoinformatics is a discipline of science and an engineering domain in which the competences of surveyors, cartographers, photogrammetry, computer science and database specialists are brought together. With the use of available tools, methods of processing can be developed, as well as spatial data analytics and, when needed, creating new tools. The scope of geoinformatics applications is extensive. It is both sophisticated spatial analysis that from raw data extracts strategic knowledge about studied phenomena and a search for the finest methods of projecting the navigation applications. Geoinformatics is applied in a various industries, with entertainment industry among them.

Spatial data describe the real world around us through certain models. Often they are assimilated with the widely understood concept of a map. Depending on adopted data collection approach, the purpose of their usage, the models representing the reality may vary. Cartographers develop models of reality where precision of objects, both spatial and semantic is selected in a way so that database is effectively used, meeting the original assumptions. Models can vary on their level of detail, complexity. However, they must be clearly embedded in space, so every spatial data model has a specific map projection. Those models can be either raster models or vector models, i.e. the information about the object of the real world can be represented through a set of pixels in a picture or in a form of data describing objects through its geometry and sets of attributes. If we have a model of vector data, the key issue is to describe the shape and position of real world objects. This can be done both in 2D or 3D space where the objects are displayed using simple geometric shapes (lines, polygons), a little more complex (Bezier curves) or specific functions (parametric modeling). Another matter is semantics, understood as selection of attributes which are describing the characteristics of real world objects. This kind of set needs to be adequate for intended application of a model – a different set will be implemented for cadaster based database and a different one for set containing model or city center in 3D.

Computer games are believed to be a future of cartography [11]. This slogan can be understood in many ways, however it can be certainly said that existing resources of cartographic data is an excellent source for the entertainment industry. Development of various field in computer science, as well as some emerging ideas (Smart City and associated Open Data) led to the moment, where in second decade of 21st century we have rich spatial data describing majority of large cities in the world. The vast amount of those data is modeled in order to create a record of different resources, thus the simplified form of geometry representation.

An attempt to use them in the process of game development requires therefore proper processing and remodeling. In case of urban agglomerations in developed countries those data are so exhaustive and detailed, that problems arise with processing, appropriate selection and generalization. The domain of geoinformatics is data processing, thus the issue of spatial data usage in games becomes to matter of appropriate design of a technique that will process raw data into the required application model and, if necessary, selection of those that are not suitable through data study.

3.2 Virtual Reality and Augmented Reality

The technology of processing data from the source to the application model is called cartographic modeling. The form of the model application depends on its functional requirements, especially its presentation methods. This particular matter is influenced by factors such as the specifics of the device, the type of the provided information, communication method or dynamics of presentation. The specificity of the device can be understood as a type of screen, performance of hardware, built-in GPS or internet

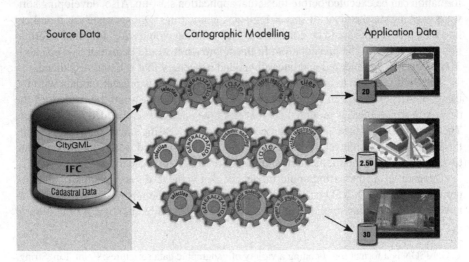

Fig. 2. Three cartographic models converting source data in a various ways in order to obtain application data. (Source: created by Gnat M.)

connection. Information provided to the device, that will be used in creation of its content can come from memory card, from GPD or from a build-in camera. The data can differ in terms of granularity depending to the actual scale of presentation or context. The method of communication can be based on image, sounds or vibrations.

Figure 2 present three examples of applications handling properly processed cartographic data as data application model.

The first scenario presents an application that displays the picture in 2D. The application works in a client–server architecture, all data are acquired from the server through the network. The data model of this application assumes that the background layer is represented by data taken from the online map service. Additional information can be dynamically added to the presentation from data in given geographic data structures format, for example GeoJSON[2]. They should be visualized based on the policy defined in cartographic composition. Source data in this process are information retrieved from municipal stocks that are available as API through service managing municipal stock of data.

The second scenario is an application that presents data in 2.5D, therefore in isometric projection (a form of axonometric projection). This application works as well in client – server architecture. In this process source data are vector 2D datasets with a cadaster service provided by the municipal administration data. In order to build such a model a selection of object that meets certain criteria needs to be conducted. Then their geometric generalization has to be adjusted to the expected level of detail. In the process of parametric modeling 2D object are converted to 3D. This process, in simplification, involves extruding data in third dimension, where the control parameter is one of the descriptive attributes. 3D objects should be used to build a scenery presenting data in isometrics. To select appropriate visualisation (colors, line thickness etc.) suitable cartographic composition should be applied. The 3D scene should be transformed into the raster. This raster has to be cut into tiles after deposition. The whole process of data formation can be executed before the actual application start-up. Also, developing software working on the layer of server logistics, generating 3D data, rasterize it and cut into tiles in real time could be considered. This solution would add additional information dynamically to the presentation. In this approach it would be useful to consider if 3D objects can be rendered and later embedded in scenery that is being visualized.

The third scenario assumes that we use application that is presenting data with the usage of AR[3]. The screen presented to the player is a picture from the camera of the device, in which there is mounted content from the databases embedded in the device. The essential part of data modeling process involves processing source data that presents the 3D city model on level of detail LOD2 (or higher), derived from CityGML model, to data model application, meaning 3D models used by adequate 3D engine.

Depending on the cartographic model we can achieve diverse results of the city representation in 2D, 2.5D or 3D. Although every one of them is possible to be implemented while building application, it is worth to reflect on technological limitation or

---

[2] GeoJSON is a format for encoding a variety of geographic data structures: Point, LineString, Polygon, MultiPoint, MultiLineString, and MultiPolygon.
[3] Augmented Reality.

possibilities. 2D model is definitely the most cost and effort – efficient one, giving appropriate representation. However, to create a connection, a sense of belonging of a player to the games environment, AR with 3D city model is the best solution. It enables to not only to develop more realistic and exciting game for players, but also to perform the tasks better (Fig. 3).

Fig. 3. Reality vs potential development scenario (Source: picture by Piotr Zbierajewski)

4   Possible Usage of the Gamification

The use of ISO standards as well as the ones offered by OGC in gathering, processing and sharing spatial data in recent years included also the area of multiscale 3D modeling of individual buildings and entire cities (CityGML using different level of detail LOD). Currently created ARML[4] standard will enable the integration of geoinformation technology and so called augmented reality. The combined use of GIS tools, AR applications and mobile devices (smartphones, tablets) equipped additionally with a GPS locating system will enable the implementation of a number of diversified social, participation or urban planning related projects:

---

[4] ARML 2.0 provides an interchange format for Augmented Reality applications to describe and interact with objects in an AR scene, with a focus on mobile, vision-based AR. The candidate standard describes the virtual objects that are placed into an AR scene, as well as the registration of the virtual objects in the real world, and allows interaction with and dynamic modification of the AR scene using ECMAScript bindings [26].

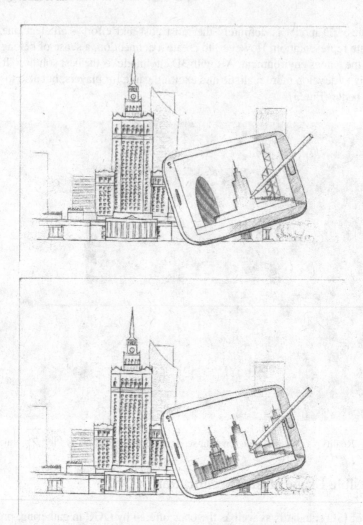

Fig. 4. Design and on-line visualization of changes in the skyline in the city center in Warsaw with use of AR technology (Source: Drawing by Agnieszka Kardaś)

- Gamification as a tool supporting the process of spatial planning. The usage of simple editing tools in 3D modeling will allow for visualization directly in the area of potential effects of the created projects (Fig. 4). The usage of gamification techniques will enable the participants of the geoinformation urban game to 'vote' for the most interesting projects. The essence of this action is to encourage citizens to the active participation and taking part in the process of shaping the urban tissue. The usage of such defined technology and field observation of augmented reality on the screen of mobile device (before it will actually become an investment - Fig. 5a) would surely prevent the construction of skyscraper that falls into the one of the most recognizable view on Belweder: the residence of polish president in Warsaw (Fig. 5b).

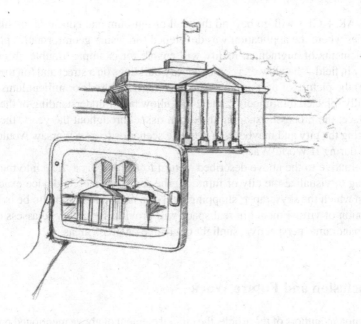

Fig. 5. a. Planning changes of spatial management around Belvedere with use of AR technology.
b. View of the Belvedere from the Royal Łazienki Park in Warsaw.

- Augmented reality as a component supporting perception of time changes in urban
  space. Using a mobile device application with embedded integrated components of

GIS + AR + GPS will go beyond the wall of museum and education on history of the cities where the application was developed for. Using geoinformatics platforms with elements of augmented reality will enable for example 'double observation' directly in field – the view of a contemporary buildings on a street and (on the mobile device) the picture of past urban tissue from tens, hundreds or millenniums before. Carefully selected multimedia content will allow better understanding of the history of a place: the processes shaping the urban tissue throughout the years, the factors impacting the city and maybe even possible scenarios (e.g. if Warsaw would not be ruined during II World War).

- An alternative to the above described 'return to the past' is a 'leap into the future' enabling to visualize the city of future on the screen of mobile device exactly in a spot, in which the skyscraper, shopping mall or a new road is about to be build. The perception of virtual model in real space will provide possibility to assess changes in the panorama, perspective, sunlight or change on green areas.

5   Conclusion and Future Work

In the opinion of authors of the article, the crucial element of above mentioned processes and approaches is the usage of an integrated geoinformation technologies and augmented reality with elements of gamification in the framework of the 'serious games'. The aim of creating the postulated by the authors of article mobile applications GIS + AR, is not an improvement of technologies, but the growth of social awareness and participation, development of participatory democracy and (geo)information society that is shaping the surrounding space in the spirit of sustainable development.

New generation geoinformation portals and gamification tools will allow us not only to collect data and to transform them into useful information; they will also enable us to acquire spatial knowledge and to use it to get to know the surrounding space and to transform it in a rational way. It is worth to remember the words of Einstein: "Imagination is more important than knowledge. For knowledge is limited to all we now know and understand, while imagination embraces the entire world, and all there ever will be to know and understand" we may believe that imagination is the only limit for the geoinformation society in the era of creating smart cities.

Smart cities are more than just modern technology. Smart city is, first of all, the human dream about a friendly city co-created by involved residents, which uses "high" technology of tomorrow and social geo-participation to become the reality of today. Smart city is more than just an idea – it is a process of competent usage of information resources concerning all spheres of activity of the city, which covers, apart from attractive, functional space, innovative economy and a network of services, also the comfort of living of its residents, manifested in form of social participation.

References

1. Andrzejewska, M., Baranowski, M., Fiedziukiewicz, K., Kowalska, A., Matuszkiewicz, J.M., Rusztecka, M., Roo-Zielińska, E., Solon, J.: O partycypacji społecznej w planowaniu przestrzennym. Zastosowania geowizualizacji w celu wzmocnienia udziału społecznego w planowaniu przestrzennym. [On social participation in spatial planning. Applications of geoimaging to strengthen social participation in spatial planning] (2007). http://pspe.gridw.pl/movies/O%20partycypacji_spolecznej.pdf
2. Bąk, M., Kulawczuk, P., Szcześniak, A.: Dobre rządzenie poprzez wkład społeczny, Najlepsze praktyki prowadzenia konsultacji z organizacjami pozarządowymi, Rekomendacje dla Polski [Best practices in consultations with NGOs, recommendations for Poland]. Warszawa: Instytut Badań nad Demokracją i Przedsiębiorstwem Prywatnym (2013). http://www.iped.pl
3. Buczek, A., Olszewski, R.: Rola bazy danych obiektów topograficznych w tworzeniu infrastruktury wiedzy przestrzennej". In: Olszewski, R., Gotlib, D. (eds.) Rola bazy danych obiektów topograficznych w tworzeniu infrastruktury informacji przestrzennej w Polsce, pp. 307–317. Główny Urząd Geodezji i Kartografii, Warszawa (2013). http://www.gugik.gov.pl/__data/assets/pdf_file/0009/22959/Monografia-GBDOT-cz.-1.pdf
4. Bunchball company homepage. http://www.bunchball.com/
5. CBOS survey conclusion: Potencjał społecznikowski i zaangażowanie Polaków w wolontariat [CBOS survey conclusion: "Social activity potential and the involvement of Polish citizens in voluntary work"]. BS 23, Warszawa (2012). http://www.cbos.pl/SPISKOM.POL/2012/K_023_12.PDF
6. CityOne Game. http://www-01.ibm.com/software/solutions/soa/newsletter/aug10/cityone.html
7. Clark, A.: Serious Games: Complete Guide to Simulations and Serious Games: How the Most Valuable Content will be Created in the Age Beyond Gutenberg to Google. Wiley, New York (2009)
8. Czekiel-Świtalska, E.: Konsultacje społeczne w planowaniu miejscowym - wybrane zagadnienia [Social consultations in local spatial planning – selected issues]. Przestrzeń i Forma 17, 325–338 (2012). http://www.pif.zut.edu.pl/pif-17_pdf/C-01_Czekiel.pdf
9. European Social Survey. http://www.europeansocialsurvey.org/
10. Fiedukowicz, A., Kołodziej, A., Kowalski, P., Olszewski, R.: Społeczeństwo geoinformacyjne i przetwarzanie danych przestrzennych. In: Olszewski, R., Gotlib, D. (eds.) Rola bazy danych obiektów topograficznych w tworzeniu infrastruktury informacji przestrzennej w Polsce, pp. 300–306. Główny Urząd Geodezji i Kartografii, Warszawa (2013)
11. Garfield, S.: On the Map: A Mind-Expanding Exploration of the Way the World Looks. Avery (2013)
12. Goodchild, M.F.: Citizens as sensors: the world of volunteered geography. GeoJournal 69(4), 211–221 (2007)
13. Kapp, K.: The Gamification of Learning and Instruction: Game-based Methods and Strategies for Training and Education, 10. Wiley, New York (2012)
14. Leszek, K.: Utilization of 3D city models at LOD1, LOD2 and LOD3 for Smart City concept development in terms of spatial planning for Plock. Master thesis (2015)
15. McGonigal, J.: Reality is Broken: Why Games Make Us Better and How They Can Change the World. Penguin, New York. (2011)
16. Olech, A.: Dyktat czy uczestnictwo? Diagnoza partycypacji publicznej w Polsce [Dictatorship or participation? Diagnosis of public participation in Poland]. Instytut Spraw Publicznych, Warszawa (2012)

17. Playable City. http://www.watershed.co.uk/playablecity/overview
18. Przewłocka, J.: Zaangażowanie społeczne Polaków w roku 2010: wolontariat, filantropia, 1 %, Raport z badań [Social involvement of Polish citizens in the year 2010: voluntary service, charity, 1 %. Study report]. Warszawa: Stowarzyszenie Klon/Jawor (2011). http:// civicpedia.ngo.pl/files/civicpedia.pl/public/raporty/zaangazowanie2010.pdf
19. Report on the participation budget for the year 2015. Urząd Miasta Warszawa, Warszawa. http://twojbudzet.um.warszawa.pl/sites/twojbudzet.um.warszawa.pl/files/raport_-_budzet_partycypacyjny_2015.pdf
20. Sintomer, Y., Herzberg, C., Rocke, A., Allegretti, G.: Transnational model of citizen participation: the case of partcipatory budgeting. J. Public Delibaration **8**(2), 70–116 (2012). http://www.publicdeliberation.net/cgi/viewcontent.cgi?article=1234&context=jpd
21. Smith A.: Civic engagement in the Digital Age (2013). http://www.pewinternet.org/2013/04/25/civic-engagement-in-the-digital-age/
22. Szlachetko, J.: Konsultacyjne formy udziału spoółeczeństwa w planowaniu przestrzennym [Consultation – based forms of social participation in spatial planning] Instytut Metropolitalny (2014)
23. Towarzystwo Urbanistów Polskich: Partycypacja społeczna w planowanie przestrzennym [Social participation in spatial planning]. TUP, Warsaw (2014)
24. Zichermann, G., Linder, J.: The Gamification Revolution: How Leaders Leverage Game Mechanics to Crush the Competition. McGraw-Hill, New York (2013)
25. PASI: http://www.ptip.org.pl/auto.php?page=Encyclopedia
26. ARML standards: http://www.opengeospatial.org/standards/requests/94

A Study of Virtual Reality Headsets and Physiological Extension Possibilities

Thitirat Siriborvornratanakul[✉]

Graduate School of Applied Statistics,
National Institute of Development Administration (NIDA),
118 SeriThai Road, Bangkapi, Bangkok 10240, Thailand
thitirat@as.nida.ac.th

Abstract. Since the worldly notable companies put serious investment in VR headsets in 2014, everything about VR has reemerged from the deep of fiction-like prototypes to the surface of actual consumer products. This reemergence is much stronger than the past with very active supports from commercial VR content creation devices like 360-degree cameras. In this paper, we study important triggers and trends related to VR headsets from 2014 to now. In details, we discuss about the timeline of important events, high potential related products together with example usage scenarios, major user experience concerns with possible solutions, and future possibilities led by VR headsets. We also propose novel usage scenarios where recently popular heart rate monitoring wearable devices are used in combination with VR headsets in order to open up a new communication channel between the headset and the wearer.

Keywords: Virtual Reality Headset · 360-degree camera · Physiology

1 Virtual Reality Headsets

In research communities, studies of Virtual Realities (VR) for immersive experiences have been investigated since decades ago as some referencing numbers shown in Fig. 1. The first head-mounted device (HMD) is believed to be [24] proposed back in 1968. Despite of the long history, VR trends were obviously missed out on mainstream consumer's attention and, in our opinion, it is not until the year 2014 that the trends of VR have truly been resurrected.

In 2014, there were significant moves from the world's technology giants towards VR headsets. In March, Facebook paid $2 billion to acquire Oculus VR—the 'Oculus Rift' VR headset tech startup that hugely succeeded in their 2012 Kickstarter campaign. In the same month during Game Developers Conference (GDC2014), Sony presented the prototype of VR headset for PlayStation 4 named 'Project Morpheus' (renamed to PlayStation VR afterwards). Three months later in June at the Google I/O developer conference for Android devices, Google introduced Google Cardboard, the cheap DIY VR headset using an Android smartphone as processor and display. Then in September at IFA technology conference, Samsung unfolded the Samsung Gear VR Innovator Edition

© Springer International Publishing Switzerland 2016
O. Gervasi et al. (Eds.): ICCSA 2016, Part II, LNCS 9787, pp. 497–508, 2016.
DOI: 10.1007/978-3-319-42108-7_38

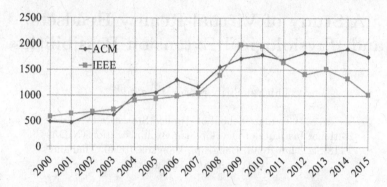

Fig. 1. The number of search results (vertical axis) using the 'virtual reality' keyword as retrieved from ACM digital library and IEEE Xplore digital library during years of 2000 to 2015 (horizontal axis). (Information retrieved on 22 January, 2016).

Fig. 2. Images from left to right are Oculus Rift, Morpheus (PlayStation VR), Google Cardboard, and Samsung Gear VR as presented in 2014.

headset running with the Galaxy Note 4 smartphone. These four VR headsets as shown in Fig. 2 share the core concept of using traditional stereoscopic vision where the third dimension (i.e. depth) is artificially generated by carefully aligning two offset 2D images. Moreover, in October 2014, there was news that Google invested \$542 million in Magic Leap, the mysterious tech startup company working on the so-called head-mounted virtual retinal display. Their Cinematic Reality, utilizing projection of a digital light field into user's eyes, was claimed to provide better, more realistic and more user friendly 3D vision than the old and bulky stereoscopic display does. This secretive technology is designed for Augmented Reality (AR) glasses which are right in the neighborhood of VR headsets.

The triggers in 2014 have effectively turned the whole world back to VR technologies. The year 2015 was an exciting year where progresses in research and development of the four leading VR headsets were continuously presented along with new VR headsets being invested by many world leading technology companies such as LG, Panasonic, HTC and Huawei. In March 2015, Apple Inc. also showed their step towards VR/AR by investing \$32 million to acquire Metaio—the German-based AR startup company behind Ferrari AR showroom app and Ikea AR catalogue app. Then in June 2015, Microsoft officially unfolded their AR/VR headset named HoloLens in Electronics Entertainment Expo (E3 2015). Besides, there was news in October that the mysterious Magic Leap was

Table 1. Overview of VR/AR headset boxes released in 2016. (Information retrieved on 17 March, 2016)

Product	Company	Release	Price	Note
Gear VR	Samsung	11/2015	$99	Free with the preorder of Samsung Galaxy S7 or S7 Edge.
Oculus VR	Facebook	3/2016	$599	Include an Xbox controller, a motion tracking sensor and a remote control but exclude Oculus touch controllers.
HoloLens	Microsoft	3/2016	$3,000	Developer edition.
HTC Vive	HTC	4/2016	$799	Include two-hand tracking controllers and full-room position sensors.
PlayStation VR	Sony	10/2016	$399	Exclude two hand batons and a motion sensing camera.

successfully attracted the Chinese e-commerce giant Alibaba to invest $200 million in their secretive AR glasses. 2016 is considered as the year of VR hardware due to many scheduled releases of consumer edition VR headsets from tech giants as partly concluded in Table 1.

Clearly, the ultimate goal of companies investing in VR headsets is to make VR headsets and VR experiences widespread and become the next common household gadgets. Unlike VR experiences in setup laboratories or well-prepared exhibitions, VR's home experiences introduce many limitations like limited physical space and obstructing furniture as discussed in [23, 26]. The work of [26] solves a conflict between unlimited virtual space and limited physical space by simply asking users to turn around when they reach a physical door. In [23], the authors focus on dealing with obstructing physical furniture or objects by replicating them in the virtual environment; detailed observation is done to find acceptable levels of mismatch between physical objects and their virtual replicas that do not break VR illusion. With the increase of consumer VR headsets, we believe that similar observations regarding home-user's VR experiences will increase and become more common in the near future.

Speaking of the true VR, it is not just about visual and audio simulation as provided by VR headsets but simulation of humans' five fundamental senses. Starting VR experiences in the visual sense is good as it's the highly informative sense we use to experience most things. Seeing and hearing the realistic virtual world is the first step that diverts us from the physical world. But in order to keep going without breaking the VR illusion, a touch experience is required in the next step. The significance of hands in VR experiences is obvious with the presence of Oculus Touch controllers of Oculus Rift, HTC Vive controllers and Wand controllers of Sony PlayStation VR; all are separated two-hand controllers for simulating and controlling two virtual hands. With these controllers, hands can be added to the virtual environment. But to complete the other side of an interactive equation, touching responses (i.e., tactile or haptic feedback) are also required and we believe that this is the next step of VR controllers as suggested by studies of passive haptics for VR experiences in [23]. According to [23], the mismatch between what seeing (virtual environment) and what touching (physical object) can result in the vanished VR illusion.

When using properly, VR experiences can help designers cross accessibility barriers of people with disabilities. An example in [1] uses the VR headset (acting like the AR see-through goggle) to help designers understand how people with different visual impairments see the world. As VR experiences progressively expand to other fundamental senses, it means more chances for reducing gaps between humans, not only the physical gap measured in kilometers but also the unmeasured physical gap of disabilities.

2   Virtual Reality Content Creation

Mass production of VR headsets alone is not enough for VR experiences to become truly widespread unless there are a sufficient number of VR contents. Creating a 3D VR environment for real-time interactions is not an easy task and it requires highly technical skills. Apart from commercial games, viewing 360° videos is an accessible and easy alternative for nontechnical consumers to taste a bit of VR immersive experiences by VR headsets. Being aware of the approaching of consumer VR, since 2015, Youtube and Facebook have allowed uploading and viewing 360° videos using standard PCs, smartphones or tablet computers either with or without VR headsets. Also live VR was tested publicly for broadcasting the US presidential debate and some sport matches, allowing owners of VR headsets to watch these broadcasts in real-time immersive 3D experiences. In the latest news in March 2016, the world first VR cinema has just been opened in Amsterdam; the cinema is currently available for 50 people; each sits on a 360° swivel chair when a 30-minute 360° movie is shown through a head-strap Samsung Gear VR.

It is Nikon that unexpectedly joined the consumer VR parade by stepping a foot in the niche market of consumer-grade 360° camera and presenting their first 360° action camera named KeyMission 360 in Consumer Electronics Show (CES2016) held in January 2016. Then, at Mobile World Congress (MWC2016), Samsung unfolded its consumer 360° camera named Samsung Gear 360 said to be released within 2016. Moreover, in the annual F8 developer conference held in April 2016, Surround 360, a 360° video camera consisting of a 17-camera array from Facebook, was introduced and its design as well as image stitching codes was said to be available as an open-source project on GitHub in summer. Besides, in order to introduce the idea of "VR for everyone (not just for gamers)," in May 2016, Facebook announced that soon it will allow uploading a 360° photo to Facebook's newsfeed. The uploaded 360° photo refers to either an actual 360° photo shot by a 360° video camera or a panoramic photo shot in panorama mode of recent smartphones. According to Facebook, viewing the uploaded panoramic photo in newsfeed will be the same as viewing the true 360° photo, meaning that the viewing angle can be freely changed by dragging a cursor or finger across the photo, tilting a smartphone, or moving a head mounted with VR headset.

Fig. 3. Images from left to right are Nikon KeyMission 360, Samsung Gear 360, Facebook Surround 360 and Lytro Immerge.

The true 360° videos or images shot by proper 360° cameras deliver 3D experiences by virtually putting us at the center of a spherical room whose wall is patched by images; we are free to change viewing directions but not allowed to wander around. The more advanced and perhaps more expensive alternative was proposed by Lytro in 2015 when they introduced Lytro Immerge whose first prototype was scheduled for 2016. With Lytro's light field technology, a focus can be freely picked at a desired depth even after the image was shot. Combining with 360° capability, Lytro Immerge is said to provide realistic 3D cinematic experiences. This means that we are not stuck at the same viewpoint as the camera, but can freely change the viewpoint and virtually wander around inside the 360° video as far as the camera can collect light field information. Figure 3 shows images of 360° cameras from Nikon, Samsung, Facebook and Lytro.

Since the resurrection of VR headsets in 2014, there are many consumer-based VR concepts proposed scatteringly, from headsets to action cameras and from 360° videos to 360° light field images. We believe that these scattering concepts will soon be united in ways that deliver better user experiences and open up more possibilities. One example of uniting is [8] where concepts of VR live streaming, 360° video camera and VR headset are combined into a single system named JackIn Head—the VR-based telepresence system with one user wearing the JackIn Head headgear and the other user wearing a normal VR headset. The headgear wearer acts like a field reporter who streams his or her 360° live experience to the audience wearing the VR headset. Note that integrating devices is not just about putting things together but about solving problems introduced by the integration. In JackIn Head, they propose the image stabilizing solution to reduce the audience's cybersickness caused by movements or unintended motions of the headgear wearer.

In conclusion, it is quite obvious that the world's technology giants have seriously put effort in bringing VR headsets and VR content creation technologies into the spotlight. Starting from the year 2016 where many consumer VR products have been scheduled to set off, it may be about time that the just-workable VR headset in the past will find its way to cross the tipping point and finally become the almost-perfect VR headset providing impressive consumer experiences.

3 Virtual Reality's User Experience Problems

While development of VR headsets and VR content creation devices seem to go well in the spotlight, the headset's user experiences have been major concerns in order for VR industry to cross the tipping point and survive in the long run. Poor 2D vision may look bad on a screen, but poor 3D vision may cause us undesired sicknesses. It was back in 1975 when this kind of sicknesses was first studied in [20]. Regarding VR systems, there are three frequently mentioned sicknesses [3]—motion sickness, simulator sickness, and cybersickness. Motion sickness (a.k.a. car sickness, sea sickness, travel sickness) causes sickness to passengers in moving, spinning or swinging vehicles. Simulator sickness as found in pilot's flight simulation, is caused by discrepancy between the simulator's physical motion and the vehicle's virtual motion. Cybersickness (a.k.a. VR sickness) occurs to users exposing in a virtual environment. These three sicknesses share some symptoms but not totally identical to one another.

In cybersickness, common symptoms include eyestrain, headache, nausea, disorientation and sweating [12]. So far the actual causes of cybersickness and its underlying mechanisms have not been completely understood. The most popular theory used to explain cybersickness is the Sensory conflict theory [11,12] saying that cybersickness occurs because of conflict between the visual and vestibular senses. For VR headsets, this refers to situations where the (virtual) motions as seen by users via the headset's display are not matched with the (physical) motions as felt by the vestibular sense. In order to provide some guidelines for VR designers, previous studies like [3,21] break down causes of cybersickness and categorize them into three groups—individual factors (age, gender, illness, posture), device factors (lag, flicker, calibration, FOV, ergonomics), and task factors (control, duration). Although individual factors are difficult to generalize in consumer VR systems with diverse users, previous studies regarding the device and task factors have greatly contributed to the latest generations of consumer VR headsets. For example, in addition to the head trackers embedded in the headsets, Oculus Rift, HTC Vive and Sony PlayStation VR include separated positional trackers for tracking headset's positions in the physical space. These separated trackers are added not only for efficiency reasons but also for user experience reasons as many previous studies including [14] discover that doing so helps reduce severity of the sickness.

Apart from the three groups of fundamental factors, recent studies in [10,25] propose a similar technique of adding a reference point in the virtual world in order to reduce the sickness and allow longer gaming experiences. Kitazaki et al. [10] conclude that either fixed or moving reference point is fine for this purpose. Venere [25] addes that their subject users simply did not notice the virtual reference while playing games. Combining these discoveries, using a virtual reference point may be considered as one technique for reducing sickness in future VR headset experiences. Also there is the latest study in [7] that observes problems of visual discomfort in VR headsets from a different perspective. They point out that it is the nature of 3D stereoscopic vision itself that causes the problem because the stereoscopic vision used in most well-known VR headsets suffers from the vergence-accommodation conflict due to lack of focus cues. Hence, [7] proposes the new affordable VR headset utilizing the light field technique so that the headset lets the eyes do their usual works of simultaneous converging (to the area of interest) and focusing. Although cybersickness evaluation is not done, their experiments show promising results in real time, allowing correct or nearly correct front and rear focuses in both computer generated 3D virtual environment and 3D cinematic reality environment captured by the stereo light field camera setup. Nevertheless, without additional side-by-side comparative experiments, it is difficult to decide which one is better—the traditional stereoscopic vision as widely used in most VR headsets right now, or the newly proposed but affordable light field technique. One thing to be certain is as long as VR headsets are still in the spotlight, the search for better user experiences will surely be continued.

These vision-based user experience concerns introduce lower age limits written in VR headsets' safety notice. For Oculus Rift and Samsung Gear VR, users should be at least 13 years old. For Sony PlayStation VR, the lower age limit is 12 years old. The reason behind this is because kids in these ages are still in "the critical period in visual development."

4   Virtual Reality Headset with Physiological Extension

In Gartner's list of top 10 strategic technology trends for 2016, the device mesh and ambient user experience are two trends listed under the digital mesh category. Combining the two trends conveys a network of collaborative devices that is capable of sensing real-time context information and creating user experiences that respond to the context accordingly, automatically and immediately. Recent high-end VR headsets with embedded motion sensors and external motion controllers are examples of how a device network enables motion-based ambient user experiences. However, in order to enhance the ambient user experience to the next step, not only user's external motions but also user's internal physiological signals should be taken into consideration.

Linking physiological signals with user's physical/emotional behaviors, a.k.a. psychophysiology, is a popular field in recent user researches, particularly game user researches. Previous study [17] in user researches experimentally confirms that taking physiological signals into account helps increase discovery rates regarding emotional user experience issues. Psychophysiological studies in [4, 16] also show how physiological signals can be correlated to specific game play experiences, allowing a noninterrupted in-depth observation of gamer's emotions regarding each stage of game. In these two works, the measured physiological signals include electrodermal activity, electroencephalography, skin-conductance, heart rate, etc. and the observed game user experiences include challenge, tension, competance, flow and immersion. Complicated discoveries as well as methodologies of psychophysiology in game user researches can be directly carried on to VR headset scenarios as well in order to quantitatively measure users' internal emotions or responses while wearing the headsets. To be more specific, in VR headset scenarios, because physiological signals are naturally emitted and free from user's intentional controls, they are exceptional user inputs that add no workload to user, do not interrupt real-time VR immersive experiences, and provide rich information regarding user's emotions that no motion sensor can measure.

For more than a decade ago before the reemergence of VR headsets in 2014, VR in medical has been an active field that frequently utilizes VR headsets and physiological signals. Common scenarios include exposing patients to a therapist-designed virtual environment in order to distract them from pains or mentally cure them by a relaxing environment. However, using VR headsets alone is a one-way communication that patients wearing VR headsets have no chance to express their real-time feedbacks to therapists. In order to create a two-way communication where patients can take part in suggesting or changing the virtual environment without interrupting their immersive VR illusion, the concept of psychophysiology is used as it requires no explicit inputs from patients. Existing VR headset systems using physiological signals for medical treatment include systems where real-time physiological signals are used to adjust or change the virtual environment to match patients' current emotional states or internal mechanisms. Proposals in [6,22] utilize GSR (Galvanic Skin Response) sensors to measure electrical conductance of the skin and use it to adaptively modify contents of the virtual environment in order to treat acute pains in [6] and chronic headaches in [22]. Another example of medical usages falls in forensic psychiatry[1] where combining VR headsets and psychophysiology enables a real-time decision making tool that allows a psychiatric to better understand and provide improved treatment for violent offenders subjectively in a highly controlled environment.

Focusing on physiological sensors available in recent consumer markets, the most popular and promising seem to be heart rate monitors as embedded in recent health and exercise tracking devices. Three popular alternatives for measuring real time heart rate are electrocardiography (ECG) electrodes, chest-strap devices and wrist-worn devices. Despite of the highest accuracy of traditional ECG electrodes, their intrusive nature of attaching electrodes on user's body limits their usages in laboratory setup environments. Chest-strap devices as available in many fitness trackers are a compromised alternative; although they can be uncomfortable to get on, they are easier to wear and provide more or less accurate heart rate tracking compared to ECG. The last alternative of wrist-worn heart rate measuring devices wins the two previous alternatives in terms of comfortable and nonintrusiveness because wrist-worn bracelets or watches have been human's accessories since centuries ago. However, according to [18], the wrist-worn alternative tends to be the least accurate for real-time heart rate measurement. Technologies behind recent commercial wrist-worn heart rate monitoring shine some special light into blood vessels and measure heart beats by sensing changes in blood volume. According to [19] and most manufacturers,

---

[1] "A specialised branch of psychiatry which deals with the assessment and treatment of mentally disordered offenders in prisons, secure hospitals and the community," https://www.healthcareers.nhs.uk/, Online; accessed: 2016-03-10.

these light-based sensing technologies when being worn on a wrist, are accurate if and only if wearers are staying still or at rest. For accurate heart rate monitoring during intense activities, the commercial chest-strap devices are recommended instead of the wrist-worn.

Utilization of heart rate has been proposed for diverse applications. During game play in order to keep gamer's heart rate inside the exercise's effective interval, [15] adjusts game's contents in real time according to heart rate data read from the chest-strap Polar S810i. Using heart rate data to create a recommendation system is another popular scenario as one example in an in-flight music recommendation system called iHeartrate [13]. In iHeartrate, the Emfit heart rate sensor hidden in the passenger seat is responsible for sensing passenger's real-time heart rate and the recommended music is selected with a purpose to keep passenger's heart rate within the normal range. Visualizing heart rate information is another interesting scenario proposed recently in TastyBeats [9] and TOBE [5]. In TastyBeats, they record user's one-day heart rate activities via the chest-strap Polar H7 heart rate monitor; at the end of the day, a personalized drink is mixed by four flavors of drinks that represent four heart rate intervals. According to the paper, TastyBeats is a prototype for creating a personalized sport drink that matches individual athletes. In TOBE, the purpose of visualizing real-time heart rate data is for tangible out-of-body experiences that represent user's inner states of physiological signals. There are many sensors used in TOBE, including ECG heart-rate sensing electrodes hidden inside the lab coat. TOBE uses heart rate data as part of their user's physiological activity visualization projected on a tangible avatar where heart rate data are also used in conjunction with measured respiration to indicate user's relaxation state. Nevertheless, to the best of our knowledge, no previous works have used wrist-worn devices, particularly smartwatches, for real-time heart rate monitoring or VR headset applications yet.

From our study focusing in consumer usage scenarios, the number of heart rate-based scenarios is small, and the number of heart rate- plus VR headset-based scenarios is extremely low and still immature. However, as suggested by previous researches, there are possibilities for future researchers to explore and combine these two popular devices in either category as following:

1. *User experience evaluation:* Collect physiological signals during the uses of VR headsets and use them later to evaluate parts of user experiences that do not appear explicitly. The evaluation may be done offline after the VR headset session finishes. Appropriate heart rate sensors for this category should be either the ECG sensor or the chest-strap sensor as both provide accurate results suitable for deep analysis.
2. *Adaptive contents:* Measure real-time physiological signals and use them to adaptively change/adjust/suggest/control contents or stories of the virtual environment in real time. Appropriate heart rate sensors for this category should depend on how each system prioritizes accuracy and convenience. Besides, for a convenient heart rate sensor like smartwatch, usage scenarios may be extended to include a 360° video camera. By using a smartwatch

and a 360° video camera in synchronization, it can record not only 360° videos but also heart rate data, enabling a new dimension of future video watching experiences.

In addition, for low-end VR headsets without two-hand controllers included, using a wrist-worn smartwatch introduces not only an additional heart rate-based monitoring feature but also an additional control pane that enables novel joint interactions with smartphone like those proposed in Duet [2]. Besides, it is also possible to use smartwatch as an optional DIY one-hand controller for low-end VR headsets.

5   Conclusion

In this paper, we review trends of commercial VR headsets. The detailed timeline showing the resurrection of VR headsets in consumer markets, is described, following by how the trend has been emphasized by multiple releases of 360° video cameras for VR content creation. We then discuss the main user experience problem of using VR headsets together with recently proposed solutions. Finally, we observe possibilities of extending VR headsets to user's internal mechanism by mean of measuring physiological signals via external sensors, particularly heart rate sensors which have been popular recently in most health monitoring wearable devices. Although the number of previous works combining VR headsets and heart rate sensors is relatively small and still immature, our study reveals two recommended strategies for future researches interesting in the combination of VR headsets and heart rate sensors.

References

1. Ates, H.C., Fiannaca, A., Folmer, E.: Immersive simulation of visual impairments using a wearable see-through display. In: Proceedings of the International Conference on Tangible, Embedded, and Embodied Interaction (TEI 2015), pp. 225–228 (2015)
2. Chen, X., Grossman, T., Wigdor, D., Fitzmaurice, G.: Duet: exploring joint interactions on a smartphone and a smart watch. In: Proceedings of the SIGCHI Conference on Human Factors in Computing Systems (CHI 2014), pp. 159–168 (2014)
3. Davis, S., Nesbitt, K., Nalivaiko, E.: A systematic review of cybersickness. In: Proceedings of the Conference on Interactive Entertainment (IE 2014), pp. 1–9 (2014)
4. Drachen, A., Nacke, L., Yannakakis, G., Pedersen, A.: Correlation between heart rate, electrodermal activity and player experience in first-person shooter games. In: Proceedings of the ACM SIGGRAPH Symposium on Video Games (Sandbox 2010), pp. 49–54 (2010)
5. Gervais, R., Frey, J., Gay, A., Lotte, F., Hachet, M.: TOBE: tangible out-of-body experience. In: Proceedings of the International Conference on Tangible, Embedded, and Embodied Interaction (TEI 2016), pp. 227–235 (2016)

6. Gromala, D., Tong, X., Choo, A., Karamnejad, M., Shaw, C.D.: The virtual meditative walk: virtual reality therapy for chronic pain management. In: Proceedings of the ACM Conference on Human Factors in Computing Systems (CHI 2015), pp. 521–524 (2015)

7. Huang, F.C., Chen, K., Wetzstein, G.: The light field stereoscope: immersive computer graphics via factored near-eye light field displays with focus cues. ACM Trans. Graph. (TOG) **34**(4), 60 (2015)

8. Kasahara, S., Rekimoto, J.: JackIn Head: immersive visual telepresence system with omnidirectional wearable camera for remote collaboration. In: Proceedings of the ACM Symposium on Virtual Reality Software and Technology (VRST 2015), pp. 217–225 (2015)

9. Khot, R.A., Lee, J., Aggarwal, D., Hjorth, L., Mueller, F.F.: TastyBeats: designing palatable representations of physical activity. In: Proceedings of the ACM Annual Conference on Human Factors in Computing Systems (CHI 2015), pp. 2933–2942 (2015)

10. Kitazaki, M., Nakano, T., Matsuzaki, N., Shigemasu, H.: Control of eye-movement to decrease VE-sickness. In: Proceedings of the ACM Symposium on Virtual Reality Software and Technology (VRST 2006), pp. 350–355 (2006)

11. Kolasinski, E.M.: Simulator sickness in virtual environments. Technical report 1027 (1995)

12. LaViola, J.J.: A discussion of cybersickness in virtual environment. ACM SIGCHI Bull. **32**(1), 47–56 (2000)

13. Liu, H., Hu, J., Rauterberg, M.: iHeartrate: a heart rate controlled in-flight music recommendation system. In: Proceedings of the International Conference on Methods and Techniques in Behavioral Research (MB 2010), p. 26 (2010)

14. Llorach, G., Evans, A., Blat, J.: Simulator sickness and presence using HMDs: comparing use of a game controller and a position estimation system. In: Proceedings of the ACM Symposium on Virtual Reality Software and Technology (VRST 2014), pp. 137–140 (2014)

15. Masuko, S., Hoshino, J.: A fitness game reflecting heart rate. In: Proceedings of the ACM SIGCHI International Conference on Advances in Computer Entertainment Technology (ACE 2006), p. 53 (2006)

16. Mirza-Babaei, P., Nacke, L., Gregory, J., Collins, N., Fitzpatrick, G.: How does it play better? exploring user testing and biometric storyboards in games user research. In: Proceedings of the SIGCHI Conference on Human Factors in Computing Systems (CHI 2013), pp. 1499–1508 (2013)

17. Mirza-babaei, P., Long, S., Foley, E., McAllister, G.: Understanding the contribution of biometrics to games user research. In: Proceedings of the International Conference: Think Design Play (DiGRA 2011), vol. 6 (2011)

18. Palladino, V.: Who has the most accurate heart rate monitor? tom's guide (2015). www.tomsguide.com/us/heart-rate-monitor,review-2885.html. Accessed 05 Mar 2015

19. Profis, S.: Do wristband heart trackers actually work? a checkup. CNET (2014). http://www.cnet.com/news/how-accurate-are-wristband-heart-rate-monitors/. Accessed 05 Mar 2015

20. Reason, J.T., Brand, J.J.: Motion Sickness, pp. 83–101. Academic Press, New York (1975)

21. Rebenitsch, L., Own, C.: Individual variation in susceptibility to cybersickness. In: Proceedings of the ACM Symposium on User Interface Software and Technology (UIST 2014), pp. 309–317 (2014)

22. Shiri, S., Feintuch, U., Weiss, N., Pustilnik, A., Geffen, T., Kay, B., Meiner, Z., Berger, I.: A virtual reality system combined with biofeedback for treating pediatric chronic headache–a pilot study. Pain Med. **14**(5), 621–627 (2013)
23. Simeone, A.L., Velloso, E., Gellersen, H.: Substitutional reality: using the physical environment to design virtual reality experiences. In: Proceedings of the ACM Conference on Human Factors in Computing Systems (CHI 2015), pp. 3307–3316 (2015)
24. Sutherland, I.E.: A head-mounted three dimensional display. In: Proceedings of AFIPS Fall Joint Computer Conference (AFIPS 1968), pp. 757–764 (1968)
25. Venere, E.: 'Virtual nose' may reduce simulator sickness in video games. Purdue University News (2015). http://www.purdue.edu/newsroom/releases/2015/Q1/virtual-nose-may-reduce-simulator-sickness-in-video-games.html. Accessed 23 Jan 2016
26. Williams, B., Narasimham, G., Rump, B., McNamara, T.P., Carr, T.H., Rieser, J., Bodenheimer, B.: Exploring large virtual environments with an HMD when physical space is limited. In: Proceedings of the symposium on Applied Perception in Graphics and Visualization (APGV 2007), pp. 41–48 (2007)

A Novel Integrated System of Visual Communication and Touch Technology for People with Disabilities

Chutisant Kerdvibulvech[✉]

Graduate School of Communication Arts and Management Innovation,
National Institute of Development Administration, 118 SeriThai Rd., Klong-chan,
Bangkok 10240, Bangkapi, Thailand
chutisant.ker@nida.ac.th

Abstract. Due to the current popularity of the internet of things (IoT), the research topic for communicating, connecting, and supporting people remotely through the internet is very popular in computational science in recent years. This paper presents a new integrated application to assist people with disabilities based on enhanced technologies of visual and touch communications in exploiting information and communication technology (ICT) innovative technologies. Our research aim is to help hearing impaired people to communicate both visually and affectingly to their loved one who may live distantly in different part of the world. By integrating an augmented reality application for visual communication and a wearable jacket for touch communication, it is able to support hard of hearing people via the human-computer interaction experience. A Google cardboard is also built for allowing people with hearing loss and deafness to have an immersive experience visually using augmented reality for geometric visualization. A hugging communication wearable tool, called T.Jacket, using sensor technology is then extended and applied to assist disabled people for hugging their loved one remotely by reproducing an artificial hug sense between two people affectingly. Experimental results have also been included to show the robustness of the proposed integrated application.

Keywords: Visual communication · Computational science · People with disabilities · Hard of hearing people · Information and communication technology · T.Jacket · Wearable tool · Touch technology

1 Background

As we currently live in a time of tremendous change technologically demographically and economically, the influence of the internet of things (IoT) plays an important role in every day of our lives in exploiting and achieving information and communication technology (ICT) innovative technologies. There are more and more people, from abled people to disabled people and from young persons to elderly persons, who are quite interested in taking advantage of smart and innovative technology for communication in a novel way. In this decade, the technology based on the internet of things in computational science, such as virtual environments, computer vision, image processing, pattern recognition, and wearable computing, extraordinarily changes and reshapes the

© Springer International Publishing Switzerland 2016
O. Gervasi et al. (Eds.): ICCSA 2016, Part II, LNCS 9787, pp. 509–518, 2016.
DOI: 10.1007/978-3-319-42108-7_39

way we communicate for people distantly. Communication between people and people, or between people and things, does not necessarily be technologically limited to only one individual human sense. This is since each human has traditionally five sense organs: sight (vision), touch, hearing, taste (gustation), and smell (olfaction). Our main purpose of the research described here in this paper is to integrate a system with two essential senses: touch and visual communications, to help disabled people with special needs. Our contribution is aimed to assist hearing impaired people to communicate both visually and affectingly to another people who live distantly in every corner of the world. Figure 1 shows some examples of our integrated system during using the presented work. In the cardboard, it shows the virtual smartphone's screen for illustrating how the users can see in the system.

Fig. 1. Examples of our integrated system between touch and visual communications using a built cardboard headset for supporting people with disabilities

First, sight is an important part of how we communicate. In recent years, several researches for communication through digital sight using computer vision, human-computer interaction and augmented reality have been proposed. For instance, an immersive telepresence work was built by Nguyen et al. [1] for providing a radical video communication experience distantly in the meetings. As shown in Fig. 2, this system is able to allow an interaction among people and collaborative contents shared remotely. Moreover, a computer vision-based algorithm [2] was created for analyzing and tracking the motion of human's organs. Therefore, by using this tracking algorithm, it is able to allow people to communicate and interact with other people from any different location remotely and automatically through digital sight. Next, a multi-touch system using augmented reality was built by Novotny et al. [3]. Their concept is to use two display units for showing two contexts of the same object for education purpose. It also aims for presenting of virtual heritage. Second, touch is also a vital part for non-verbal communication. Digital touch communication based on the innovative technologies from any location has been recently a very popular research topic for connecting people to people. For example, Huisman et al. [4] studied a collaborative game with virtual agents for simulating social touch, such as human hugs and handshakes. This study revealed that the social touch through virtual agents is able to influence the people's perception positively. Furthermore, [5] presented recent innovative technologies used for touch communication such as the therapeutic robot for affective touch from Massachusetts Institute of Technology (MIT) and the Pet Internet through digital touch from National University Singapore (NUS). Robots with a human touch was also discussed in [6] by Mone. In addition, a transient microbot framework was built by Chen et al. [7] to transport drugs inside the human body for touch communication. This framework uses a micro-to-macro cross-scale communication model that is able to create a touch-communication (TouchCom) paradigm. However, most of the aforementioned works are different from ours presented in this paper. This is because, to the best of our knowledge, no previous research paper in computational science has integrated a system with two essential senses: touch and visual communications, to help people with disabilities. This paper presents a new integrated system to solve this challenging robustly and accurately.

In fact, there are many recent research works attempting to achievably build the systems for people with disabilities using multimedia and virtual reality. For instance, a system for simulating and identifying the signs of rapid patient deterioration was presented by Wu et al. [8] for medical purposes using multimedia and virtual reality approaches as an assistive technology. So this system used virtual reality applied to medicine. The analyzed influence of virtual human animation was then studied in 2014 virtually and interactively on the sensational responses of patients with disabilities and special rehabilitation for medical study and assessment in the conditions of patient deterioration. This research study uses sensors of psycho-physical Electro Dermal Activity (EDA), Positive and Negative Affect Survey (PANAS), and Differential Emotions Survey (DES). By enhancing social presence with interactive virtual humans in each situation synthetically, it is able to elicit a response in non-positive emotions (e.g., suffering, irritation and dread) as the disabled patient's medical symptom declined virtually in the simulated interpersonal experience. Also more recently, Hornero et al. [9] in August 2015 proposed a wireless system using an augmentative and alternative communication instrument. The augmentative and alternative communication application is aimed for supporting people with impairments such as children born with development disability, young people with physical disabilities and cerebral palsy who somehow may have severely speech limitations, and elders with language impairments. It has two different blocks connected by a network system of communication board sheets separately, so that the interface for people with disabilities in the mentioned

Fig. 2. Examples of a real-time immersive telepresence work, called Immersive Telepresence system for Entertainment and Meetings (ITEM), implemented by Nguyen et al. in [1] for providing a radical video communication experience distantly

system is designed for simple use economically. It is also scalable and changeable adaptively for the number of conditions and the vocabulary of speech disabled people. Furthermore, a learning system was built by Saad et al. [10] in 2015 for supporting young people with intellectual development disorder in intellectual functioning and adaptive behavior using multimedia technology. This is achieved by integrating Skinner's Operant Conditioning Model with Mayer's Cognitive Theory of Multimedia Learning. This system provides some interactive multimedia contents educationally such as images and short clips for teaching young disabled people for both group and individual activities. This system has many benefits by enhancing learning relative performance and higher scores for young people with intellectual disabilities. Besides, Kbar and Aly [11] created a set of applications designed to aid elderly people and people with disabilities to access to information and communication technology related-workplace. In this system, a scenario for the ambient intelligence-based assistive technology is designed similarly as the ambient and assisted living (AAL) for disabled and older people. By using ambient intelligence technology, people with special rehabilitation and needs are able to interact with smart and intuitive interfaces that are embedded in all kinds of objects naturally. However, in [8–11], they do not aim to use two main important senses of human (i.e., visual and touch senses) together to support disabled people with special needs. In other words, their research works only focus on an individual sense which is not somehow convenient for some specific groups of people.

Moreover, Neto and Fonseca [12] in 2014 presented a photo-to-speech application, so-called Camera Reading for Blind People using a smartphone. This application uses a technological set of frameworks of Text to Speech Synthesis (TTS) and Optical Character Recognition (OCR) for recognizing words in the photos. An integration of filters (i.e., CIColorControls and CIColorMonochrome) is utilized for improving the quality of the results. It is able to allow people with visual impairments to take a photo and then hear the text that appears in each taken photo interactively. In addition, Magnenat et al. [13], proposed recently in 2015, from Disney Research Zurich developed a printed coloring book method for children. It is done by texturing and displaying the color characters in three dimensions using a smartphone in a children's coloring book. In their work, a texturing algorithm was also presented in real-time for transforming the input texture from a two-dimensional colored drawing to a three-dimensional character. Furthermore, Hornero et al. [14] in August 2015 proposed a wireless system using an augmentative and alternative communication instrument. The augmentative and alternative communication application is aimed for supporting people with impairments such as children born with development disability, young people with physical disabilities and cerebral palsy who somehow may have severely speech limitations, and elders with language impairments. It has two different blocks connected by a network system of communication board sheets separately, so that the interface for people with disabilities in the mentioned system is designed for simple use economically. It is also scalable and changeable adaptively for the number of conditions and the vocabulary of speech disabled people. Moreover, a learning system was built by Saad et al. [15] in 2015 for supporting young people with intellectual development disorder in intellectual functioning and adaptive behavior using multimedia technology. This is achieved by integrating Skinner's Operant Conditioning Model with Mayer's Cognitive Theory of Multimedia Learning. This system provides some interactive educational multimedia contents such as images and short clips for teaching young disabled people for both group and individual activities. This system has many benefits by enhancing learning relative performance and higher scores for young people with intellectual disabilities. Nevertheless, in every mentioned system in [12] for recognizing words, [13] for texturing and displaying the color characters, [14] for an augmentative and alternative communication instrument and [15] for multimedia-based adaptive behavior, they have different goals for our presented work in this paper. In other words, our research works do not focus on just only an individual sense. In this way, it will be more convenient for some specific groups of people such as disabled people and/or people with special needs. Our main purpose of the research described here is to integrate a system with two essential senses: touch and visual communications, to help disabled people with special needs simultaneously.

In the rest of the presented paper, it is organized as follows. The next section in Sect. 2 will explain about our proposed system integrated between vision and touch and how does it work. We then show our representative experimental results in Sect. 2. Finally in the last section (Sect. 3), we conclude the paper and points out the directions of our future work.

2 The Proposed Integrated System Between Vision and Touch

This paper integrates the interaction for supporting deaf people via the human-computer interaction experience from an augmented reality application [2] for vision and a wearable T.Jacket [16] for touch using geometric visualization and sensor technology, respectively, in achieving ICT innovative technologies. This concept is somehow similar to assistive technology device that is able to help disabled people. First, for visual communication, we use a Google cardboard [17] we built for allowing people with disabilities to have an immersive experience about the scene and story we created using augmented reality technology. In computational science, computer science and computer engineering, the internet of things does not mean only devices, but it also means about the communications and connections between people. In this paper, we present a new integrated application between touch and visual communications, especially for people with disabilities. Figure 3 shows some examples of our integrated system while the user is using the system. The user can both sit and stand while wearing both instruments. Our contribution is aimed to assist hearing impaired people to communicate both visually and affectingly to another people who live distantly in every corner of the world. An augmented reality application for vision and a wearable jacket for touch are integrated in the proposed system. We first use the Vidinoti image application [18] to create the augmented reality environments. This QR-code is used to recognize the pose of each camera from the smartphones we use in the system. Representative information for visual communication will be then overlaid onto the background for helping disabled people with special needs. After that by using a sensor method in a hugging

Fig. 3. The integrated system of visual communication and touch technology for assisting disabled people

communication wearable tool, called T.Jacket. This idea is also similar to assistive technology device.

For visual communication, we use a new cardboard headset we built. By using this cardboard headset, it is able to allow people with disabilities to have an immersive experience about the scene and story we created using augmented reality technology. This cardboard headset we designed is extended from the Google cardboard. The difference is that ours is more suitable for people with disabilities in term of how to use it. Figure 4 illustrates our created cardboard and the T.Jacket we used. For touch communication, we utilize the extended T.Jacket, a hugging communication wearable tool for allowing people with hearing loss to be able to hug their loved one remotely by reproducing a hug sense between two people. By integrating augmented reality with the T.Jacket and touch technology, it is able to simulate the feeling of a hug with laterally used air pressure to help people with disabilities to communicate touchingly. In this way, our system works well for people who are living remotely in different parts and places of the world. Figure 4 depicts some additional experimental results during testing the system with hearing impaired people. Finally, we conduct some experiments to confirm

Fig. 4. A cardboard headset we created and integrated for allowing people with special needs to have an immersive experience about the scene and story using augmented reality technology and our extended wearable T.Jacket

the feasibility and usability of our proposed integrated system. Additional experimental results were conducted during testing our proposed system with the real hearing impaired person who is wearing our huggable T.Jacket we extended and the cardboard headset we built. We test our set-up on two sets of example data (both abled people and disabled people). Each set of example data is composed of eight people: eight abled people and eight disabled people. There are six abled people and seven disabled people who gave positively compliments when using the proposed application. There are two abled people and one disabled people who gave neutral feedback and comment to the presented system. After conducting both qualitative and quantitative evaluations, our experimental results have shown that the integrated system we presented in this paper is positively beneficial for disabled people with special needs.

3   Conclusions

We presented an integrated application to support disabled people with special needs based on enhanced technologies of visual and touch communications in achieving ICT innovative technologies. For geometric visualization and augmented environments, we use augmented reality and a smartphone setup to provide multiple scenarios. By using the Vidinoti image application, we are able to create the augmented reality environments. This QR-code is also used for recognizing the position of each camera from the mobile-phones in the system. A cardboard headset is then built for assisting people with disabilities to have an immersive experience visually using augmented reality. After that, we integrate the interaction for supporting deaf people via the human-computer interaction experience between an augmented reality application and a hugging communication wearable jacket for visual and touch, respectively. This improves the understanding of communications and helps the people with special needs understand the new way of communication of the 21st century. We tested our set-up on two sets of example data (abled people and disabled people). After testing, we found out that the proposed system is quite positively received by the users.

Even though we produce an accurate result that, we believe, is generally acceptable, there are several things that we should still improve the current system. Part of the future research work is intentionally aimed at designing, creating and implementing an integrated application for supporting more challenging human senses such as hearing, taste (gustation), and smell (olfaction). Moreover, we plan to cope with the ease of the user interface while using the system. In the current system, it does not work well with totally disabled people in every human sense organ: sight, touch, hearing, taste, and smell. In the future, we intend to make technical improvements for every human sense, such as smelling screen methodology [19] and temporal characteristics of designed taste experiences [20], to additionally refine these mentioned issues in a better way uniquely.

Acknowledgments. This research presented herein was partially supported by a research grant from the Research Center, NIDA (National Institute of Development Administration).

References

1. Nguyen, V., Lu, J., Zhao, S., Vu, D., Yang, H., Jones, D., Do, M.N.: ITEM: immersive telepresence for entertainment and meetings – a practical approach. IEEE J. Sel. Topics Signal Process. (J-STSP) **9**(3), 546–561 (2015). Special Issue on Interactive Media Processing for Immersive Communication
2. Kerdvibulvech, C: A methodology for hand and fingers motion analysis using adaptive probabilistic models. EURASIP J. Embed. Syst. (JES) (18), 9 (2014). Springer Publisher
3. Novotnya, M., Lacko, J., Samuelcika, M.: Applications of multi-touch augmented reality system in education and presentation of virtual heritage. Procedia Comput. Sci. **25**, 231–235 (2013). Elsevier Publishing, Science Direct
4. Huisman, G., Kolkmeier, J., Heylen, D.: Simulated social touch in a collaborative game. In: Auvray, M., Duriez, C. (eds.) EuroHaptics 2014, Part I. LNCS, vol. 8618, pp. 248–256. Springer, Heidelberg (2014)
5. Kerdvibulvech, C.: Vision and virtual-based human computer interaction applications for a new digital media visualization. In: The Proceeding of the 23rd International Conference on Computer Graphics, Visualization and Computer Vision (WSCG), Eurographics Association & ACM SIGGRAPH, Plzen, pp. 247–254, 8–12 June 2015
6. Mone, G.: Robots with a human touch. Commun. ACM **58**(5), 18–19 (2015). CACM Homepage archive
7. Chen, Y., Kosmas, P., Anwar, P.S., Huang, L.: A touch-communication framework for drug delivery based on a transient microbot system. IEEE Trans. Nanobiosci. **14**(4), 397–408 (2015). IEEE Computational Intelligence Society
8. Wu, Y., Babu, S.V., Armstrong, R., Bertrand, J.W., Luo, J., Roy, T., Daily, S.B., Cairco Dukes, L., Hodges, L.F., Fasolino, T.: Effects of virtual human animation on emotion contagion in simulated inter-personal experiences. IEEE Trans. Visual Comput. Graphics **20**(4), 626–635 (2014)
9. Hornero, G., Conde, D., Quilez, M., Domingo, S., Pena Rodriguez, M., Romero, B., Casas, O.: A wireless augmentative and alternative communication system for people with speech disabilities. IEEE Access **3**, 1288–1297 (2015). IEEE Publishing
10. Saad, S., Dandashi, A., Alja'am, J.M., Saleh, M.: The multimedia-based learning system improved cognitive skills and motivation of disabled children with a very high rate. Educ. Technol. Soc. **18**(2), 366–379 (2015)
11. Kbar, G., Aly, S.: SMART workplace for persons with DISABiLitiEs (SMARTDISABLE). In: 2014 IEEE International Conference on Multimedia Computing and Systems (ICMCS), pp. 996–1001. IEEE Publishing, 14–16 April 2014
12. Neto, R., Fonseca, N.: Camera reading for blind people. Procedia Technol. **16**, 1200–1209 (2014). Elsevier Publishing
13. Magnenat, S., Ngo, D.T., Zund, F., Ryffel, M., Noris, G., Röthlin, G., Marra, A., Nitti, M., Fua, P., Gross, M., Sumner, B.: Live texturing of augmented reality characters from colored drawings. In: The Proceeding of the IEEE International Symposium on Mixed and Augmented Reality (IEEE ISMAR 2015), Fukuoka, 10 p., 29 September 2015
14. Hornero, G., Conde, D., Quilez, M., Domingo, S., Pena Rodriguez, M., Romero, B., Casas, O.: A wireless augmentative and alternative communication system for people with speech disabilities. IEEE Access **3**, 1288–1297 (2015). IEEE Publishing
15. Saad, S., Dandashi, A., Alja'am, J.M., Saleh, M.: The multimedia-based learning system improved cognitive skills and motivation of disabled children with a very high rate. Educ. Technol. Soc. **18**(2), 366–379 (2015)
16. Cheok, A.D., Pradana, G.A.: Virtual touch. Scholarpedia **10**(4), 32679 (2015)

17. Yoo, S., Parker, C.: Controller-less interaction methods for Google cardboard. In: The Proceeding of the ACM Symposium on Spatial User Interaction (SUI), p. 127. ACM, New York (2015)
18. Perdikakis, A., Araya, A., Kiritsis, D.: Introducing augmented reality in next generation industrial learning tools: a case study on electric and hybrid vehicles. In: The Proceeding of the Procedia Engineering, the Manufacturing Engineering Society International Conference (MESIC), vol. 132, pp. 251–258. Elsevier Publishing, ScienceDirect (2015)
19. Matsukura, H., Yoneda, T., Ishida, H.: Smelling screen: presenting a virtual odor source on a LCD screen. In: The Proceeding of the IEEE Virtual Reality (VR), IEEE Computer Society, Lake Buena Vista, pp. 167–168, 18–20 March 2013
20. Obrist, M., Comber, R., Subramanian, S., Piqueras-Fiszman, B., Velasco, C., Spence, C.: Temporal, affective, and embodied characteristics of taste experiences: a framework for design. In: The Proceeding of the ACM CHI Conference on Human Factors in Computing Systems (CHI), Toronto, pp. 2853–2862, 26 April–1 May 2014

A Multi-classifier Combination Method Using SFFS Algorithm for Recognition of 19 Human Activities

Feng Lu[1], Danfeng Wang[2], Haoying Wu[1(✉)], and Wei Xie[2]

[1] Key Laboratory of Fiber Optic Sensing Technology and Information Processing, Ministry of Education, College of Information Engineering, Wuhan University of Technology, Wuhan, Hubei, China
lufengwut@163.com, why_dd@126.com
[2] College of Information Engineering, Wuhan University of Technology, Wuhan, Hubei, China
{wdf168168,xiewei0916}@163.com

Abstract. In order to investigate the human activity recognition and classification, which is significant for human-computer interaction (HCI), a multi-classifier combination method using Sequential Forward Feature Selection (SFFS) algorithm is proposed in this paper for recognition of 19 human daily and sports activities. The dataset collected by wearable sensor units is obtained from UCI Machine Learning Repository. The main contents of this method include: (1) extracting features from the raw sensor data after preprocessing; (2) reducing features by SFFS algorithm; (3) classifying activities by 10-fold cross validation with a multi-classifier combination algorithm based on the grid search for parameter optimization. The experimental results indicate that, compared with other traditional activity recognition methods, which use principal component analysis (PCA) to reduce features or use a single classifier to classify activities, the multi-classifier combination method using SFFS achieves the best recognition performance with the average classification accuracy of 99.91 %.

Keywords: Activity recognition and classification · Feature extraction and reduction · Parameter optimization · Multi-classifier combination · Sequential Forward Feature Selection (SFFS)

1 Introduction

With the rapid development of artificial intelligence, human activity recognition, which helps robots or other intelligent devices to recognize human activities and understand the behavioral intentions by monitoring human daily activities, has become a hot spot that numerous scholars are working on in recent years. And human activity recognition contributes to human-computer interaction (HCI), which improves human living standards by providing humanistic service. Some studies have shown that the human activity recognition can be used in monitoring the activities of the elderly and the

O. Gervasi et al. (Eds.): ICCSA 2016, Part II, LNCS 9787, pp. 519–529, 2016.
DOI: 10.1007/978-3-319-42108-7_40

young activities remotely for security, fall detection [1], medical diagnosis, and computer games in virtual reality, and so on.

In earlier research, visual perception is the main way to obtain the motion information in activity recognition, and it is still going on today [2]. However, vision-based motion capture system which has higher requirements of environmental parameters can only be used in a finite space such as a house or a laboratory.

With the rapid advances in MEMS (Micro-electro Mechanical System) technology, the weight, size, cost and energy of motion sensors have reduced considerably. Thus, wearable motion sensor system has becoming a emerging branch of research in human activity recognition [3]. For example, Kerem Altun et al. [4–6] used five Xsens MTx units fixed on body to collect human motion signals and recognize 19 types of daily and sports activities; Khan et al. [7] used a single accelerometer fixed in the chest to classify seven activities, including lying, sitting, standing, walking, running, up and down stairs; Zhang et al. [8] used a accelerometer, a gyroscope and a magnetometer fixed on human's right hip to classify nine activities.

The classification techniques used in human activity recognition include Decision Trees, K-Nearest Neighbor (KNN), Artificial Neural Networks (ANNs) [9], SVM [10], Gaussian mixture model [11], Multi-Classifier Adaptive-Training algorithm (MCAT) [12], etc. The MCAT is a semi-supervised learning algorithm which using four classifiers to utilize unlabelled data and adapting the activity recognition classifier model to a particular person. And this method achieves the average classification accuracy of 82.70 % in the recognition of eight atomic activities.

In this paper, a multi-classifier combination recognition method using SFFS algorithm is proposed to classify 19 types of human daily and sports activities, and the experimental dataset comes from [4–6, 13], which the motion signals collected by a wearable motion sensor system. In this method, SFFS algorithm is used for feature reduction in data processing to eliminate redundancy and reduce the computational complexity, grid search algorithm is applied to optimize the parameters of multi-classifier combination model, and 10-fold cross validation is adopted in the classification process to verify the recognition performance of this activity recognition method. This method achieves the average recognition accuracy of 99.91 %.

This paper consists of five sections, and the structure is as follows: the Sect. 2 gives a brief introduction of the human activity dataset in the experiments; the Sect. 3 describes the human activity recognition method in detail; Sect. 4 compares and discusses the experimental results of different activity recognition methods; and Sect. 5 prospects the conclusion.

2   The Human Activity Dataset

The human activity dataset in this paper is the Daily and Sports Activities Data Set from UCI Machine Learning Repository, and there are 19 types of activities in the dataset, as shown in Table 1. These activities are performed at the Bilkent University Sports Hall, in the Electrical and Electronics Engineering Building, and in a flat outdoor area on campus [13].

Table 1. 19 types of human daily and sports activities

No.	Activity description	No.	Activity description
A1	Sitting	A11	Walking on a streadmill with a speed of 4 km/h in 15 deg inclined position
A2	Standing	A12	Running on a treadmill with a speed of 8 km/h
A3	Lying on back	A13	Exercising on a stepper
A4	Lying on right side	A14	Exercising on a cross trainer
A5	Ascending stairs	A15	Cycling on an exercise bike in horizontal position
A6	Descending stairs	A16	Cycling on an exercise bike in vertical position
A7	Standing in an elevator still	A17	Rowing
A8	Moving around in an elevator	A18	Jumping
A9	Walking in a parking lot	A19	Playing basketball
A10	Walking on a streadmill with a speed of 4 km/h in flat position		

Each activity listed above is performed by eight healthy experimenters (four males and four females, whose ages range from 20 to 30) for 5 min in their own styles with no restriction on how the activities should be performed. For this reason, there are some differences in the speeds and amplitudes for each activity among experimenters.

The motion signal duration in this dataset is collected by a wearable system [4–6, 13] which consists of five Xsens MTx units (MTx, a 3DOF orientation tracker, manufactured by Xsens Technologies). And each unit contains a tri-axial accelerometer, a tri-axial gyroscope and a tri-axial magnetometer, which provides drift-free 3D motion data: 3D acceleration, 3D rate of turn and 3D earth-magnetic field. Five units are fixed on the torso (T), right arm (RA), left arm (LA), right leg (RL) and left leg (LL) separately. The sampling frequency of the wearable motion sensor system is 25 Hz.

In general, there are 7500 ($= 5 * 60 * 25$) data vectors for each activity per experimenter, and 1140000 ($= 19 * 8 * 7500$) data vectors in the total dataset. The dimension of vectors is 45 ($= 5 * 3 * 3$), including the x-, y-, z-axis signals acquired from the accelerometers, gyroscopes and magnetometers in five MTx units.

3 The Human Activity Recognition Method

3.1 The Overall Process of the Method

The overall process of human activity recognition in this method is shown as Fig. 1. At first, the total human activity signal duration in the dataset are converted to a data sample set with the corresponding activity labels from 1 to 19 after data preprocessing; then features are extracted from the activity samples for classification; and then SFFS algorithm is used for feature reduction to eliminate redundancy and simplify the

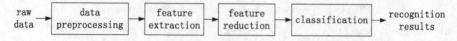

Fig. 1. The overall process of the activity recognition method

computation; finally, the selected features act as the input vectors of the multi-classifier combination model to classify 19 types of human daily and sports activities.

3.2 Data Preprocessing Using Slide Time Window Technology

The slide time window technology is applied to the raw sensor signal duration after data normalization, and the signal duration is divided into a plurality of equal length of time slices (observation windows), and the corresponding activity category of each observation window is defined as the numerical labels from 1 to 19. In this paper, the length of each window is set to 4 s, in order words, there are 100 (= 4 * 25) data vectors in each window; and the overlap rate of the adjacent windows is set as 50 % based on the experience to prevent the loss of useful information in the window edges. Regarding each time window as a activity sample, and there are 149 (= 7500/100 * 2 − 1) samples for each activity per experimenter and 22648 (= 19 * 8 * 149) activity samples in the whole dataset.

3.3 Feature Extraction in Temporal and Frequency Domains

The data in each time window can be seen as a matrix $S_{100 * 45}$, and there are 29 features extracted in each column of $S_{100 * 45}$: the maximum value (max), the minimum value (min), the mean value (mean), the average rectified value (arv), variance (var), skewness (skew), kurtosis (kurt), root mean square (rms), the peak to peak value(pp), 10 values ($A(i)$, $i = 1, 2, ..., 10$) chosen at regular intervals in the auto-correlation sequences, 5 peak values ($B(j)$, $j = 1, 2, ..., 5$) and corresponding frequencies ($F(j)$, $j = 1, 2, ..., 5$) in the Discrete Fourier Transform (DFT) sequences. Therefore, there are 1305 (= 45 * 29) features at all and the raw signal duration $D_{1140000 * 45}$ has been transformed into the feature sample set $F_{22648 * 1305}$.

3.4 Feature Reduction Using SFFS Algorithm

Because not all features work for activity recognition and the number of them is too large to result in redundancy. Moreover, too much data would lead to high computational complexity, and even "the curse of dimensionality". Therefore, it is necessary to reduce the dimensionality of feature samples, in other words, reduce the number of features. The common method of feature reduction is Principal Component Analysis (PCA), which is also known as Karhunen-Loeve (KL) Transform. This method produces uncorrelated features and achieves the linear transformation from the high input vectors to the low output vectors. However, PCA is a linear method which is not very helpful on extracting the nonlinear structure of motion data.

Therefore, the feature reduction technique adopted in this activity recognition method is the Sequential Forward Feature Selection (SFFS) algorithm, a sub-optimal search technology, and the main contents of the algorithm are as follows:

- Calculating the criterion value of each feature (each column of the feature sample set) and selecting one with the best value based on a certain classification criterion;
- Combing the selected feature with each other feature to get two-dimensional feature vectors;
- Calculating the criterion value of each 2-D feature vector, and selecting the 2-D vector with the best criterion value;
- Repeating the steps that combining the n-D feature vector with each other feature and selecting the best one until the criterion value of the final feature vector is optimal.

Finally, there are 52 features selected in those 1305 features after SFFS algorithm, and the selected features can be seen in Table 2.

Table 2. The selected features after SFFS

		x-axis	y-axis	z-axis
T	Accelerometer	min mean pp	rms	A(1) min
	Gyroscope	F(2)	A(4) B(1) F(5)	
	Magnetometer	rms	B(3)	
RA	Accelerometer	kurt	B(5) kurt	
	Gyroscope	rms	rms	A(5)
	Magnetometer		A(2)	arv
LA	Accelerometer	min		
	Gyroscope		skew pp	
	Magnetometer		max	A(3) min
RL	Accelerometer	B(5) min	B(3) mean skew	A(3) kurt
	Gyroscope	A(3) A(5)	A(3)	arv
	Magnetometer	A(3) max	rms pp	B(5)
LL	Accelerometer	A(5) max	B(3) mean	max
	Gyroscope			A(2)
	Magnetometer	var	A(3) A(5)	A(4)

3.5   Classification by the Multi-classifier Combination Model

Support Vector Machine (SVM), which is a supervised learning algorithm, is usually used in pattern recognition, data mining, regression analysis and classification for linear problems. SVM also can solve the nonlinear problems because of its nonlinear mapping ability which mapping the nonlinear input space which is low dimension to the linear output space which is high dimension.

AdaBoost, which is short for "Adaptive Boosting", is a adaptive algorithm which can be used in conjunction with weak classifiers to improve their classification performance. The results of the weak classifiers is combined into a weighted sum which represents the final result of the AdaBoost classifier. Researches have proven that as long as the performance of each weak classifier is slightly better than the result of random guessing, the AdaBoost classifier can be a strong classifier.

Because the performance of a single classifier is lower than the combination of multiple classifiers in most cases, the latter strategy is adopted in this paper, which connecting Quadratic SVM (QSVM) and AdaBoost Tree ensemble learning algorithm (AT, AdaBoost classifier using Decision Tree as the weak classifier) in series as a multi-classifier combination model to classify the 19 types of human activities.

The classification algorithm of multiple classifiers combination is shown as follow.

Algorithm.

- Training the **QSVM** classifier model with the training sample set **T1**;
- Classifying the testing sample set **t1** using **QSVM** and finding the activity labels **WL** (from 1 to 19) which have lower accuracy according to the classification results **QR**;
- Extracting the training samples from **T1** which with the same activity labels in **WL** to compose a new training set **T2**;
- Training the **AT** classifier model with **T2**;
- Composing the new testing set **t2** by extracting the samples from **t1** which have the same activity labels in **WL** based on **QR**;
- Classifying **t2** using **AT** and getting the classification results **AR**;
- Replacing the classification results in **QR** with the corresponding results in **AR** to form the final result labels **QAR** of the multi-classifier combination model;
- Comparing **QAR** with actual activity labels **L** and calculating the classification accuracy.

In order to set up the optimal parameters of the multi-classifier combination model, the grid search technology is adopted to avoid blindness and randomness in model parameters selection, and the classification precision is improved. Grid search algorithm which finds the best parameters of classification model by brute force, is widely used in machine learning.

4 Results and Discussions

4.1 Experimental Setup

This experiment was carried out on MATLAB platform with the 2.26 GHz Intel core i5 dual core processor 430 M. And 10-fold cross validation was used to verify the recognition performance of 19 human daily and sports activities.

In 10-fold cross validation method, the activity sample set is randomly partitioned into ten equal sized sub-sample sets. For the ten sub-sample sets, a single sub-sample set is regarded as the validation set for testing the classifier model, and other nine sub-sample sets are used as the training sample set. And then repeat ten times, with each sub-sample set used exactly once as the validation set. The classification results from ten tests are averaged to produce the final estimation.

In this experiment, there are 149 samples of each activity per experimenter, and so 15 samples in each sub-sample (15 * 10 = 150, and one sample is repeated in two sub-samples to make the same size of each sub-sample). Therefore, the size of training sample set is 20520 (= 15 * 9 * 19 * 8) and the size of testing sample set is 2280 (= 15 * 1 * 19 * 8) in each test.

After parameter optimization by grid search algorithm, the best parameters of the multi-classifier combination model are set as: for QSVM classifier model, KernelScale = 5, BoxConstraint = 2; for AT classifier model, MaxNumSplits = 27, AdaBoostM1 (the number of weak learners) = 500.

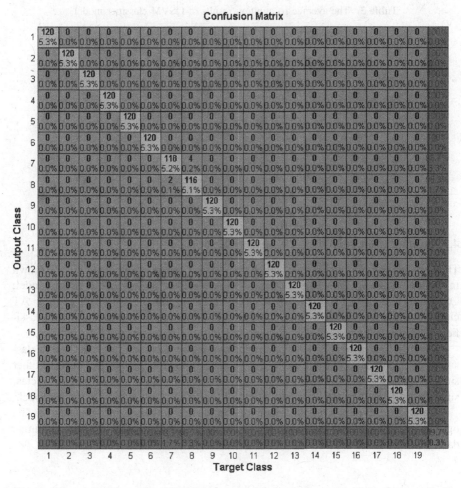

Fig. 2. The confusion matrix of QSVM classification result (in a test) (Color figure online)

The confusion matrix of QSVM classifier model (in a test) is shown in Fig. 2. It can be found that, the samples which with the activity labels of 7 and 8 have a lower classification accuracy compared with those which with other activity labels. And it is also be proved in Table 3, which shows the average recognition accuracy of the QSVM classifier model in each activity.

As can be seen in Table 3, for QSVM classifier model, the recognition accuracy rate of each activity isn't lower than the average recognition accuracy rate of all activities (99.75 %), except the activity 7 and activity 8, which the recognition accuracy rate is 99.09 % and 96.83 % respectively.

Therefore, based on the classification result labels of QSVM classifier model, the samples which with the result labels of 7 and 8 are taken separately out of the training sample set to be classified again by the AT classifier model for improving the recognition accuracy.

Table 3. The average recognition results of QSVM classifier model

Activity	Accuracy	Activity	Accuracy
A1	100 %	A11	100 %
A2	99.84 %	A12	99.92 %
A3	100 %	A13	100 %
A4	100 %	A14	100 %
A5	100 %	A15	100 %
A6	100 %	A16	100 %
A7	**99.09 %**	A17	100 %
A8	**96.83 %**	A18	99.90 %
A9	100 %	A19	99.75 %
A10	100 %	**average**	99.75 %

4.2 Comparison on Different Classifier Models

The average recognition results of different classifier models in all activities are shown in Table 4. In there, QSVM + AT represents the multi-classifier combination model; QSVM, AT, Tree, QD and KNN respectively represent the QSVM classifier, AT classifier, Decision Tree classifier, Quadratic Discriminant Analysis and KNN classifier. SFFS algorithm are used for feature reduction in all recognition methods.

It can be seen in Table 4, the average recognition accuracy rates of QSVM classifier (99.75 %) and AT classifier (99.68 %) are significantly higher than those of other single classifier models. In fact, this is the reason that QSVM classifier and AT classifier are chosen to be the basic classifiers of the multi-classifier combination model.

Moreover, the multi-classifier combination method proposed in this paper has the optimal recognition performance with 99.91 % average accuracy for 19 types of human daily and sports activities.

Table 4. The recognition results of different classifier models

No.	QSVM + AT	QSVM	AT	Tree	QD	KNN
Test 1	99.96 %	99.87 %	99.65 %	69.56 %	96.71 %	98.51 %
Test 2	99.96 %	99.82 %	99.61 %	69.08 %	96.80 %	98.51 %
Test 3	99.91 %	99.65 %	99.52 %	69.30 %	96.10 %	98.33 %
Test 4	99.82 %	99.56 %	99.65 %	69.65 %	96.49 %	98.11 %
Test 5	99.87 %	99.61 %	99.61 %	69.65 %	95.05 %	98.16 %
Test 6	99.91 %	99.87 %	99.69 %	69.61 %	96.80 %	98.38 %
Test 7	99.96 %	99.74 %	99.74 %	69.74 %	96.36 %	98.38 %
Test 8	99.91 %	99.82 %	99.82 %	70.13 %	96.67 %	98.07 %
Test 9	99.91 %	99.82 %	99.74 %	69.39 %	95.88 %	98.60 %
Test 10	99.91 %	99.78 %	99.82 %	69.52 %	95.75 %	98.07 %
average	**99.91 %**	99.75 %	99.68 %	69.56 %	96.36 %	98.31 %

4.3   Comparison on Different Feature Reduction Technologies

In addition, in order to evaluate the effects of different feature reduction technologies on activity recognition, for each activity, the recognition results of the multi-classifier combination methods using SFFS and using PCA are shown in Table 5. Because of 52 features selected after SFFS, there are also the first 52 principal components remained after PCA to reconstruct the sample set.

It's easy to be found that, for every activity, the recognition accuracy rate of the method using SFFS is higher than those using PCA, and the average recognition accuracy rate of the method using SFFS is 99.91 %, while it is only 97.06 % for the method using PCA. Therefore, compared with the traditional methods using PCA, SFFS algorithm improves the recognition accuracy of 19 human daily and sports activities by selecting more useful features in feature reduction.

Table 5. The average recognition results of methods using SFFS and PCA

Activity	SFFS	PCA	Activity	SFFS	PCA
A1	100 %	100 %	A11	100 %	91.51 %
A2	99.84 %	97.49 %	A12	99.92 %	99.59 %
A3	100 %	100 %	A13	100 %	99.92 %
A4	100 %	100 %	A14	100 %	100 %
A5	100 %	100 %	A15	100 %	100 %
A6	100 %	97.50 %	A16	100 %	99.27 %
A7	99.43 %	91.34 %	A17	100 %	100 %
A8	99.51 %	83.25 %	A18	99.90 %	97.25 %
A9	100 %	99.75 %	A19	99.75 %	96.16 %
A10	100 %	91.08 %	**average**	**99.91 %**	**97.06 %**

5   Conclusion

In this paper, a multi-classifier combination activity recognition method using SFFS algorithm for feature reduction is presented to classify 19 types of human daily and sports activities. Grid search technology is adopted to optimize the multi-classifier combination model's parameters, and 10-fold cross validation is applied in classification to evaluate the method's recognition performance. The experimental results indicate that the multi-classifier combination model has better recognition performance than other common single classifiers, and for feature reduction, SFFS algorithm selects the features with the best classification ability, while some useful information is lost in PCA. Therefore, it does seem fair to conclude that the multi-classifier combination recognition method using SFFS algorithm proposed in this paper improves the activity recognition performance of 19 human activities compared with traditional methods.

Because of the respective activity styles among the experimenters, the activity recognition accuracy would be slightly lower when the participants in testing sample set don't participate in the training sample set, so this problem need further research in the future.

Acknowledgments. This study was supported by National Natural Science Foundation of China (Grant No. 61403289). The authors would like to thank the assistance of UCI Machine Learning Repository for the dataset support.

References

1. Ozdemir, A., Barshan, B.: Detecting falls with wearable sensors using machine learning techniques. Sensors **14**(6), 10691–10708 (2014)
2. Guan, D., Ma, T., Yuan, W.: Review of sensor-based activity recognition systems. IETE Tech. Rev. **28**, 418 (2014)
3. González, S., Sedano, J., Villar, J.R., et al.: Features and models for human activity recognition. Neurocomputing **167**(C), 52–60 (2015)
4. Altun, K., Barshan, B., Tunçel, O.: Comparative study on classifying human activities with miniature inertial and magnetic sensors. Pattern Recogn. **43**(10), 3605–3620 (2010)
5. Barshan, B., Yüksek, M.C.: Recognizing daily and sports activities in two open source machine learning environments using body-worn sensor units. Comput. J. **57**, 1649–1667 (2014)
6. Altun, K., Barshan, B.: Human activity recognition using inertial/magnetic sensor units. In: Salah, A.A., Gevers, T., Sebe, N., Vinciarelli, A. (eds.) HBU 2010. LNCS, vol. 6219, pp. 38–51. Springer, Heidelberg (2010)
7. Adil Mehmood, K., Young-Koo, L., Lee, S.Y., et al.: A triaxial accelerometer-based physical-activity recognition via augmented-signal features and a hierarchical recognizer. IEEE Trans. Inf. Technol. Biomed. **14**(5), 1166–1172 (2010)
8. Mi, Z., Sawchuk, A.A.: Human daily activity recognition with sparse representation using wearable sensors. IEEE J. Biomed. Health Inf. **17**(3), 553–560 (2013)
9. Ermes, M., Parkka, J., Mantyjarvi, J., et al.: Detection of daily activities and sports with wearable sensors in controlled and uncontrolled conditions. IEEE Trans. Inf. Technol. Biomed. **12**(1), 20–26 (2008)

10. Wu, H., Liu, H., Liu, D.: Two dimensional direction recognition using uniaxial tactile arrays. IEEE Sens. J. **13**(12), 4897–4903 (2013)
11. Bruno, B., Mastrogiovanni, F., Sgorbissa, A.: Analysis of human behavior recognition algorithms based on acceleration data (2013)
12. Cvetković, B., et al.: Adapting activity recognition to a person with multi-classifier adaptive training. J. Ambient Intell. Smart Environ. **7**(2), 171–185 (2015)
13. Lichman, M.: UCI Machine Learning Repository, University of California, School of Information and Computer Science, Irvine (2013). [http://archive.ics.uci.edu/ml]

Embedded Implementation of Template Matching Using Correlation and Particle Swarm Optimization

Yuri Marchetti Tavares[1(✉)], Nadia Nedjah[2], and Luiza de Macedo Mourelle[3]

[1] Departamento de Armas, Diretoria de Sistemas de Armas da Marinha,
Rio de Janeiro, Brazil
yurimtavares@yahoo.com.br
[2] Departamento Engenharia Eletrônica e Telecomunicações,
Universidade do Estado do Rio de Janeiro, Rio de Janeiro, Brazil
[3] Departamento de Engenharia de Sistemas e Computação,
Universidade do Estado do Rio de Janeiro, Rio de Janeiro, Brazil

Abstract. The template matching is an important technique used in pattern recognition. The goal is find a given pattern, from a prescribed model, in a frame sequence. In order to evaluate the similarity of two images, the Pearsons Correlation Coefficient (PCC) is widely used. This coefficient is calculated for each of the image pixels, which entails a computationally very expensive operation. This paper proposes the implementation of Template Matching using the PCC based method together with Particle Swarm Optimization as an embedded system. This approach allows for a great versatility to use this kind of system in portable equipment. The results indicate that PSO is up to 158x faster than the brute force exhausted search. So, the thus obtained co-design with PCC computation implemented in hardware, while the PSO process in software, is a viable way to achieve real time template matching, which is a pre-requisite in real-word applications.

Keywords: Embedded systems · Co-design · Particle swarm optimization · Template matching · Correlation · Tracking

1 Introduction

With development and enhancement of sensors and intelligent equipment capable of capturing, storing, editing and transmitting images, the acquisition of information, which is extracted from images and videos, became an important research area. In the defense and security areas, this kind of research is very relevant to recognition and tracking targets in image sequences. It can provide solutions for development of surveillance systems [1], monitoring, fire control [2], guidance [3], navigation [4], remote biometric [5], guided weapons [6], among many others.

A pattern is an arrangement, or a collection of objects that are similar, and is identified by its elements disposition. One of the most used techniques

© Springer International Publishing Switzerland 2016
O. Gervasi et al. (Eds.): ICCSA 2016, Part II, LNCS 9787, pp. 530–539, 2016.
DOI: 10.1007/978-3-319-42108-7_41

for finding and tracking patterns in images is generally identified as template matching [7,8]. Among the methods used for evaluate the matching process, the correlation is very known and widely used. The task consists basically to find out a small image, considered as a template, inside a larger image. This task is computationally very expensive, especially when using large templates and extensive image sets [9].

In [10] the authors use and compare Genetic Algorithms and Particle Swarm Optimization (PSO) to improve the time processing of template matching. The PSO performed better in terms of time processing and robustness. The system was implemented in MATLAB and took advantages from PSO method like high convergence, great simplicity and easy organization [11]. This approach provides a good solution of the targets tracking problem. However, in many cases of security and defense applications, the surveillance system needs to be embedded in portable hardware equipment.

This paper proposes an implementation of the template matching using correlation and PSO in an embedded system. Beyond this, a comparison with the exhaustive search and an evaluation of the time results are performed. For this, the Sect. 2 present some related works; in Sect. 3 is presented the correlation concept; in Sect. 4 is briefly described the hardware used; in Sect. 5 are described the methodology and the how the system was implemented; the Sect. 6 presents the results and, finally, the Sect. 7 concludes the work.

2   Related Works

In [12] the co-design methodology is used to implement an automatic video surveillance system in a Field Programmable Gate Array (FPGA) device. Like most part of image processing applications, the system must have a high performance to support real time applications. The most computationally intensive operations were implemented in the FPGA and the control and user interface were developed in software and executed by a soft processor. For object tracking, the background subtraction was used. Although efficient, this technique requires that the background remains static and is not applicable when the targets and the background change in subsequent frames.

In [8], several techniques of template matching are discussed. An architecture that is based on spectral correlation, together with its FPGA based implementation, is proposed. The correlation is calculated in the frequency domain using the Fast Fourier Transform in two dimensions and complex multiplication.

In [13], a technique for generation of panoramic images from image combinations is proposed. The set is composed by two or more digital images from different sensors or angles. The method is divided into three steps, which were all implemented in FPGA, with efficient hardware resources utilization and expected results.

In this paper, we attempt to evaluate the impact that the utilization of PSO would introduce regarding the efficiency of the object tracking process. To do so, we implemented both PSO and the exhausted search in an embedded platform.

Differently form the cited works, the main purpose is to develop a portable system, capable of processing images in real-time, allowing changes in both the target and background. Moreover, in contrast with existing work, we use the correlation in time domain.

3   Correlation

Template matching is used in images processing to determine the similarity of two entities (pixels, curves or forms) of the same type. The pattern to be recognized in an image is compared to a pre-defined template. After the similarity evaluation considering all the possibilities, the matching that provided the highest correlation value, above a given threshold, is identified as the location of the pattern inside the image.

Among the template matching techniques for tracking targets, the correlation technique is well known and often used. The Pearsons Correlation Coefficient (PCC) [14] is used as a measure of similarity between two variables and can be interpreted as dimensionless index, having values in $[-1, 1]$. Coefficient equal to 1 means a perfect positive correlation between the two variables. Coefficient equal -1 means a perfect negative correlation between the two variables. Coefficient equal to 0 means that the two variables do not have linear dependency. For imaging applications, the Pearson's Correlation Coefficient can be calculated as defined in Eq. 1:

$$corr = \frac{\sum\limits_{i=1}^{N}(p_i - \overline{p})(a_i - \overline{a})}{\sqrt{\sum\limits_{i=1}^{N}(p_i - \overline{p})^2}\sqrt{\sum\limits_{i=1}^{N}(a_i - \overline{a})^2}} \tag{1}$$

wherein p_i is the intensity of the pixel i in the template; $\overline{p}$ is the average intensity of the pixels of the template; a_i is the intensity of the pixel i in the patch of the image A; $\overline{a}$ is the average intensity of the pixels in the patch of the image A. The template and the image A must have the same dimensions.

The ideal correlation utilization of the Eq. 1 considers that the target appearance remains the same in all frames [15]. A possible solution to this limitation consists of updating the template with each new frame.

4   Hardware Platform

In order to achieve the purpose of this work, the template matching using PCC is implemented on the Smart Vision Development Kit (SVDK) evaluation board (rev 1.2) from Sensor to Image [16]. This board together with a Zynq XC7Z015 based PicoZED module provides a hardware environment for developing and evaluating designs targeting machine vision applications. The SVDK board provides features common to many embedded processing systems, including a DDR3 memory, tri-mode Ethernet PHYs, HDMI video encoder, general purpose I/O,

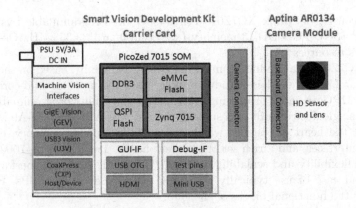

Fig. 1. SVDK block diagram [16]

and a UART interface, but also components to realize machine vision interface standards like USB3 Vision, GigE Vision and CoaXPress. Machine vision sensors can be connected using a standard PCI Express connector. Figure 1 shows the block diagram of SVDK.

The PicoZed 7Z015 System-On Module (rev. C) is a low cost evaluation board targeted for broad use in many applications. The features provided by the

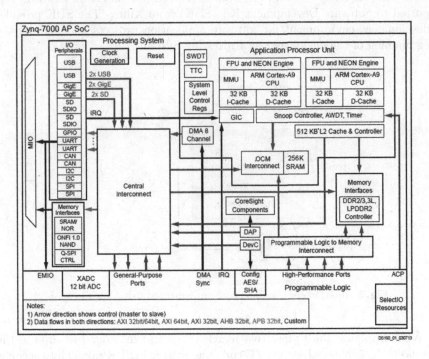

Fig. 2. Zynq-7000 AP SoC functional blocks [18]

PicoZed consist of Xilinx XC7Z015-1CLG485C All Programmable System on Chip (AP SoC), 1 GB DDR3 memory (x32), On-board 33.333 MHz Oscillator, and others resources [17].

The XC7Z015 is a product from Xilinx Zynq-7000 family, based on the Xilinx AP SoC architecture. These products integrate a feature-rich dual-core ARM Corte-A9 MPCore based processing system (PS) and Xilinx programmable logic (PL) in a single device, built on a state-of-the-art. The ARM Cortex-A9 MPCore CPUs are the heart of the PS, which also includes on-chip memory, external memory interfaces, and a rich set of I/O peripherals [18]. The Zynq-7000 family offers the flexibility and scalability of an FPGA, while providing performance, power, and ease of use typically associated with ASIC and ASSPs. Figure 2 illustrates the functional blocks of the chip.

5   Implementation Methodology

In this section, the methodology and the steps exploited to implement the exhausted search and the template matching with PSO in the FPGA are described.

First, the Zynq hardware platform is developed using the software tool Vivado from Xilinx. The hardware interface is shown in Fig. 3. The ZYNQ7 Processing System IP was used and the in/out pins and peripherals were configured.

Once the embedded system has been developed, the hardware platform was exported to Software Development Kit (SDK) from Xilinx. The SDK provides a development environment for software application projects and permits the programming of the device and its Processing System (PS).

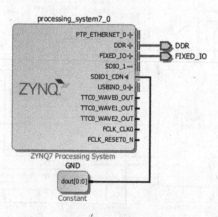

Fig. 3. Interface of the hardware implemented using Vivado tool

When applying the Exhaustive search (ES), and in order to compute the correlation by exhausted search, a C code was written in SDK taking into account the following steps:

1. First, obtain the main image and the template, bytes in gray scale, by serial communication RS232.
2. Then, extract of a patch, of the same size of template, for each pixel. Edges of the main image were completed with zeros.
3. For each pixel of the main image, compute the corresponding PCC.
4. Finally, identify the point where the correlation is the highest, which should be the target object center.

During the application of the Particle Swarm Optimization (PSO), and in order to compute the correlation while using PSO, another C code is developed in SDK considering the following steps:

1. First, obtain the main image and the template, bytes in gray scale, by serial communication RS232.
2. Then, generate the initial particle swarm, with random positions and velocities.
3. For each particle, extract a patch of the same size as the template, and compute the corresponding PCC at its center. It is noteworthy to point out that the limits of the main image are completed with zeros.
4. After that, store the best value of PCC found for each particle and also that related to the whole the swarm of particles.
5. Repeat the steps 3 and 4 until an acceptable PCC value is reached or a maximum limit for the iteration number is exceeded.

To receive the images and results in both cases (ES and PSO), the software MATLAB is used via a serial port with a 115200 baud rate.

6   Results

Initially, an aircraft video was downloaded from [19]. From this video, we extracted 2 frames, identified hereafter as $aircraft_1$ and $aircraft_2$. Figure 4 shows each frame and its corresponding correlation behavior. The frame $aircraft_1$ has a resolution of 361×481 pixels while $aircraft_2$ has 355×479 pixels.

Moreover, the first frame of the video benchmark EgTest02, downloaded from [20], is used in the performance evaluation. Hereafter, this frame is identified as *cars* and shown in Fig. 5 together with the corresponding PCC behavior. It is possible to see in the correlation graphic that it shows many local maxima and thus entails a more difficult optimization process in order to identify the target.

Both algorithms are used to find out the pixel localization of the template center within the main image. All the considered templates have a resolution of 64×64 pixels. The canonical algorithm of PSO is used and the parameters were configured, empirically after many executions. The parameters were set as follows: 100 particles, inertial coefficient $w = 1$, cognitive coefficient $c_1 = 1.5$, social coefficient $c_2 = 2$, maximum velocity set to 10. As stopping criteria of the PSO, we combined an acceptable correlation coefficient of 0.95 or maximum of 30 iterations.

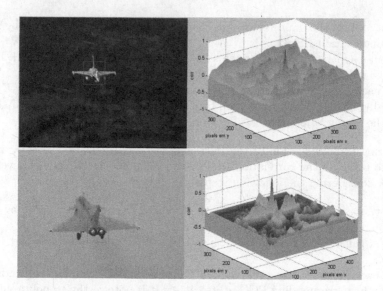

Fig. 4. Frames *aircraft₁* (top) and *aircraft₂* (bottom)

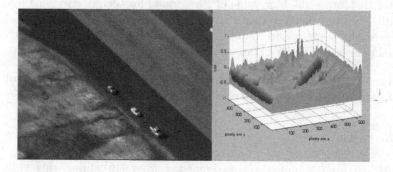

Fig. 5. Frame *cars*

The memoization is an optimization technique used to speed up programs by storing the results of expensive function calls and returning the stored result when the same inputs occur again. In order to improve the processing time, this technique is implemented and the correlation values computed for a particle in a given pixel are stored in a local memory. If a particle comes to appear in a position that at least another particle has been located previously, then the correlation value is reused instead of recomputed, and thus accelerating the computational process of PCC.

Some of the results, together with the processing time and number of iterations as well as the maximum correlation value found for the three images are presented in Tables 1, 2 and 3, respectively.

Note that the execution of the PSO algorithm was repeated 100 times and the average results are shown in Table 4. These results confirm that PSO is an

Table 1. Results for $aircraft_1$

	ES	1st PSO	2nd PSO	3rd PSO
Position (l,c)	(154,215)	(154,215)	(54,321)	(154,216)
Correlation	0.999	0.999	0.554	0.964
Iterations	-	7	30	22
Time (ms)	392064	1568	5854	4771

Table 2. Results for $aircraft_2$

	ES	1st PSO	2nd PSO	3rd PSO
Position (l, c)	(222,231)	(221,231)	(222,232)	(222,232)
Correlation	0.999	0.952	0.959	0.959
Iterations	-	8	11	13
Time (ms)	387489	1820	2483	2926

Table 3. Results for cars

	ES	1st PSO	2nd PSO	3rd PSO
Position (l, c)	(323,389)	(252,310)	(324,390)	(252,310)
Correlation	0.999	0.758	0.962	0.758
Iterations	-	30	20	30
Time (ms)	768001	6113	4411	6124

Table 4. Average results

	$aircraft_1$	$aircraft_2$	cars
Iterations	16.71	13.35	22.53
Time (ms)	3600	2907	4842

excellent tool to decrease the processing time during template matching when PCC is used.

For real-time applications, the results still need to be improved. In this purpose, the PSO based solution was evaluated and some tests were performed to identify the code sections that are more expensive computationally. Table 5 shows the results for $aircraft_1$ for the five different experiments that were evaluated.

In Table 5, time duration T_1 corresponds to the time spent in the execution a full iteration; T_2 corresponds to the time spent in an iteration without the computation of the PCC; T_3 is the time spent in the execution of an iteration without the computation of the PCC and without patch extraction; T_4 provides the time spent in an iteration without the computation of the PCC, patch extraction and also without performing the best particles verification; and finally T_5 corresponds

Table 5. Processing time for one iteration of $aircraft_1$

	T_1	T_2	T_3	T_4	T_5
Time (μs)	226200	24570	95.9	77.8	36.9

to the time spent in an iteration without computing the PCC, patch extraction and also without computing the particles position and velocity update.

It is possible to identify that the PCC computation is the most expensive part of the algorithm, followed by the memory access done while extracting the patches of the main image. One way that can be done to improve the processing time is to implement the PCC computation into a dedicated efficient hardware, exploiting any possible opportunity to parallelize the entailed process. The PSO can still be executed by the soft processor, but the PCC computation must be done by the hard IP that can be implemented in Programmable Logic part of the chip. This approach, usually called co-design, is a methodology to develop an integrated system using hardware and software components, to satisfy performance requirements and cost constraints [21]. The final target architecture usually has software components executed by the soft processor that is assisted by some dedicated hardware components developed especially for the application. It is noteworthy to point out that this is the next step in the development of this work.

7   Conclusion

The PSO is an excellent tool for optimization in the template matching by correlation, achieving a good performance and robustness. The processing time for used benchmarks was improved in 108x, 133x and 158x.

The most expensive part of the algorithm is the correlation computation, followed by the memory access for extract the patch from the main image. For real-time applications, the processing time should be further improved by implementing the computation of the correlation coefficients in hardware, organizing the image memory so as to accelerate the memory access necessary during patch extraction. Furthermore, increasing the clock frequency could be one extra trick to shorten the overall processing time.

For future works, the authors intend to implement the correlation coefficient computation into Programmable Logic part of the Zynq chip, as dictated by the co-design approach.

References

1. Narayana, M.: Automatic tracking of moving objects in video for surveillance applications. Kansas University (2007)
2. Ali, A., Kausar, H., Khan, M.I.: Automatic visual tracking and firing system for anti aircraft machine gun. In: International Bhurban Conference on Applied Sciences & Technology, Islamabad, Pakistan, January 2009

3. Choi, H., Kim, Y.: UAV guidance using a monocular-vision sensor for aerial target tracking. Control Eng. Pract. **22**, 10–19 (2014)
4. Forlenza, L.K., Fasano, G., Accardo, D., Moccia, A.: Flight performance analysis of an image processing algorithm for integrated sense-and-avoid systems. Int. J. Aerosp. Eng. **2012**, 1–8 (2012). Hindawi Publishing Corporation
5. Benfold, B., Reid, I.: Stable multi-target tracking in real-time surveillance video. In: Proceedings of Computer Vision and Pattern Recognition (CVPR), Colorado Springs (2011)
6. Olson, T.L.P., Sanford, C.W.: A real-time multistage IR image-based tracker. In: Proceedings of Acquisition, Tracking and Pointing XIII, SPIE, vol. 3692, July 1999
7. Ahuja, K., Tuli, P.: Object recognition by template matching using correlations and phase angle method. Int. J. Adv. Res. Comput. Commun. Eng. **2**(3), 1368–1373 (2013)
8. Mahalakshmi, T., Muthaiah, R., Swaminathan, P.: An overview of template matching technique in image processing. Res. J. Appl. Sci. Eng. Technol. **4**(24), 5469–5473 (2012)
9. Sharma, P., Kaur, M.: Classification in pattern recognition: a review. Int. J. Adv. Res. Comput. Sci. Softw. Eng. **3**(4), 2013
10. Tavares, Y.M., Nedjah, N., Mourelle, L.M.: Utilização de otimização por enxame de partículas e algoritmos genéticos em rastreamento de padrões. In: XXII Congresso Brasileiro de Inteligência Computacional (2015)
11. Siciliano, A.V.: Algoritmos genéticos e particle swarm optimization e suas aplicações problemas de guerra eletrônica. In: IX Simpósio de Guerra Eletrônica, Instituto Tecnológico da Aeronáutica, São José dos Campos, SP (2007)
12. Ngo, H.T., et al.: Real-time video surveillance on an embedded, programmable platform. Microprocess. Microsyst. **37**(6), 562–571 (2013). Elsevier
13. Vinod, G., Anitta, R.: Implementation of FFT based automatic image mosaicing. Int. J. Adv. Res. Electr. Electron. Instrum. Eng. **2**(12), 6002–6009 (2013)
14. Miranda, A.N.: Pearsons correlation coefficient: a more realistic threshold for applications on autonomous robotics. Comput. Technol. Appl. **5**, 69–72 (2014)
15. Matthews, I., Ishikawa, T., Baker, S.: The template update problem. IEEE Trans. Pattern Anal. Mach. Intell. **6**, 810–815 (2004)
16. SVDK Hardware User Guide: Document Revision 1.1, May 2015
17. PicoZed 7Z015 / 7Z030 System-On Module Hardware User Guide, Version 1.3, July 2015
18. Zynq-7000 AP SoC Technical Reference Manual - UG585, version 1.10, February 2015
19. Rafale - Avião Caça de alta Tecnologia (Brasil) (2015). http://www.youtube.com/watch?v=e3wi-i_hDVQ
20. Collins, R.T.; Zhou, X.; Teh, S.K.: An open source tracking testbedand evaluation web site. In: IEEE International Workshop on PerformanceEvaluation of Tracking and Surveillance (PETS) (2005). http://vision.cse.psu.edu/data/vividEval/datasets/datasets.html
21. Nedjah, N., Mourelle, L.M.: Co-design for System Acceleration: A Quantitative Approach. Springer, Berlin (2007)

Visualizing High Dimensional Feature Space for Feature-Based Information Classification

Xiaokun Wang and Li Yang[✉]

Department of Computer Science, Western Michigan University,
Kalamazoo, MI, USA
{xiaokun.wang,li.yang}@wmich.edu

Abstract. Feature-based approaches represent an important paradigm in content-based information retrieval and classification. We present a visual approach to information retrieval and classification by interactively exploring the high dimensional feature space through visualization of 3D projections. We show how grand tour could be used for 3D visual exploration of high dimensional feature spaces. Points that represent high dimensional feature observations are linearly projected into a 3D viewable subspace. Volume rendering using splatting is used to visualize data sets with large number of records. It takes as input only aggregations of data records that can be calculated on the fly by database queries. The approach scales well to high dimensionality and large number of data records. Experiments on real world feature datasets show the usefulness of this approach to display feature distributions and to identify interesting patterns for further exploration.

Keywords: Data clustering · Information categorization · Information visualization · Interactive information retrieval

1 Introduction

Advances in multimedia data processing, data management, information retrieval and World Wide Web have resulted in content-based retrieval and classification to emerge as active areas of research with many important applications. Typical content-based retrieval systems allow users to specify queries by providing examples of objects similar to the ones they wish to retrieve. An important search paradigm is the feature-based approach, which maps a data entity to a record of numeric features extracted from the entity. Examples of features include color, texture and shape for images, and motion parameters for videos. Searching a large information archive then corresponds to searching over the high-dimensional feature space. Feature-based similarity search has been extensively studied in information retrieval [8]. Due to the subjective nature of information retrieval, it is unlikely that answers to a user's query will completely satisfy the user's need. Among retrieved answers, instead, the user may find one or more objects that are closer to what she has in mind and then refine the query. Unlike that for conventional database queries, the criterion used here is "likelihood" rather than "yes or no".

© Springer International Publishing Switzerland 2016
O. Gervasi et al. (Eds.): ICCSA 2016, Part II, LNCS 9787, pp. 540–550, 2016.
DOI: 10.1007/978-3-319-42108-7_42

The inaccuracy of query answers and, furthermore, the difficulty and subjectivity in capturing the contents of multimedia objects suggest that we should incorporate subjective human judgments into the query and classification process. Therefore, it seems that interactive information visualization may help to support feature-based content-based classification. In the feature-based paradigm, features extracted from a data entity are supposed to fully represent the data entity for search and classification. By visually inspecting the data distributions in high dimensional feature space, we may expect that, in addition to the retrieval of information, it is possible to find interesting patterns and abnormities in large information archives.

In this paper, we assume that feature data are organized into a large relational table, where each feature is a domain in the table and all features extracted from an entity are represented as a record in the table. By large relational table, we imply that the table is very wide (up to a few hundred features) or very long (a huge number of records). For a small table with a few domains and a few records, scatterplot is an excellent means for visualization. However, scatterplot loses its effectiveness as either data dimensionality or the number of data records becomes large. To deal with high dimensionality, a natural extension to scatterplot is to project data orthogonally onto 2D or 3D subspaces that allow us to see. There are an infinite number of possibilities to project high dimensional data onto a lower-dimensional space. The grand tour method [1,4] provides an overview of the projections through sampling all possible planes randomly. By displaying a series of intermediate projections obtained by interpolation, the entire process creates an illusion of continuous smooth motion. A tour could be further guided by various data distribution measurements [3,7,12]. In this paper, we use the location of data clusters to guide the path of a tour, based on the fact that four non-coplanar points can determine a 3D linear projection, where the four points are displayed as apart as possible and the distances between them are preserved. Interesting patterns in data clusters will most likely surface during the journey of a cluster-guided tour. This way of viewing higher dimensional data helps understanding of the data clusters and allows viewers to grouping data entities for information classification and retrieval.

Large data archives are major data sources in the big data era. Modern digital libraries and data archives have often archived so many data entities that it is impractical to display each entity as a point in a scatterplot. Even if the display of many points were possible, its usefulness would be diminished by overwhelming appearance and excessive occlusion. To handle large number of records, one improvement to scatterplot is to draw density map instead of individual data points. The way to draw 3D density map is volume rendering. The idea behind volume rendering is that data points are grouped into multidimensional cubes. Each cube represents a range of values for each of the variables. Data points belonging to each range are then categorized into each cube. Volume rendering takes as input these cubes and displays them onto screen as voxels. One way for volume rendering of 3D volume data is splatting. It takes 3D volume data as input and contributes a 2D footprint for each 3D voxel. These footprints are then blended together to generate the final image. Although splatting was designed for 3D volume data, the same principle applies to the rendering of high dimensional data where footprints of high-dimensional "voxels" are directly tossed to a 2D display.

Notice that volume rendering takes as input only a set of data "voxels" and, in each voxel, the density or number of data records. The size of this aggregated data is expected much smaller than that of the individual data records, making possible remote visual exploration of features of distributed data archives in a way that features are extracted and aggregated locally on each remote site before the aggregation results are transferred back in an ad hoc manner. Furthermore, data voxels and the number of records in each voxel are directly obtainable from databases by the SQL GROUP BY query. To further accelerate the process, data cubes can be pre-calculated and stored as summary tables in databases. This is in the same way as summary information of data in large databases is gathered to answer On-Line Analytical Processing (OLAP) queries. Volume rendering takes as input either materialized cubes in data warehouse or aggregations being calculated on the fly. The database support enables fast rendering of very large datasets.

We are better off thinking the visual exploration of features extracted from large distributed data archives as multidimensional data cubes volume rendered by grand tour. A typical scenario of visual exploration is like this: Initially, cube sizes are chosen, data are aggregated and an overview of the dataset is presented through grand tour and volume rendering. Once viewers identify interesting sub-parts of the dataset and decide to drill down, records of data will be selected and aggregated with smaller cube sizes in an ad hoc manner. This drill down process could continue until the extreme case where individual data records are retrieved.

A useful visualization tool has to deal with data of various types, such as numerical data, categorial data and missing values. Categorical values can be arranged along an axis in scatterplot in the same way as continuous values once they are explicitly ordered. We are lucky that, unlike real world relational data, most extracted features are generated by automatic techniques and thus contain no missing values.

Visual exploration of general-purpose relational data has been extensively studied. Common techniques include scatterplot, scatterplot matrices, and parallel coordinates. 3D visual exploration of large relational datasets has been reported [10]. Nonlinear techniques have also be studied to preserve certain properties or maximize certain criteria. These include the distance-preserving approach [11] and recent developments [5,13] using data embedding techniques. These techniques have rarely been reported to be applied to feature space exploration for information retrieval and classification.

This paper applies the visualization approach to feature-based information classification and studies its usefulness in finding patterns and data entities of concern. In the rest of this paper, we will briefly introduce the idea of grand tour and continue our discussion on volume rendering. We think grand tour together with volume rendering as a new way to retrieve interesting elements from large information archives. Finally, we will present experiment results when the approach be applied to real world feature datasets to demonstrate its usefulness.

2   3D Grand Tour

Grand tour is a way to visualize high dimensional space by extending data rotation to high dimensions. Given a source frame and a destination frame in a high dimensional Euclidean space, grand tour moves continuously projections along the geodesic path from the source to the destination. It does this by computing explicitly a sequence of interpolated frames. The high dimensional data is then projected onto the low dimensional subspace spanned by each frame. For 3D projections, a geodesic path is simply a rotation in the (at most) 6-dimensional subspace containing both the current and the target 3D subspaces. Moving along geodesic path is a way of assuring that the sequence of projections is smooth and thus comprehensible.

There are various ways of choosing the source and the destination frames of a tour. For clustered datasets, we have a way, what we call the *cluster-guided tour* [10], to show the structure of data clusters. As we know, a nontrivial linear combination of four non-coplanar points determines a unique 3D subspace of n-space. Suppose that we have partitioned an n-dimensional dataset into k $(k \geq 4)$ clusters, and let $\{c_j\}_{j=1}^k$ denote the centroids of these clusters. We can then compute a 3D *cluster-guided projection* by projecting the n-dimensional dataset onto a 3D subspace determined by four chosen cluster centroids c_a, c_b, c_c and c_d in $\{c_j\}$. This projection preserves the inter-cluster distances. The Euclidean distance between any two of the four cluster centroids $\{c_a, c_b, c_c, c_d\}$ will remain unchanged after the projection. This projection is a good perspective to view these four clusters since the four clusters are visualized as separate as possible. Cluster-guided projection gives a visual way to check a data cluster against other clusters.

Cluster-guided tour enables a viewer to move among all cluster-guided projections and, thus, is a way to quickly check the distribution of all data clusters. Given k cluster centroids, there are at most $\binom{k}{4}$ combinations of unique 3D cluster-guided projections. The basic idea behind cluster-guided tour is simple: Choose a target projection from these cluster-guided projections, move from the current projection to the target projection, and continue. We illustrate the 3D cluster-guided tour on the Boston housing data set [6]. The dataset was clustered into 6 clusters. Among 15 possible 3D cluster-guided projections, one is plotted in Fig. 1. The 3D cluster-guided tour reveals significant information about the position of data clusters.

3   Volume Rendering of High Dimensional Feature Space

Scatterplot is incapable of visualizing a large number of data points. One problem is that the cost of drawing each record as a point on screen becomes prohibitively

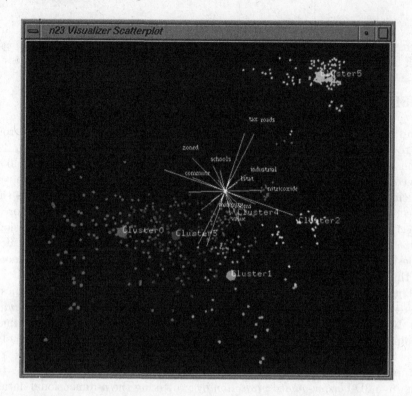

Fig. 1. A 3D cluster-guided projection where the 3D subspace is determined by centroids of 4 clusters, Clusters 0, 1, 3, 5.

high for interactive rendering. Another problem is that records with the same values in some features would be drawn in the exact same location, preventing the viewer from seeing more than one data point. More importantly, a scatterplot of a large dataset would result in many data points cluttered together, making it nearly impossible for a viewer to understand (Fig. 2). An improvement to scatterplot is to draw a density map instead of individual data points. The way to draw density map in 3D graphics is volume rendering. It is the direct display of volume data sampled in 3D. It helps the user perceive the density of data points in each location of a 3D scatterplot.

High dimensional space is usually quite sparse. A lot of storage space would be wasted on regions with no data if we store the data for volume rendering in the volumetric format. Relational table is a better data structure. Using relational table as the data structure, each high dimensional data "voxel" is represented as a record in the table. The weight of each voxel is an aggregation of data records within the voxel. In most cases, it is the number of data points. In some other cases, it may be the sum or even the weight of the values of some other variables. Since relational table is used as the underlying data structure, no storage is wasted on regions with no data.

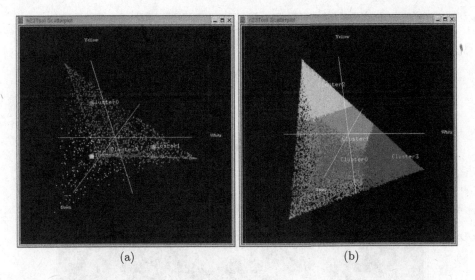

(a)   (b)

Fig. 2. A scatterplot of (a) 2000 data points; and (b) 68,040 data points cluttered together. (Color figure online)

The first step in getting data for volume rendering is to aggregate data values to a desired resolution. A great advantage of using the relational table as the internal data structure is that data voxels can be calculated easily from relational database by using data aggregation operations supported by the database, for example, by issuing an SQL GROUP BY query. To further speed up data aggregation, a summary table could be created in advance in database to store the aggregation results. The trade-off between accuracy and rendering speed is determined by the voxel resolution.

There are many methods of implementing volume rendering. Since the result is expected to represent an amorphous cloud of colored data points, it is felt that texture splatting using Gaussian footprints [9] is the most natural choice. 3D splatting takes a 3D volumetric data as input and projects each 3D voxel onto the 2D viewing plane. The projection of each voxel is approximated by a Gaussian footprint, whose opacity and color depend on those of the voxel. Splatting was proposed to improve the calculation speed at the price of less accurate rendering. For our purpose, a high degree of accuracy is not necessary as compared to the rendering speed. The only problem we have is that the aggregated dataset is high dimensional and traditional splatting takes 3D volume data as input. In our approach, footprint splatting is combined with 3D projection of multidimensional voxels so that each multidimensional voxel is directly splatted onto the 2D viewing plane.

Opacity of each footprint is assigned as a function of weight of the corresponding voxel. The function being used is identical to the function used in [2], that is, $\alpha = 1 - e^{-\mu \cdot weight}$, where α is the opacity of the footprint of a voxel, *weight* is the weight of aggregated data points in the voxel, and μ is a scale

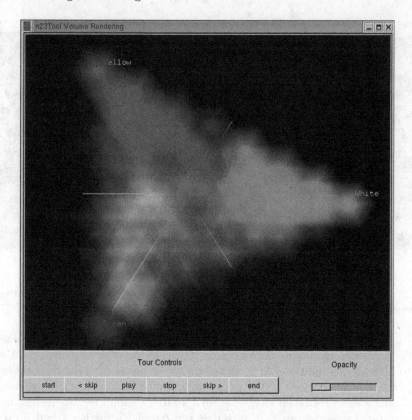

Fig. 3. Volume rendering of a simple projection.

factor. This exponential opacity function is effective in modelling light propagation through clouds. Footprints are blended together at each image pixel to form
a final image that approximates the appearance of a scatterplot density map of
the original data (Fig. 3). A slider is used to vary the value of the scale factor
μ. This allows global scaling of the opacity for each footprint to make an entire
image of rendered footprints more or less transparent.

Volume rendering takes as input only aggregation of data records in each
voxel. This makes it possible to interactively explore very large datasets if backend databases are fast enough to answer data aggregation queries. The less aggregated part of data will never be retrieved until you select an interesting subset
of data and drill down. This pick-and-drill-down process could continue until
you reach the record level when sets of individual data points are retrieved. In
the case that data archive is distributed across remote sites, aggregation queries
could be sent to each remote site where features are extracted and aggregated
locally. Only results of data aggregation, the sizes of which are much smaller
than that of the original, need to be transferred back. In this way, our approach
is capable of interactive visual exploration of very large distributed information
archives.

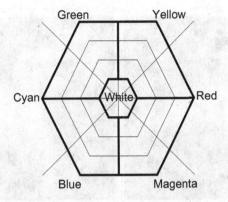

Fig. 4. Preprocessing of the color histogram dataset. Original data dimensions are combined into only five dimensions.

4   Experimental Results

Now we present the results of applying the approach to a real world image feature dataset. The experiment demonstrates the effectiveness of our visual techniques. We use the Corel Image Features dataset from the UCI KDD Archive [6]. The dataset contains image features extracted from a Corel image collection. Four sets of features are available based on color histogram (32 dimensions), color histogram layout (32 dimensions), color moments (9 dimensions) and co-occurence texture (16 dimensions). The datasets use Hue-Saturation-Value (HSV) to encode color. Each dataset has 68,040 records. In the color histogram feature set, each dimension is the percentage of image pixels for a certain Hue-Saturation values. There are 8 bins for Hue and 4 bins for Saturation, resulting in a total number of 32 dimensions (Fig. 4). We combined some dimensions and got five dimensions, white, yellowish, greenish, blueish, and purplish, eventually from the color histogram dataset. These five dimensions are marked by five regions with thick boundaries in Fig. 4.

The color histogram feature set contains no brightness information. We added two dimensions, the mean and standard derivation of the V (brightness) component, from the color moment dataset. Values for these dimensions are normalized as a Gaussian sequence. The texture feature dataset has 16 dimensions which are co-occurrences in 4 directions (horizontal, vertical and two diagonal directions) of 4 texture measures: Angular Second Moment, Contrast, Inverse Difference Moment and Entropy. We averaged the values for four directions of each measure. Finally, we joined all these dimensions and got a dataset with 11 dimensions.

Figure 5 shows some screen snapshots of the dataset with 11 dimensions volume rendered. In the figure, each color represents a data cluster. The screen snapshots are the visualization against color component features. The overall shape of data distribution is like a tetrahedron because of the fact that the sum of all color histogram values in a record must be 1. Some areas contain

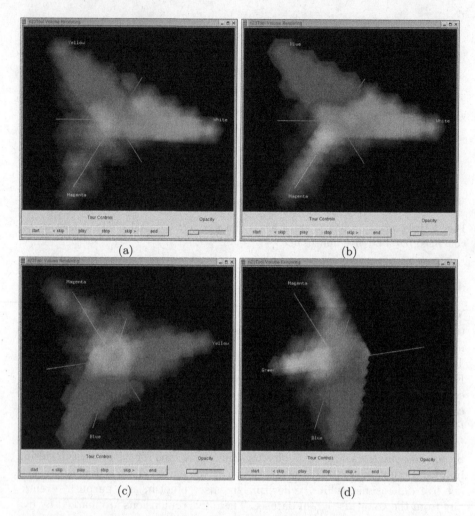

Fig. 5. Volume rendering of the Corel image features. (Color figure online)

no points and are fully transparent. This means there is no image in the Corel image collection that has such combination of features. Some areas have light transparent color because they contains less data points than other parts in the space, for example, the yellow areas along the positive Magenta axis. This means that few images in the collection have Magenta as the major color.

The visualization of feature space may disclose many interesting patterns. Figure 6(a) shows a scatterplot against three features: the angular second moment of texture and the mean and standard derivation of the V values. At center of the distribution is a big hole. The cave-shaped data distribution also suggests some inherent relations among these three features. Figure 6(b) shows a scatterplot against three texture measures: angular second moment, contrast

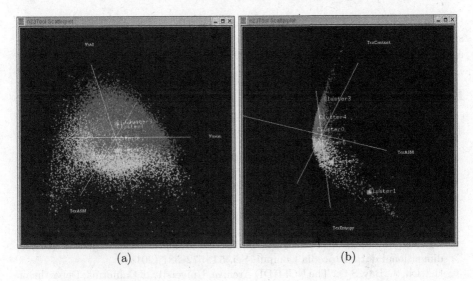

(a)   (b)

Fig. 6. Scatterplot of the Corel image features. (Color figure online)

and entropy. The data distribution shows the relationship among entropy, contrast and angular second moment. Remember that all texture measures we use here are averages of those on four directions.

5 Conclusion

Because of the rapid growth of the World Wide Web, content-based information classification together with other information seeking tasks such as information extraction, summarization and retrieval is becoming more and more important and is valuable in all disciplines that require access to textual and multimedia information. This paper discusses visual exploration of large feature datasets. We hope that this visual exploration technique would become a useful tool for information classification.

In summary, we have used grand tour to deal with high dimensionality of data and proposed high dimensional volume rendering to deal with huge number of records. Grand tour creates an illusion of smooth travel through high dimensional space. Cluster-guided tour preserves distances between cluster centroids. Volume rendering enables the visual exploration of very large datasets. Only aggregations or parts of data needed in rendering are retrieved from databases. This approach scales well to large datasets. Making projections and scatterplots has a complexity linear to the number of data points. This is in the sense that all data points have to be projected one by one. With volume rendering which takes as input the binned and aggregated relational data, this complexity can be greatly reduced.

This approach represents a new strategy that uses visual exploration techniques for information classification. We have shown in the experiment how the technique can be used to visualize a large feature dataset. We hope that this strategy will be useful to visually explore data in other areas. Potential new applications remain as part of the future work.

References

1. Asimov, D.: The grand tour: a tool for viewing multidimensional data. SIAM J. Sci. Stat. Comput. **6**(1), 128–143 (1985)
2. Becker, B.G.: Volume rendering for relational data. In: IEEE Symposium on Information Visualization (InfoVis 1997), Phoenix, Arizona, October 1997
3. Cook, D., Buja, A.: Manual controls for high-dimensional data projections. J. Comput. Graph. Stat. **6**(4), 464–480 (1997)
4. Cook, D.R., Buja, A., Cabrera, J.: Grand tour and projection pursuit. J. Comput. Graph. Stat. **4**(3), 155–172 (1995)
5. Dzwinel, W., Wcisło, R.: Very fast interactive visualization of large sets of high-dimensional data. Procedia Comput. Sci. **51**, 572–581 (2015)
6. Hettich, S., Bay, S.D.: The UCI KDD Archive, University of California, Department of Information and Computer Science, Irvine, CA (1999). http://kdd.ics.uci.edu
7. Hurley, C., Buja, A.: Analyzing high-dimensional data with motion graphics. SIAM J. Sci. Stat. Comput. **11**(6), 1193–1211 (1990)
8. Jones, K.S., Willett, P.: Readings in Information Retrieval. Morgan Kaufmann Publishers, San Francisco (1997)
9. Westover, L.: Footprint evaluation for volume rendering. ACM Comput. Graph. **24**(4), 367–376 (1990)
10. Yang, L.: Visual exploration of large relational datasets through 3D projections and footprint splatting. IEEE Trans. Knowl. Data Eng. **15**(6), 1460–1471 (2003)
11. Yang, L.: Distance-preserving mapping of patterns to 3-space. Pattern Recogn. Lett. **25**(1), 119–128 (2004)
12. Young, F.W., Rheingans, P.: Visualizing structure in high-dimensional multivariate data. IBM J. Res. Dev. **35**(1/2), 97–107 (1991)
13. Zhang, Z., Chow, T.W.S., Zhao, M.: Trace ratio optimization-based semi-supervised nonlinear dimensionality reduction for marginal manifold visualization. IEEE Trans. Knowl. Data Eng. **25**(5), 1148–1161 (2013)

Online Appearance Manifold Learning for Video Classification and Clustering

Li Yang$^{(\boxtimes)}$ and Xiaokun Wang

Department of Computer Science, Western Michigan University,
Kalamazoo, MI, USA
{li.yang,xiaokun.wang}@wmich.edu

Abstract. Video classification and clustering are key techniques in multimedia applications such as video segmentation and recognition. This paper investigates the application of incremental manifold learning algorithms to directly learn nonlinear relationships among video frames. Video frame classification and clustering are performed to the projected data in an intrinsic latent space. This approach has avoided partitioning video frames into arbitrary groups. It works even when the input video frames are under-sampled or unevenly distributed. Experiments show that video classification and clustering give better results in the latent space than in the original high dimensional space.

Keywords: Dimensionality reduction · Manifold learning · Video classification · Video data mining

1 Introduction

The ultimate challenge in video analysis is how to bridge the semantic gap between the formative information represented by a video sequence and the corresponding cognitive information. Formative information refers to pixel values and measures (e.g. color, shape, length, etc.). Cognitive information refers to human being's interpretation of the corresponding formative information. Video analysis has a broad coverage of topics [3]. It has important applications in industrial inspection, remote sensing, transportation, and video surveillance.

Video data contain huge amount of information and require more succinct presentations for analysis. A widely accepted methodology defines four hierarchical levels to represent a video sequence: *clip*, *scene*, *shot*, and *key frame*. From the highest level to lower levels, video clips are segmented into scenes and scenes are further segmented into shots each of which consists of a set of key frames. A scene normally represents an abstract concept such as a wedding event, a conversation, etc. A shot consists of a set of continuous video frames, some of which are defined as key frames which capture the semantic essence of the corresponding shot. Given a video shot, it is thus important to extract key frames and to represent the shot by the extracted key frames for further analysis. Usually, key frames are extracted by clustering all video frames into a few groups and then choosing one frame from each group as the key frame. Researches on

© Springer International Publishing Switzerland 2016
O. Gervasi et al. (Eds.): ICCSA 2016, Part II, LNCS 9787, pp. 551–561, 2016.
DOI: 10.1007/978-3-319-42108-7_43

shot detection, shot clustering [8,9], and key frame extraction [7,10] have been broadly reported in the literature.

The formative information of a video frame is usually represented by a high dimensional vector. Therefore, techniques for dimensionality reduction would be useful to get more succinct representations of video data for further processing. A classical method for dimensionality reduction is principal component analysis (PCA). PCA is a linear model and assumes that the data distribute on a hyperplane in high dimensional space. Therefore, it is not directly applicable to video data. Video data are commonly believed [4] to be distributed on nonlinear manifolds in high dimensional space.

To solve this problem, researchers have focused on partitioning the video data in high dimensional space and approximating the distribution of data by applying linear models to each partition. For example, [2] reports an online learning algorithm for constructing probabilistic appearance manifolds. The algorithm works by first constructing a generic appearance manifold from multiple video sequences of different persons. For online processing, the manifold is updated each time a new video frame is inserted. The generic appearance manifold is learned by partitioning the video frames into different clusters using the k-means clustering algorithm. PCA is then applied to each cluster to create a low dimensional linear approximation. The low dimensional clusters are connected by a transition matrix whose elements capture the likelihood that successive frames make transitions between the clusters. When a new video frame comes, the online learning process has two steps to follow: (1) estimate the pose of the new coming frame by Bayesian probability; (2) minimize the Euclidean distance between the appearance manifold and the key frames. In the first step, the new frame is assigned into a cluster and approximated by a weighted linear combination of its k nearest neighbors. The weights in the combination are then used to further approximate the frames in other poses of the same person. In the second step, the linear model is updated by the new video frame. Finally, the transition matrix is updated incrementally by counting the actual transitions between different linear models. The success of such an approach for video classification depends on the quality of clustering of video frames. In the approach, the number of clusters are arbitrarily assigned by the user.

With these observations, we believe that methods in manifold learning may play an important role in video analysis. These methods are able to learn nonlinear manifolds without partitioning the data into groups. For video classification, therefore, a promising idea is to use one of these methods to project the data to the data's intrinsic latent space where each dimension corresponds to a degree of freedom of the data. We expect that video classification and clustering can be done more effectively on the projected data in the latent space.

Manifold learning methods work in two phases: the first phase collects local information and the second phase derives global coordinates from the local information. Both phases are linear although the entire method is nonlinear. For example, Isomap [5] constructs a neighborhood graph of all data points and then use the graph to estimate geodesic distances between every pair of data points.

It then applies multidimensional scaling to the estimated geodesic distances to derive the low dimensional data configuration. Clearly, its success depends on whether the first step can successfully build a connected neighborhood graph. Isomap assigns each point neighbors as its k nearest neighbors (the k-NN approach) or all points within a fixed radius from the point. The approach may not be directly applicable to video data. This is because video data are continuous streams and are often under-sampled or unevenly distributed, in which cases these methods do not guarantee the connectivity of the constructed neighborhood graph. Data embedding methods need to work in an incremental way in order to process video streams. When a new frame is added or an old frame is dropped, there should be mechanisms to update the neighborhood graph, to update the estimated geodesic distances, and to adjust the resulting data configurations in the low dimensional space.

To overcome this problem, we have developed an algorithm named k-VC [6] to build k-connected neighborhood graphs. Furthermore, we have developed an algorithm, incremental k-VC [11], to incrementally update k-connected neighborhood graphs. The algorithm minimizes the maximal edge length of the neighborhood graph, which means that only short edges would remain on the graph. The geodesic distance between every pair of data points can be updated when a new data point is inserted or an existing data point is deleted.

Following this idea, we have studied online learning of appearance manifolds from continuous video streams. This paper reports our work in video data classification and clustering by applying incremental manifold learning algorithms. For dimensionality reduction of video data, video frames in high dimensional space can be projected incrementally into low dimensional space. Such an incremental algorithm for dimensionality reduction enables online learning of appearance manifolds of video sequences. Using our algorithms to preprocess the high dimensional video data, classification and clustering algorithms are expected to provide better results for video classification and key frames extraction.

2   Online Video Classification and Clustering

This section describes our approach for online video classification and key frame extraction using incremental algorithms for data embedding we developed previously. Our approach takes two stages. The first stage incrementally projects the high dimensional data to a low dimensional space. The second stage applies classification or clustering techniques to the data in the low dimensional space. The two stages and steps we used are listed in Table 1.

In Step 1 of the first stage, a newly inserted data point introduces two sets of edges: a set of edges to be added and a set of edges to be deleted. In Step 2, the edges to be deleted can break existing shortest paths whereas the edges to be added can create new shortest paths. The lengths of all shortest paths that go through the changed edges need to be updated. In implementation, the incremental algorithm initially sorts the n edges incident to the new vertex in a non-decreasing order of edge length. It then inserts in sequence k edges to the

Table 1. Major steps of video classification and clustering.

1st Stage: Incremental projection of high dimensional data
    Step 1: incrementally constructs k-connected neighborhood graphs when a new data point comes in;
    Step 2: updates the length of the shortest path between every pair of data points;
    Step 3: estimates the coordinates of the new data point and updates the coordinates of the existing data points.
2nd Stage: Video classification and clustering
    Classification: applying decision tree algorithm to the low dimensional data configuration;
    Clustering: applying k-means clustering algorithm to the low dimensional data configuration;

neighborhood graph. If the longest one of the k new edges is longer than the longest existing edge in the graph, we are all done. Otherwise, we add more new edges incident to the new vertex in a non-decreasing order of edge length until the length of the edge to be inserted is not shorter than the longest existing edge in the graph. Of course, some of these inserted edges are redundant with respect to k-connectivity at this moment. The next step then tries iteratively to drop the longest edges from the neighborhood graph. In each iteration, if two end vertices of the longest edge are still k-connected after deletion of the edge, the edge is then deleted. This procedure continues until the next longest edge cannot be deleted and it remains the longest edge in the finally updated neighborhood graph. Note that, in each step of the iteration, the neighborhood graph is guaranteed to be k-connected. Therefore, the finally updated neighborhood graph keeps k-connected.

When a vertex is deleted, the edges incident to it are also deleted which would affect the connectivity of some vertex pairs. Our idea to solve this problem is to insert new edges between every pair of vertices to which the deleted vertex directly connects. In this way, these vertices construct a clique. Certainly, such a graph may contain redundant edges in the clique. This problem can be solved by testing the edges in the clique from long to short whether its removal will still keep the graph k-connected. Such a process is repeated until an edge cannot be deleted.

We explain a little more in detail the third step of Stage 1. It deals with the issue on how to calculate the projection of the new data point and how to update the low dimensional configurations of existing data points. Let $\mathbf{X} = (\mathbf{x}_1, \ldots, \mathbf{x}_n)^T$ denote the existing n data points in the low dimensional space and g_{ij} denote the estimated geodesic distance, i.e. length of the shortest path, between the corresponding original data points of $\mathbf{x}_i$ and $\mathbf{x}_j$. Suppose we are projecting data to a d-dimensional space and the projected data points are centered, that is, $\sum_i \mathbf{x}_i = \mathbf{0}$, inner product matrix $\mathbf{X}\mathbf{X}^T$ should approach $\mathbf{B} = -(I - 1/n)\mathbf{G}(I - 1/n)/2$ where $\mathbf{G} = \{g_{ij}^2\}$, I is the identity matrix and

1 is a matrix with all elements as 1. In fact, classical multidimensional scaling seeks $\mathbf{X}\mathbf{X}^T$ to be as close to $\mathbf{B}$ as possible in the least square sense. This is achieved by assigning $\mathbf{X} = (\sqrt{\lambda_1}\mathbf{v}_1, \ldots, \sqrt{\lambda_d}\mathbf{v}_d)$, where $\lambda_1, \ldots, \lambda_d$ are the d largest eigenvalues of $\mathbf{B}$ and $\mathbf{v}_1, \ldots, \mathbf{v}_d$ are the corresponding eigenvectors.

We follow the idea of Law and Jain [1] to estimate the coordinate $\mathbf{x}_{n+1}$ of a new data point. Denote $d_{ij}^2 = \|\mathbf{x}_i - \mathbf{x}_j\|^2$. Because $\sum_i \mathbf{x}_i = \mathbf{0}$, we have

$$\sum_j \|\mathbf{x}_j\|^2 = \frac{1}{2n}\sum_{ij} d_{ij}^2. \tag{1}$$

Similarly, by defining $d_i^2 = \|\mathbf{x}_i - \mathbf{x}_{n+1}\|^2$, we have

$$\|\mathbf{x}_{n+1}\|^2 = \frac{1}{n}(\sum_{i=1}^n d_i^2 - \sum_{i=1}^n \|\mathbf{x}_i\|^2), \tag{2}$$

$$\mathbf{x}_{n+1}^T\mathbf{x}_i = -\frac{1}{2}(d_i^2 - \|\mathbf{x}_{n+1}\|^2 - \|\mathbf{x}_i\|^2). \tag{3}$$

If we approximate d_{ij}^2 by g_{ij}^2 and d_i^2 by $g_{i,n+1}^2$, the inner product f_i between $\mathbf{x}_{n+1}$ and $\mathbf{x}_i$ is estimated as

$$f_i \approx \frac{1}{2}(\frac{\sum_j g_{ij}^2}{n} - \frac{\sum_{lj} g_{lj}^2}{n^2} + \frac{\sum_l g_{l,n+1}^2}{n} - g_{i,n+1}^2). \tag{4}$$

The coordinates of $\mathbf{x}_{n+1}$ in the low dimensional configuration can then be obtained by solving $\mathbf{X}\mathbf{x}_{n+1} = \mathbf{f}$ in the least square sense, where $\mathbf{f} = (f_1, \ldots, f_n)^T$. The solution is given by

$$\mathbf{x}_{n+1} = (\frac{1}{\sqrt{\lambda_1}}\mathbf{v}_1^T\mathbf{f}, \ldots, \frac{1}{\sqrt{\lambda_d}}\mathbf{v}_d^T\mathbf{f})^T. \tag{5}$$

The low dimensional coordinates of existing data points $\mathbf{x}_i$, $i = 1, \ldots, n$ should also be adjusted in view of the modified geodesic distance matrix $\mathbf{G}_{new}$. This can be viewed as an incremental eigenvalue problem, as $\mathbf{x}_i$'s are obtained by eigen decomposition, where eigenvalues and eigenvectors of $\mathbf{B}_{new}$ can be found iteratively. A good initial guess of dominant eigenvectors of $\mathbf{B}_{new}$ are the column vectors of $\mathbf{X}$. In each iteration, we first apply QR-decomposition to $\mathbf{B}_{new}\mathbf{X}$ and let $\mathbf{Q}$ denote the Q-factor, we then apply eigen-decomposition of $\mathbf{Q}^T\mathbf{B}_{new}\mathbf{Q}$ and let $(\lambda_i, \mathbf{u}_i)$ be the i-th eigenpair, where $i = 1, \ldots, d$. The improved set of eigenvectors of $\mathbf{B}_{new}$ is then obtained as $\mathbf{Q}[\mathbf{u}_1, \ldots, \mathbf{u}_d]$.

In the second stage, we apply decision tree algorithm to the low dimensional data for the purpose of video classification. In particular, we use a subset of video sequences as the training set to construct a decision tree. The decision tree is then used to classify new coming video frames. Since the projected space is low dimensional, the rule used in each non-leaf node of the decision tree is restricted to a small number of variables. Although the physical meanings of these variables may be opaque, they reflect the semantic variances of the video frames.

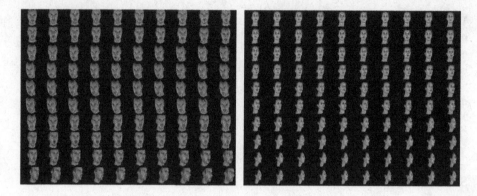

Fig. 1. Video sequences of two subjects, male and female.

The results of classification yielded by decision tree on the projected data is expected to be more accurate and resistant to noises.

We apply the classical k-means clustering algorithm to the projected data in the low dimensional space for the purpose of video clustering and key frame extraction. The k-means clustering algorithm first defines k centroids, one for each cluster. The algorithm runs iteratively and, in each step, associates each data point to the nearest centroid. The k cluster centroids are recalculated as barycenters of the clusters after each iteration. This process continues until there is no more significant change to cluster centroids.

3   Implementation and Experiments

We have implemented the proposed approach in Matlab and the graph-theoretic parts using the C programming language. This section presents results of three experiments. The first shows how our approach incrementally projects the video frames to 2D space. The second and third experiments apply the approach to video classification and key frame extraction respectively.

Our concern in the first experiment is the relationship between the high dimensional video frames and the corresponding projected low dimensional data points. We take video sequences of facial images of two subjects in this experiment. Figure 1 shows the original data set[1]. Each video sequence consists of 100 continuous frames each of which is an image with 31×46 pixels and 256 gray levels. We used the first 70 images of each subject as a training set and the rest 30 images as a test set. Batch version of the k-VC algorithm was first applied to the 140 training samples (70 for each subject). Incremental k-VC is then applied to update the neighborhood graph when new frames from the two video sequences come in. Consecutive frames of one subject's video sequence comes first and are followed by another's. The three pictures in the first column in Fig. 2 show the changes of the low dimension configuration when one video

---

[1] Available at http://www.ee.surrey.ac.uk/personal/a.hilton/research.html.

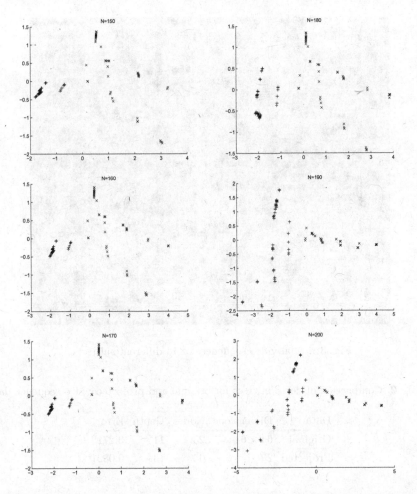

Fig. 2. 2D projections of two subjects' video sequences using incremental 4-VC.

sequence of one subject ('×') comes in. The second column shows the changes of low dimensional configuration when a video sequence of another subject ('+') comes in. We can see from Fig. 2 that consecutive frames of new coming video sequence are classified into the areas formed in the training process.

Our second experiment is to demonstrate how the proposed algorithm helps to improve the quality of video classification. The experimental data set[2] contains 13 video sequences of facial images, one sequence is recorded for each of the 13 subjects. Each video sequence contains 65 consecutive frames representing a consecutive motion of facial impressions of a specific subject. Figure 3 shows the first 13 frames of each subject. Each frame is represented as an image of 64×64 pixels and 256 gray levels.

---

[2] Available at http://amp.ece.cmu.edu/projects/FaceAuthentication/Default.htm.

Fig. 3. Facial video sequences of 13 different subjects.

Table 2. Comparison of decision trees for original and projected video sequence data

Data	Dimensions	Nodes	Depth	Error
Original	64×64	25	11	0.2718
Projected	2	65	13	0.0821

We use the first 50 frames in each video sequence as a training set to build decision tree. The built decision tree is then used to classify the rest consecutive 15 frames of each subject. For comparison, we have built two decision trees, one directly from the original video frames and the other from the 2D projections. The first decision tree is used to classify the rest 15 consecutive frames of each subject and the second decision tree is used to classify the 2D projections of the 15 consecutive frames of each subject. Figure 4 visualizes the two decision trees. The leaf nodes are denoted with different capital letters indicating different subjects. Splitting criteria are displayed at intermediate nodes of the tree. The two classification trees are summarized in Table 2. We can see significant improvement of classification error rate on the projected data.

The third experiment reports results for video frame clustering and key frame extraction. The data set[3] used in this experiment is a video sequence of 1000

---

[3] Available at http://www.cs.toronto.edu/~roweis/data.html.

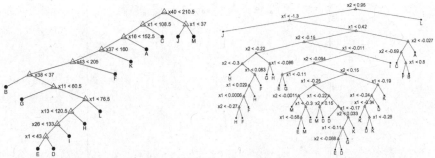

(a) Decision tree from the original data   (b) Decision tree from the 2D projections

Fig. 4. Decision trees for 13 subjects' video sequences.

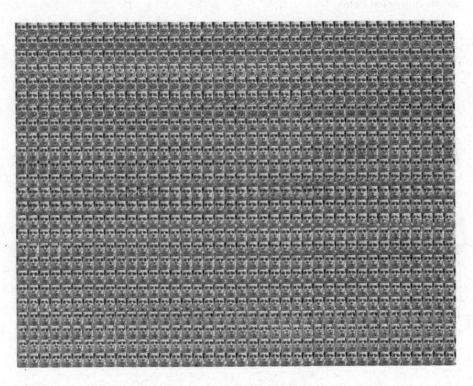

Fig. 5. A video sequence of 1,000 continuous frames.

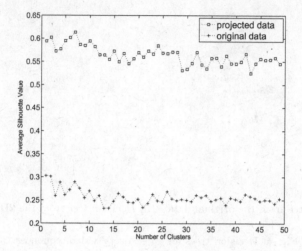

Fig. 6. Average silhouette values of results of k-means clustering on the original data and the projected data.

consecutive frames of one subject. The data set is shown in Fig. 5. We measure the quality of k-means clustering using silhouette values[4].

We choose the number of clusters as $k = 2, \ldots, 50$. For each value of k, we calculate the average of 1000 silhouette values. Figure 6 shows two plots of silhouette values of the results of the k-means clustering algorithm, one on the original video frames and the other on the projected data points. Higher average silhouette value indicates better performance. Figure 6 clearly shows that the average silhouette values of clusters of the projected data are consistently higher than those of the original video frames. Such results demonstrate the effectiveness of our proposed approach to preprocess video data for the purpose of video frame clustering.

4   Conclusion and Discussion

In this paper, we proposed an approach of using incremental data embedding techniques as a preprocessing step for video frame classification and clustering. The approach works by first projecting the video frames from high dimensional space to low dimensional space and then perform classification or clustering to data in the low dimensional space.

---

[4] Silhouette value is a measure of how similar a data point is to data points in its own cluster versus data points in other clusters. Let a_i denote the average distance of the i-th point to all other points in the same cluster. Let b_i denote the minimum of average distances of the i-th point to all points in other clusters, that is, the average of distances to all points in the next closest cluster. The i-th point's silhouette value is defined as $s_i = (b_i - a_i)/\max(a_i, b_i)$. Silhouette values range from -1 to $+1$.

It should be noted that the proposed approach heavily depends on efficient implementations of underlying algorithms and therefore would benefit from modern research on algorithms. For example, the operation for testing the k-connectivity between two vertices in our approach is a direct use of network flow algorithm, which is fairly costly. In addition, the presented method for updating lower dimensional configurations is based on iterative approximation. In essence, the updating process is actually updating the results of singular value decomposition (SVD). Our approach would benefit from modern research of efficient algorithms in these classical areas.

Machine learning for multimedia data analysis is a research area that has enormous applications in the big data era. Specific to the technical approaches in this paper, future work will be on the development of useful software systems. We also plan to apply the incremental algorithm for manifold learning to other applications such as speech recognition and text-based information retrieval.

References

1. Law, M.H.C., Jain, A.K.: Incremental nonlinear dimensionality reduction by manifold learning. IEEE Trans. Pattern Anal. Mach. Intell. **28**(3), 377–391 (2006)
2. Lee, K.C., Kriegman, D.: Online learning of probabilistic appearance manifolds for video-based recognition and tracking. In: Proceedings of the IEEE Conference on Computer Vision and Pattern Recognition, Washington, DC, vol. 1, pp. 852–859 (2005)
3. Roach, M., Mason, J., Xu, L., Stentiford, F.: Recent trends in video analysis : a taxonomy of video classification problems. In: Proceedings of the International Conference on Internet and Multimedia Systems and Applications, Honolulu, HI, pp. 864–871, August 2003
4. Seung, H.S., Lee, D.D.: The manifold ways of perception. Science **290**(5500), 2268–2269 (2000)
5. Tenenbaum, J.B., de Silva, V., Langford, J.C.: A global geometric framework for nonlinear dimensionality reduction. Science **290**, 2319–2323 (2000)
6. Yang, L.: Building k-connected neighborhood graphs for isometric data embedding. IEEE Trans. Pattern Anal. Mach. Intell. **28**(5), 827–831 (2006)
7. Yeo, B.L., Liu, B.: Rapid scene analysis on compressed videos. IEEE Trans. Circ. Syst. Video Technol. **5**(6), 533–544 (1995)
8. Yeung, M., Yeo, B.L.: Time-constrained clustering for segmentation of video into story units. In: Proceedings of the International Conference on Pattern Recognition, Vienna, Austria, pp. 375–380, August 1996
9. Yeung, M., Yeo, B.L., Liu, B.: Extracting story units from long programs for video browsing and navigation. In: Proceedings of the International Conference on Multimedia Computing and Systems, Hiroshima, Japan, pp. 296–305, June 1996
10. Zhang, H.J., Low, C.Y., Smoliar, S.W.: Video parsing and browsing using compressed data. Multimedia Tools Appl. **1**(1), 89–111 (1995)
11. Zhao, D., Yang, L.: Incremental isometric embedding of high dimensional data using connected neighborhood graphs. IEEE Trans. Pattern Anal. Mach. Intell. **31**(1), 86–98 (2009)

Patch Based Face Recognition via Fast Collaborative Representation Based Classification and Expression Insensitive Two-Stage Voting

Decheng Yang, Weiting Chen$^{(\boxtimes)}$, Jiangtao Wang, and Yan Xu

Computer Science and Software Engineering Institute,
East China Normal University, Shanghai, China
yang_dc@outlook.com, {wtchen,jtwang}@sei.ecnu.edu.cn,
51141500088@ecnu.cn

Abstract. Small sample size (SSS) is one of the most challenging problems in Face Recognition (FR). Recently the collaborative representation based classification with l2-norm regularization (CRC) shows very effective face recognition performance with low computational cost. Patch based CRC (PCRC) also could well handle the SSS problem, and a more effective method is conducted PCRC on different scales with various patch sizes (MSPCRC). However, computation of reconstruction residuals on all patches is still time consuming. In this paper, we devote to improve the performance for SSS problem in face recognition and decrease the computational cost. First, fast collaborative representation based classification (FCRC) is proposed to further decrease the computational cost of CRC. Instead of computing reconstruction residual on all classes, FCRC computes the residual on a small subset of classes which has a big coefficient, such a category full make use of the discrimination of representation coefficients and decrease the computational cost. Our experiments results show that FCRC has a significantly lower computational cost than CRC and slightly outperforms CRC. FCRC is especially powerful when it is applied on patches. To further improve the performance under varying expression, we use a two-stage voting method to combine the recognition outputs of all patches. Extended experiments show that the proposed two-stage voting based FCRC (TSPFCRC) outperforms many state-of-the-art face recognition algorithms and have a significantly lower computational cost.

Keywords: Face recognition · Small sample size · Ensemble learning · Collaborative representation

1 Introduction

Small sample size problem is one of the most fundamental and challenging issues in FR. The performance of traditional appearance based FR methods, such as the classical Eigenface [1], Fisherface [2], and LPP [3], degrades much with the decrease of training samples.

© Springer International Publishing Switzerland 2016
O. Gervasi et al. (Eds.): ICCSA 2016, Part II, LNCS 9787, pp. 562–570, 2016.
DOI: 10.1007/978-3-319-42108-7_44

Inspired by the sparse coding mechanism of human vision system [4], the sparse representation based classification (SRC) [5] scheme shows very interesting FR results, in SRC the query face image is coded as a sparse linear combination of all the training samples via l1-norm minimization, and is classified to the class which has the least representation residual. However, recent researches have shown that the costly l1-norm sparse regularization on the representation vector in SRC is not necessary, and using the non-sparse l2-norm to regularize the representation coefficients can lead to similar FR results with a significant decrease in computational cost. The collaborative representation based classification (CRC) was then proposed in [6] by representing the query sample with non-sparse l2-regularization. However, both CRC and SRC cannot well handle the SSS problem due to the query sample cannot be well represented when the training sample size is very small.

To solve the SSS problem, many methods have been proposed. In [7], virtual samples and generic training set were used. However, such a category can lead the trained classifiers unstable and have poor generalization ability when the available samples are insufficient. A more stable and effective way is ensemble learning. These ensemble learning based methods can be roughly divided into three categories. The first category of methods is patch (or block) based methods, which usually involve steps of local region partition, local feature extraction and classification combination [8–10], in [11], to solve the SSS problem, CRC was conducted on patches (PCRC). To further improve the performance, a multi-scale PCRC (MSPCRC) was proposed. MSPCRC considers PCRC on each scale as a base classifier and learns scale weights to fuse multi-scale decisions. The second category considers the fact that the global and local features can provide complementary information [12, 13]. Third, different types of facial features are combined for face verification [14, 15], such as SIFT, LBP, Gabor response and gray values.

Among these methods, patch based methods are very popular because of its effectivity. However, most of patch based methods are complicate and time-consuming. In this paper, we focus on the decrease of computational cost and improvement of the recognition rate for SSS problem. Although the computational cost of CRC is much lower than l1-norm regularized SRC, patch based CRC is still time-consuming because of we must compute reconstruction residuals of all classes on all patches. To address it, we propose a fast collaborative representation based classification (FCRC) which exploits the discrimination of coding coefficients. After choosing several classes with big coefficients as candidates, FCRC compute the reconstruction residuals on these candidate classes, then classify the query sample by checking which class yield minimum reconstruction residual. Operating FCRC on patches is much more efficient than PCRC because we only compute reconstruction residuals on a small subset of classes. To further improve the performance under varying expression, a novel two-stage voting category is presented to combine recognition outputs of all patches. We find the samples difficult to classify after first voting, then using the subset of patches which insensitive to expression to vote again.

The rest of this paper is organized as follows. Section 2 describes fast collaborative representation based classification (FCRC). Section 3 presents the two-stage voting classification combination. Section 4 conducts experiments and conclusions are made in Sect. 5.

2 Fast Collaborative Representation Based Classification

The collaborative representation based classification (CRC) with l_2 -norm regularization was presented in [6]. Denote by $A = [A_1, A_2, \ldots, A_c]$ the matrix formed by original training samples, where A_i is the sub-set of training samples from class i, and c is the number of classes. Let y be a query sample to be classified. In CRC, the collaborative representation of it is

$$\hat{a} = \arg\min_a \left\{ \|y - Aa\|_2^2 + \lambda \|a\|_2^2 \right\} \tag{1}$$

The l_2-regularization term $\|a\|_2$ not only makes the least square solution stable but also introduces a certain amount of sparsity to $\hat{a}$, yet this sparsity is much weaker than that by l_1-regularization.

To further analysis, the Eq. (1) can be analytically derived as $\hat{a} = Py$, where $P = (A^T A + \lambda \cdot I)^{-1} A^T$. Clearly, P is independent of y so that it can be pre-calculated. This makes the computation of coding coefficients very fast because we just simply project query sample y onto P via Py. The classification of CRC is performed by checking which class yields the minimal regularized reconstruction error. The recognition output of the query sample y is $identity(y) = \arg\min_i\{r_i\}$, where $r_i = \|y - A_i \cdot \hat{a}_i\|_2 / \|\hat{a}_i\|_2$ and $\hat{a} = [\hat{a}_1; \hat{a}_2; \ldots; \hat{a}_c]$.

Reconstruction residual based CRC is effective for face recognition, considering the coding coefficients also has some discrimination information, some heuristics can be used for classification. To make full use of the discrimination of coding coefficients, we can classify the query sample by checking which class has the maximum class-coefficient (MCC), suppose that each class has m training samples, so the coding coefficients of class i can be denoted as $\hat{a}_i = [a_{i1}, a_{i2}, \ldots, a_{im}]$, the class-coefficient is computed by summing the coding coefficients of class i, it can be denoted as $cc_i = \sum_{j=1}^m a_{ij}$, the class which has the max class-coefficient will be the identity of y. Such a method is much simpler and more efficient than reconstruction residual based CRC, the comparison of recognition rate between MCC and CRC can be seen in Fig. 1, the experiment setting is described in Sect. 4, Fig. 1 shows that the performance of MCC is lower than CRC especially when the training sample size is very small.

To balance performance with efficiency, a novel classification method combining the advantage of MCC and CRC is proposed. We compute reconstruction residual on a small subset of classes with big values of class-coefficients. The detail of the proposed FCRC algorithm is summarized in Table 1.

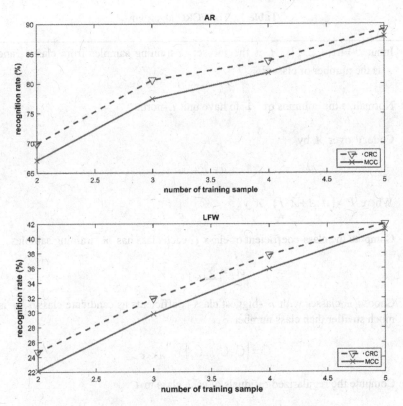

Fig. 1. Recognition rates of MCC and CRC on AR (up) and LFW (down).

3 Expression Insensitive Two-Stage Voting

After operating FCRC on patches, how to combine the classification outputs of all patches is a critical problem. Majority voting [9], linear weighted combination [16], kernel plurality [8] and probabilistic model [12] can be employed for the combination. In [9, 13], the authors show that the weighted combination leads to little improvement compared to the simple majority voting. To improve the performance against the variations of expression, we use a two-stage majority voting method to combine the recognition output of all patches. In first voting, all patches are used for voting, the class which has largest frequency will be the final identity of the query sample. If the frequency of final identity is lower than the preset threshold, the sample is difficult to classify, then the query sample will be voted again by using the patches insensitive to expression. The flowchart of the proposed method is given in Fig. 2.

Table 1. The FCRC algorithm

---

1. Input: $A = [A_1, A_2, ..., A_c]$, A_i is the sub-set of training samples from class i, and c is the number of classes.

2. Normalize the columns of A to have unit l_2-norm.

3. Code y over A by

$$\hat{a} = Py \tag{2}$$

Where $P = (A^T A + \lambda \cdot I)^{-1} A^T y$.

4. Compute the class-coefficient of class i, each class has m training samples.

$$cc_i = \sum_{j=1}^{m} a_{ij} \tag{3}$$

5. Choose n classes with n-biggest class-coefficients as candidate classes, n is much smaller than class number c.

$$C = [C_1, C_2, ...C_n] \quad n << c \tag{4}$$

6. Compute the regularized residuals of each class in C

$$r_k = \|y - A_k \hat{\alpha}_k\|_2 / \|\alpha_k\|_2 \tag{5}$$

7. Output the identity of y as

$$\text{identity}(y) = \arg\min_k \{r_k\} \tag{6}$$

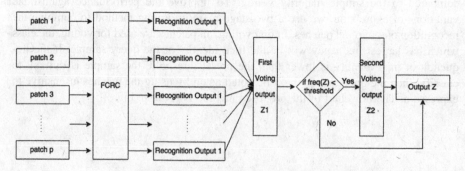

Fig. 2. Flow chart of TSPFCRC for face recognition

4   Experiment Analysis

In this section, the Multi-PIE [17] and AR [18] databases in controlled environments together with the LFW database [19] in uncontrolled environments were used to test the FR performance of the proposed method.

We use CRC, SRC and NN methods as baseline, and use the state-of-the-art patch based methods including BlockFLD [24], Patch based CRC (PCRC), and Multi-scale patch based CRC (MSPCRC) for comparison. The detail parameter setting of these method are described in [11].

For our proposed two-stage voting patch based FCRC (TSPFCRC), the candidate class number in FCRC was set as one-tenth of class number, the parameter was set as 0.001, and the patch size for Multi-PIE, AR, and LFW are set as 10×10 (32×32 image), 10×10, and 20×20, respectively.

4.1   Multi-PIE Database

In our experiments, a subset of Multi-PIE [17] database that contains 10 images with neutral expression and 10 images with smile expression per person from session 3 was chosen. For the training set, we used 2–5 samples per subject from images with neutral expression. For the testing sets, 3 samples from images with smile expression were used. The face images are resized to 32×32. The FR results are listed in Table 2. We can see that the proposed TSPFCRC outperform all the other methods.

4.2   AR Database

The AR face database [18] contains over 4,000 color frontal face images of 126 people with simultaneous variations in expressions, lighting conditions and occlusions. As in [5], a subset of AR contains two-session data of 50 male subjects and 50 female subjects was chosen in our experiments. For each subject, 2–5 samples from session 1 were selected for training and 3 samples were selected for testing. The face images are resized to 32×32.

Table 2. Recognition accuracy on the Multi-PIE database (%)

Method	2	3	4	5
CRC	62.6 ± 13.8	74.3 ± 6.3	78.5 ± 5.2	80.4 ± 3.7
SRC	61.9 ± 14.0	73.2 ± 8.9	78.6 ± 6.5	80.8 ± 4.2
NN	54.9 ± 14.5	64.7 ± 12.1	71.9 ± 9.9	74.5 ± 8.8
BlockFLD	66.1 ± 6.9	71.1 ± 5.7	76.4 ± 4.6	79.2 ± 3.2
PCRC	68.8 ± 10.9	76.0 ± 6.2	79.4 ± 4.8	81.3 ± 3.7
MSPCRC	72.4 ± 10.5	79.6 ± 5.9	83.6 ± 4.0	84.6 ± 2.6
FCRC	63.0 ± 11.9	74.8 ± 6.8	79.1 ± 4.9	80.8 ± 4.2
TSPFCRC	**75.9 ± 5.9**	**81.0 ± 4.1**	**84.8 ± 4.1**	**86.0 ± 3.3**

Table 3. Recognition accuracy on the AR database (%)

Method	2	3	4	5
CRC	69.9 ± 12.6	80.6 ± 10.4	83.8 ± 9.6	89.1 ± 6.2
SRC	69.7 ± 14.8	79.0 ± 10.6	83.5 ± 8.9	88.2 ± 5.7
NN	48.5 ± 9.5	54.7 ± 9.0	58.5 ± 9.1	63.2 ± 7.0
BlockFLD	71.5 ± 11.5	78.6 ± 9.8	84.2 ± 8.7	87.6 ± 4.2
PCRC	82.2 ± 11.3	87.7 ± 9.4	89.9 ± 8.5	92.9 ± 6.7
MSPCRC	82.3 ± 11.5	87.8 ± 10.5	90.2 ± 9.1	93.6 ± 7.6
FCRC	74.4 ± 11.1	80.7 ± 10.6	84.0 ± 11.7	91.1 ± 9.6
TSPFCRC	**83.2 ± 10.3**	**89.4 ± 8.0**	**92.3 ± 7.6**	**94.4 ± 6.1**

Table 4. Recognition accuracy on the LFW database (%)

Method	2	3	4	5
CRC	24.7 ± 2.1	31.9 ± 2.4	37.8 ± 2.6	42.0 ± 3.2
SRC	24.4 ± 2.4	32.7 ± 3.2	38.7 ± 2.4	44.1 ± 2.6
NN	9.3 ± 1.7	11.4 ± 1.8	13.0 ± 1.7	14.3 ± 1.9
BlockFLD	18.0 ± 2.1	22.3 ± 2.1	26.2 ± 2.6	28.4 ± 2.5
PCRC	32.0 ± 1.9	37.0 ± 2.8	40.2 ± 2.5	42.9 ± 2.6
MSPCRC	35.0 ± 1.6	41.1 ± 2.8	46.0 ± 3.0	49.0 ± 2.9
FCRC	24.9 ± 2.2	33.0 ± 2.4	38.7 ± 2.9	43.6 ± 2.2
TSPFCRC	34.7 ± 2.6	**43.0 ± 3.0**	**48.5 ± 2.5**	**53.8 ± 2.9**

The recognition accuracy on the AR database is shown in Table 3. The results validate that TSPFCRC is the best in accuracy, although the improvement is slightly.

4.3 LFW Database

The LFW database [19] contains images of 5,749 different individuals in unconstrained environment. LFW-A is a version of LFW after alignment using commercial face alignment software. We gathered the subjects including no less than ten samples and then get a dataset with 158 subjects from LFW-a. For each subject, 2–5 samples are randomly chosen for training and another 2 samples for test. The images are firstly cropped to 121 × 121 and then resized to 32 × 32. The comparison of competing methods is given in Table 4. From the results, we can see that the proposed category of FCRC and expression insensitive two-stage voting are also suit for uncontrolled situation.

4.4 Efficiency

In this section, we compare the efficiency between the FCRC and CRC, we test the running time on a query sample of AR database which contains 126 subjects. For the proposed FCRC, the candidate class number is 10. The comparison results can be seen

Table 5. The comparison of running time between CRC and FCRC on AR database

Method	Running time (ms)
CRC	6.5
FCRC	1.4
PCRC	122.5
TSPFCRC	20.9

in Table 5. The result clearly show that FCRC has a much lower computational cost than CRC. Such a method is very useful when operating it on patches, it could significantly decrease the computational cost compared to patch based CRC (PCRC).

5 Conclusion

In order for a more effective and efficient face recognition when the number of training samples per class is small, we propose a fast collaborative representation based classification which could significantly decrease the computational cost, especially when operating it on patches. To further improve the FR performance, an effective classification combination method by using two-stage voting category was used to handle the variation of expression. The proposed method is simple but effective, experimental results on controlled and uncontrolled face databases showed that the proposed TSPFCRC outperform many state-of-the-art patch based methods such as MSPCRC and BLDA and with much lower computational cost.

Acknowledgement. The research is supported by the National Natural Science Foundation of China (Grant No. 81101119, Grant No. 61340036), the Open Project of Software/Hardware Co-design Engineering Research Center MoE, and National Key Basic Research Program (Grant No. 2011CB707104).

References

1. Zhang, J., Yan, Y., Lades, M.: Face recognition: eigenface, elastic matching, and neural nets. Proc. IEEE **85**(9), 1423–1435 (1997)
2. Belhumeur, P.N., Hespanha, J.P., Kriegman, D.J.: Eigenfaces vs. fisherfaces: recognition using class specific linear projection. IEEE Trans. Pattern Anal. Mach. Intell. **19**(7), 711–720 (1997)
3. He, X., Yan, S., Hu, Y., Niyogi, P., Zhang, H.-J.: Face recognition using laplacianfaces. IEEE Trans. Pattern Anal. Mach. Intell. **27**(3), 328–340 (2005)
4. Vinje, W.E., Gallant, J.L.: Sparse coding and decorrelation in primary visual cortex during natural vision. Science **287**(5456), 1273–1276 (2000)
5. Wright, J., Yang, A.Y., Ganesh, A., Sastry, S.S., Ma, Y.: Robust face recognition via sparse representation. IEEE Trans. Pattern Anal. Mach. Intell. **31**(2), 210–227 (2009)
6. Zhang, L., Yang, M., Feng, X.: Sparse representation or collaborative representation: which helps face recognition? pp. 471–478

7. Su, Y., Shan, S., Chen, X., Gao, W.: Adaptive generic learning for face recognition from a single sample per person, pp. 2699–2706

8. Kumar, R., Banerjee, A., Vemuri, B.C., Pfister, H.: Maximizing all margins: pushing face recognition with kernel plurality, pp. 2375–2382

9. Kumar, R., Banerjee, A., Vemuri, B.C.: Volterrafaces: discriminant analysis using volterra kernels, pp. 150–155

10. Lu, J., Tan, Y.-P., Wang, G.: Discriminative multimanifold analysis for face recognition from a single training sample per person. IEEE Trans. Pattern Anal. Mach. Intell. **35**(1), 39–51 (2013)

11. Zhu, P., Zhang, L., Hu, Q., Shiu, S.C.K.: Multi-scale patch based collaborative representation for face recognition with margin distribution optimization. In: Fitzgibbon, A., Lazebnik, S., Perona, P., Sato, Y., Schmid, C. (eds.) ECCV 2012, Part I. LNCS, vol. 7572, pp. 822–835. Springer, Heidelberg (2012)

12. Lin, D., Tang, X.: Recognize high resolution faces: from macrocosm to microcosm, pp. 1355–1362

13. Su, Y., Shan, S., Chen, X., Gao, W.: Hierarchical ensemble of global and local classifiers for face recognition. IEEE Trans. Image Process. **18**(8), 1885–1896 (2009)

14. Wolf, L., Hassner, T., Taigman, Y.: Effective unconstrained face recognition by combining multiple descriptors and learned background statistics. IEEE Trans. Pattern Anal. Mach. Intell. **33**(10), 1978–1990 (2011)

15. Guillaumin, M., Verbeek, J., Schmid, C.: Is that you? Metric learning approaches for face identification, pp. 498–505

16. Tan, X., Chen, S., Zhou, Z.-H., Zhang, F.: Recognizing partially occluded, expression variant faces from single training image per person with SOM and soft k-NN ensemble. IEEE Trans. Neural Netw. **16**(4), 875–886 (2005)

17. Sim, T., Baker, S., Bsat, M.: The CMU pose, illumination, and expression database. IEEE Trans. Pattern Anal. Mach. Intell. **25**(12), 1615–1618 (2003)

18. Martinez, A.M.: The AR face database, CVC Technical report, vol. 24 (1998)

19. Huang, G.B., Ramesh, M., Berg, T., Learned-Miller, E.: Labeled faces in the wild: a database for studying face recognition in unconstrained environments, Technical report 07–49, University of Massachusetts, Amherst (2007)

A Video Self-descriptor Based on Sparse Trajectory Clustering

Ana Mara de Oliveira Figueiredo[1]([✉]), Marcelo Caniato[1],
Virgínia Fernandes Mota[2], Rodrigo Luis de Souza Silva[1],
and Marcelo Bernardes Vieira[1]

[1] Universidade Federal de Juiz de Fora, Juiz de Fora, Brazil
{anamara,marcelo.caniato,rodrigoluis,marcelo.bernardes}@ice.ufjf.br
[2] Colégio Técnico, Universidade Federal de Minas Gerais, Belo Horizonte, Brazil
virginiaferm@dcc.ufmg.br

Abstract. In order to describe the main movement of the video a new motion descriptor is proposed in this work. We combine two methods for estimating the motion between frames: block matching and brightness gradient of image. In this work we use a variable size block matching algorithm to extract displacement vectors as a motion information. The cross product between the block matching vector and the gradient is used to obtain the displacement vectors. These vectors are computed in a frame sequence, obtaining the block trajectory which contains the temporal information. The block matching vectors are also used to cluster the sparse trajectories according to their shape. The proposed method computes this information to obtain orientation tensors and to generate the final descriptor. The global tensor descriptor is evaluated by classification of KTH, UCF11 and Hollywood2 video datasets with a non-linear SVM classifier. Results indicate that our sparse trajectories method is competitive in comparison to the well known dense trajectories approach, using orientation tensors, besides requiring less computational effort.

Keywords: Block matching · Human action recognition · Self-descriptor · Sparse and dense trajectories · Trajectory clustering

1 Introduction

Human action recognition is a challenging problem in Computer Vision, and it has been an active area of research for over a decade. An important part of the recognition process is the representation of motion information. This was approached in different ways over the course of the years. Recently, using video descriptors for representing this kind of information have become a trend among researchers. In this context, orientation tensors were used in [1–3] for describing video information. The tensor keeps data that represent the relationship between vectors coefficients associated with the original motion information. Likewise,

M.B. Vieira—The authors thank FAPEMIG, CAPES and UFJF for funding.

O. Gervasi et al. (Eds.): ICCSA 2016, Part II, LNCS 9787, pp. 571–583, 2016.
DOI: 10.1007/978-3-319-42108-7_45

our work also employs tensors. Our descriptor is the result of the concatenation of separate cluster vectors obtained from cross products between two different motion informations.

Two different approaches are combined here in order to detect motion displacement. The first one is block matching. The other one is the calculation of the brightness gradient between two consecutive frames. The cross product between the vectors of these two methods is calculated, and then this resulting vector is clustered according to the mean angle of the block matching trajectory vectors. Following other works in the literature, we use spatial-temporal motion information for better representing the motion.

In order to evaluate the video descriptor obtained by the proposed method, we used well-known datasets, such as KTH, UCF11 and Hollywood2. Besides, a Support Vector Machine (SVM) was employed for classification purposes.

2   Related Works

In [4], we have used a block matching method to extract motion information from a video. In this method, each frame is divided into blocks of a predetermined size, and each block is matched to a correspondent block in a subsequent frame. The matching can occur in a sequence of frames in order to obtain the trajectory of a block within the sequence. The method calculates displacement vectors for each block and uses a histogram to quantize these vectors. This histogram is also used to calculate a tensor capable of describing a video.

The idea of using trajectories for extracting human activities has become increasingly popular. A work that uses the space-time trajectories idea is presented in [5]. The motion is decomposed into dominant and residual parts. These parts are used in the extraction of space-time trajectories and for the descriptor computation. The resulting descriptor, named DCS, is based on differential motion scalar quantities, as well as on divergence, curl and shear features (hence the acronym DCS). It uses the Vector of Local Aggregated Descriptor (VLAD) encoding technique to perform action recognition.

The residual motion discussed above also plays an important role in the method proposed by [6]. In their work, two methods are combined to estimate camera motion: SURF descriptors and dense optical flow. Feature points are matched and used to estimate a homography with RANSAC. A human detector is also employed to remove inconsistent matches originated from human activity and trajectories generated from camera motion.

Finally, [1] presented a tensor motion descriptor for video sequences using optical flow and HOG3D information. In that work, they use an aggregation tensor-based technique, combining two descriptors. One of them carries polynomial coefficients which approximate optical flow, and the other one carries data from HOGs. This descriptor is evaluated by a SVM classifier using KTH, UCF11 and Hollywood2 datasets. Orientation tensors are also used as motion descriptors in [7] in conjunction with dense optical flow trajectories. For each optical flow point, he computes the cross product between the trajectory displacement vector and the 3D gradient vector in a window around that point. The optical flow displacement vectors are also used for clustering based on trajectories shape. According to this clustering, the resulting motion vectors are grouped and represented by a tensor.

In the new method presented in this work, block matching is employed as in [4], but a Variable Size Block Matching Algorithm (VSBMA) is used instead to obtain a better division of the frame. In our work, the approach of using block matching vectors considerably reduces the effort needed for tracking motion, as compared to the use of HOG3D and dense trajectories.

3   Proposed Method

The basis for our proposed method will be presented in the following sections. The method presentation will be divided in five stages. In general terms, we calculate the block displacement using the 4SS block matching. The displacement vector is used in the computation of the cross product with the brightness gradient in the block area. The resulting three-dimensional vectors are then clustered according to the average of the angles of trajectory vectors.

Two different variations of the method were used during experiments. These differences are entirely concentrated in the first stage, presented in Sect. 3.1, where the sparse trajectories are computed using block matching. Both the method of block division and the trajectory formation are switched in each variation. The subsequent stages remain unchanged.

3.1   Computing Sparse Trajectories

The input of our method is a video, i.e., a set of frames $V = \{F_k\}$, where $k \in [1, n_f]$ and n_f is the number of frames. The goal is to characterize a video sequence action, so the movement is described for $k - 1$ frames, because at least one successor frame is needed for the calculation. We also need to calculate the displacement block between the first two frames of the sequence. For that purpose, a Variable Size Block Matching Algorithm (VSBMA) is used. The frame k of the video is subdivided into $n_x \times n_y$ non-overlapping blocks of exactly $s_0 \times s_0$ pixels, where s_0 is an initial block size fixed for the first frame. If the frame dimension is not a multiple of s_0, the remaining right and bottom pixels do not form blocks, as can be seen in Fig. 1(a). The algorithm searches for each block from the reference frame in the target frame based on any BMA search strategy. If the match error found is greater than a fixed threshold, the block is split into four half-sized blocks until the error is below the threshold or it reaches a predetermined smallest permitted size s_s. A quadtree is used for this purpose with leaf nodes corresponding to blocks of varying sizes. The objective is to make the edge of the blocks coincide with the border of objects in the scene, forming regions with uniform displacement. After this block division, a final number of blocks n_b is reached, where $n_b \geq n_x \cdot n_y$, as can be seen in Fig. 1(b).

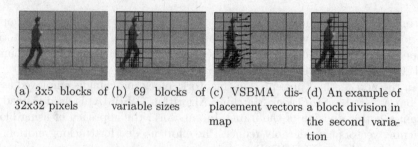

(a) 3x5 blocks of (b) 69 blocks of (c) VSBMA dis- (d) An example of
32x32 pixels   variable sizes   placement vectors a block division in
                                        map   the second varia-
                                                           tion

Fig. 1. An example of motion estimation using VSBMA

VSBMA is used here with a 4SS strategy to generate the displacement vectors map. For each block B_i, displacement vectors $v_i{}^k = (x, y) \in \mathbb{R}^2$ are calculated, where $i \in [1, n_b]$ is the block index. In this way, the most representative size is selected for each region of the frame. Specifically, the frame is divided into blocks with a predetermined initial size, which are split only when the lowest Block Distortion Measure (BDM) is still above the established threshold. Thus, a single frame can have blocks of various sizes, and their displacement vectors are all in the same map.

In order to explore the information of object trajectory through the video, we use t pairs of adjacent frames from the sequence, where t is the size of the trajectory generated from $t + 1$ frames. The block in the reference frame k is matched with a block in the 'target' frame, finding the correspondent block in frame $k + 1$. The correspondent block found is used as a reference block for the next match. The subsequent pairs use the same block size configuration as the first pair in the sequence.

An initial frame k and t successor frames are used to generate t vectors for each block trajectory, starting on the original grid. The number of frames used in the trajectories are predetermined and equal for all video.

The vector describing the displacement of a block between two consecutive frames is defined as $v_{i,j}^k = (x, y) \in \mathbb{R}^2$, where k is the index of the initial frame of the sequence and $j \in [1, t]$ is the index of the reference frame used to compute the vector. The set of these t vectors of all blocks form a map of motion vectors referent to k-th frame, as can be seen in Fig. 1(c), where $t = 4$.

3.2 Generating Histograms Using Cross Product Vectors

The same reference frames used to compute the BM are selected as references to compute the brightness gradient. These brightness gradient vectors are used to calculate the cross product with BM displacement vectors. As these displacement vectors have only the spatial components (x, y), a third component representing the time displacement is added, in order to make possible the cross product computation. As only a single frame displacement is considered, the temporal coordinate is set to 1.

These calculated cross products are represented by a vector set defined as:

$$C_{i,j}^k = \{c_q{}^k \mid q \in B_{i,j}^k,\; c_q{}^k = v_{i,j}^k \times g_q{}^k\}, \tag{1}$$

where q is a point in the block $B_{i,j}^k$ and $g_q{}^k$ is the brightness gradient in that point.

The motion vector set obtained from each block is represented by a motion estimation histogram. All the resulting vectors in $C_{i,j}^k$ are converted to equivalent spherical coordinates and quantized into histogram $h_{i,j}{}^k$. Then, we normalize these histograms, as is common in computer vision, to make it invariant to brightness magnitude. We have tested the L^1 and L^2 norm for histogram normalization and the first one obtained better results.

3.3 Clustering the Trajectories of a Frame

In order to address the problem of camera motion, trajectories are clustered based on their shape. As this clustering groups trajectories with similar angle sequences, the camera motion trajectories tend to be in the same cluster.

For each trajectory an angle vector is associated, which has a length equal to $t - 1$. The angles are calculated between two consecutive displacement vectors of the trajectory. The vector $a_i{}^k$ of the trajectory i is created using $a_i{}^k = (a_{i,1}^k, ..., a_{i,t-1}^k)$, as the following equation

$$a_{i,j}^k = \cos^{-1}\left(\frac{v_{i,j}^k \cdot v_{i,j+1}^k}{\| v_{i,j}^k \| \cdot \| v_{i,j+1}^k \|} \right), \tag{2}$$

where $j \in [1, t - 1]$ is the index of the vector element.

The result of this equation is a number between 0 and 180, because it gives us the smallest angle between two connected vectors in the trajectory. This angle vector is then used in k-means clustering, and the mean of all angles of each cluster is used to sort the cluster.

The number of clusters n_c is predetermined and stay the same for the whole video. Each cluster $X_c = \{i \mid a_i{}^k$ was assigned to the cluster $c\}$ tends to have very similar trajectories.

3.4 Generating the Frame Descriptor

The histograms, $h_{i,j}{}^k$, are represented by an orientation tensor $\mathbf{T}_{i,j}^k = h_{i,j}^k \cdot h_{i,j}^{k\ T}$ where $\mathbf{T}_{i,j}^k \in R^{n_\theta}$.

Individually, these tensors have the same information as $h_{i,j}{}^k$, but several tensors can be combined to find component correlations. In order to determine an orientation tensor that describes the block displacement in the i-th trajectory we sum all tensors along the same trajectory into a single tensor $\mathbf{T}_i^k$:

$$\mathbf{T}_i^k = \sum_{j=0}^{t} \mathbf{T}_{i,j}^k. \tag{3}$$

For the frame k, we generate c tensors $\mathbf{T}_c^k$, which together contains all information about the frame trajectories. These tensors are composed by tensors corresponding to trajectories associated to this cluster position c:

$$\mathbf{T}_c^k = \sum_{i \in X_c} \mathbf{T}_i^k. \tag{4}$$

As proposed by [7], the clusters c of each frame are ordered using the ascending angles ϕ and θ. The tensors $\mathbf{T}_c^k$ follow the same ordering. It ensure that the vectors with similar angles tend to be in the same cluster number for all frames.

3.5   Generating the Final Video Descriptor

At this point, all frames have c tensors in ascending order. In order to obtain the information about the whole video, we accumulate the tensors of all frames to obtain the descriptor for the video. To make it possible, for each cluster c, we generate a tensor $\mathbf{T}_c$ for the video by accumulating the tensors $\mathbf{T}_c^k$:

$$\mathbf{T}_c = \sum_{k=0}^{n_f} \mathbf{T}_c^k, \tag{5}$$

respecting the ascending order given by the histogram angles. In other words, the tensors corresponding to smaller angles of all frames will be added together, the tensors corresponding to second smaller angles of all frames will be added together and so on.

Finally, we concatenate the tensors $\mathbf{T}_c$ to obtain a vector $\boldsymbol{d}$ that describes the video:

$$\boldsymbol{d} = (T_1^k, ..., T_c^k)$$

This descriptor encapsulates all information about the motion within the video.

3.6   Variations of the Method

In the first variation of the proposed method the process of block division occurs between the first two frames of the sequence. The second variation modifies the block division process, by allowing it to occur in every frame of the sequence.

The first variation is a result of the steps outlined in the previous sections. First, VSBMA displacement vectors are computed using t pairs of adjacent frames, as shown in Sect. 3.1. In this phase, we use the pair of frames t_0 in the block division process, and this final division defines the size of each block. This size remains the same for the processing in the next frames.

In the second variation, a modification is introduced in the block division part of VSBMA. We use all trajectory frames in this division. We initially divide frame f_0 into blocks of a predetermined initial size. Each block is searched in the subsequent frame f_1 and can be divided according to the match error, as explained in Sect. 3.1. In the first variation, we had two conditions for ending the division of a block: the match error should be smaller than the threshold or the division should result in four blocks smaller than the smallest predetermined size. In this second variation, if we reach the first condition, we try to continue the process using the next frame f_2. The division process proceeds until the block size reaches the smallest size. In Fig. 1(d), we can notice that the number of blocks increased considerably compared to the previous method. This happens because all frames contribute to the division process, and frames in the end of the sequence produce additional divisions when compared to the other variations.

4   Experimental Results

Our proposed method makes use of six parameters. Three of them are important for VSBMA: the initial block size, the minimum allowed size for blocks and the threshold of block division. These parameters directly affect the final division of the blocks and, consequently, the descriptor performance. Plus, we have the trajectory size t, which is the amount of frame pairs used in the computation of displacement vectors. This number has to be large enough to satisfactorily catch the temporal information. The number of bins is another parameter. It affects the histogram, since this is uniformly divided into a specified number of bins. Finally, the number of clusters c is used by the k-means algorithm for grouping samples. All clustering steps are done using the stopping criterion of 500 iterations or less than 0.01 % of change in the cluster membership from last iteration.

The main goal of this experimentation is to obtain the best parameter configurations, in order to compare our results with similar works. The proposed method was then tested with different values for each parameter in the KTH dataset. In order to test our descriptor in more realistic scenarios, we also used the UCF11 and Hollywood2 datasets, which have more action classes and colorful videos. We performed these tests using the best parameter configuration found in KTH tests. As the videos in these datasets require more computational resources, varying the parameters would not be viable.

In order to measure the recognition rate achieved using our descriptors, we use an SVM classifier, which takes the descriptors for the whole database. We follow the same classification protocol as [8]. We use both Triangle and Gaussian kernel and a one-against-all strategy for multiclass classification. We produce 6 recognition rates for each parameter combination, 3 using a triangular kernel, and 3 using a Gaussian kernel. In the following sections, we present the highest result achieved from these ones.

4.1   Results for the First Variation

In order to determine the best parameter configuration for our tests, we decided to run experiments by incrementally fixing each parameter, based on empirical knowledge acquired during the development of this work. We started by evaluating the effect of varying the first parameter, i.e., the initial block size. By analyzing the results, we fixed this parameter with the size value that produced the best recognition rate. Then, we moved on to the second parameter and initiated a new set of experiments to determine, out of a set of predetermined test values, which one yielded the best result. As before, once tests were finished, the best value was permanently assigned to the parameter, initiating, then, tests for determining a value for the third parameter, and so on. This process continued until we were able to fix all six parameters. This set of values formed our intended best parameter configuration. Below we present the set of values tested for each parameter:

- Initial block size: 16×16, **32×32** and 48×48
- Minimum allowed block size: **4×4** and 8×8
- Threshold: 5, 7, **10**, 15, 20, 25 and 30
- Number of bins: 12, 24, **26** and 30
- Number of clusters: 1, 2, **3** and 4
- Trajectory size: 2, 4, **6**, 8 and 10

The values highlighted in the list above are the ones that compose the final configuration. Figure 2 shows the results obtained for each parameter variation. Notice that the trajectory size was also varied during each experiment. In the majority of tests, the best result were achieved with a trajectory size of 6. We tried to find a combination of parameters that would provide good results with a satisfactory number of trajectories. Ideally, this number has to be small enough so as to not require too much computing and large enough to accommodate important information.

Although we empirically determined the best values for each parameter, we can interpret the results to provide some insights on what might actually be happening behind the scenes. For example, during the initial block size tests, we noticed a significant increase in recognition rate when a 32×32 size was used. This is depicted by the red curve in Fig. 2(a). We believe this is related to having some blocks in the region without major movements when their size is 16×16, besides describing redundant information. We can also observe the effects of varying the smallest permitted block size, shown in Fig. 2(b). One can observe that the rates for 8×8 size (shown in blue) tend to worsen as the trajectory size increases. Besides, its best result is still much below the 4×4 curve. It occurs because blocks of smaller sizes can correctly represent regions of important movement.

In Fig. 2(c), we can note the threshold of 5 is associated with the worst curve of the graph. This value causes the formation of many trajectories, and these may have noisy information. The best results of these tests was obtained for two threshold values: 10 and 25. A threshold of 10 still causes the generation of many trajectories, consequently worsening performance. However, considering a larger span of trajectory sizes, a threshold of 10 could maintain a better rate than when using 25. This is the main reason why the value of 10 was chosen for the subsequent tests.

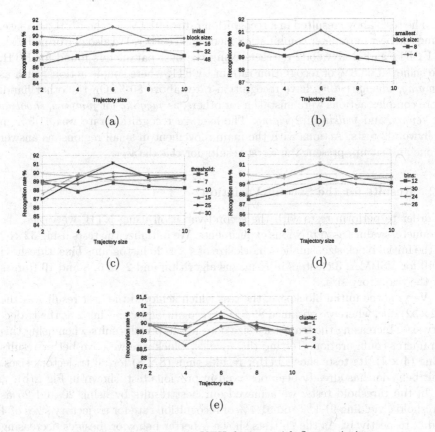

Fig. 2. Experiments on KTH dataset with first variation

Figure 2(d) shows the influence of the number of bins in the results. 30 and 26 bins both achieved good recognition rates. Although experiments with 30 bins achieved better results than those with 26 bins for trajectories of size 4, 8 and 10, we decided to use 26 bins in the subsequent tests, mainly because its result was better when 6 frames were set as the trajectory size. Small values of bins implies to larger ranges of angles and, consequently, groupings of vectors with different information. On the other hand, big values can generate empty bins, increasing the descriptor size with no information gain. With respect to the number of clusters used in the k-means algorithm (Fig. 2(e)), we use 1 cluster in order to show that the absence of trajectory clustering results in a worse recognition rate. If we observe the results for the trajectory size equal to 6, we can see that, as the number of clusters is incremented, the recognition rate only increases until a certain point, where it begins to fall. This shows a tendency for the best number of clusters for this dataset to be 3. As our method uses sparse trajectories, we achieve better results using a small number of clusters, in contrast to [7], which uses 5 clusters.

The tests above resulted in a recognition rate of 91.1 % for the KTH database, using the best parameter combination mentioned earlier. We also performed tests in UCF11 and Hollywood2 datasets using the best parameters found for KTH. We achieved 65.8 % of recognition rate in UCF11, where simple actions, such as *jumping, diving, riding*, have recognition rates above 80 %. On the other hand, more complex actions were mistaken for others, as *juggling* for *jumping*, *shooting* for *tennis*, and *walking* for *riding*. The average recognition rate was 41.5 % in Hollywood2 tests. Actions with the main movement in small regions, as answer phone and sit up, present the worst results for this dataset.

4.2   Results for the Second Variation

In order to perform tests with the second variation using KTH, we varied the parameters values as in previous experiments. We initialize the tests using 32×32 as the initial block size, smallest block size of 4×4, 26 histograms bins, threshold of 10 for VSBMA, 3 clusters in K-means algorithm and 2, 4, 6, 8 and 10 frames for the trajectory size.

Varying the initial block size, the test which achieved the best result was the (32×32) one, where we obtained 89.9 % of recognition rate, using 2 as the trajectory size. Increasing the trajectory size does not improve results when using this parameter configuration. Testing the smallest block size we have better results using (4×4). Its tests shows better results than (8×8) for all trajectory sizes. This behavior has already been observed in previous test, shown in Fig. 2(b).

In the threshold tests, we achieved our best results by using 20 and 25 as thresholds, yielding 91.1 % and 91.4 % of recognition rate for trajectory sizes of 4 and 6, respectively. As the 25 tests shown a better behavior, besides decreasing computational cost, this value was chosen for subsequent tests. The threshold value directly influences the number of trajectories: the higher the value, less trajectories are formed. As this variation form more trajectories than the previous one, the threshold 25 maintains a reasonable amount of trajectories, despite being a high value.

The tests for 26 bins present the best results for all trajectory sizes tested. It achieved 91.4 % of recognition rate using a trajectory of size 6. This was also the best rate obtained for number of clusters tests. These experiments showed a behaviors similar to the ones observed in the first variation tests, shown in Fig. 2(d) and (e). Once again, we observed that 6 is an appropriate value for the trajectory size in this dataset.

In Table 1, we show the confusion matrix of our best result in KTH. The main diagonal of this matrix represents the percentage of correctly classified videos. In this test, we achieved a recognition rate of 91.4 % using 32×32 as initial block size, 4×4 as smallest block size, 25 as threshold, 26 bins, 3 clusters and 6 as trajectory size. In *boxing, clapping, waving* and *running* classes, we achieved better recognition than using our previous method variations. Our video descriptor recognized the *boxing* action in 100 % of boxing-videos. However, we had a small worsening in recognition for *jogging* and *walking* classes, comparing

Table 1. Confusion matrix of the best result on KTH dataset with third variation. The average recognition rate is 91.4 %.

	Box	HClap	HWav	Jog	Run	Walk
Box	100	0.0	0.0	0.0	0.0	0.0
HClap	5.6	93.1	1.4	0.0	0.0	0.0
HWav	0.7	2.8	96.5	0.0	0.0	0.0
Jog	0.0	0.0	0.0	83.3	5.6	11.1
Run	0.0	0.0	0.0	21.5	76.4	2.1
Walk	0.0	0.0	0.0	0.7	0.0	99.3

with first variation results. These are the two more complicated actions of this dataset.

Concluding our experiments, we tested our best parameter configuration for this second variation against UCF11 and Hollywood2 datasets. For the UCF11 dataset, we achieved 65.6 % of recognition rate. In the Hollywood2 dataset experiment we were able to achieve a recognition rate of 40.5 %. This dataset has more complex videos than KTH. Hollywood2 videos usually present more than one person. The movement of these secondary persons cause wrong matchings, with bad impact in long trajectories.

4.3 Comparison with State-of-the-art

In Table 2, we show a comparison of our best result with state-of-the-art methods. Although it is clear that our results can not reach the state-of-the-art, that was not our initial goal. Our focus was on developing a method capable of obtaining reasonable results using sparse trajectories, which are computationally less expensive than dense trajectories.

In that matter, we were able to obtain satisfactory results. Even though we did not achieve recognition rates as high as those of dense trajectories, our

Table 2. Comparison with state-of-the-art for KTH, UCF11 and Hollywood2 datasets.

	KTH	UCF11	Hollywood2
Klaser et al. [9] (2008)	91.0		24.7
Perez et al. [3] (2012)	92.0		34
Mota et al. [1](2013)	93.2	72.7	40.3
Wang et al. [6] (2013)	**95.3**	**89.9**	59.9
Wang and Schmid [10] (2013)			64.3
Figueiredo et al. [4] (2014)	87.7	59.5	34.9
Caetano [7] (2014)	94.1		46.5
Our method	**91.4**	**65.8**	**41.5**

method uses considerably fewer trajectories and, therefore, requires less computational effort. Despite rates being a bit lower, our results are very close to those obtained by other self-descriptor methods.

It is important to emphasize that most of these methods shown in Table 2 are not restricted to the self-descriptor constraint. Their video final descriptor are calculated using information of all videos of the dataset. Our descriptor, on the other hand, does not have any dependency on other videos. So, comparing our results with other self-descriptor methods, our results are better than those of [1,3] in the Hollywood2 dataset, which is, as mentioned before, the most complex set of videos used for testing in several works.

Some state-of-the-art methods shown in Table 2 present rates of 92 % and above in KTH dataset, while our have shown a rate of 91.4 %. Their results are very close to the limit of this dataset. When the recognition rate in a database reaches that point, it is very difficult to improve it further even by some tenths. Comparing our results against our first self-descriptor method, presented in [4], this method yields better results.

We perform some tests in order to estimate the computational effort of our method. We use a machine with Intel Xeon E5-4607, 2.20 GHz, 32 GB of RAM using only 1 thread. We select 10 videos per dataset to perform this test quickly. The best parameter configuration previously found was used and our descriptors were computed with an average of 8.48, 1.66, 0.72 frames per second for the KTH, UCF11 and Hollywood2 videos, respectively. Tests in the same conditions were performed using the [7] method, which were computed with an average of 1.57, 0.61 frames per second for the KTH and Hollywood2 videos. This work does not provide results for the UCF11 database. As our method had higher processing rate, we can assume that our method requires lower computational cost than the dense trajectory method presented in [7].

5   Conclusion

This work presented a method for generating video descriptors based on sparse trajectories. Our descriptor is considered a self-descriptor, since it depends solely on the input video. It is computed by extracting and accumulating information from each frame of the video. Basically, the frame is divided into blocks and their displacement vectors are computed. These vectors are represented by a histogram, which is represented by a tensor.

We presented two variants of our method in this work. The first one uses only the first two frames of a sequence to perform the block division. After this division, the size of each block remains the same for the other frames of a sequence. This first variation achieved 91.1 %, 65.8 % and 41.5 % of recognition rate in KTH, UCF11 and Hollywood2 datasets, respectively. In our second variation the block division was allowed in all frames of the sequence. Our best result in KTH dataset was achieved using this method variation, producing a recognition rate of 91.4 %. In UCF11 and Hollywood2 datasets we achieved 65.6 % and 40.5 % of recognition rate.

This work shows that it is possible to achieve good recognition rates using sparse trajectories. We use a smaller number of trajectories compared to the method of dense trajectories. Due to this characteristic, our data storing needs are reduced. It is important to note, though, that this amount of data is enough to represent the motion of the video.

References

1. Mota, V.F., Souza, J.I.C., de Albuquerque Araújo, A., Vieira, M.B.: Combining orientation tensors for human action recognition. In: Conference on Graphics, Patterns and Images (SIBGRAPI), pp. 328–333. IEEE (2013)
2. Sad, D., Mota, V.F., Maciel, L.M., Vieira, M.B., de Albuquerque Araújo, A.: A tensor motion descriptor based on multiple gradient estimators. In: Conference on Graphics, Patterns and Images (SIBGRAPI), pp. 70–74. IEEE (2013)
3. Perez, E.A., Mota, V.F., Fernandes, V., Maciel, L.M., Sad, D., Vieira, M.B.: Combining gradient histograms using orientation tensors for human action recognition. In: 21st International Conference on Pattern Recognition (ICPR), pp. 3460–3463. IEEE (2012)
4. Figueiredo, A.M.O., Maia, H.A., Oliveira, F.L.M., Mota, V.F., Vieira, M.B.: A video tensor self-descriptor based on block matching. In: Murgante, B., et al. (eds.) ICCSA 2014, Part VI. LNCS, vol. 8584, pp. 401–414. Springer, Heidelberg (2014)
5. Jain, M., Jégou, H., Bouthemy, P.: Better exploiting motion for better action recognition. In: 2013 IEEE Conference on Computer Vision and Pattern Recognition (CVPR), pp. 2555–2562. IEEE (2013)
6. Wang, H., Schmid, C., et al.: Action recognition with improved trajectories. In: International Conference on Computer Vision (2013)
7. Caetano, F.A.: A video descriptor using orientation tensors and shape-based trajectory clustering, Universidade FederaL De Juiz De Fora (2014)
8. Schuldt, C., Laptev, I., Caputo, B.: Recognizing human actions: a local SVM approach. In: Proceedings of the 17th International Conference on Pattern Recognition (ICPR), pp. 32–36. IEEE (2004)
9. Kläser, A., Marszałek, M., Schmid, C.: A Spatio-temporal descriptor based on 3D-gradients. In: British Machine Vision Conference (BMVC), pp. 995–1004 (2008)
10. Wang, C., Wang, Y., Yuille, A.L.: An approach to pose-based action recognition. In: 2013 IEEE Conference on Computer Vision and Pattern Recognition (CVPR), pp. 915–922. IEEE (2013)

Facial Expression Recognition Using String Grammar Fuzzy K-Nearest Neighbor

Payungsak Kasemsumran[1], Sansanee Auephanwiriyakul[1,2(✉)], and Nipon Theera-Umpon[2,3]

[1] Computer Engineering Department, Faculty of Engineering,
Chiang Mai University, Chiang Mai, Thailand
payungsak2517@gmail.com
[2] Biomedical Engineering Center, Chiang Mai University, Chiang Mai, Thailand
{sansanee,nipon}@ieee.org
[3] Electrical Engineering Department, Faculty of Engineering,
Chiang Mai University, Chiang Mai, Thailand

Abstract. Facial expression recognition can provide rich emotional information for human computer interaction. It has become more and more interesting problem recently. Therefore, we propose a facial expression recognition system using the string grammar fuzzy K-nearest neighbor. We test our algorithm on 3 data sets, i.e., the Japanese Female Facial Expression (JAFFE), the Yale, and the Project- Face In Action (FIA) Face Video Database, AMP, CMU (CMU AMP) face expression databases. The system yields 89.67 %, 61.80 %, and 96.82 % in JAFFE, Yale and CMU AMP, respectively. We compare our results indirectly with the existing algorithms as well. We consider that our algorithm provides comparable results with those existing algorithms but we do not need to crop an image beforehand.

Keywords: Facial expression recognition (FER) · String grammar · K-nearest neighbor · String grammar fuzzy K-nearest neighbor · JAFFE face expression database · Yale face expression database · CMU AMP face expression database

1 Introduction

Nowadays, facial expression recognition has become more and more interesting problem in the human-computer interaction, security, or social interactions, etc. [1, 2]. This is not an easy problem since there is a variation of facial expression across human population or from the same person with different context-dependent face expression. There are many research works involving with the facial expression recognition. However, most of the algorithms need to detect face location first using either manually cropping image or automatic face detection [3–19]. In the real time application, the manually cropping image might not be preferable, since human need to be presented at the time. Or an automatic face detection might select wrong face area, hence the algorithm might provide a miss facial expression classification. Another approach is by using template matching [20]. Unfortunately, template matching does not usually perform well in general. Moreover, all of the mentioned algorithms are developed

© Springer International Publishing Switzerland 2016
O. Gervasi et al. (Eds.): ICCSA 2016, Part II, LNCS 9787, pp. 584–596, 2016.
DOI: 10.1007/978-3-319-42108-7_46

based on quantitative information not structural information. It has been known that structural information is more preferable in the facial expression recognition since it is closer to human perception.

In this paper, we utilize the string grammar fuzzy k-nearest neighbor (sgFKNN) [21] without either manual or automatic image cropping pre-processing. In particular, string grammar extracted in the form of a string of symbols can represent the face structural information. Then the sgFKNN is utilized to classify each facial expression. We implement this algorithm on 3 datasets, i.e., the Japanese Female Facial Expression (JAFFE) Database [22], the Yale facial expression database [23] and the Project- Face In Action (FIA) Face Video Database, AMP, CMU (CMU AMP) database [24]. We also provide the indirect comparison between results from our method with those from the existing methods in each data set.

2   System Description

We generate a string from 1 image using the scheme in [21, 25]. There are 5 steps as follows:

1. Each image in the data set is resized to 200×200. If the image is a color image, it will be converted into gray-scale image using luminance (Y) component [26].
2. The difference between each image (Ori_f_i) in the data set and the average of all training images in the data set (Ave_f) is calculated as

$$Dif\_f_i = Ori\_f_i - Ave\_f \text{ for } i = 1, \ldots, N \tag{1}$$

   where N is the total number of images in the data set. Figures 1 and 2 show example images from the JAFFE data set [22] and its Ave_f. Whereas the corresponding Dif_f_i of each image in Fig. 1 is shown in Fig. 3.
3. The Ori_f_i is convolving with the Gaussian kernel with $\sigma = 1$ to provide the ith blurred image ($Blur\_f_i$). Then Dif_f_i is divided by the $Blur\_f_i$ (resulting in Fi_f_i) to reduce the effect of the variation of illumination [27] as shown in Fig. 4.
4. A symbol of each nonoverlapping subimage of Fi_f_i is created and concatenated into a string. For example, if we divide Fi_f_i into 100 subimages with the size of 20×20, then a string of this image will have 100 symbols.
5. The orientation of each pixel (x,y) in the rth subimage according to the gradient direction is calculated as

$$\theta_r(x,y) = 360 - \tan^{-1}\left(\frac{Fi\_f_{ir}(x, y+1) - Fi\_f_{ir}(x, y-1)}{Fi\_f_{ir}(x+1, y) - Fi\_f_{ir}(x-1, y)}\right) \tag{2}$$

The Histogram of Gradients (HoG) [28–30] with 8 bins as shown in Table 1 is implemented in each subimage. The bin with maximum frequency will be a representative of that subimage. The bin number was utilized as a character in the string

anger   disgust   fear   happiness   neutral   sadness   surprise

Fig. 1. Examples of original images in the JAFFE data

Fig. 2. *Avg_f* from the JAFFE data

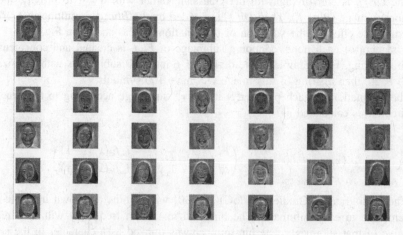

Fig. 3. *Dif_fi* between the original images in Fig. 1 and *Avg_f* in Fig. 2

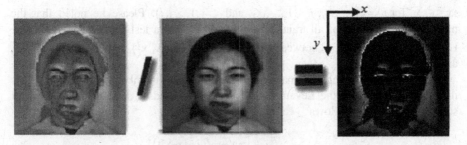

Fig. 4. An example of step 3.

Table 1. Bin orientation

Orientation	Bin No.
$0 <= \theta_r(x,y) < 45$	1
$45 <= \theta_r(x,y) < 90$	2
$90 <= \theta_r(x,y) < 135$	3
$135 <= \theta_r(x,y) < 180$	4
$180 <= \theta_r(x,y) < 225$	5
$225 <= \theta_r(x,y) < 270$	6
$270 <= \theta_r(x,y) < 315$	7
$315 <= \theta_r(x,y) < 360$	8

representing the image, e.g., within the 10th subimage, the bin number 6 has the maximum frequency, then the character of that subimage will be 6 as well. This step is repeated for all subimages to produce a string of that image. The time complexity of this step is approximately O(N). An example of this step is shown in Fig. 5.

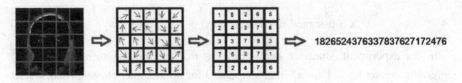

Fig. 5. String generation process.

Now, we will briefly describe the string grammar fuzzy K-nearest neighbor (sgFKNN) [21]. Suppose the training data set, i.e., strings of images in all classes, is $\mathbf{X} = \left\{ \mathbf{x}_1^1, \ldots, \mathbf{x}_{N_1}^1, \mathbf{x}_1^2, \ldots, \mathbf{x}_{N_2}^2, \ldots, \mathbf{x}_1^C, \ldots, \mathbf{x}_{N_C}^C, \right\}$ where $\mathbf{x}_i^j$ is a training string of image i in class j, and N_j is the number of training strings of images in class j. Each string $(\mathbf{x}_i^j)$ is a sequence of symbols (primitives). For example, $\mathbf{x}_i^j = x_{i1}^j x_{i2}^j \ldots x_{il}^j$, a string with length l, where each x_{ir}^j is a member of a set Σ of defined symbols or primitives

$(x_{ir}^j \in \sum$ for $i = 1, \ldots, N_j, j = 1, \ldots, C$, and $r = 1, \ldots, l)$. Please be noted that the number of the strings of all train images is still N. For a test string of image $\mathbf{x}$, the Levenshtein distances [31] between strings $\mathbf{x}$ and $\mathbf{x}_i^j$, $Lev(\mathbf{x}, \mathbf{x}_i^j)$, for $1 \leq j \leq C$ and $1 \leq i \leq N_j$ are calculated.

Then K closest strings of images are identified. The membership value in [32] is modified to cope with the Levenshtein distance and used as a membership value of string $\mathbf{x}$ in class i as follows

$$u_i(\mathbf{x}) = \frac{\sum_{j=1}^{K} u_{ij}\left[\frac{\exp\left(-m\sqrt{C}Lev\left(\mathbf{x}-\mathbf{x}_j^q\right)\right)}{\beta}\right]}{\sum_{j=1}^{K}\left[\frac{\exp\left(-m\sqrt{C}Lev\left(\mathbf{x}-\mathbf{x}_j^q\right)\right)}{\beta}\right]} \tag{3}$$

where u_{ij} is the membership value of training string of image $\mathbf{x}_j^q$ in class i, m and C are the fuzzifier and the number of subjects in the training data set, respectively. Also, β is modified from [32] and calculated as

$$\beta = \frac{\sum_{q=1}^{C} \sum_{j=1}^{N_q} Lev\left(\mathbf{x}_j^q, \mathbf{x}_{med}\right)}{\sum_{q=1}^{C} N_q} \tag{4}$$

where the median string $\mathbf{x}_{med}$ in a set of strings $\mathbf{X}$ can be calculated as [33, 34]

$$\mathbf{x}_{med} = \arg \min_{j \in \mathbf{X}} \sum_{k=1}^{N_1 + N_2 + \ldots + N_C} Lev(\mathbf{x}_j, \mathbf{x}_k) \tag{5}$$

Then, the decision is as follow:

$$\mathbf{x} \text{ is assigned to class } i \text{ if } u_i(\mathbf{x}) > u_j(\mathbf{x}) \text{ for } j \neq i \tag{6}$$

In our experiment, since we know the class that the training string of image $\mathbf{x}_j^q$ represents, we set $u_{qj} = 1$ for $\mathbf{x}_j^q$ in class q and 0 for all the other classes. Also, we set $m = 2$ in our experiment. The time complexity of this step is approximately $O(N^2)$. Hence, the time complexity of the face expression system is around $O(N^2)$.

3 Experimental Results

We implement sgFKNN on 3 data sets, i.e., the Japanese Female Facial Expression (JAFFE) Database [22], the Yale facial expression database [23], and the Project- Face in Action (FIA) Face Video Database, AMP, CMU (CMU AMP) database [24]. The JAFFE Database consists of 213 images of Japanese female facial expression, where each image corresponds to one of the seven categories of expression, i.e., anger, disgust, fear, happiness, neutral, sadness, and surprise. The Yale database contains 165 gray scale images in GIF format from 15 individuals. There are 11 images per subject, one per different facial expression or configuration, i.e., center-light, with glasses, happy, left-light, without glasses, normal, right-light, sad, sleepy, surprised, and wink. There are 3 categories of illumination, i.e., center-light, left-light, and right-light. The facial expression part is divided into 6 expressions, i.e., neutral, happiness, sadness, sleepiness, surprise, and wink. There are 88 images in this part. Since in this paper, we are only interested in facial expression, we use only these 88 facial expression images for this data set. For CMU AMP database, there are 13 subjects in the database, each with 75 images, and all of the face images are collected in the same lighting condition, allowing only human expression changes. There are 5 facial expressions in this data set, i.e., happy, normal, wink, surprise, and disgust.

We implemented the leave-one-out cross validation (LOOCV) and used the same setting for all of the data sets. The best result from the JAFFE data set was when we set K-nearest neighbor to the minimum number of images of each expression minus 1, i.e., 28 with 1600 symbols of each image. In the Yale data set, the best result was when we set K-nearest neighbor to 14 with 1600 symbols of each image. For the CMU AMP data set, 1600 symbols of each image with 5-nearest neighbor yielded the best result.

The result in term of confusion matrix from the JAFFE data set is shown in Table 2. The correct classification on average is 89.67 %. The reason of this mis-classification might be because one face expression class from some subjects is almost similar with the other class. For example, the membership value of subject number 6 shown in Fig. 6 in anger, disgust, fear, happiness, neutral, sadness, and surprise classes (all 28 nearest neighbors are shown in Fig. 7) are 0.118, 0.179, 0.128, 0.134, 0.170, 0.142, and 0.128, respectively. The decision from the algorithm is disgust while the desired class is anger. We can see from the face of subject 6 and all 28 nearest neighbors that this face is very similar to neutral or disgust emotion. Hence the membership value of this subject to these 2 classes are higher than the other classes.

The indirect comparison with the results from the Neural-AdaBoost algorithm [18] and the multiple Gabor filter with SR and SVM algorithm [14] is shown in Table 3. We can see that the correct classification rates from [14, 18] are 89.28 % and 96.81 %,

Table 2. Confusion matrix of seven expression classes of our method in the JAFFE Database using LOOCV.

		Program output						
	Expression	Anger	Disgust	Fear	Happiness	Neutral	Sadness	Surprise
Desired output	Anger	26	2	0	0	1	1	0
	Disgust	2	26	1	0	0	0	0
	Fear	0	1	29	0	0	0	2
	Happiness	0	0	0	28	2	0	1
	Neutral	1	0	0	0	27	0	2
	Sadness	1	1	0	0	2	26	1
	Surprise	0	0	0	0	0	1	29

Fig. 6. Subject number 6 who is misclassified into disgust class

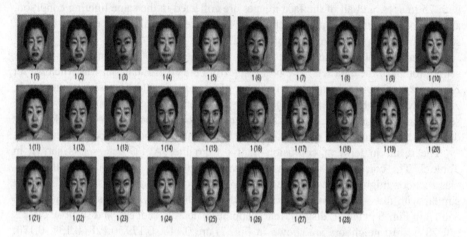

Fig. 7. 28-nearest neighbor of face expression in Fig. 6

respectively. Our algorithm is better than the algorithm in [14] but the algorithm in [18] is far better than our algorithm. This might be because the algorithm in [18] cropped the whole image into a subimage containing only face area while our algorithm does not do that.

Table 4 shows the result from the Yale data set. The average classification rate is approximately 61.80 %. The reason of misclassification might be because we utilize the HoG in the process of string generation. The HoG might create a string from one emotion that is similar to the other emotions because of background or a person's shadow in the background. Examples of the original image with the string created versus its corresponding cropped image with the string created are shown in Fig. 8. We can see that strings from both cases are different and since we do not crop the image, hence, our algorithm will assign this image to a wrong emotion class. The indirect comparison between our results with that from the algorithm in [18] is shown in Table 5. Since the algorithm in [18] only implemented for the data set in 4 classes, i.e., normal, sad, happy, and surprise, we compare our results from these 4 classes only. We can see that the neural-AdaBoost [18] yielded around 91.52 % classification rate, whereas our algorithm yielded 78.85 %. Although, our algorithm does not yield an accuracy as high as that from [18], we do not crop any image beforehand.

Table 3. Indirect comparison of the results from the JAFFE Database.

Method	Anger	Disgust	Fear	Happiness	Neutral	Sadness	Surprise	Average
Multiple Gabor filter & SR + SVM [14]	90 %	95 %	80 %	90 %	90 %	80 %	100 %	89.28 %
Neural-AdaBoost [18]	96.10 %	99.99 %	96.08 %	99.72 %	92.23 %	93.9 %	99.87 %	96.81 %
Our method	86.67 %	89.66 %	90.63 %	90.32 %	90 %	83.87 %	96.67 %	89.67 %

Table 4. Confusion matrix of six expression classes of our method in the YALE Database.

		Program output					
	Expression	Happy	Normal	Sad	Sleepy	Surprise	Wink
Desired output	Happy	11	1	0	0	0	1
	Normal	1	10	1	2	0	1
	Sad	2	2	8	2	1	1
	Sleepy	0	3	1	8	0	2
	Surprise	1	1	1	0	12	0
	Wink	3	3	0	3	0	6

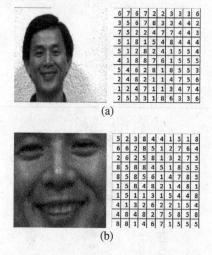

(a)

(b)

Fig. 8. (a) original image of a person in happy class with its string and (b) cropped image of (a) with its string

Table 5. Indirect comparison of the results from the Yale Database.

Method	Normal	Sad	Happy	Surprise	Average
Neural-AdaBoost [18]	86.16 %	86.79 %	97.60 %	98.33 %	91.52 %
Our algorithm	83.33 %	61.54 %	91.67 %	80.00 %	78.85 %

Table 6 shows the result from our algorithm on the CMU AMP data set. Our algorithm yields 96.82 % correct classification in this case. Again, the misclassification occurs when the subjects have similar faces in different classes. For example, the subject 85 who should be in happy class but is classified into normal class because the membership value of this subject in happy and normal classes are 0.40 and 0.59, respectively. Figure 9 shows this subject in the happy face and 5-nearest neighbor. Table 7 shows the indirect comparison between our result and the nonlinear decomposable generative model [19]. In this case, the results from [19] is from 9 persons with only 3 face expressions, i.e., happy, disgust, and surprise. We then consider only these

Table 6. Confusion matrix of six expression classes of our method in the CMU AMP Database.

		Program output				
	Expression	Happy	Normal	Wink	Surprise	Disgust
Desired output	Happy	193	2	1	1	0
	Normal	2	277	3	1	1
	Wink	1	9	172	1	1
	Surprise	0	3	1	172	0
	Disgust	0	1	3	1	129

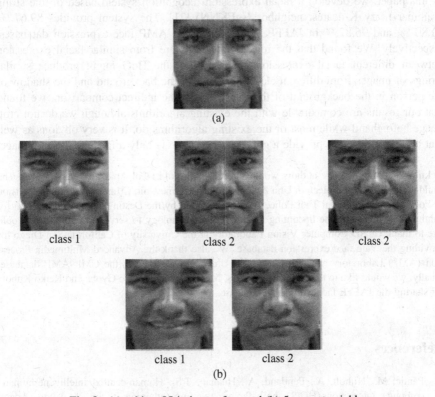

(a)

class 1   class 2   class 2

class 1   class 2

(b)

Fig. 9. (a) subject 85 in happy face and (b) 5-nearest neighbor

Table 7. Indirect comparison of the results from the CMU AMP Database.

Method	Happy	Disgust	Surprise	Average
Nonlinear decomposable generative model [19]	≈95 %	≈98 %	100 %	N/A
Our algorithm	99.48 %	99.23 %	100 %	99.60 %

face expression classes for comparison. Since we do not know who are the 9 persons used in [19], we report the result from our algorithm implemented on all 13 persons. However, our algorithm provides a better correct classification than the method in [19].

From all 3 experiments, we can see that for some data sets that have images with background, our algorithm provides comparable results with the existing algorithms, although, those algorithms have cropping preprocessing step. For the data set that comes only face without background (CMU AMP), our algorithm outperforms the existing algorithm significantly.

4   Conclusion

In this paper, we develop a facial expression recognition system based on the string grammar fuzzy K-nearest neighbor (sgFKNN) [21]. The system provides 89.67 %, 61.80 %, and 96.82 % in JAFFE, Yale, and CMU AMP face expression databases, respectively. We found that the misclassifications are from similar facial expression between different facial expression classes. Also, the HoG might produce similar strings of images from different classes because of the background and the shadow of the person in the background of the image. From the indirect comparison, we found that our results are comparable with the existing algorithms although we do not crop image beforehand while most of the existing algorithms do. It is very obvious as well that our algorithm will provide a good result if there is only a face area in the image.

Acknowledgements. The authors would like to thank Pol.Lt.Col Apichart Hattasin from The Children and Women Protection Unit & Transnational Crime Unit, Crime Investigation Division of Police Region 5, Royal Thai Police who is supported by the Destiny Rescue Foundation for valuing and foreseeing the upcoming of computing technology to serve the public. We would like to thank UCSD Computer Vision Laboratory at the University of California-San Diego for providing the Yale face expression database. We also thank the Advanced Multimedia Procession (AMP) Laboratory at Carnegie Mellon University for sharing the CMP AMP database. Finally, we would like to thank Michael Lyons, Miyuki Kamachi, Jiro Gyoba and Reiko Kubota for sharing the JAFFE face expression database.

References

1. Pantic, M., Nijholt, A., Pentland, A., Huanag, T.S.: Human-centred intelligent human? computer interaction (HCI2): how far are we from attaining it? Int. J. Auton. Adapt. Commun. Syst. **1**(2), 168–187 (2008)
2. Vinciarelli, A., Pantic, M., Bourlard, H.: Social signal processing: survey of an emerging domain. Image Vis. Comput. **27**(12), 1743–1759 (2009)
3. Zhang, Z., Lyons, M., Schuster, M., Akamatsu, S.: Comparison between geometry-based and gabor-wavelets-based facial expression recognition using multi-layer perceptron. In: The Third IEEE International Conference on Automatic Face and Gesture Recognition, pp. 454–459, April 1998
4. Sarode, N., Bhatia, S.: Facial expression recognition. Int. J. Comput. Sci. Eng. (IJCSE) **02** (05), 1552–1557 (2010)
5. Song, K.T., Chien, S.C.: Facial expression recognition based on mixture of basic expressions and intensities. In: IEEE International Conference on Systems, Man, and Cybernetics, COEX, Seoul, Korea, 14–17 October 2012

6. Kotsia, I., Pitas, I.: Facial expression recognition in image sequences using geometric deformation features and support vector machines. IEEE Trans. Image Process. **16**(1), 172–187 (2007)
7. Sadeghi, H., Raie, A.A., Mohammadi, M.R.: Facial expression recognition using geometric normalization and appearance representation. In: 8th Iranian Conference on Machine Vision and Image Processing (MVIP) (2008)
8. Wu, Y., Liu, H., Zha, H.: Modeling facial expression space for recognition. In: 2005 IEEE/RSJ International Conference on Intelligent Robots and Systems (2005)
9. Shan, C., Gong, S., McOwan, P.W.: Robust facial expression recognition using local binary patterns. In: 2005 IEEE International Conference on Image Processing (2005)
10. Bashyal, S., Venayagamoorthy, G.K.: Recognition of facial expressions using Gabor wavelets and learning vector quantization. Eng. Appl. Artif. Intell. **21**, 1056–1064 (2008)
11. Xia, H., Xu, R., Song, S.: Robust facial expression recognition via sparse representation over overcomplete dictionaries. J. Comput. Inf. Syst. **8**(1), 425–433 (2012)
12. Zhang, S., Zhao, X., Lei, B.: Facial expression recognition using sparse representation. Wseas Trans. Syst. **11**(8), 440–452 (2012)
13. Wong, J.J., Cho, S.Y.: Facial emotion recognition by adaptive processing of tree structures. In: SAC 2006 Proceedings of the 2006 ACM Symposium on Applied Computing, vol. 8, pp. 23–30. ACM, New York (2006)
14. Rania, S.E.S., Kholy, A.E., Nahas, M.Y.E.: Robust facial expression recognition via sparse representation and multiple gabor filters. Int. J. Adv. Comput. Sci. Appl. **4**(3), 82–87 (2013)
15. Bartlett, M.,S., Littlewort, G., Frank, M., Lainscsek, C., Fasel, I., Movellan, J.: Recognizing facial expression: machine learning and application to spontaneous behavior. In: Computer Vision and Pattern Recognition (2005)
16. Shan, C., Gong, S., McOwan, P.W.: Facial expression recognition based on local binary patterns: a comprehensive study. Image Vis. Comput. **27**, 803–816 (2009)
17. Jabid, T., Kabir, M.H., Chae, O.: Robust facial expression recognition based on local directional pattern. ETRI J. **32**(5), 784–794 (2010)
18. Owusu, E., Zhan, Y., Mao, Q.R.: A neural-adaboost based facial expression recognition system. Expert Syst. Appl. **41**(7), 3383–3390 (2014)
19. Lee, C.-S., Elgammal, A.: Facial expression analysis using nonlinear decomposable generative models. In: Zhao, W., Gong, S., Tang, X. (eds.) AMFG 2005. LNCS, vol. 3723, pp. 17–31. Springer, Heidelberg (2005)
20. Perveen, N., Gupta, S., Verma, K.: Facial expression recognition using facial characteristic points and gini index. In: 2012 Students Conference on Engineering and Systems (2012)
21. Kasemsumran, P., Auephanwiriyakul, S., Theera-Umpon, N.: Face recognition using string grammar fuzzy K-nearest neighbor. In: 2016 8th International Conference on Knowledge and Smart Technology (2016)
22. Japanese Female Facial Expression (JAFFE) Database. http://www.kasrl.org/jaffe.html
23. Georghiades, A.: Yale face database, Center for computational Vision and Control at Yale University (2002). http://cvc.cs.yale.edu/cvc/projects/yalefaces/yalefaces.html
24. Advanced Multimedia Processing Lab. http://chenlab.ece.cornell.edu/projects/FaceAuthenti cation/download.html
25. Kasemsumran, P., Auephanwiriyakul, S., Theera-Umpon, N.: Face recognition using string grammar nearest neighbor technique. J. Image Graph. **3**(1), 6–10 (2015)
26. Gonzalez, R.C., Woods, R.E.: Digital Image Processing, 3rd edn. Pearson Education Inc, New Jersey (2008)
27. Phillips, P.J., Beveridge, J.R., Draper, B.A., Fivens, G., O'Toole, A.J., Bolme, D., Dunlop, J., Lui, Y.M., Sahibzada, H., Weimer, S.: The good, the bad, and the ugly face challenge problem. Image Vis. Comput. **30**, 177–185 (2012)

28. Chandrasekhar, V., Takacs, G., Chen, D., Tsai, S., Grzeszczuk, R., Girod, B.: CHoG: Compressed histogram of gradients a low bit-rate feature descriptor. IEEE Conference on Computer Vision and Pattern Recognition, pp. 2504–2511 (2009)
29. Psyllos, A., Anagnostopoulos, C.N., Kayafas, E.: Vehicle model recognition from frontal view image measurements. Comput. Stand. Interfaces **33**, 142–151 (2011)
30. Dalal, N., Triggs, B.: Histograms of oriented gradients for human detection. In: IEEE Computer Society Conference on Computer Vision and Pattern Recognition, pp. 886–893 (2005)
31. Bezdek, J.C., Keller, J.M., Krishnapuram, R., Pal, N.R.: Fuzzy Models and Algorithms for Pattern Recognition and Image Processing. Kluwer Academic Publishers, USA (1999)
32. Yang, M.S., Wu, K.L.: Unsupervised possibilistic clustering. Pattern Recogn. **39**, 5–21 (2006)
33. Kohonen, T.: Median strings. Pattern Recogn. Lett. **3**, 309–313 (1985)
34. Klomsae, A., Auephanwiriyakul, S., Theera-Umpon, N.: A novel string grammar fuzzy C-medians. In: Proceedings of the 2015 IEEE International Conference on Fuzzy Systems, Istanbul, Turkey, August 2015

Living in the Golden Age of Digital Archaeology

Rosa Lasaponara[1](✉) and Nicola Masini[2]

[1] CNR-IMAA, C.da S. Loja, 85050 Tito Scalo, PZ, Italy
rosa.lasaponara@imaa.cnr.it
[2] CNR-IBAM, C.da S. Loja, 85050 Tito Scalo, PZ, Italy

Abstract. The aim of this work is to provide a short overview on the most commonly digital technologies today available and used for historical landscape investigations as well as for archaeological and palaeo-environmental studies. One of the main advantage of these techniques is their capability to provide a huge amount of information in non invasive, non-destructive way, also protecting and preserving cultural heritage. The impact of digital technologies for archaeology regards researchers, professionals as well as end-users and enables us not only to improve knowledge and documentation, but also fruition and sustainable touristic exploitation as well as management and monitoring.

Keywords: Digital archaeology · UAV · Satellite · SAR · GIS · GPS · Prediction model

1 Introduction

The field of digital archaeology is stepping in its golden age characterized by an increasing growth of both classical and emerging multidisciplinary methodologies, addressed to the study and conservation of archaeological heritage. The availability of the new digital technologies have opened, for the cultural heritage, new infinite possibilities, unthinkable only a few years ago.

A wide availability of 3D technologies, available from active and passive satellite, aerial and ground sensors, including laser scanning, GIS mapping tools, virtual 4D modeling, augmented reality etc. enables us to address manifold strategic challenges. Thus opening a new horizon in Archaeology. The impact of digital technologies for archaeology regards researchers, professionals as well as end-users. This is clearly evident thinking about, for example, the new portable devices, as tablets and smartphones, nowadays equipped with integrated (Global Position System) GPS, very powerful processors and video cards, which permit us to enjoy virtual reconstructions and an increasing amount of information available "exactly on site and on time".

The digital tools nowadays available for archaeology enable us to get extremely precise results speeding up the work during the diverse phases of archaeological investigations ranging from survey, mapping, excavation, documentation, exploitation and monitoring at diverse scales of interest, moving from small artifacts to architectural structures and landscape reconstruction. Also it is possible to integrate ancient environment reconstruction, obtainable for satellite data, with mapping of past flora and fauna

© Springer International Publishing Switzerland 2016
O. Gervasi et al. (Eds.): ICCSA 2016, Part II, LNCS 9787, pp. 597–610, 2016.
DOI: 10.1007/978-3-319-42108-7_47

and anthropological aspects having all of them, as in museum's showcases. virtually linked in the screen of a tablet. AS an example Fig. 1 shows results from satellite based investigations used as a basis information layer to set up a virtual reconstruction of mediaeval historical landscape in S, Giovanni in Fiore – Calabria-a site of South of Italy.

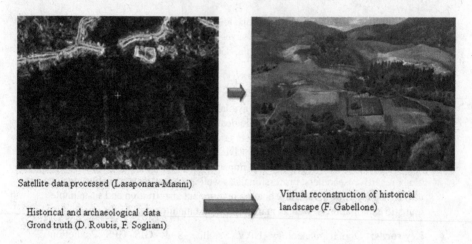

Satellite data processed (Lasaponara-Masini)

Historical and archaeological data
Grond truth (D. Roubis, F. Sogliani)

Virtual reconstruction of historical
landscape (F. Gabellone)

Fig. 1. Virtual reconstruction of a Medieval landscape S Giovanni in Fiore Calabria (Italy).

One of the most important points is that all these technologies are available at different costs for different purposes and needs [1, 2], even with a small budget it is possible to implement a very effective solution. For example, satellite optical imagery are available from google Earth and using a small "drone", equipped with GPS and digital camera remotely controlled, it is possible to carry on detailed survey and documentation, dynamic monitoring and robust engineering analysis even for complex archaeological structures. This simple photographic survey can provide a reliable 3D reconstruction with small margin of error ranging from a few centimeters to a few millimeters. Moreover, the calculation may be managed in cloud computing sending the shootings to a remote powerful server that elaborate a model preview in a few minutes of post-processing.

Moreover, we already live in an age of a growing availability of free data and open access software tools that enhance a powerful link between in situ investigations and computer-based analysis thus offering a new opportunity for the exploitation of archaeological results. Therefore, the new and emerging challenges are the dissemination of data, the interoperability, the costs, simplicity in use and speed of applications, to make them open and understandable to everybody.

One of the main greatest advantage of digital technologies (available from air, space, ground) is that they gather an immense amount of information, as those obtained from Lidar [3] on archaeological remain even buried in a non-invasive, non-destructive way, also protecting and preserving them (Fig. 2).

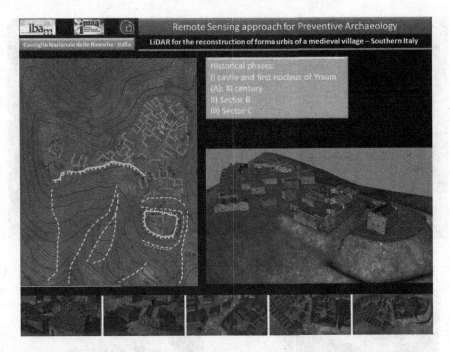

Fig. 2. Lidar survey and virtual reconstruction of a Medieval castle and village

2 Aerial Digital Archaeology

2.1 Overview

Historically, aerial photography has been the first remote sensing tool extensively used for digital archaeology for surveying emerging archaeological remains as well as for detecting underground archaeological structures through the reconnaissance of the so-called "soil", "crop", "marks" and "shadow" marks [4–6].

Crop-marks are linked to the presence of buried walls and/or filled ditches in vegetated areas. They produce local variations in moisture and nutrients content and, consequently, in the growth of vegetation, that can be revealed by differences of height or colour in crops Using multispectral images crop-marks can be detected by spectral variations in specific channels more sensitive to vegetation (as near infrared) or spectral indices (i.e. mathematical combinations of different spectral channels) as NDVI, etc.

Damp-marks occur when the presence of archaeological deposits, such as buried walls, filled ditches and pit etc., They induce local changes in the drainage capability of the soil and can be revealed by changes in colour or texture or using multispectral data by spectral variations in specific channels more sensitive to moisture or spectral indices (i.e. mathematical combinations of different spectral channels) as NDWI etc.

Finally, shadow marks are micro/medium-microtopographic relief linked to archaeological remains, as artworks, platforms, ditches and shallow remain, and they can be revealed by the presence of shadow (Fig. 3).

Fig. 3.   3 D model obtained from a low cost drone

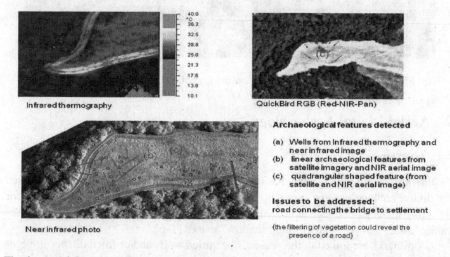

Infrared thermography

QuickBird RGB (Red-NIR-Pan)

Archaeological features detected

(a) Wells from Infrared thermography and
     near infrared image
(b) linear archaeological features from
     satellite imagery and NIR aerial image
(c) quadrangular shaped feature (from
     satellite and NIR aerial image)

Issues to be addressed:
road connecting the bridge to settlement

Near infrared photo

(the filtering of vegetation could reveal the
presence of a road)

Fig. 4.   Aerial thermography survey conducted for the Etruscan site S. Giovenale (Viterbo) Italy
(Color figure online)

The visibility of the subtle traces linked to archaeological marks (as those present n
desert area [7–9], appearing as crop/weed, soil and shadow marks, has a great intra- and
inter year variability due to changes in crop types and phenology, soil moisture content
and other surface parameters. Additionally, the visibility of micro-relief depends on
many factors, such as off-nadir viewing angle of the collected imagery, time of image
acquisition, view geometry, sun angle (micro-topographic relief variations are more
visible in early morning or late evening) and surface characteristics, in particular, the
presence of vegetation can completely cover the micro-relief. These limitations can be
overcome by two technologies that have strongly improved the performance of aerial

remote sensing in archaeology: (i) UASs and (ii) airborne laser scanner (ALS), also referred to as Light Detection And Ranging (LiDAR). the airborne laser scanning.

2.2 UAVs System

Undoubtedly, compared to traditional aerial archaeological, the UAVs offer several advantages, particularly low cost and ability to cover large in a short window of time. There are currently a wide range of) UAVs. A classification of diverse UAV system based on size, weight, endurance, range, and flying altitude, is in [1, 2].

A typical UAV platform for geomatics purposes [1, 2] have varying costs ranging from 1,000 to 50,000 Euro, depending on (i) payload, (ii) instrumentation, (iii) degree of automation (iii) autonomy in terms of battery.

Low-cost solutions generally offer lower payload and usually require human assistance in the take-off and landing phases More expensive system systems offer higher payload, with the capability to jointly host large format cameras systems (see Table 1) and thermocamera, hypespectral imaging. LiDAR or SAR instruments onboard see [1, 2].

2.3 Lidar System

Currently, a LiDAR [3] survey could be carried out by using two different types of ALS sensor system: (i) conventional scanners or discrete echo scanners and (ii) full-waveform (FW) scanners. The first, generally, delivers only the first and last echo, thus losing many other reflections. The second is able to detect the entire echo waveform for each emitted laser beam, thus offering improved capabilities especially in areas with complex morphology and/or dense vegetation cover (Fig. 5).

Fig. 5. Airborne laser scanning survey aerial photo (up) and DTM (bottom) obtained after the removal of vegetation cover

Airborne laser scanning (ALS) systems are generally composed of different components which contribute to the final accuracy of the range data. All the components should
be accurately calibrated and integrated.

Nowadays ALS is regarded as a well-established tool used for archaeological and
cultural landscape. The majority of published studies are based on data collected by
conventional ALS, for the management of archaeological monuments, landscape studies
and archaeological investigations to depict micro-topographic earthworks in bare
ground sites and in forested areas. Recently ALS has been also used for underwater
archaeological even if the majority archaeological studies make use of LiDAR for land
investigations throughout the world in Europe, Central America, Canada and limited
locations in North America including the United States. Finally several researches have
been also focalized on data manipulation. In absence of vegetation the use of aerial
photos may be adequate for the detection of archaeological marks and the reconstruction
of detailed DTMs. Digital cameras must be accurately calibrated, and later elaborated
with powerful processing chain as the SfM.

Hyperspectral and Thermal sensors have been also exploited in archaeology (see for
example) even if less than multispectral data. Up to now, Hyperspectral and Thermal
data application and impact in archaeological investigations have been still limited due
to the high cost of the sensors compared to the improvement achievable compared to
other lower cost technologies.

3   Satellite Digital Archaeology

3.1   Passive Data

Presently, the great amount of multispectral VHR satellite images, even available free
of charge in Google Earth, opened new strategic challenges in the field of remote sensing
in archaeology. The importance of applying space technology [4–9] to archaeological
research has been paid great attention worldwide, due to the following aspects:

(i)   the improvement in spectral and spatial resolution reveals increasing detailed
      information for archaeological purposes;
(ii) the synoptic view offered by satellite data helps us to understand the complexity
      of archaeological investigations at a variety of different scales;
(iii) satellite-based digital elevation models (DEMs) are widely used in archaeology
      for several purposes to considerably improve data analysis and interpretation;
(iv) the availability of long satellite time series allows the monitoring of hazard and
      risk in archaeological sites;
(v)   remotely sensed data enable us to carry out both inter and intra site prospection
      and data analysis.

The availability of high resolution satellite data has been so rapidly growing that
new problems have arisen mainly linked with methodological aspects of data analyses.
In this context, the main concern is the lack of correspondence between the great amount
of remote sensing image and effective data processing methods capable to reliably

enhance and automatic extract the subtle traces of archaeological remains still present in the modern landscape.

The availability of VHR satellite data has determined an increasing use of satellite data in archaeology. As an example, Fig. 4 shows the increasing number of papers based on the use of satellite data which were recently published in the Journal of Archaeological Science. This increasing trend can be also observed in other journals see, for example, the special issues published focused on the topic in a number of specialized peer review journals, such as Journal of Archaeological Science, (vol. 38, issue 9, 2011), Archaeological Prospection (vol. 16, issue 13, 2009), Journal of Cultural Heritage vol. 10S, 2009., Photo Interpretation European Journal of Applied Remote Sensing, vol. 46, 2010, (http://www.ibam.cnr.it/earsel/Editorial-activity.htm).

The access to VHR satellite images is different, depending on the satellites owners, in the case of private companies such as IKONOS, QuickBird and OrbView images are well distributed. A good distribution network also exists for SPOT, the IndianSatellites and EROS (Fig. 6).

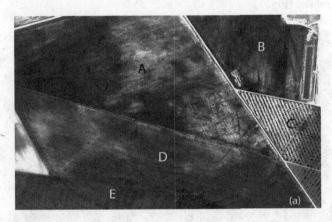

Fig. 6.   Google Earth picture: a view of the Neolithic Village in Schifata (Foggia) Italy

3.2   Active Data

Recently the use of SAR data in archaeology [10] has shown the great potential of these technologies for site detection (buried or emerging archaeological remains) and monitoring as well as for landscape archaeology. SAR enables us to overcome some limitations of optical imaging providing (i) all weather acquisitions (ii) at any time of day or night and (iii) capable to 'penetrate' (to some extends) vegetation and/or soil depending on the antenna wavelength, surface characteristics (ice, desert sand, close canopy, etc.) and conditions (moisture content).

Only in recent years [see 11, 12 and references therein quoted] some investigations have been carried out using aerial and space SAR to detect traces of archaeological interest. In particular, the NASA project attempted to devise a systematic method and protocol to survey large areas merging aerial SAR and satellite multispectral data.

Although several achievements have been made in recent years, the exploitation of SAR remote sensing to detect archaeological marks is still in its infancy, especially for X-band data such as TERRA-SAR and COSMO-SkyMed (for additional details see http://www.e-geos.it/products/pdf/csk-user_guide.pdf).

Satellite Synthetic Aperture Radar (SAR) has entered into a golden age with a rich availability of data from both historical archives and numerous operative satellite plat-forms, which, compared to the past, offer advanced imaging mode capability available at diverse (L,C and X) bands.

The currently available satellite SAR systems provide data with a greater flexibility in the selection of incidence angles and polarizations even in the scale of one meter and less. These advanced technical characteristics make the use of SAR data very attractive for numerous application fields, including archaeology.

A correct identification and interpretation of archaeological features (marks) on the basis of radar images is not a straightforward task and requires knowledge about ground surface conditions as well as about the interaction mechanisms between radar waves and surface sensed (Fig. 7).

Fig. 7. Cosmo Sky med view of Sabratha (Lybia)

4   Ground Digital Archaeology

Geophysical prospecting is a non-destructive technique for subsoil investigation, which consists of the measurement of some physical properties of soil (as anomalies in magnetic, electric or radio signal) that can reveal its structure, as well as the presence of buried objects. In particular, magnetometry detects magnetic [see 13–15 and refer-ences therein quoted] anomalies in the vertical component of the earth's magnetic field providing useful information for the detection of archaeological deposits as burnt areas, kilns and hearths, building remains, and even pits and ditches,. Resistivity is based on a small electrical current which is passed through the earth. The differences in relative resistance is used to map features including ditches, pits, voids and structural features such as wall footings, garden features, platforms, paths, tracks and roads. Compared to magnetometry, resistivity is more time consuming, GPR is based on radio signals. subsurface features and objects can be profitable identified on the basis of the recording reflections. The time of the transmitted and received signal provides information on the presence and depth of structural features.

Magnetometry, resistivity, ground penetrating radar are very popular thecnologies for archaeological investigations, today, considered as high powerful subsurface imaging tools that can provide detailed evidence of past occupation and activities not visible in surface. Obviously Magnetometry, resistivity, ground penetrating radar survey must be implemented using adequately configured survey equipment, with properly data sampling strategies and appropriate post survey data processing.

4.1 Magnetometry

The magnetic survey is one of the most important non-destructive investigation techniques widely used in archaeological research. The magnetic method is based on the measurement of changes in Earth's Magnetic Field or its gradient. Variations or anomalies up to 50 000 times weaker than the local magnetic can be measured. These small local variations of the earth's magnetic field are useful to describe and retrieve the subsoil of a given area. Magnetic methods have been largely used for archaeological prospection since the 1950s, being that many archaeological features show a contrast in magnetic properties compared to the surrounding materials, thus making possible their identification.

Nowadays, there is a large variety of sophisticated sensors (Gaffney & Gater 2003). The most used devices can be categorized in two groups (i) magnetometric devices and (ii) gradiometric devices on the basis of the method used to perform the measurements. The magnetometric devices are based on an single sensor that measures the 'total magnetic field', (i.e. the field resulting from the combination of an anomaly with the ambient earth's field). The gradiometric devices are not able to refers about the diurnal variations of the magnetic field. To cope with this issue, the gradiometric devices provide measurements from at least two opposed magnetic sensors, which are calibrated in a same location. In this way the value of earth's magnetic field is considered as conventional 0 value and, therefore, the sensors of the gradiometer measure the variations from this reference value. The distance between the sensors in the case of gradiometers, or the distance between the ground and the sensor in the case of magnetometry determines the depth of investigation and the resolution of acquisitions.

Applied to archaeological investigations, magnetometry makes it is possible to detect magnetic anomalies produced by the baked bricks or rocks used in the building of a buried house or by the sedimentary alteration induced by an excavated ditch. The identification of these structures depend on the contrast between the magnetic proprieties of the archaeological materials and the surroundings areas. Some remains as, for example, wells, tombs, deposits of materials as well as roads, alignments of walls, pits and ditches, can be identified more easily than others. As expected, iron objects or bricks or ceramic materials can be easily because they tend to exhibit high contrast anomalies, due to their constituent material or coherent magnetic fields they acquired during the firing.

Moreover, it is interesting to highlight that also cultivated areas exhibited in their upper layer (about 30 cm) a soil magnetic susceptibility higher compared to the one found in the deeper layers. This phenomenon is induced by several process as for example, the main are (i) fermentation resulting from the degradation of organic waste, (ii) common practices to burn to clear (from dry grass, shrubs and cultivation residuals) the land before cultivating it.

As a general role, a good result is easier in uniform ground, with smooth surface and moderate magnetic susceptibility, whereas the presence of disturbing elements, as volcanic or recent metallic material provokes noise.

4.2 Resistivity

(https://www.ipfw.edu/centers/archaeology/Geophysics/resistivity.html) is an active technique based on the exploitation of electrical current which is introduced into the soil by one probe in order to measure by a second probe the resistance of the soil through a voltage measurement. The value of the resistance of the soil between the two probes is obtained considering the ratio of voltage (which is the parameter measured) to current which is injected and therefore a known quantity. The distance between the mobile probe array determines the depth at which the instrument can penetrate and see beneath the surface. In other word, features smaller than the probe separation distance are not likely to produce a distinct anomaly. The modern resistivity devices enable us to take multiple measurements in the same location by activating sequentially the measurement of electrodes with different spacing. In this way it is possible to obtain several maps at diverse resolutions resulting from the different spacing of the electrode configuration.

One of the major limitations of resistivity in archaeology is the quite limited capability in creating adequate vertical profiles and in turn in discriminate depth.

Applied to archaeological investigations, magnetometry makes it is possible, for example, to detect archaeological features which area characterized by higher or lower resistivity than their surroundings. For example, a stone foundation tends to impede the flow of electricity, whereas the presence of organic deposits tends to conduct electricity more easily than surrounding soils. As a general role, soil resistance is linked to several parameters among them "porosity, permeability, saturation, and chemical nature of entrapped fluids". Therefore, high resistance can be associated to soils characterized by dry condition, coarse grain composition, and low salinity. On the other side, low resistance soils can be associated to soils characterized by moist conditions, fine grain composition, and high salinity.

Therefore it is expected that building materials will tend to be more resistive than sedimentary soils as well as cavities, ashes or paved floors. Resistivity survey are useful to identify buried buildings, walls, foundations walkways, etc. as well as areas previously excavated and later filled as pits and ditches.

4.3 Ground Penetrating Radar

The Ground Penetrating Radar (GPR) is based on the use of high frequency electronic signals into the earth to detect buried features. High frequency electromagnetic (EM) pulses are generated and transmitted into the ground with known proprieties. Transmitted wavelength systems from 200 to 400 MHz are typical used for archaeological prospection. Lower frequency systems have superior soil penetration which is critical in certain conditions (GPR is useful to sense until a depth of around 10 meters); whereas higher frequency systems have a higher resolution (being that the resolution capability is a function of the transmitted wavelength). The returning signals generate as reflections enable us, knowing the velocity of the pulses into the ground, to calculate the depth of the potential buried objects that produced the reflections. From the archaeological point of view it is important to remind that buried artefacts or sedimentary strata are characterized by physical and chemical properties that affect significantly electrical conductivity and magnetic permeability which are important parameters that determine the velocity of the EM wave. Given soil properties or objects buried interrupt the signal provoking backscattering that is elaborated in 2D or 3D mapping providing therefore information on potential underlying archaeological features such as architecture or artefacts are able to induce significant differences or variations of the backscattered signals. Obviously, as for other geophysical survey, if the archaeological features have a similar material composition of neighboring areas and in turn exhibit, close or identical physical and chemical properties, then the objects may be 'invisible' to GPR equipment.

Applied to archaeological investigations, GPR makes it is possible, for example, to detect archaeological features as ancient roads, house floors, architectural features such as walls and wells; geophysical features such as riverbeds; and even smaller objects such as tools and other artefacts. As a general role, GPR is very valuable tool on urban sites where the use of magnemetry is limited or impossible by the presence of modern metallic material. GPR is very valuable tool for detecting ditches, cave structures, ancient mines or large landscape features such as dry river channels.

One of the major limitations of GPR is the detection of softer and more finely-differentiated types of deposit encountered on many rural archaeological sites. For example, damp clay soils may present critical problems enabling only poor penetration below a few centimeters.

5 Data Integration

The full exploitation of data provided by diverse sensors (from aerial, space and ground acquisition) the usage of 3D models, acquired from airborne and terrestrial laser scans, impose the integration of all the available information (in digital and non digital format) within a GIS environment and software technologies which provide effective solutions for the management, integration, elaboration, full exploitation and publication of heterogeneous data sources provided by excavation reports, geophysical prospection, cartography, aerial and satellite photogrammetry.

GIS environment or web-based GIS environment tools allow a new and more effective way to conduct archaeological research, storing handling and sharing geospatial data from heterogeneous sources in a collaborative way.

The huge amount of data (big data), the increasing needs of data integration, archiviation and processing along with the necessity to make cost effective and easier available these GIS based technologies require new approaches and concepts in the development of infrastructures.

These issues can be reliable and effectively addressed by a WebGIS platform, based on and built with open source components, i.e. open standards, metadata and open source (OSS) architectures.

In the mid and late nineties, the terms GIS and Internet indicated two distinct and separate fields. Today the combination and the increasing use of these systems in regards to archeological applications is clear in the rapid spread of archaeological webGIS, leading to the creation of many platforms with interfaces and functionality oriented at both specialist and non-specialist audience".

A webGIS architecture provides flexible tools for the diverse needs, applications and "usage phases" ranging from data collection phase to the system fruition. In fact, in recent years the development of open webGIS source tools has played an important role with regard to different aims as, for example, (i) publication of the results of an excavation, (ii) placement of archaeological evidence in the territory,(iii) inclusion of archaeological data in broader national geoportals aimed at landscape protection (iv) the inclusion in projects for dissemination also to an audience of non-specialists.

The critical point is therefore to create easy access tools that can suitably face (i) domain expert needs, such as archaeologists, remote sensing community, manager, museums, and laypersons (ii) interested in the cultural assets of the area for educational or tourist purpose (https://rometheimperialfora19952010.wordpress.com/tag/descriptio-romae-webgis/) (iii) effective interchange among computer platforms, i.e., interoperability i.e. "the capability to communicate, execute programs, or transfer data among various functional units in a manner that requires the user to have little or no knowledge of the unique characteristics of those units".

6   Final Remarks

Digital data source from aerial, satellite and ground platform devices provided from active and passive sensors integrated with excavation reports, geophysical prospection, cartography and other documentary sources can suitable support study, documentation, management and systematic monitoring activities for archaeological and historical landscape. Landscape is an integral part of our archaeological heritage being that it preserves the main features that identity and exhibit the evolutionary history of civilization over time. The analysis and interpretation of these features have a strategic importance for the promotion and preservation of the archaeological landscape, which today, unfortunately, is more and more exposed to degradation phenomena. The identification of traces of past human activities, still fossilized in the modern landscape, is the first important step to preserve these "proof" of the landscape history from "extinction processes".

Cultural heritage management and preservation is a strategic priority to assure cultural treasure and evidences of the human past to future generations but, at the same time, it also represents a strategic and valuable economic asset, if inspired to sustainable development strategies. This is an extremely important key factor for the countries which are owners of an extraordinary cultural legacy, which is today particular fragile due to many reasons including industrial risk, pollution impacts and degradation factors. Considering the potential adverse impact of numerous threats to cultural heritage and landscape it is important to ensure the preservation and enjoyment of legacy as unique resources (non -renewable) to be protected not only for the future generations but also for a sustainable economic exploitation coupled with social and cultural developments.

Moreover the analysis could be improved also from the modelling point of view. For example the use of the right cell size is not an easy choice, because a smaller cell size returns a result too disaggregated and probably unuseful, while a bigger cell size is not good to map environmental phenomena or particular existing structure that could be useful to determine settlement pattern and choices and so to determine better the sensibility areas.

Another example is that the model could reveal in real time which parameters are not useful for the analysis. For example, in this study, at the start four parameters were used, while, during the process it appeared clear that the landform one was not indicative of settlement choices, so it was not used to derive the sensibility maps.

Acknowledgments. This paper has been carried out in the framework of the project "Smart Cities and Communities and Social Innovation" (Avviso MIUR n.84/Ric 2012, PON 2007 – 2013 del 2 marzo 2012) Misura IV.1, IV.2, 2013 – 2015.

References

1. Nex, F., Remondino, F.: UAV for 3D mapping applications: a review. Appl. Geomatics **6**(1), 1–15 (2014)
2. Remondino, F., Barazzetti, L., Nex, F., Scaioni, M., Sarazzi, D.: UAV photogrammetry for mapping and 3D modeling–current status and future perspectives. Int. Arch. Photogrammetry, Remote Sens. Spat. Inf. Sci. **38**(1), C22 (2011)
3. Lasaponara, R., Masini, N.: On the processing of aerial LiDAR data for supporting enhancement, interpretation and mapping of archaeological features. In: Murgante, B., Gervasi, O., Iglesias, A., Taniar, D., Apduhan, B.O. (eds.) ICCSA 2011, Part II. LNCS, vol. 6783, pp. 392–406. Springer, Heidelberg (2011)
4. Lasaponara, R., Masini, N.: Satellite remote sensing in archaeology: past, present and future perspectives. J. Archaeol. Sci. **38**(9), 1995–2002 (2011)
5. Lasaponara, R., et al.: Towards an operative use of remote sensing for exploring the past using satellite data: the case study of Hierapolis (Turkey). Remote Sens. Environ. **174**, 148–164 (2016)
6. Lasaponara, R., Masini, N.: Remote sensing in archaeology: an overview. J. Aeronaut. Space Technol. **6**(1), 7–17 (2013)
7. Masini, N., Lasaponara, R., Orefici, G.: Addressing the challenge of detecting archaeological adobe structures in Southern Peru using QuickBird imagery. J. Cult. Herit. **10**, e3–e9 (2009)
8. Lasaponara, R., Masini, N., Scardozzi, G.: New perspectives for satellite-based archaeological research in the ancient territory of Hierapolis (Turkey). Adv. Geosci. **19**, 87–96 (2008)
9. Lasaponara, R., et al.: New discoveries in the Piramide Naranjada in Cahuachi (Peru) using satellite, Ground Probing Radar and magnetic investigations. J. Archaeol. Sci. **38**(9), 2031–2039 (2011)

10. Lasaponara, R., Masini, N.: Satellite synthetic aperture radar in archaeology and cultural landscape: an overview. Archaeol. Prospection **20**(2), 71–78 (2013)
11. Chen, F., Lasaponara, R., Masini, N.: An overview of satellite synthetic aperture radar remote sensing in archaeology: from site detection to monitoring. J. Cult. Herit. (2015)
12. Stewart, C., Lasaponara, R., Schiavon, G.: Multi-frequency, polarimetric SAR analysis for archaeological prospection. Int. J. Appl. Earth Observ. Geoinf. **28**, 211–219 (2014)
13. Gaffney, C.: Detecting trends in the prediction of the buried past: a review of geophysical techniques in archaeology*. Archaeometry **50**(2), 313–336 (2008)
14. Gaffney, C.F., Gater, J.: Revealing the Buried Past: Geophysics for Archaeologists. Tempus Publishing, Stroud (2006)
15. Aspinall, A., Gaffney, C., Schmidt, A.: Magnetometry for Archaeologists. Rowman Altamira, Lanham (2009)

Low Cost Space Technologies for Operational Change Detection Monitoring Around the Archaeological Area of Esna-Egypt

Rosa Lasaponara[1(✉)], Abdelaziz Elfadaly[2,3], and Wael Attia[3]

[1] CNR-IMAA, C.da S. Loja, 85050 Tito Scalo, Pz, Italy
rosa.lasaponara@imaa.cnr.it
[2] UNIBAS University of Basilicata, Potenza, Italy
[3] NARSS, National Authority for Remote Sensing and Space Sciences,
Cairo, Egypt

Abstract. Cultural sites are being continuously threatened by natural hazardous processes and human intervention. There is a general agreement on the need for their protection for present and future human generations. The approach to do is unclear technologically inadequate and (or) lacks financing. Pollution, urban encroachment, population pressure and major development projects are seriously impinging on the precious heritage material values of man in innumerable cases.

Remote sensing and GIS provide a historical database from which hazard maps may be generated, indicating which areas are potentially dangerous. The zonation of hazard must be the basis for any environmental risks management project and should supply planners and decision-makers with adequate and understandable information. The objective of this paper is the detection and mapping urban sprawl and agricultural areas around Esna city in order to assess their impact on the Esna temple and to propose mitigation measures.

Keywords: Esna temple · Remote sensing · Operative monitoring · Urban sprawl · Mitigation strategies

1 Introduction

In the last two decades, the increasing development of ground, aerial and space remote sensing techniques and the tremendous advancement of Information and Communication Technologies (ICT) have focused a great interest in the use of remote sensing and ICT for supporting cultural heritage applications. In particular, the improved spatial and spectral capability of active and passive sensors, the availability of data free of charge as Sentinel 1 and 2 along with open source friendly software have opened new challenging prospectives for the use of EO (Earth Observation) and ICT not only for the investigation [1–10] but also, for the management and valorisation, monitoring [11–13] and preservation of cultural resources, as well as for educational purposes.

These challenges substantial deal with the exploitation of such data as much as possible, and, in turn, with the setting up of effective and reliable automatic and/or

O. Gervasi et al. (Eds.): ICCSA 2016, Part II, LNCS 9787, pp. 611–621, 2016.
DOI: 10.1007/978-3-319-42108-7_48

Table 1. An overview of satellite sensors and missions

	Launch	Country	Pan	Ms
High and very high resolution satellite sensors.				
SPOT 1	1986	France	10 m	20 m
SPOT 2	1990	France	10 m	20 m
SPOT 3	1993	France	10 m	20 m
MOMS 02	1993	Germany	4.5 m	13.5 m
IRS-1C	1995	India	5.8 m	23.5 m
MOMS-2P	1996	Germany	6 m	18 m
ADEOS	1996	Japan	8 m	16 m
IRS-1D	1997	India	5.8 m	23.5 m
SPOT 4	1998	France	10 m	20 m
IKONOS 2	1999	USA	0.8 m	2.4 m
KITSAT 3	1999	S. Korea	15 m	15 m
UoSAT 12	1999	UK	10 m	30 m
Kompsat 1	1999	S. Korea	6.6 m	
EROS A1	2001	Israel	1.8 m	
QuickBird	2001	USA	0.6 m	2.4 m
TES	2001	India	1 m	
SPOT 5	2002	France	5 (2.5) m	10 m
OrbView 3	2003	USA	1 m	4 m
Resourcesat	2003	India	5.8 m	5.8 m
Bilsat	2003	Turkey	12 m	28 m
ROCSat	2004	RO China	2 m	4 m
Cartosat 1	2005	India	1 m	2.5 m
Kompsat 2	2006	S. Korea	1 m	4 m
Topsat	2005	UK	2.5 m	5 m
ALOS	2006	Japan	2.5 m	10 m
Resurs DK2	2006	Russia	1 m	2.3 m
EROS B	2006	Israel	0.7 m	
WorldView	2007	USA	0.5 m	2 m
Cartosat 2	2007	India	0.8 m	
RapidEye	2008	Germany	5 m	5 m
GeoEye-1 (Former name OrbView 5)	2008	USA	0.4 m	1.6 m
THEOS	2008	Thailand	2 m	15 m
RazakSat	2009	Malaysia	2.5 m	5 m

semiautomatic data processing strategies for an operative use of satellite technologies for documentation, management, monitoring and preservation (Table 1).

Additional challenges to this field of research are related to the crucial importance of the integration of remote sensing with other traditional archaeological data sources, such as aerial photos, in situ analysis, historical documentation, etc. This integration requires great efforts aimed at creating a strong interaction among archaeologists, scientists and managers interested in using remote sensing and ICT for supporting and preserving cultural heritage.

In this paper we focus on the operational monitoring based on free of charge satellite images set up for Esna in Egypt. The city of Esna (see Fig. 1) is 50 km south of the Valley of Kings.

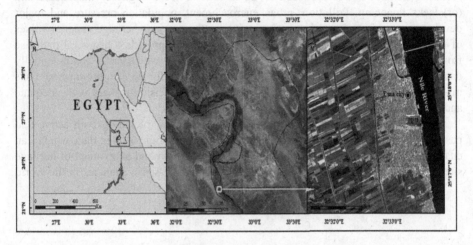

Fig. 1. Study area

Esna is an agricultural village located on the west bank of the Nile, just south of a dam built on the river in 1906. It was the ancient city of Senat, called Latopolis by the Greeks. Today it is a very famous touristic destination for its temple of Khnum, which was buried for many centuries and river barrage. Tuthmosis III laid the foundations of the Temple in the 18th Dynasty, that was later completed by Ptolemaic and Roman Emperors, from 40–250 A.D. The temple is in a huge pit in the center of the city, 10 m deep below the level of the main road. The roof is still intact and is located at the level of the foundations of the surrounding houses. Although the Khnum Temple is one of the last built in Egypt, is less intact than other monuments built centuries before, and retains only the pillared hall, which, however, is well preserved. Its ceiling is supported by 24 columns with floral capitals. An astronomical decoration, with a large zodiac, decorates the ceiling.

For the investigated area, satellite data has been used for setting up operation monitoring tools for the identification and quantification of changes to identify threats to the archaeological area linked to landscape degradation due to anthropogenic activities as agriculture (that can induce pollution), urban sprawl etc.

2 Methodology

Cultural sites are being continuously threatened by natural hazardous processes and human intervention. This paper aims at using low cost Remote sensing and GIS techniques to assess the current conditions in the archaeological area of Esna, analyze the ongoing trend comparing and predict the situation in the future. This aim was achieved by (i) performing a multitemporal analysis of satellite images (ii) assessing the current environmental status (iii) forecasting and infering situation in the future in a GIS environment, (iv) finding solutions for mitigation strategies (Fig. 2).

There is a general agreement on the need for their protection for present and future human generations. The approach to do is unclear technologically inadequate and (or) lacks financing. Pollution, urban encroachment, population pressure and major development projects are seriously impinging on the precious heritage material values of man in innumerable cases. Remote sensing and GIS provide a historical database from which hazard maps may be generated, indicating which areas are potentially dangerous. The zonation of hazard must be the basis for any environmental risks management project and should supply planners and decision-makers with adequate and understandable information.

Recent improvements in earth observation techniques (including both active and passive sensors from satellite, aerial and in situ technologies) offer data which can enable new applications specifically for the documentation and assessment of heritage sites. In this context, UNESCO in partnership with some space agencies in the world

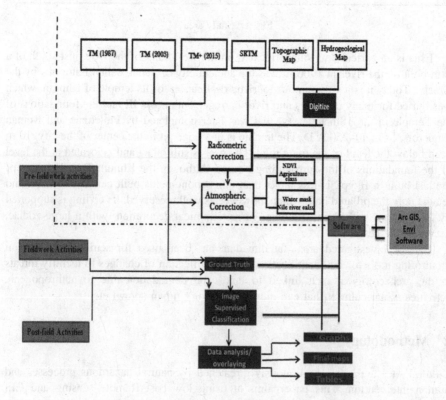

Fig. 2. Methodology

(NASA, ESA, DLR, ASI, CNES, Chinese, etc.) had and has strongly supported the spreading and promoting the use of space technologies to assess the state of conservation of UNESCO-inscribed cultural and natural heritage sites.

Satellite time series offer great potential for a quantitative assessment of urban expansion, urban sprawl and the monitoring of land use changes and soil consumption. Nowadays, remotely sensed data with an acceptable spatial and spectral resolution, together with GIS software tools, have become increasingly available. The data processing adopted for the estimation of urban encroachment, agriculture pressure are mainly based on the use of free of charge data and open source software. The data processing has been mainly based on the classification carried out using free of charge data acquired for diverse years. The comparison of the outputs obtained from the multitemporal classification provided information on the ongoing changes in the study area. The multitemporal analysis was mainly focused on the identification of negative impact linked to urban expansion and agriculture activities.

2.1 Data Characteristics

For the purposes of our investigation free of charge were used for setting up an operational tool for change detection monitoring. The satellite we used belong to the Landsat satellite program that was designed mainly for vegetation monitoring. Therefore the five spectral bands (see Table 2) were selected from green to thermal infrared to be applied in forestry, agriculture, geology, and land-use planning. The spectral bands were selected to study specific resources on Earth and specifically: (i) band 1 for green reflectance from healthy vegetation, (ii) band 2 for chlorophyll absorption in vegetation, (iii) bands 3 and 4 for recording near-infrared reflectance peaks in healthy green vegetation and for detecting water-land interfaces.

Table 2. Spectral range of MSS bands

Band	Resolution	Wavelength μm	Description
1	79 m	0.5–0.6	Green
2	79 m	0.7-0.7	Red
3	79 m	0.7–0.8	Near infrared
4	79 m	0.8–1.1	Near infrared
5	240 m	10.4–12.6	Thermal infrared

For practical use of MSS data, it is important to remind us that for technical reasons MSS bands on the first 3 Landsat satellites were labeled as 4, 5, 6, and 7. Later, these bands were relabeled to 1, 2, 3, and 4 for MSS onboard Landsat 4 and 5 satellites. For Landsat satellite details see Table 3.

The four bands from visible to near infrared have a spatial resolution of 79 m × 79 m, whereas the thermal infrared (only present on Landsat 3) has a spatial resolution of 240 m × 240 m.

Landsat's satellite program was designed to have regular acquisition schedule (revisits each spot on the earth every 16 to 18 days) and long-term data archive with

Table 3. Satellites carrying the MSS sensor

Satellite	Launched	Decommissioned
Landsat 1	July 23, 1972	January 6, 1978
Landsat 2	January 22, 1975	February 25, 1982
Landsat 3	March 5, 1978	March 31, 1983
Landsat 4	July 16, 1982	June 15, 2001
Landsat 5	March 1, 1984	MSS sensor failed in 1992, TM sensor still operational

data available from 1972–1992. The MSS has been one of the first sensors at high spatial resolution onboard Landsat satellites, but of course today it is classified as moderate-resolution image source and its spatial resolution is considered its major limitation.

MSS has been operative until 1992 when it stopped acquiring images because it was replaced by Thematic Mapper (TM) with improved technical characteristics namely higher spectral and spatial capability (7 spectral channels among them 6 at 30 m and one in the thermal range at 90 m).

LANDSAT TM multispectral data are acquired from a nominal altitude of 705 km (438 miles) in a near-circular, sun-synchronous orbit at an inclination of 98.2 degrees, imaging the same 185-km (115-mile) swath of the Earth's surface every 16 days. The Landsat-TM consist of seven spectral bands whose spatial resolution is 30 meters for bands 1–5 and 7 band, while the band 6 (thermal infrared) is acquired at 120 meters but is resampled to 60 meters. The size of a scene are approximately 170 km from north to south and 183 km east-west. All the TM spectral bands (1 to 5 and 7) along with their typical applications are listed in Table 4.

Table 4. TM and ETM + band designations.

Spectral bands	Wavelength (micron)	Resolution (meters)	Use
B 1–blue-	0.45–0.52	30	Bathymetric mapping; distinguishes soil from vegetation; deciduous from coniferous vegetation.
B 2–green	0.52–0.61	30	Emphasizes peak vegetation, which is useful for assessing plant vigor.
B 3–red	0.63–0.69	30	Emphasizes vegetation slopes.
B 4– reflected IR	0.76–0.90	30	Emphasizes biomass content and shorelines.
B 5– reflected IR	1.55–1.75	30	Discriminates moisture content of soil and vegetation; penetrates thin clouds.
B 6– thermal	10.40–12.50	120	Useful for thermal mapping and estimated soil moisture.
B 7– reflected IR	2.08–2.35	30	Useful for mapping hydrothermally altered rocks associated with mineral deposits.

3 Results

TM data acquired in 1987, 2003 and 2015 were used for the investigations. The Landsat data were georeferenced to UTM coordinate system (zone 36 North) and classified using supervised and unsupervised classification.

The main land uses of the study area include urban and agriculture, both of them regarded as deteriorating factors on archaeological sites, especially in the study area where the two land cover tend to facilitate the deterioration of foundation due to the seepage of the drainage water underneath the temples. Both agriculture and urbanization tend to increase in the period of investigation, with a linear increasing trend which achieve the greatest in 2015. In particular, Fig. 3 shows the evolution and enlargement of the urban encroachment affected the site: green, red and yellow pixels clearly depict the area where the urbanization increased from 6000 square kilometer at year 1985 up to 8000 at year 2015. A similar trend is also observed for the agricultural area (see Fig. 6).

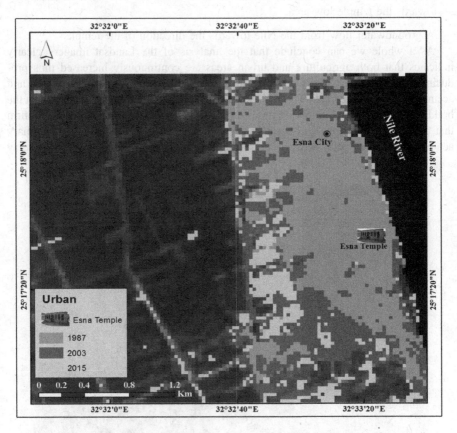

Fig. 3. Change detection of urban areas for (1987, 2003 and 2015) (Color figure online)

The total area of the agricultural lands have reduced but at the same time new agricultural lands appeared in the vicinity of Esna temple and new urban areas are built on the old agricultural lands in the Nile Valley. Actually, after the construction of Aswan High Dam a new reclaimed areas has been added to the cultivated lands. The agricultural expansion toward the monuments areas such as Esna Temple which is located at the fringe of the desert and having nearly the ground elevation of the newly cultivated lands particularly at the eastern parts of these temples. The farmers use excess water to irrigate the Sugarcane crop who consumes more water to grown up. The sub-soil water level increase cause deterioration and salinization of soil around the temple area (dry sabkha) and causing aggressive deterioration of the sandstone rock foundation (Figs 4 and 5).

The unplanned urban encroachments around the temple have negative impacts on the temple for the following reasons:

- The sightseeing of the area become very limited and become invisible.
- There are no sewage systems, so all the untreated sewage domestic water drains directly to the ground which causes an increase of the sub-soil water level flowing towards the temple low area.

A groundwater flow from the Nile towards the direction of the temple.

As a whole we can conclude that the analysis of the Landsat imagery clearly indicates that both agriculture and urban areas are continuously increased thus producing an increasing threat the temple area because favoring the seepage water and accumulation of wastewater. This causes a deterioration to the foundation being that the building stone of the temple tend to absorb water. Situ investigations clearly confirm that the temple is seriously affected by weathering factors. Water rising by capillary action in the pore space between stone foundations vertically and horizontally may

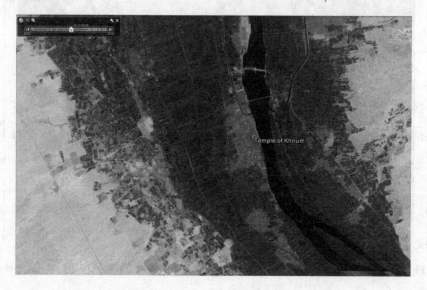

Fig. 4. Google Earth pictures 2013

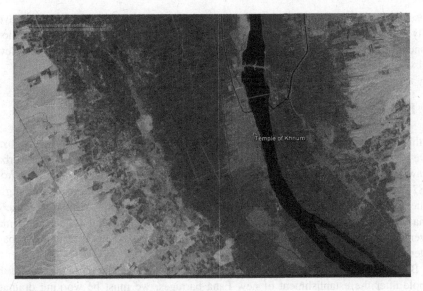

Fig. 5. Google Earth pictures 2015 which clearly show the increasing of urban area

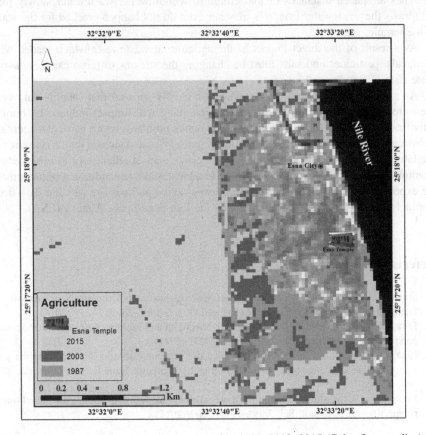

Fig. 6. Change detection for agricultural area in 1987, 2003, 2015 (Color figure online)

cause deterioration as far the sub-soil water table is very close to the surface. Moreover, daily and seasonal dry and wet cycles, acting on the surface and inside the stone block foundation, cause aggressive deterioration of the sandstone foundation. The water level and wind fluctuation are the main factors eroding the temple foundation.

4 Final Remarks

In this study satellite Landsat data available free of charge from the NASA website have been used to set up an operation low cost tool to detect change and estimate urban sprawl and agricultural areas. The increasing trend of both has been quantified from 1987 to 2015. And also verified by in situ analysis and Google Earth. According to some studies conducted in Temple and environmental conditions similar to those of Esna archaeological area, the increasing in both agriculture and urbanization may adversely impact the archaeological due to the action they make in favoring the seepage water and accumulation of wastewater.

Because the level of the groundwater has become a huge danger on the area of the temple after the establishment of new Esna barrages, we must be working drainage systems covered with a layer of sponge to withdrawal the groundwater or digging up Trenches at spaced distances of the temple to withdraw the wastewater slowly then withdraws the wastewater from this trencher even do not happen cracked for the walls of the temple.

As a result of the direct impact of the agricultural wastewater which loaded with chemicals, pesticides and salts must be changing the streams of these canals to avoid these risks.

As a whole, results from our investigation clearly showed that satellite data very useful tools for (i) capturing land use changes along with impact induced by human activities at a site level (ii) monitoring environmental problems with a particular attention addressed to the urbanization and agricultural areas. We can conclude that: (i) the current availability of long term satellite time series provide an excellent tool to observe and monitor changes from a global view down to a local scale, (ii) additional improvement are expected to be obtained in the future using active and passive satellite data from Sentinel 1 and 2 provided free of charge by the Europena Space Agency (ESA).

References

1. Stewart, C., Lasaponara, R., Schiavon, G.: Multi-frequency, polarimetric SAR analysis for archaeological prospection. Int. J. Appl. Earth Obs. Geoinf. **28**, 211–219 (2014)
2. Lasaponara, R., Masini, N.: Satellite remote sensing in archaeology: past, present and future perspectives. J. Archaeol. Sci. **38**(9), 1995–2002 (2011)
3. Lasaponara, R., et al.: Towards an operative use of remote sensing for exploring the past using satellite data: the case study of Hierapolis (Turkey). Remote Sens. Environ. **174**, 148–164 (2016)
4. Lasaponara, R., Masini, N.: Satellite remote sensing in archaeology: past, present and future perspectives. J. Archaeol. Sci. **38**(9), 1995–2002 (2011)

5. Lasaponara, R., Masini, N.: On the processing of aerial LiDAR data for supporting enhancement, interpretation and mapping of archaeological features. In: Murgante, B., Gervasi, O., Iglesias, A., Taniar, D., Apduhan, B.O. (eds.) ICCSA 2011, Part II. LNCS, vol. 6783, pp. 392–406. Springer, Heidelberg (2011)

6. Lasaponara, R., Masini, N.: Remote sensing in archaeology: an overview. J. Aeronaut. Space Technol. **6**(1), 7–17 (2013)

7. Masini, N., Lasaponara, R., Orefici, G.: Addressing the challenge of detecting archaeological adobe structures in Southern Peru using QuickBird imagery. J. Cult. Heritage **10**, e3–e9 (2009)

8. Lasaponara, R., Masini, N., Scardozzi, G.: New perspectives for satellite-based archaeological research in the ancient territory of Hierapolis (Turkey). Adv. Geosci. **19**, 87–96 (2008)

9. Lasaponara, R., et al.: New discoveries in the Piramide Naranjada in Cahuachi (Peru) using satellite, Ground Probing radar and magnetic investigations. J. Archaeol. Sci. **38**(9), 2031–2039 (2011)

10. Chen, F., Lasaponara, R., Masini, N.: An overview of satellite synthetic aperture radar remote sensing in archaeology: from site detection to monitoring. J. Cult. Heritage (2015)

11. Lasaponara, R.: Estimating interannual variations in vegetated areas of Sardinia island using SPOT/VEGETATION NDVI temporal series. IEEE Geosci. Remote Sens. Lett. **3**(4), 481–483 (2006)

12. Lanorte, A., Manzi, T., Nolè, G., Lasaponara, R.: On the use of the Principal Component Analysis (PCA) for evaluating vegetation anomalies from LANDSAT-TM NDVI temporal series in the Basilicata region (Italy). In: Gervasi, O., Murgante, B., Misra, S., Gavrilova, M. L., Rocha, A.M.A.C., Torre, C., Taniar, D., Apduhan, B.O. (eds.) ICCSA 2015. LNCS, vol. 9158, pp. 204–216. Springer, Heidelberg (2015)

13. Nolè, G., Murgante, B., Calamita, G., Lanorte, A., Lasaponara, R.: Evaluation of urban sprawl from space using open source technologies. Ecolog. Inf. (2014). doi:10.1016/j.ecoinf. 2014.05.005

Satellite Based Monitoring of Natural Heritage Sites: The Case Study of the Iguazu Park

Antonio Lanorte, Angelo Aromando, Gabriele Nolè,
and Rosa Lasaponara[⌗]

CNR-IMAA, C.da S. Loja, 85050 Tito Scalo, PZ, Italy
rosa.lasaponara@imaa.cnr.it

Abstract. Up to nowadays, satellite data have become increasingly available, thus offering a low cost or even free of charge unique tool, with a great potential for operational monitoring of vegetation cover, quantitative assessment of urban expansion and urban sprawl, as well as for monitoring of land use changes and soil consumption. This growing observational capacity has also highlighted the need for research efforts aimed at exploring the potential offered by data processing methods and algorithms, in order to exploit as much as possible this invaluable space-based data source. The work herein presented concerns an application study on the monitoring of vegetation cover with the use of multi-temporal (2010–2014) satellite Modis data. The selected test site is the Iguazu park highly significant, being it one of the most threatened global conservation priorities (http://whc.unesco.org/en/list/303/). In order to produce synthetic maps of the investigated areas to monitor the status of vegetation and ongoing subtle changes, satellite data were processed using Principal Component Analysis (PCA). Results from our investigations pointed out a n ongoing degradation trend.

Keywords: Remote sensing · Change detection · PCA · Iguazu park

1 Introduction

Earth Observation (EO) is being successfully used for providing on-going monitoring of natural and cultural site, for assessing natural and man-made risks including emerging threats to natural and cultural resources. EO technologies are allowing the international scientific community to better understand the global environment and how it is changing natural and cultural site including details ranging from a global down to a local scale analysis.

Nevertheless, it is necessary to improve the capacity in the use of remote sensing technologies and data to monitor the state of conservation of natural and cultural heritage including UNESCO-inscribed sites, moving from scientific exercises to operational applications and framework.

In this paper we focus on the use of free of charge satellite data for the monitoring of the Iguazu park selected because highly significant, being it one of the most threatened global conservation priorities (http://whc.unesco.org/en/list/303/). In order to produce synthetic maps of the investigated areas to monitor the status of vegetation

© Springer International Publishing Switzerland 2016
O. Gervasi et al. (Eds.): ICCSA 2016, Part II, LNCS 9787, pp. 622–631, 2016.
DOI: 10.1007/978-3-319-42108-7_49

and ongoing subtle changes, satellite data were processed using Principal Component Analysis (PCA) [1, 2]. It is a highly powerful statistical technique for processing multidimensional data and reducing variance and dimensionality of the data set by projecting the data along new non-correlated axes in a new component space. The PCA technique is applied in almost all disciplines of physical sciences and engineering and it is extensively used in the field of remote sensing data processing.

Like all other techniques, it also has its set of advantages and disadvantages. The foremost advantage of Principal component analysis technique is the ability to identify patterns in the data emphasizing their similarities and differences. Also, a PC transform isolates those data bands which are most informative but this does not imply any loss of information contained in a dataset and the original data can be recovered from the principal components.

The main disadvantage of PCA technique is related to its scene dependence. This means that the representation of the same object acquired by remote sensing images at different times varies depending on temporal scene conditions or data in other pixels in the scene and not only on the specific property of the same object.

Moreover the PCA outputs are software dependent. [3] warn that the computed PC image displays may appear different depending upon the way the computations are made in different software with the same input data set.

Finally, the interpretation procedure for PC images may be difficult and not immediate.

PCA transformation produces new principal components which are uncorrelated with one another and ordered in terms of the amount of variance they represent with respect to the set of the original bands. These bands are often more interpretable than the source data.

PCA technique extracts and displays the greatest variance in a data on the first axis (called the first principal component), the second greatest variance on the second axis (which is orthogonal to the first) and so on. It is also used to reduce the dimensionality of a data set consisting of a large number of interrelated variables [4].

Satellite remote sensing digital images are numeric; therefore, their dimensionality can be reduced using PCA [5]. After PC transformation, new pixel values are computed and stored but the coordinates of the pixel remains the same.

PCA is widely used in detecting changes in time series data and has become one of the most popular techniques because of its simplicity and capability of enhancing the information on change.

In our investigations, PCA was used to emphasize the areas that present significant changes in multi-temporal data sets (i.e. by using NDVI multi-temporal data sets). This is a direct result of the high correlation that exists among images for regions that do not change significantly and the relatively low correlation associated with regions that change substantially.

The major portion of the variance in a multi-temporal image data set is associated with constant cover types and is represented in PC1, while the regions with localized change will be enhanced in later components. In particular, each successive component contains less of the total data set variance. In other words, the first component contains the major portion of the variance, whereas, later components contain a very low proportion of the total data set variance.

PCA is called unstandardized if the principal components are calculated using the covariance matrix, but if it is calculated using the correlation matrix then it is called standardized PCA.

Generally the correlation matrix is used when the data range differs greatly between bands and normalization is needed.

In multispectral remote sensing, the standardized PCA is reported to have improved signal to noise ratio (SNR) as compared to the unstandardized PCA for the same data set [6–10].

2   Methods

Principal Component Analysis decorrelates multivariate data by translating and/or rotating the axes of the original feature space, so that the data can be represented without correlation in a new component space. In order to do this, it is first computed: (i) the covariance matrix (S) among all input spectra bands (each element of S is calculated by using formula 1), then (ii) eigenvalues and eigenvectors of S in order to obtain the new feature components.

$$\text{cov} k1, k2 = 1/nm \sum_{i=1}^{n} \sum_{j=1}^{m} \left(\text{MVC }_{i,j,k1} - \mu_{k2} \right) \left(\text{MVC }_{i,j,k2} - \mu_{k2} \right) \tag{1}$$

Where k1, k2 are two input time series dates, MVCij the annual maximum NDVI value in row i and column j, n the number of row, m the number of columns and μ is the mean of all pixel MVC values in the subscripted input dates.

The percent of total dataset variance explained by each component is obtained by formula 2:

$$\% i = 100 * \lambda_i / \sum_{i=1}^{k} \lambda_i \tag{2}$$

where λi are eigenvalues of S.

Finally, a series of new image layers (called eigenchannels or components) are computed (using formula 3) by multiplying, for each pixel, the eigenvector of S for the original value of a given pixel in the input bands

$$P_i = \sum_{i-1}^{n} P_k \times u_{k,i} \tag{3}$$

where Pi indicates a spectral channel in component i, uki eigenvector element for component i in input band k, Pk spectral value for channel k, number of input band.

A loading, or correlation R, of each component i with each input band k can be calculated by using formula 4.

$$R_{k,i} = u_{k,i} \times (\lambda_i)^{\frac{1}{2}} \times (\text{var}_k)^{1/2} \tag{4}$$

where vark.is the variance of input date k (obtained by reading the kth diagonal of the covariance matrix).

The PCA transforms the input multispectral bands in new components that should be able to make the identification of distinct features and surface types easier. This is a direct result of two facts: (i) the high correlation existing among channels for areas that do not change significantly over the space; and (ii) the expected low correlation associated with higher presence of noise.

The major portion of the variance in a multi-spectral data set is associated with homogeneous areas, whereas localised surface anomalies will be enhanced in later components. In particular, each successive component contains less of the total dataset variance. In other words, the first component contains the major portion of the variance, whereas, later components contain a very low proportion of the total dataset variance. Thus they may represent information variance for a small area or essentially noise and, in this case, it must be disregarded. Some problems can arise from the fact that eigenvectors cannot have general and universal meaning since they are extracted from the series itself.

3 Study Area and Data Set

3.1 Study Area

Located in Misiones Province in the Northeastern tip of Argentina and bordering the Brazilian state of Parana to the north, Iguazú National Park was inscribed on the World Heritage List in 1984. The micro-climate is extremely humid thus favouring lush and dense sub-tropical vegetation harbouring a diverse fauna.

Iguazu National Park and the neighbouring property constitute a significant remnant of the Atlantic Forest, one of the most threatened global conservation priorities. Around 2000 plant species, including some 80 tree species have been suggested to occur in the property along with around 400 bird species, including the elusive Harpy Eagle. The parks are also home to some several wild cat species and rare species such as the broad-snouted Caiman.

"Today, the parks are mostly surrounded by a landscape that has been strongly altered due to heavy logging, both historically and into the present, the intensification and expansion of both industrial and small-scale agriculture, plantation forestry for pulp and paper and rural settlements. Jointly, the two sister parks total around 240,000 hectares with this property's contribution being c. 67,000 hectares." (Fig. 1).

3.2 Data Set

This paper uses 16-day interval MODIS NDVI (MOD13Q1) time series from 2010 to 2014. MODIS NDVI MOD13Q1 is a 16-day composite product, which contains 12 layers, including NDVI, EVI, and pixel reliability layers. We obtained the results for the study area from the Principal Component Analysis applied to a temporal series (2010–2014) of the yearly Maximum Value Composite (MVC) of MODIS NDVI maps (Fig. 2).

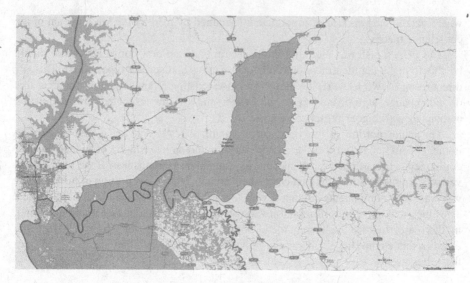

Fig. 1. Iguazu park from Open Street Map

Fig. 2. NDVI max 2010 (Color figure online)

The temporal series 2010–2014 made up of the MVCs of NDVI maps were combined to produce one singular MVC of NDVI map for each year giving a total of 5 NDVI maps. This choice was performed because the yearly MVC of NDVI was found to be a reliable indicator of variations that can affect the state of vegetation cover.

4 Results

Results obtained from the application of unstandardized PCA to the NDVI temporal series are shown in Figs. 3, 4 and 5. These results are obtained by using GRASS-GIS software. Figure 3 shows the first PC component, Fig. 4 the second PC component and Fig. 5 the anomalies map obtained from the second PC component. The Fig. 6 show the loadings of the second component. The loadings representing the correlation of each component (in this case the second component) with each input date (NDVI). In this case the loadings show in general a decreasing trend in all three series. This trend justifies the presence of negative anomalies.

Figure 7 shows the map of NDVI anomalies obtained from the second standardized PC component by using PCI-Geomatica software. Figure 8 show the loadings of the second component with a decreasing trend.

Preliminary results obtained in this study allow to make the following considerations:

1. Results from both standardized and un-standardized PCA, obtained with different software, tend to fit each other in the indication of decreasing trend of NDVI during the reporting period, with the presence of negative anomalies.
2. The main difference that emerges between the PCA made using the covariance matrix and PCA made using the correlation matrix, refers to the year 2011 of the considered time series. This aspect should be investigated with further analysis.

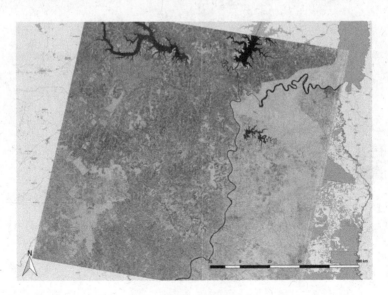

Fig. 3. PCA1

Fig. 4. PCA2

Fig. 5. PCA2 - NDVI anomalies (Color figure online)

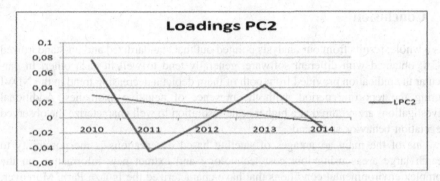

Fig. 6. Loadings of PC2 (unstandardized PCA) (Color figure online)

Fig. 7. PCA2 (PCI-Geomatica) – Anomalie (Color figure online)

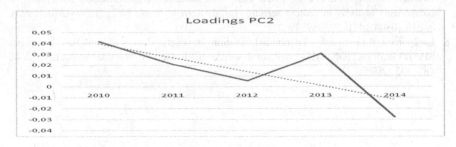

Fig. 8. Loadings of PC2 (standardized PCA)

5   Conclusion

As a whole, results from our analysis pointed out that standardized and un-standardized PCA, obtained with different software, generally tend to well fit each other. In particular the indication provided from both of them depict a decreasing trend in the NDVI during the reporting period, with the presence of negative anomalies. Additional investigations are required at a higher scale resolution to well characterize the observed vegetation behavior and trends.

One of the main advantages of satellite based observations is the possibility to search large areas using low cost technology and extract new information on the complex environmental conditions that have characterized the Iguazu Park. Moreover, results from satellite based investigations, along with all the other information available from previous studies, can be used to create predictive modeling. They are useful tools for setting up mitigation strategies to reduce the ongoing degradation trend.

Acknowledgments. The authors thank the Ministero degli Affari Esteri e Cooperazione Internazionale (MAECI) for supporting this activity in the framework of the Project "Smart management of cultural heritage sites in Italy and Argentina: Earth Observation and pilot projects" PGR00189.

References

1. Johnson, R.A., Wichern, D.W.: Applied Multivariate Statistical Analysis. Prentice Hall, New Jersey (2001)
2. Lattin, J.M., Caroll, J.D., Green, P.E.: Analyzing Multivariate Data. Brooks/Cole, Thomson Asia Pte. Ltd., Singapore (2004)
3. Gupta, R., Tiwari, R., Saini, V., Srivastava, N.: A simplified approach for interpreting principal component images. Adv. Remote Sens. **2**(2), 111–119 (2013). doi:10.4236/ars.2013.22015
4. Jolliffe, I.T.: Principal Component Analysis. Springer, New York (2002)
5. Munyati, C.: use of principal component analysis (PCA) of remote sensing images in wetland change detection on Kafue Flats, Zambia. Geocarto Int. **19**(3), 11–22 (2002). doi:10.1080/10106040408542313
6. Fung, T., LeDrew, E.: Application of principal components analysis to change detection. ISPRS J. Photogrammetry Remote Sens. **53**(12), 1649–1658 (1987)
7. Eklundh, L., Singh, A.: A comparative analysis of standardised and unstandardised principal components analysis in remote sensing. Int. J. Remote Sens. **14**(7), 1359–1370 (1993). doi:10.1080/01431169308953962
8. Lasaponara, R.: Estimating interannual variations in vegetated areas of Sardinia island using SPOT/VEGETATION NDVI temporal series. IEEE Geosci. Remote Sens. Lett. **3**(4), 481–483 (2006)

9. Lanorte, A., Manzi, T., Nolè, G., Lasaponara, R.: On the use of the principal component analysis (PCA) for evaluating vegetation anomalies from LANDSAT-TM NDVI temporal series in the Basilicata Region (Italy). In: Gervasi, O., Murgante, B., Misra, S., Gavrilova, M.L., Rocha, A.M.A.C., Torre, C., Taniar, D., Apduhan, B.O. (eds.) ICCSA 2015. LNCS, vol. 9158, pp. 204–216. Springer, Heidelberg (2015)

10. Nolè, G., Murgante, B., Calamita, G., Lanorte, A., Lasaponara, R.: Evaluation of urban sprawl from space using open source technologies. Ecol. Inf. (2014). doi:10.1016/j.ecoinf. 2014.05.005

Author Index

Printed in the United States
By Bookmasters